D0422991

MANAGEMENT OF RADIOACTIVE WASTES
FROM THE NUCLEAR FUEL CYCLE

Vol. II

PROCEEDINGS SERIES

MANAGEMENT OF RADIOACTIVE WASTES FROM THE NUCLEAR FUEL CYCLE

PROCEEDINGS OF A SYMPOSIUM
ON THE
MANAGEMENT OF RADIOACTIVE WASTES
FROM THE NUCLEAR FUEL CYCLE
JOINTLY ORGANIZED BY THE
INTERNATIONAL ATOMIC ENERGY AGENCY
AND THE
OECD NUCLEAR ENERGY AGENCY
AND HELD IN VIENNA, 22–26 MARCH 1976

In two volumes

Vol. II

INTERNATIONAL ATOMIC ENERGY AGENCY
VIENNA, 1976

MANAGEMENT OF RADIOACTIVE WASTES FROM THE NUCLEAR FUEL CYCLE
IAEA, VIENNA, 1976
STI/PUB/433
ISBN 92–0–020376–0
Printed by the IAEA in Austria
September 1976

FOREWORD

Seven international symposia covering various aspects of radioactive waste management have been held by the International Atomic Energy Agency, either alone or jointly with the Nuclear Energy Agency of the Organization for Economic Co-operation and Development (OECD/NEA). The first two of these symposia, held in 1959 and 1962, dealt respectively with the disposal of radioactive wastes and with the treatment and storage of high-level radioactive wastes. Subsequent symposia discussed the treatment of low- and intermediate-level radioactive wastes (1965), disposal of radioactive wastes into the ground (1967) and treatment of airborne radioactive wastes (1968). The two most recent symposia on the subject were held in Aix-en-Provence (France) in 1970 on developments in the management of low- and intermediate-level radioactive wastes and in Paris in 1972 on the management of radioactive wastes from fuel reprocessing.

From these symposia, it is clear that suitable technology and processes have been conceived and have been and are being developed for managing the present-day amounts of radioactive wastes and effluents from nuclear facilities. But with the increasing emphasis that is being placed on nuclear power a continuing expansion in nuclear fuel cycle facilities in many countries is inevitable. Important policy issues are involved in this expansion, especially with regard to the radioactive waste management requirements. Meanwhile, a concerned public is questioning the effectiveness of this management. In spite of the capability of present technology and development work to cope satisfactorily with present needs, long-term safe disposal has not yet been fully demonstrated for the larger quantities that will arise in the future.

In view of this, the Agency and OECD/NEA felt it was timely to hold a symposium to review the current situation in the management of radioactive wastes generated by nuclear fuel cycle facilities, to identify those areas where important advances have been made, and to indicate where further technological development is needed. It was hoped that the exchange of information and conclusions gained from the symposium would guide the research and development efforts for national nuclear programmes as well as foster international co-operation.

The symposium, held in Vienna on 22–26 March 1976, was attended by over 350 participants from 32 countries and five international organizations. The 62 papers presented covered essentially all aspects of managing nuclear fuel cycle wastes. The major topic of interest was the technology for incorporating high-level radioactive liquid waste from fuel reprocessing operations into solid forms such as calcines, glasses or ceramics for safe interim storage and eventual disposal. Many countries are now examining the possibilities of disposing of the solidified, high-level waste products and plutonium-contaminated waste into suitable geological formations, including rock salt, clays and crystalline rock.

The symposium underlined that considerable progress has indeed been made in the development of radioactive waste management technology and flow-schemes, to the extent that they now are workable and available for collecting, treating, packaging and storing safely all hazardous radioactive wastes evolving from the nuclear fuel cycle. There is in short no lack of appropriate methods, but much of the technology for putting these methods into effect remains in its developmental stage. It is hoped that these Proceedings, which include the papers and the discussions, will assist and guide national and international efforts in those areas of radioactive waste management where the technology is yet to be taken through the demonstration phase.

EDITORIAL NOTE

The papers and discussions have been edited by the editorial staff of the International Atomic Energy Agency to the extent considered necessary for the reader's assistance. The views expressed and the general style adopted remain, however, the responsibility of the named authors or participants. In addition, the views are not necessarily those of the governments of the nominating Member States or of the nominating organizations.

Where papers have been incorporated into these Proceedings without resetting by the Agency, this has been done with the knowledge of the authors and their government authorities, and their cooperation is gratefully acknowledged. The Proceedings have been printed by composition typing and photo-offset lithography. Within the limitations imposed by this method, every effort has been made to maintain a high editorial standard, in particular to achieve, wherever practicable, consistency of units and symbols and conformity to the standards recommended by competent international bodies.

The use in these Proceedings of particular designations of countries or territories does not imply any judgement by the publisher, the IAEA, as to the legal status of such countries or territories, of their authorities and institutions or of the delimitation of their boundaries.

The mention of specific companies or of their products or brand names does not imply any endorsement or recommendation on the part of the IAEA.

Authors are themselves responsible for obtaining the necessary permission to reproduce copyright material from other sources.

CONTENTS OF VOLUME II

MANAGEMENT OF ALPHA-BEARING WASTE (Session VIII)

GEOLOGIC DISPOSAL (Session IX)

SEA DISPOSAL (Session X)

BURIAL OF RADIOACTIVE WASTE (Session XI)

EVALUATION OF SOLIDIFIED, HIGH-LEVEL WASTE PRODUCTS
(Session VI)

Chairman:

F. GERA, Italy

DEVELOPMENT AND RADIATION STABILITY OF GLASSES FOR HIGHLY RADIOACTIVE WASTES

A.R. HALL, J.T. DALTON,
B. HUDSON, J.A.C. MARPLES
Atomic Energy Research Establishment,
Harwell, Didcot, Oxfordshire,
United Kingdom

Abstract

DEVELOPMENT AND RADIATION STABILITY OF GLASSES FOR HIGHLY RADIOACTIVE WASTES.

The variation of formation temperature, crystallizing behaviour and leach resistance with composition changes for sodium-lithium borosilicate glasses suitable for vitrifying Magnox waste are discussed. Viscosities have been measured between 400 and 1050°C. The principal crystal phases which occur have been identified as magnesium silicate, magnesium borate and ceria. The leach rate of polished discs in pure water at 100°C does not decrease with time if account is taken of the fragile siliceous layer that is observed to occur. The effect of 100 years' equivalent α- and β-irradiation on glass properties is discussed. Stored energy release experiments demonstrated that energy is released over a wide temperature range so that it cannot be triggered catastrophically. Temperatures required to release energy are dependent upon the original storage temperature. Helium release is by Fick's diffusion law up to at least 30% of the total inventory, with diffusion coefficients similar to those for comparable borosilicate glasses. Leach rates were not measurably affected by α-radiation. β-radiation in a Van de Graaff accelerator did not change physical properties, but irradiation in an electron microscope caused minute bubbles in lithium-containing glasses above 200°C.

1. INTRODUCTION

In this paper we describe the choice of suitable glass compositions for retaining waste from Magnox fuel, the measurement of some of the properties of these glasses and the results of accelerated radiation damage experiments. Since the experience gained in the Fingal 1 plant [1] new compositions have become essential because the levels of magnesium (25 wt %) and aluminium (20 wt %) in current Windscale wastes increase the founding temperature of the original silica/sodium tetraborate formulation to well above the original 1050°C. Increasing the sodium content for such formulations reduces the founding temperature back to 1050°C, but the resulting glass has poor leaching resistance and forms a separate molybdate/chromate phase very readily.

2. COMPOSITIONS OF GLASSES STUDIED

The new compositions studied have been chosen so that the effects of composition changes can be compared. (Table I) Thus, in glasses M1, 2, 3, waste is increased at the expense of silica, M1, 5, 9, show variations in the alkali/silica ratio; in M20, 21, 22 and M23, 24, 25, the effect of changing alkali/silica ratio at discrete boric oxide levels is demonstrated. Most compositions contain mixed alkalis, but as a comparison with M1, M18 (lithia) and M19 (soda) are included to demonstrate the mixed-alkali effect.

The feed for glass making consisted of oxides and carbonates. After melting, the cast blocks were stress relieved at 500°C for half an hour before cooling to room temperature. For comparison, a few melts have been made from nitrate feed; so far tests have shown that the resulting glasses do not differ

Table I. Glass Compositions, Formation Temperatures and Leach Rates

No.	COMPOSITION						Formation Temp. (°C) (a)	Leach rate at 100°C (b) ($g \cdot cm^{-2} \cdot d^{-1}$) x$10^3$
	Waste		mol %					
	(wt%)	(mol%)	SiO_2	Li_2O	Na_2O	B_2O_3		
M1	25.2	19.00	45.70	7.50	7.50	20.30	950(950)	1.18(0.86)
M2	31.2	24.00	40.70	7.50	7.50	20.30	1000(975)	1.27(0.95)
M3	37.0	29.00	35.70	7.50	7.50	20.30	1000(1025)	1.52(1.16)
M5	25.3	19.00	44.70	8.00	8.00	20.30	950(925)	1.78(1.20)
M6	31.3	24.00	39.70	8.00	8.00	20.30	950(950	2.00
M7	37.1	29.00	34.70	8.00	8.00	20.30	950(975)	2.30
M9	25.3	19.00	43.70	8.50	8.50	20.30	900(900)	2.85
M10	31.3	24.00	38.70	8.50	8.50	20.30	950(950)	2.22
M11	37.1	29.00	33.70	8.50	8.50	20.30	950(975)	2.46
M18	26.2	19.00	45.70	15.00	-	20.30	950(925)	1.91(2.56)
M19	24.3	19.00	45.70	-	15.00	20.30	1050(1050)	2.55(1.90)
M20	25.6	19.00	55.85	7.50	7.50	10.15	1000(1000)	0.221(0.214)
M21	25.6	19.00	54.85	8.00	8.00	10.15	1000(1000)	0.233(0.224)
M22	25.7	19.00	53.85	8.50	8.50	10.15	1000(975)	0.264(0.233)
M23	25.8	19.00	60.92	7.50	7.50	5.08	1050(1050)	0.140(0.167)
M24	25.8	19.00	59.92	8.00	8.00	5.08	1050(1050)	0.160(0.168)
M25	25.9	19.00	58.92	8.50	8.50	5.08	1000(1000)	0.168(0.240)

Notes

(a) The first figure is the temperature used. The figure in brackets is the estimated temperature necessary for formation.

(b) The first figure is the value for the glass. The figure in brackets is for crystallised glass.

from those made from oxides and carbonates. Some small particles of undissolved oxide are occasionally found in the glass and deposits containing palladium metal and probably ruthenium oxide occur on the glass surface and at the crucible/melt interface.

Samples containing 5 wt % PuO_2 for α-damage investigations were made by preparing a parent glass in which the rare earths were omitted to accommodate the PuO_2; this glass was granulated by casting into water. The plutonia was dissolved in the resulting frit at 1050°C in Pt-10% Rh-3% Au crucibles. Samples were cast in silver foil lined steel molds preheated to 450°C. The molds were then held at 450°C for one hour and cooled to room temperature over half an hour.

3. CRYSTALLISING BEHAVIOUR OF GLASSES

Slow cooling from 950°C produces crystals large enough to be identified and analysed. At temperatures at which crystallisation occurs rapidly (600-900 the glass matrix is so soft that containment in silica crucibles is necessary.

To reveal and identify optically the crystals formed, thin sections were prepared by conventional petrographic techniques. With so many elements present many crystalline phases are possible. In practice only some five or six species have been observed and of these Harrison et al. [2] have identified Enstatite ($MgSiO_3$), magnesium borate ($Mg_2B_2O_5$) and Cerianite (CeO_2). In the plutonia containing glass the Cerianite is replaced by plutonia (PuO_2).

In glasses containing 25 wt % waste oxide the maximum volume of crystalline phase formed was 20%. When the waste level was raised to 43 wt % as much as 60 vol % of the sample could be crystallised. At the higher waste concentrations crystallisation is occurring as the glass is being formed.

4. THE FORMATION OF MOLYBDATES IN MELTS

Between 600 and 800°C the fission product molybdenum reacts with the alkalis to form a phase of the type $R_2O.MoO_3$. This phase, coloured yellow by dissolved chromate, is water soluble and contains Cs, Ce and Sr, the most toxic long lived isotopes. On the laboratory scale when the melts are entirely free of sulphate ions, re-dissolution of this phase readily takes place at temperatures above 900°C, particularly if the alkali contains 50 or more mol % lithium. Glasses based solely on sodium tend to form greater amounts of molybdate which takes higher temperatures to re-dissolve. Morris [3] has pointed out that the molybdate phase avidly collects sulphate ions introduced as feed contaminants. When the phase contains a substantial proportion of sulphate it will not dissolve in the glass unless heated to 1100°C and even when dissolved it re-precipitates on cooling as stringers and globules.

5. METHODS OF MEASURING GLASS VISCOSITIES

Viscosity data is needed over a wide temperature range from the strain point to the processing temperature ($10^{13.5}$ to 10^1 $N \cdot s \cdot m^{-2}$)*. A penetrometer was used to determine viscosities in the range 10^2 to 10^7 $N \cdot s \cdot m^{-2}$. For use at higher temperatures a method has been developed using nickel crucibles with accurately drilled orifices in their bases. These are first calibrated at room temperature by measuring the rate of flow of fluids of known viscosity. The rate of flow of glass through the orifice is then measured at temperatures between 850°C and 1050°C. Since the flow rate is proportional to the 4th power of the orifice radius it is necessary to correct for thermal expansion in the calculation. Viscosities at low temperatures were readily obtained from compressive creep measurements. Tests were made in the range 400 to 450°C, giving viscosities from 10^{12} to 10^{16} $N \cdot s \cdot m^{-2}$. An example of the results obtained by the three different methods is shown in Fig. 1.

A smooth curve can be drawn through the points obtained by the different methods which each cover a part of the temperature range. The most difficult range of temperature in which to measure viscosity is between 700 and 900°C where crystallisation occurs rapidly in some compositions and interferes with the flow properties of the glass.

Viscosities for compositions 1-3 are shown in Fig. 2. Rapid crystallisation caused anomalous behaviour in the 31 and 37 wt % waste glasses.

6. LEACHING METHODS AND RESULTS

For routine tests on a large number of experimental compositions, some sort of accelerated leach test is required. Several tests have been compared,

* $10\ N \cdot s \cdot m^{-2} = 100$ poises.

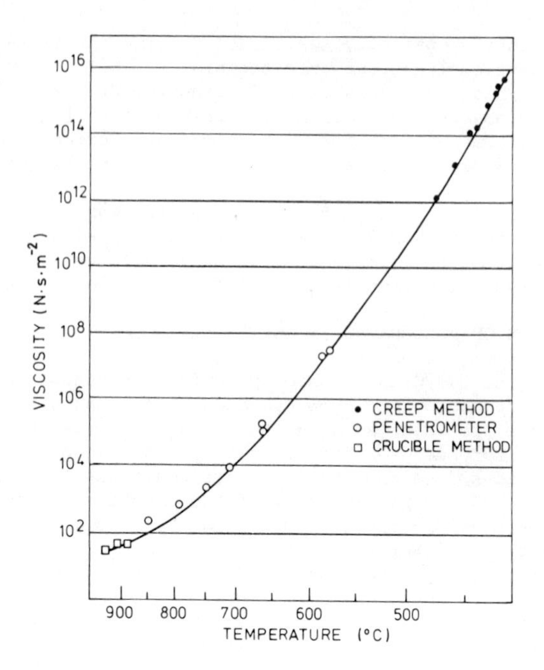

FIG.1. The viscosity of glass M5.

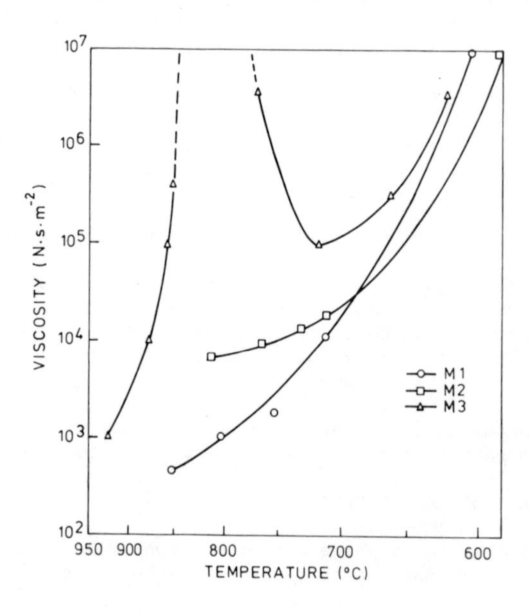

FIG.2. The viscosity of glasses M1, M2 and M3.

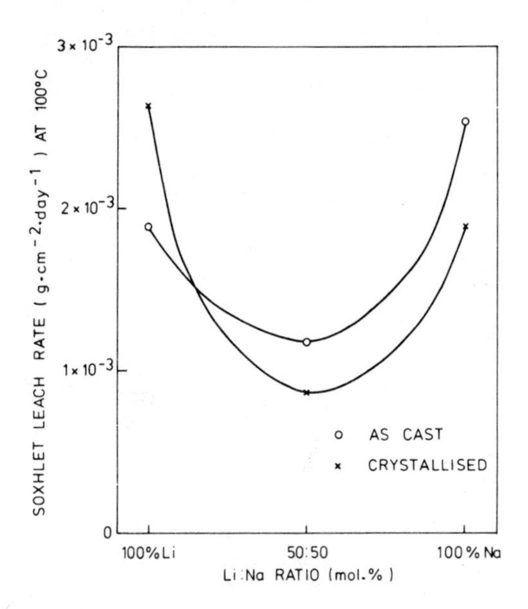

FIG.3. The effect of mixed alkalis on the leach rate.

i.e. powdered samples agitated in hot water, discs in water filled sealed tubes shaken at various temperatures, and a Soxhlet technique similar to that described by Mendel [4]. The use of solid discs or slabs of glass leached in the Soxhlet apparatus at 100°C has been adopted as the standard test.

Leaching tests on solid polished discs showed that the glass surface becomes depleted in Li, Na and B and retains most of the Si, Al, Mg, Fe, Ba, Sr, Y, Cr, La, and Ni. On drying, this layer shrinks to a granular powder which, if of sufficient thickness, becomes detached. We consider that the depleted surface layer should be included as part of the total weight loss; the leach rates quoted in Table 1 were calculated in this way. Examination of samples of leached glass grains shows evidence of the depleted surface layer. It is clear that long term leach tests on powdered glass samples give a falsely optimistic result, (a) because of the exclusion of this weak surface layer from the weight loss and (b) because the surface area available to attack decreases more rapidly than calculation from weight loss would show. For these reasons leach tests on small glass grains have shown rates which decrease with time, whereas on massive samples the rates are constant.

It is well known that mixed-alkali glasses tend to have properties superior to comparable glasses containing either single species. Fig. 3 shows a comparison of leach rates for three glasses before and after crystallisation. As can be seen, the durability is improved by a factor of two for the mixed-alkali glass. We have also observed a decrease in viscosity due to the mixed-alkali effect and replacement of half the sodium atoms by lithium reduced the tendency to form the separate molybdate-chromate phase.

Glasses with high alkali content and correspondingly low melting points are inevitably less durable than those formed at high temperatures. However, a particularly encouraging series of glasses was that shown in Table 1 as

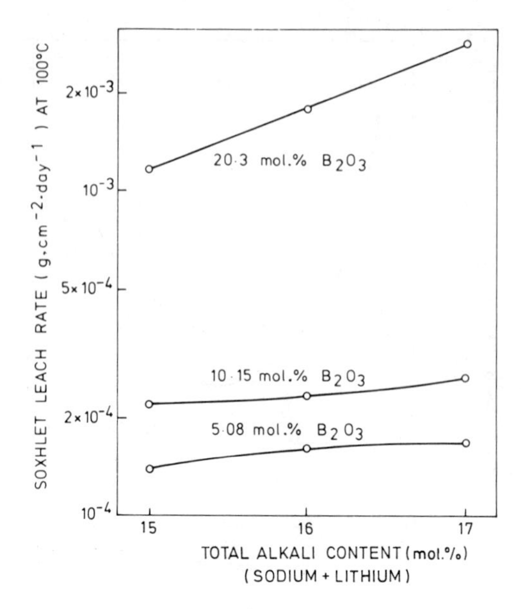

FIG.4. The effect of alkalis and B_2O_3 on the leach rate.

glasses 20-25. The formation temperatures lie between 1000 and 1050°C but the durabilities are better by a factor of 10 than a comparable sodium-borosilicate glass. These good qualities appear to be due to a balance between the presence of mixed alkali plus the benefit gained from lowering the boric oxide content. The crystals which grew in these glasses on slow cooling were predominantly magnesium silicate and the remaining glass phase seemed stable and most reluctant to crystallise further. Fig. 4 shows the effect of varying B_2O_3 and alkali contents on leach rates.

7. RADIATION DAMAGE

The number of α, β and γ decays per ml of glass in the first 100 years are given in Table II. The numbers will vary within the ranges given depending on the reactor, the burn up, the cooling time and other details of the fuel cycle. Assuming a value of 25eV for the displacement energy, the number of atoms displaced from their 'lattice' sites has been estimated for each type of decay and this value and the resulting total number of displacements is given in the Table II: α decays cause the most displacements followed by β and then γ. The number of displacements caused by neutrons, (n, α) reactions and fission fragments has also been estimated and shown to be relatively insignificant.

Experiments have been carried out to simulate the damage due to α and β decays on an accelerated time scale. γ rays cause damage only by giving rise to secondary electrons and so are covered by the β simulations.

7.1. α damage experiments

Because 90% of the displacements caused by an α decay are due to the recoil nucleus it is not possible to simulate the damage using accelerated α particles. A realistic simulation can however be made by adding a short half

Table II. Displacements in the glass in the first 100 years

Radiation	Decays per ml (a)	Displacements per decay (b)	Displacements per ml
α	3×10^{17} to 1.5×10^{19}	140 (α) 1500 (recoil)	5×10^{20} to 2.5×10^{22}
β	2×10^{19} to 1×10^{20}	0.13	3×10^{18} to 1×10^{19}
γ	10^{11}-10^{12} rads		$\sim 10^{18}$

Notes

(a) Assumes 12 wt. % fission products in the glass.

(b) Assumes E_d = 25 eV.

life α emitter to the glass. In the present experiments 5 wt % $^{238}PuO_2$ was added replacing an equal molecular fraction of the rare earth elements in the list of simulated fission products. In a year this gave 2×10^{18} decays per ml, typical of the dose the actual glass will receive in the first 100 years.

Deleterious effects that the α damage might cause include (a) volume changes (b) stored energy which would be released later giving a dangerous temperature rise (c) helium storage (d) changes in leach rate and (e) cracking of the glass. Each of these effects has been studied.

The samples were prepared as described in Section 2. Eight samples containing ^{239}Pu were also made as controls. The samples were immediately placed in two storage ovens at 50°C and 170°C, respectively. Each oven was equipped with two dilatometers, one sample of ^{238}Pu glass and one ^{239}Pu glass were monitored for length changes at each storage temperature. The lengths of the ^{239}Pu samples did not vary throughout the test, the ^{238}Pu sample at 50°C increased in length by 0.02%, whereas the ^{238}Pu sample at 170°C showed a length increase of 0.4% in the first 50 days, followed by a small steady decrease. Unfortunately the densities were not measured before storage. Densities were measured after 1 year's storage, and again after heating in the stored energy release measurements described in Section 7.2. Samples stored at 50°C and subsequently heated to 460°C increased in density from 2.630 to 2.634 $g \cdot cm^{-3}$. Densities of samples stored at 170°C were 2.636 $g \cdot cm^{-3}$. On heating to 480°C this decreased to 2.630, and on heating to 520°C to 2.627. Thus there is a disparity between dilatometer and density measurements which must be resolved.

7.2. Stored energy

The α-damage stored energy was released and measured in a D.T.A. apparatus at a constant heating rate of 1°C per minute. The total stored energy released from samples stored at 50°C for 1 year was 97 $J \cdot g^{-1}$; the release began at ~70°C and was complete at 460°C. The energy release from

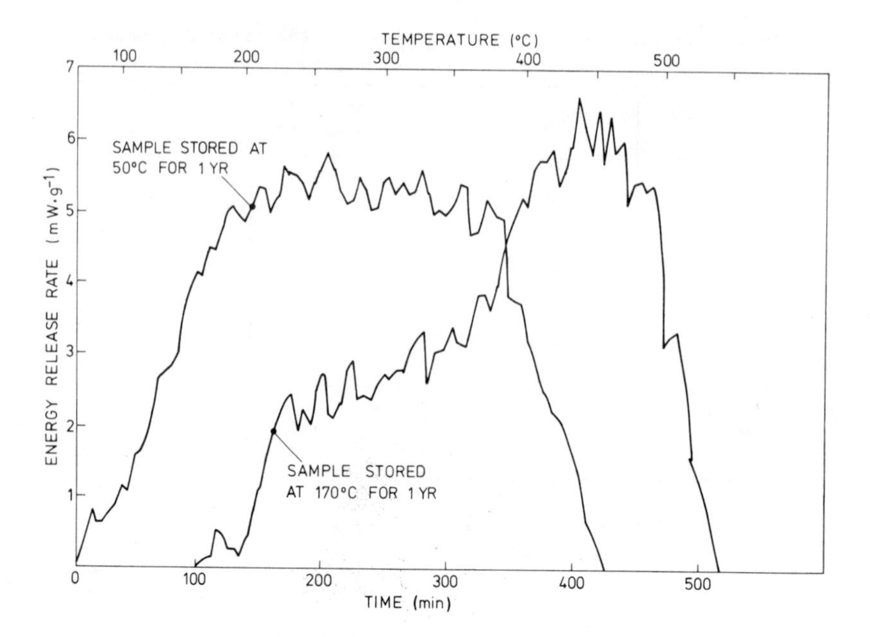

FIG.5. Stored energy release rates.

samples stored at 170°C was 82 $J \cdot g^{-1}$, commencing at ~190°C and reaching completion at 520°C. Fig. 5 shows the rates of energy release. Since the specific heat of the glass is about 1 $J \cdot g^{-1} \cdot {}^{\circ}C^{-1}$ these total stored energies even if triggered, could only lead to a temperature rise of less than 100°C. However, the width of the release peak is so large that triggering is unlikely to occur.

The higher temperature required for complete release from the 170°C samples indicates that at the higher storage temperature additional, different, defect configurations have been formed: thus the energy required for the formation of any specific type of 'lattice' defect is temperature dependent. This leads to the conclusion that the stored energy is quantised as in crystalline materials, although in a glass so many types of 'lattice' displacement are possible that the appearance is of a continuous spectrum. Thus the small peaks seen in Fig. 5 may be real and not due to experimental scatter.

7.3. Helium generation

Two pairs of samples wrapped in aluminium foil were sealed in two evacuated Pyrex capsules. The capsules were stored for a year at 50°C and 170°C respectively, they were then broken under vacuum and the helium content measured. One sample from the 50°C capsule was slit into blocks measuring 1x1.5x3 mm for diffusion coefficient measurements. The theoretical amount of helium, calculated from the number of α-particles, was 0.077 cm^3 per cm^3 of glass, and blocks heated to ~800°C released exactly this amount. Diffusion coefficients were measured on blocks heated to temperatures between 330°C and 480°C, a typical experimental run being shown in Fig. 6. No helium had been

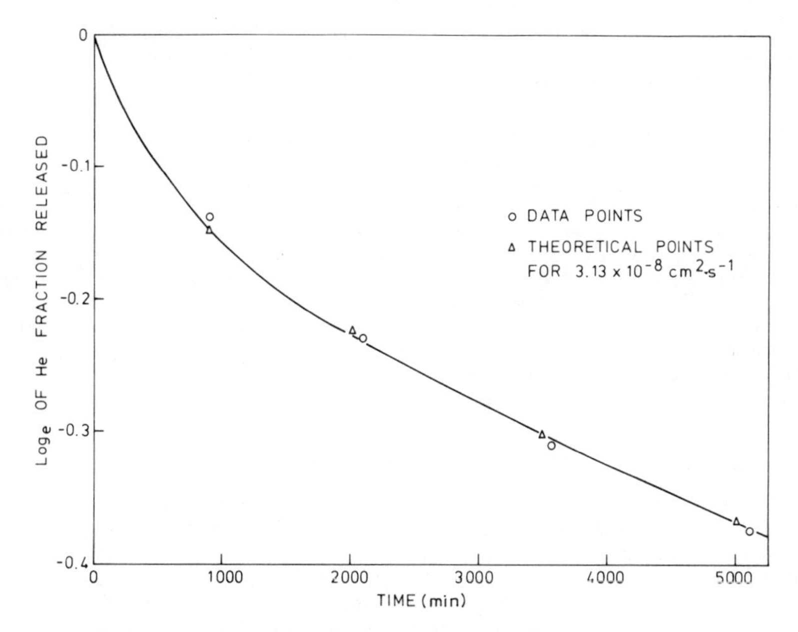

FIG.6. Helium diffusion at 480°C.

released into the capsule from the samples stored at 50°C, 0.01 cm^3 of helium per cm^3 was released from the samples into the capsule stored at 170°C. This is equivalent to a diffusion coefficient of ~4×10^{-11} $cm^2 \cdot s^{-1}$. This value, taken in conjunction with values measured at 400-480°C, gives an activation energy of ~36 kJ·$mole^{-1}$, compared to ~25 kJ·$mole^{-1}$ for most borosilicate glasses, and a diffusion coefficient at 200°C comparable to that predicted for a glass containing 65% of network formers [5].

7.4. Leaching of plutonia glasses

Hot Soxhlet leaching tests were made on polished flat slabs cut from samples of both ^{238}Pu and ^{239}Pu glasses stored at 50°C and 170°C. After 10 days leaching, all specimens showed leach rates of ~1.5×10^{-3} $g \cdot cm^{-2} \cdot day^{-1}$, with no measurable trends to show that either storage temperature or α-damage had affected the glass.

7.5. Microscopic examination of plutonia glasses

Transmitted light samples of ^{238}Pu glasses measuring ~10 μm thick were examined. Autoradiographs showed that some of the Pu had separated out from the glass, and optical microscope examination showed cubes of ~5 μm side which were PuO_2 or a phase very rich in plutonium. No cracks or other defects could be seen, although it was observed that the ^{238}Pu glass stored at 50°C was more difficult to grind to very thin sections without cracking.

8. β-DAMAGE EXPERIMENTS

The average β-energy is about 0.5 MeV although since the more energetic β-particles will be more damaging it is probably better to simulate the β-particles using electrons accelerated to a somewhat higher voltage than this.

Two experiments have been carried out (a) using a Van de Graaff to irradiate samples over a period of about 50 hours and (b) using a high voltage microscope when the appropriate dose is accumulated in seconds.

8.1. Van de Graaff irradiations

Samples 1cm square and about 1mm thick were irradiated with electrons accelerated to 0.5 MeV in a Van de Graaff accelerator. The total dose was 10^{19} electrons per cm^2 and since the range will be about 1mm this is equivalent to about 10^{20} β-decays per ml, the 100 year dose for the actual glass. The irradiation took a total of 45 hours at 10 μA and the beam heated the sample to 240°C. The temperature was raised to 300°C using the sample heater. All the samples irradiated appeared to be completely unaffected as judged by inspection, by electron microscope examination and by density measurements before and afterwards. Small changes were observed in the absorption spectrum. It has not yet been possible to test samples for leach resistance.

8.2. Using the high voltage electron microscope

The microscope can give beam voltages up to 1 MeV and, even when reduced in intensity to the minimum necessary for viewing, the focussed beam still gives a very high dose rate and power density in the 1 μm thick specimen. Relating the electron flux through the microscope foil to the equivalent dose in a thick block of glass involves some assumptions since the β-particles will all be absorbed in the glass whereas most of the electron beam of the microscope passes through the specimen without displacing any atoms.

At room temperature all the glasses tested were completely unaffected by the electron bombardment even when subjected to doses calculated to be over 100x that which will be received in practice.

At 200°C and above, however, some of the glasses were found to form minute bubbles 0.02 μm in diameter over all the irradiated area. The minimum dose to form bubbles was only about 4x the lifetime β-dose. That they are bubbles and not voids was shown when some, formed at 300°C, expanded when the specimen was heated to 700°C without further irradiation: voids would be expected to collapse.

The tendency to form these bubbles was strongly dependent on composition, the glasses containing lithium being very susceptible. Indeed in the absence of lithium, but with an equivalent amount of sodium, the glasses were remarkably radiation resistant even at high temperature. The gas filling the bubbles has not been positively identified but is almost certainly oxygen which constitutes about 50% of the glass.

The damage appears to be caused by a direct displacement mechanism rather than an ionisation mechanism since the total volume of the bubbles for a given dose is almost unaffected by voltage between 100 and 1000keV. Bubbles do not form near surfaces or near previously damaged regions suggesting that gas atoms or molecules preferentially diffuse to a nearby sink rather than nucleate a fresh bubble.

It is possible that the electron beam heats the specimen to a very high temperature locally, causing it to decompose. Calculations of the temperature rise suggest that this is not so and some experiments with much lower beam currents produced bubbles after about 30 minutes: it seems unlikely that so small a specimen would take so long to heat up but would reach equilibrium quickly.

It is difficult to explain the difference between this experiment and the Van de Graaff irradiations. A feasible explanation is that free oxygen atoms produced by collisions from the high energy electrons will diffuse through the glass and can either be trapped at a 'vacancy' in the network or combine with another free oxygen atom to form a gas molecule. If the glass contains a concentration of 'vacancies' apart from those caused by irradiation then the chances of formation of a gas molecule will be much higher at the high dose rate in the electron microscope than at the lower dose rates in the Van de Graaff.

9. ACKNOWLEDGEMENTS

The authors would like to thank K.A. Boult, H.E. Chamberlain, R.K. Dawson, E.A. Harper, Dr. L. Hobbs, A. Hough, R.H. Smith and G.J. Weldrick for their contributions to the work reported here.

10. REFERENCES

[1] GROVER, J.R., CHIDLEY, B.E., UKAEA Rep. AERE-R 3178 (1960).
[2] HARRISON, R.K., et al., U.K. Institute of Geological Sciences Report No. 84 (1975).
[3] MORRIS, J.B., AERE Harwell, unpublished work (1975).
[4] MENDEL, J.E., WARNER, I.M., USAEC Rep. BNWL 1741 (1973).
[5] DOREMUS, R.M., Glass Science, J. Wiley and Sons (1973).

DISCUSSION

H.W. LEVI: As a general comment on irradiation experiments, I would say that the experimental conditions usually applied simulate a very unfavourable situation. The dose that a solidified product receives in reality over a very long period of time is delivered to it experimentally in one year. The radiation defects therefore have less chance to anneal than they would have under real conditions. Hence we can be sure that the radiation effects found under test conditions are worse than those that would occur in the stored vitrified waste.

J.A.C. MARPLES: Yes, I agree. Since, in general, the accelerated experiments give negative results, we can be doubly sure that our glasses will be stable under the much slower rate of irradiation that is found in practice.

W. LUTZE: Your temperature range for energy release tallies well with the range we found in borosilicate glasses after irradiating small powdered samples with thermal neutron fluxes of up to 10^{19} n/cm^2 at temperatures below 100°C. Energy for storage is delivered by recoil from the reaction $^{10}B(n, \alpha)^7Li$. The stored energy was found, however, always to be $\leqslant$ 10 cal/g.

There are two questions I would like to ask you. How did you measure the He release, and what was the temperature dependence of the release rate? This could provide an extrapolated value for room temperature.

J.A.C. MARPLES: The use of the $^{10}B(n, \alpha)^7Li$ reaction is not completely equivalent to actinide α-damage, since most of the latter is caused by the heavy recoil nucleus; the lithium-7 nucleus has more similarity to an α-particle in its effects. Perhaps this, or slight composition differences, or different degrees of crystallization, affect the quantity of stored energy.

The helium release was found by collecting the helium in a capsule surrounding the sample and measuring the quantity with a mass spectrometer at the end of the storage period.

We have measured the diffusion coefficient at various temperatures by determining the helium release rate. When the experiments are finished, we shall certainly be able to find a room-temperature value by extrapolation.

Nina V. KRYLOVA: How much molybdenum can be incorporated into the glass, in weight per cent, without the yellow phase occurring?

J.A.C. MARPLES: The only molybdenum present in the glass during our experiments was that used to simulate the fission product levels (~ 1 wt%). In the absence of sulphate this is certainly not enough to produce the yellow phase.

W. HILD: I can perhaps add a comment on the yellow phase formation. As pointed out in the paper that my co-author, Mr. Kaufmann, presented (paper IAEA-SM-207/79, these Proceedings, Vol. 1), we have carried out tests on this yellow phase and on light phase formation. For borosilicate glass formulations similar to those investigated by you we determined conditions under which up to 3.5 wt% molybdate can be incorporated into the glass without phase separation. This figure is also valid for mixtures of molybdates and chromates, in addition to 1.5 wt% sulphate.

J.A.C. MARPLES: Our experiments on the occurrence of this phase under our conditions have not been as extensive as this, but our experience would be consistent with the values you report.

R.W. BARNES: Is the yellow phase, i.e. sulphate-chromate-molybdate, likely to separate out in later years?

J.A.C. MARPLES: The yellow phase forms during the heating cycle and then floats on the glass. Under equilibrium conditions it will usually redissolve, but equilibrium may not be attained under production schedules. It is therefore important to prevent the phase forming in the first place. We do not think the phase will separate out after the glass has formed, especially under storage conditions where the temperature will be well below 400°C.

EVALUATION OF PRODUCTS FOR THE SOLIDIFICATION OF HIGH-LEVEL RADIOACTIVE WASTE FROM COMMERCIAL REPROCESSING IN THE FEDERAL REPUBLIC OF GERMANY

E. EWEST, H.W. LEVI
Nuclear Chemistry and Reactor Division,
Hahn-Meitner Institut für Kernforschung Berlin GmbH,
Berlin,
Federal Republic of Germany

Abstract

EVALUATION OF PRODUCTS FOR THE SOLIDIFICATION OF HIGH-LEVEL RADIOACTIVE WASTE FROM COMMERCIAL REPROCESSING IN THE FEDERAL REPUBLIC OF GERMANY.

The main purpose of the solidification product within the waste management scheme is to serve as a second barrier against release of radioactivity from the final disposal site. When defining the stability requirements quantitatively, the general philosophy is to make the product 'as stable as practical'. The stability categories to be considered are, in order of increasing significance: irradiation stability, thermal stability, mechanical stability, and chemical stability. The product groups to be analysed are calcines, glasses, ceramics and composites. The conclusion is that glasses are at present the first choice for waste solidification with the long-term emphasis on borosilicate glasses. Glass ceramics and metal matrix composites may be important advances to meet the 'as stable as practical' requirement.

1. INTRODUCTION

The safe disposal of highly radioactive waste from fuel reprocessing is a key issue in the long-term exploitation of nuclear energy. The risks associated with radioactive waste are different from other nuclear risks in that:

(1) its magnitude will increase rather than approach a steady state as long as nuclear power is generated
(2) it will remain for many generations after nuclear power has been replaced by other sources of energy.

An adequate concept of high-level radioactive waste management has been developed in the Federal Republic of Germany that includes the following basic steps:

(a) Interim tank storage of liquid waste
(b) Solidification after five years of liquid storage
(c) Interim storage of solidified waste in surface facilities
(d) Final disposal in a salt mine.

Of these steps solidification has received most attention over the last ten years so that a range of options now exists, though at different stages of development. A comprehensive study on the entire radioactive waste problem is in progress in the Federal Republic of Germany, initiated and funded by the Minister of Research and Technology and carried out by eight institutions actively engaged in waste research or management. This paper is a result of this study and will present an approach to a comparative evaluation of alternative solidification products and to the further policy of waste solidification.

The safety philosophy of the final disposal concept is based on the idea that the underground disposal facility is designed to guarantee isolation of radioactivity from man's environment. Solidification of the waste serves two purposes. It is an additional safety factor in all handling and storage operations. Primarily, however, the solidification product is considered a second release barrier in the final disposal concept. In the event of the geological barrier failing it has to be assumed that the waste container has corroded away and that the long-term safety rests mainly in the inherent stability of the product.

The general requirement to be met by a solidification product is stability against any destructive influence to which a highly radioactive solid may be exposed, i.e.:

(a) Irradiation stability
(b) Thermal stability
(c) Mechanical stability
(d) Chemical stability.

As absolute stability cannot be achieved, the question is how stable is stable enough. At least at present, this question can hardly be answered because of the lack of reliable information on long-term failure modes of disposal facilities like salt mines and on the extremely small failure probabilities. Therefore, bearing in mind the particular quality of the risk associated with the waste, the product will have to be made 'as stable as practical'.

2. PRODUCT ALTERNATIVES

The present product alternatives, classified according to their basic nature, are:

(i) Calcines
(ii) Glasses
    borosilicate glass
    phosphate glass
(iii) Ceramics
    borosilicate glass ceramic
    supercalcine
(iv) Composites
    coated calcine particles
    matrix embedded calcine or glass particles.

Groups (i) and (ii) include well-known standard products. Different processes have been developed in different countries to obtain these products. Some of them have been demonstrated with full-scale radioactivity or even run on a routine basis for some time. There is no work on calcine development in the Federal Republic of Germany. Research and development work in this country is concentrated on two borosilicate processes, VERA [1] and FIPS [2], and on a phosphate glass process in two versions, PHOTHO and PAMELA [3].

Groups (iii) and (iv) include advanced products, each a follow-up development of one of the standard products.

The ceramics approach stems from the idea of incorporating fission product and actinide ions into chemically stable crystalline species that are suitable as host phases. Glass ceramic formation is essentially a controlled heat treatment of a borosilicate glass of appropriate composition, resulting in partial crystallization. A glass ceramic process has been elaborated in the Federal Republic of Germany [4]. Supercalcine is a ceramic obtained by adding appropriate chemicals to the waste solution, to enable the formation of the desired crystalline species, and firing the dry residue at a temperature higher than that used for ordinary calcines. Interest in this development is concentrated in the USA [5].

The idea of composites is to add properties to the basic product that it does not or not sufficiently possess. Coatings are to furnish calcine particles with chemical stability. Embedding of particles in an inactive and stable matrix is in the first place to make a granular product monolithic. In the process PAMELA [3], a German/Eurochemic co-development, phosphate glass particles are embedded in a metal, e.g. lead or aluminium, to improve the heat conductivity of phosphate glasses and its mechanical resistance.

In the following discussion a reference solidification product is assumed with a specific volume of 70 litres per tonne of irradiated fuel of 30 000 MW · d/t burn-up, 3% original enrichment, no Pu-recycle and 1100 days residence time in the reactor.

3. IRRADIATION STABILITY

With regard to radiation damage, α-decay is the most important type of radioactive decay. Figure 1 shows the total α-dose as a function of time; during the first 100 years it is of the order of 10^{18} cm^{-3}. Experimental approaches to simulate such α-doses in a relatively short time have been the incorporation of ^{244}Cm into the product and the irradiation of borosilicate glasses with thermal neutrons yielding B(n, α) reactions.

Among radiation effects energy storage received most attention. Figure 2 shows the temperature rise upon the release of energy in glass for the worst case, that is under adiabatic conditions. Irradiation experiments on glasses with a dose of 10^{18} α-particles per cm^3 at room temperature from ^{244}Cm led to about 20 cal/g stored energy, which was already in the saturation range [6]. The same dose from the B(n, α) reaction at 60°C yielded about 5 cal/g [7]. No studies on energy storage in crystalline solidification products have been published yet. Recent results from USA, however, indicate that there is no significant difference between glass and calcine [8]. No data exist for glass ceramics.

It should be borne in mind that the dose experimentally applied in hours or days will in reality be absorbed in 100 years. This gives a good deal of the defects a chance to anneal. In conclusion, the quantities of stored energy to be expected in any solidification product are probably not a serious risk.

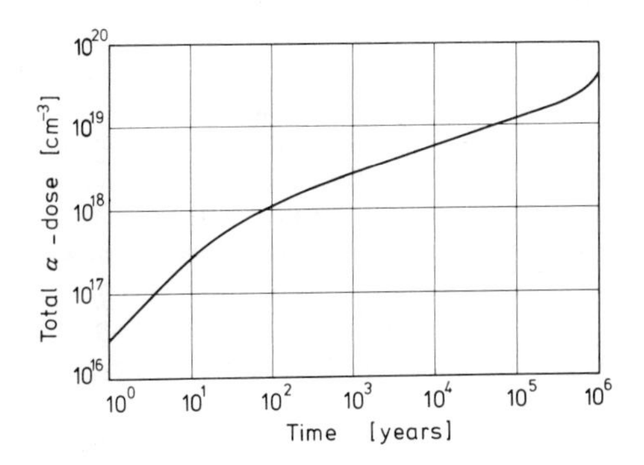

FIG.1. Total α-dose per cm³ of solidified waste versus time (70 l of solidified waste per tonne of heavy metal from LWR with 30 000 MW · d, 1100 days cycle time, 3% original enrichment, no Pu-recycling).

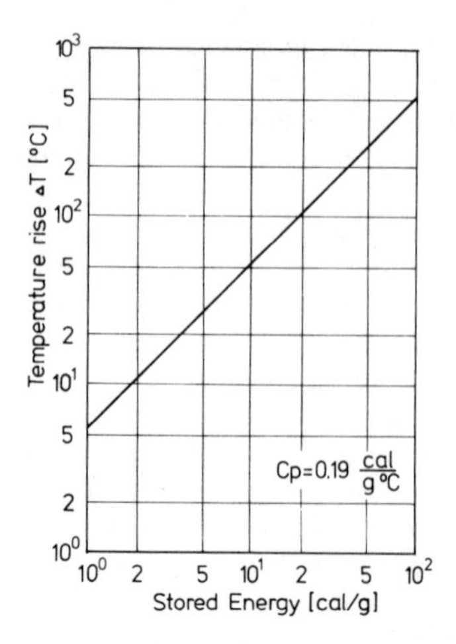

FIG.2. Temperature rise upon release of stored energy in glass under adiabatic conditions.

As yet, it is not quite clear whether the formation of helium by α-decay will affect the stability of the product. Bubble formation may occur at high concentration of helium, which is the consequence of low diffusivity and of the presence of defects like vacancies trapping the helium. The problem is still under study in several laboratories.

Radiation, as the most typical factor possibly affecting the long-term stability of solidified waste, has attracted remarkably little interest over many years of solidification research. Besides the still existing gaps in our information on energy storage and He effects, no experimental data are available on whether radiation affects the integrity or chemical stability of solidified waste. More experimental evidence is required to prove that radiation will have no serious effect on the long-term stability of solidified waste.

4. THERMAL STABILITY

The solidified waste cylinder is a homogeneous heat source with a temperature gradient along the radius. The difference between the centreline and the surface temperature is a function of the rate of heat generation and the thermal conductivity of the solidification product. Table I shows the thermal conductivities of several products. The surface temperature of the waste cylinder depends on the storage conditions and not on product properties. As yet the storage conditions have not been specified and so only temperature differences can be estimated – shown in Fig. 3 – as a function of the age of the waste. It is evident from the plot that as long as standard conditions (0.234 m cylinder diameter, 5 years ageing prior to solidification) are maintained there can be no significant benefit from improving the thermal conductivity beyond that typical for glasses.

As a result of the temperature gradient thermal stresses will build up in the solidified waste and may crack the material. In our calculations a maximum permissible stress of 500 kgf/cm^2 was assumed, which in turn leads to a maximum permissible temperature gradient of

$$\Delta T_{max.} \approx 5 \times 10^{-4} \frac{1}{\alpha}$$

(α = linear expansion coefficient)

TABLE I. THERMAL CONDUCTIVITIES, λ, OF DIFFERENT SOLIDIFICATION PRODUCTS

Product	λ ($W \cdot m^{-1} \cdot deg^{-1}$)
Borosilicate glass	1.2
Phosphate glass	1.0
Calcine	0.25
Borosilicate glass ceramic	1.5
Phosphate glass in Pb matrix	10.0

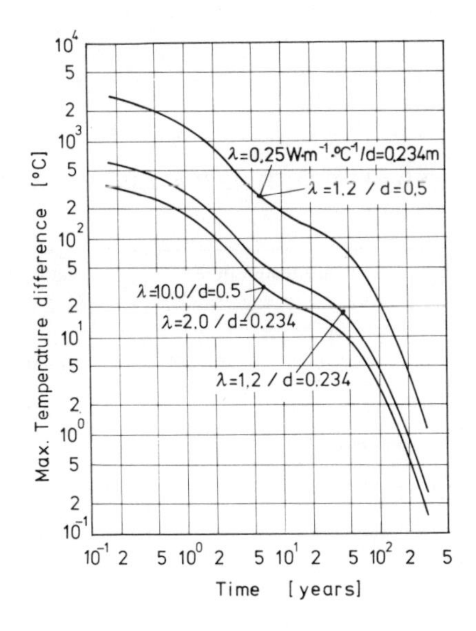

FIG.3. Temperature differences in a waste cylinder versus age of the waste.
γ = thermal conductivity ($W \cdot m^{-1} \cdot deg^{-1}$), d = diameter of the waste cylinder (for waste data see Fig.1).

Table II shows expansion coefficients of certain solidification products and the corresponding maximum permissible temperature gradients. Clearly they can be maintained under standard solidification conditions and with a thermal conductivity of about 1.2 $W \cdot m^{-1} \cdot °C^{-1}$.

From various safety aspects a softening and melting point as high as technology permits is desirable. Table III shows the relevant values. Very little is known on the volatility of fission products from various solidification products.

The long-term temperature effect most likely to occur is conversion into a thermodynamically more stable form. Glasses are well known to be a metastable form of matter and therefore likely to devitrify. This was confirmed experimentally for both borosilicate glasses [9] and phosphate glasses [10]. Elevated temperature and the presence of impurities favour the devitrification process.

TABLE II. EXPANSION COEFFICIENTS FOR VARIOUS MATERIALS

Product	Laboratory	$\alpha(\times 10^{-5})$	$\Delta T_{max.}$
Borosilicate glass			
VG 38	GfK	1.80	} 30
VG 98		1.56	
VC 15		0.89	60
Eucryptite	HMI	0.75	70
Glas ceramic			
VC 15	GfK	0.99	50
Eucryptite	HMI	0.50	100
Perovskite		1.00	50
Phosphate glass	Gelsenberg	1.10	50
Bottle glass		0.90	60
Fused Silica		0.05	1000
Steel		1.00–1.40	

TABLE III. SOFTENING AND MELTING TEMPERATURES OF SOLIDIFICATION PRODUCTS

Product	Softening temperature (°C)	Melting temperature (°C)
Borosilicate glass	500–600	1000–1200
Phosphate glass	350–450	800–1000
Glass ceramic	~750	~1200
Pb matrix	–	327
Calcine	very high	very high

Phosphate glasses are more readily devitrified than borosilicate glasses. Both types of devitrified products are more leachable than the parent glasses. The effect is more drastic with phosphate glasses. Studies on the devitrified products have shown that the responsibility for the increased leachability rests with the crystalline phases in the case of phosphate glasses and with the residual glass phase in the case of borosilicate glasses. Furthermore, it could be shown that upon devitrification of borosilicate glasses fission products may be enriched in very stable crystalline phases, depending on the composition of the glass.

5. MECHANICAL STABILITY

When a block of solidified waste is crushed by mechanical impact, fragments of various sizes will emerge. Two consequences are to be considered: the increase in the surface area will favour the leaching process; the formation of very small particles of the order of 100 μm and less will possibly enable radioactive material to be spread by air. Even larger particles may be spread by water.

Non-monolithic calcines do not have any mechanical stability. Preliminary results of impact tests have shown that metal matrix products are the most stable ones. Glass ceramics are more stable than glasses. However, the experimental data at present available do not permit a reliable comparison of all alternative solidification products. Further tests of products to be compared under equal experimental conditions are necessary.

6. CHEMICAL STABILITY

The only chemical attack on the solidification products deserving serious consideration is water leaching. Chemical interaction between rock salt and any of the solidification products under consideration is not to be expected unless the temperature rises above the melting point of the salt. A rough estimate shows that diffusion of fission products into the salt is not a significant safety factor.

The leaching behaviour is the most thoroughly studied property of solidification products. As a consequence, there is a wide spectrum of procedures and results. In the majority of the experiments an integral leaching rate R ($g \cdot cm^{-2} \cdot d^{-1}$) is determined in order to compare the leach resistance of different solidification products. In addition, there have been a number of experiments to study the time dependence of the leaching process. They should provide the necessary information to calculate the total radioactivity release upon accidental contact with water over a long period of time.

Leach rates measured on borosilicate glasses are in the range of 10^{-6} to $10^{-4} g \cdot cm^{-2} \cdot d^{-1}$, largely depending on experimental conditions. Almost identical results were obtained in some simultaneous tests with water and a NaCl solution. A careful study of the spectrum of data and procedures leads to the conclusion that $10^{-5} g \cdot cm^{-2} \cdot d^{-1}$ is a reasonable average suitable for characterizing the leach resistance of borosilicate glasses. Leach rates of phosphate glasses are within the range of borosilicate glasses, with a trend to somewhat lower values. Devitrification increases the leaching rates, in the case of phosphate glasses by up to three orders of magnitude.

TABLE IV. FRACTIONAL RELEASES FROM A SOLIDIFIED WASTE CYLINDER OVER 1000 YEARS

Calculated according to a 144-day leach experiment on borosilicate glass. The leach rate R was obtained from the 144-day release, the other coefficients were obtained by fitting the respective equations to the experimental curve. M and M', and A and A' are mass and surface area of the block and the experimental sample, respectively.

Leaching period (a)	Fractional release			
	$F = \frac{A}{M} \cdot R \cdot t$ ($R = 1.9 \times 10^{-6}$ $g \cdot cm^{-2} \cdot d^{-1}$) (%)	$F = \frac{A}{M} \cdot \frac{M'}{A'} \cdot a\sqrt{t}$ (%)	$F = \frac{A}{M} \cdot \frac{M'}{A'} (b\sqrt{t} + ct)$ (%)	$F = \frac{A}{M} \cdot \frac{M'}{A'} d \cdot t^{2/3}$ (%)
10	0.041	0.0074	0.021	0.014
20	0.083	0.011	0.039	0.022
50	0.208	0.016	0.091	0.041
100	0.416	0.024	0.177	0.065
1000	4.168	0.075	1.670	0.303

Leaching rates of the order of 10^{-5} g·cm^{-2}·d^{-1} have been found for borosilicate glass ceramics. Surprisingly, the controlled crystallization process, as opposed to the spontaneous one, does not increase the leachability relative to that of the parent glass. The same leachability as for the basic product is generally assumed for matrix composites.

Calcines are well known to be readily leachable in water. Hot pressed supercalcine shows a leach resistance similar to that of the other chemically stable products.

In general, it may be concluded that a leach rate in the range of 10^{-6} g·cm^{-2}·d^{-1} is probably something like a lower limit, unless a substantially different solidification technology is employed. Less leachable products may be very high melting glasses, high temperature ceramics or very sophisticated composites. It would be worth checking systematically whether materials science offers product alternatives substantially more stable and still realistic in terms of technology.

An important task in evaluating the risk of radioactivity release by water leaching is the mathematical description of the process. All efforts to do this on the basis of a physical understanding of the process have not led very far. This leaves us with empirical approaches where the physics of the process is packed into coefficients and exponents obtained by fitting experimental leach curves. Table IV shows a sample calculation of fractional releases from a long 0.234 m diameter waste cylinder over a period of 1000 years. They are obtained by fitting several mathematical expressions to experimental data from a 144-day column leach experiment with a powdered borosilicate glass and recalculation to the specific surface of the glass block. It may be noted that the results, except that with R = const. which is certainly unrealistic, are within a relatively narrow range. The mass fractions which are leached over a period of 1000 years are in any reasonable extrapolation below 2% of the original mass. This demonstrates that an up-to-date solidification product provides in fact an effective release barrier. It should be further pointed out that the experimental results and consequently the real leach process is highly dependent on several parameters like flow rate, pH value of the leachant and temperature. Bearing this in mind, it may be suggested that further efforts to penetrate the problem of leach kinetics more deeply are of no use. Instead, an $F = f(t^n)$ fit to experimental leach curves is considered the best and a good enough approximation to evaluate the radioactivity release from a waste block.

It is most important, however, to carry out a series of carefully designed leach experiments on relevant solidification products to supply reliable information on fractional release as a function of time, flow rate, chemical composition of the leachant and temperature. Furthermore, it remains to be proved how the leach behaviour of a large waste block fits to the experimental results obtained with small samples, even with powders. Finally, international agreement on test procedures should be reached to facilitate comparison of results. Only if this is done can a complete set of reliable data on the radioactivity release to be expected from different solidification products over relevant periods of time be made available.

7. PRODUCT EVALUATION

Table V lists the stability criteria in order of increasing weight relative to the long-term release barrier function of the solidification product. The stability categories are of different significance. Categories (3) and (4) are directly relevant for the release barrier function of the solidification product. Radioactivity release from the final disposal site may occur by a leaching process or, less likely, via spreading by air. Categories (1) and (2) are relevant mainly for maintaining the original properties of the solidification product.

Radiation stability was assigned the least weight because the dose to be expected is moderate and smeared out over a long period of time. Serious effects on the stability are unlikely. Moreover, the available experimental information is meager and not very helpful for a comparison of products.

Properties relating to the thermal stability have been characterized in more detail. They are considered important for maintaining the product stability over intermediate periods of time.

TABLE V. STABILITY CRITERIA

Rank	Stability category	Individual criteria
1	Radiation	Energy storage Effects on other stability categories
2	Thermal	Heat conductivity (max. temperature and thermal stresses) Softening or melting temperature thermodynamics
3	Mechanical	Degree of fragmentation
4	Chemical	Leachability

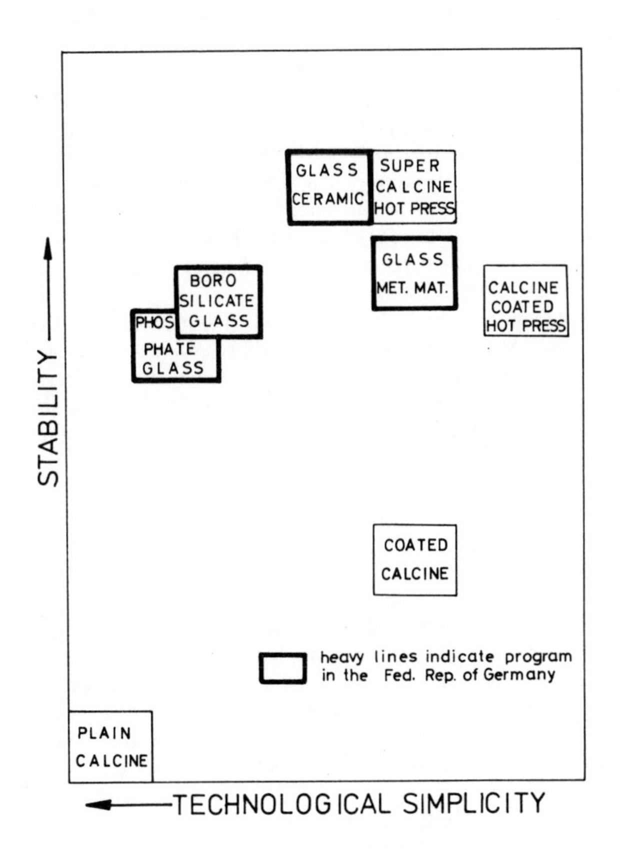

FIG.4. Schematic presentation of product evaluation.

As leaching is the most likely path by which radioactivity may be released, chemical stability has been assigned the highest rank.

In the final evaluation the product technology has to be considered as well. If one is selecting a product for use in the near future, this will have a high weight. It might be better to have the waste solidified in an acceptable form, even if not in the very best one. In the long run, however, stability criteria should clearly be paramount.

Figure 4 is a semi-quantitative approach to a rating of solidification products according to stability and technological simplicity. Such an attempt is obviously never free of personal judgement and should not be over-interpreted. Nevertheless, it is felt that the general picture is correct.

8. POLICY CONSIDERATIONS

The following considerations on waste solidification policy in the Federal Republic of Germany are derived:

Calcines are not suitable as final solidification products. They have no favourable mechanical and chemical stability properties, which are the categories considered most important.

Coating and matrixing the calcine particles may overcome the stability drawbacks but will replace them by severe technological difficulties. Calcine may only be discussed as a non-final solidification product for interim storage, provided the safety concept does definitely not require inherent stability of the product. In this case it may be an advantage that the calcine can be converted into a final product with some advanced technology possibly available at a later time.

Glass is among the solidification products with the highest mechanical, chemical and irradiation stability currently known. As far as the thermal stability of glass is concerned the possibility of devitrification and its not very high softening temperature are disadvantages, causing some uncertainty about its long-term performance. Glass solidification technology is in an advanced state of development and is likely to be available in the Federal Republic of Germany in the mid-eighties. In view of their properties and the state of technology, glasses are at present the first choice for high-level waste solidification.

There is still competition between borosilicate and phosphate glasses. Under the conditions of the present German HAW reference scheme they will be essentially equivalent in terms of stability. Therefore the decision should be made according to the availability of the technology.

In the long run, however, borosilicate glass should be given preference. Its higher thermal stability provides more flexibility in terms of waste management schemes, it leaves the option of a proper ceramic product, and it will probably be a world-wide solution to the solidification problem with all the benefits from exchange of experience.

A ceramic product, thermodynamically more stable than glass, will reduce the uncertainty about the state of solidified waste after a period of storage or disposal. It may become desirable in order to meet the 'as stable as practical' requirement.

Among the alternative ceramic products, glass ceramic is favoured in the Federal Republic of Germany. This is mainly because the glass ceramic process probably requires only slight modifications of the borosilicate process. It is also believed that this process offers the best chance of achieving with a reasonable effort a ceramic product with at least the same chemical and mechanical stability as glass. The technological penalty for any ceramic product with tailored crystalline host phases is a reduced flexibility towards the chemical composition of the waste. Current experience, however, indicates that this penalty will be tolerable in the case of glass ceramics.

The metal matrix process PAMELA, originally developed for phosphate glass particles, may be suitable also for borosilicate glass and even for glass ceramic. This should provide a product of excellent impact stability and extremely high thermal conductivity. The latter may be useful if less ageing time prior to solidification or cylinders of larger diameters are desired. Drawbacks are the lower overall melting point of the product due to the metal, uncertainties about the long-term chemical stability of the metal and the relatively complex technology.

The above spectrum of products and processes covers a sufficient number of reasonable alternatives to be sufficiently sure that the problem of waste solidification can be solved. Besides continued effort to develop the technology, further research in the area of product characterization and evaluation is required. On the one hand, more and precise data on irradiation behaviour, volatilization of radioactive species, relevant mechanical properties and leachability are needed. Collection of those data should be an internationally co-ordinated effort. On the other hand, the problem has to be approached to replace the 'as stable as practical' principle by defining stability requirements that have to be met.

REFERENCES

[1] GUBER, W., et al., these Proceedings (Paper IAEA-SM-207/79), Vol. 1.
[2] HALASZOWICH, S., et al., these Proceedings (Paper IAEA-SM-207/19), Vol. 1.
[3] VAN GEEL, J., et. al., these Proceedings (Paper IAEA-SM-207/83), Vol. 1.
[4] DE, A.K., et. al., these Proceedings (Paper IAEA-SM-207/11), Vol. 2.
[5] McCARTHY, G.J., COO-2510-5 USERDA Rep.
[6] MENDEL, J.E., McELROY, J.L., 77th Annual Meeting of the American Ceramic Society, Washington 1975.
[7] REENTS, D., personal communication.
[8] ROBERTS, F.P., personal communication.
[9] DE, A.K., et. al., 77th Annual Meeting of the American Ceramic Society, Washington 1975 and European Nuclear Conference, Paris 1975 (in press).
[10] BLASEWITZ, J.G., et. al., Management of Radioactive Wastes from Fuel Reprocessing (Proc. IAEA/NEA Symp. Paris, 1972), OECD/NEA, Paris (1973).

DISCUSSION

R.W. BARNES: I am pleased to see that ceramics are also being considered as appropriate materials. It was reported by Mr. Isaacson at the Paris Conference organized by OECD in 1972 that by forming caesium aluminosilicate the caesium ion is locked in and cannot be exchanged. The solubility of caesium, normally characterized by high elemental volatility and relatively high solubility of the compounds, is reduced to the solubility of the aluminosilicate lattice, which is virtually insoluble in neutral or alkaline water. Both strontium and plutonium likewise form virtually insoluble silicates. This would seem to be an admirable back-up, if required. The use of ceramics would appear to open up a possibility of producing such compounds.

H.W. LEVI: Thank you for the comment. As I have stated, I regard a ceramic product as a useful improvement compared to glasses. The question whether a ceramic product will be required for final disposal is still open.

THE LEACHING OF RADIOACTIVITY FROM HIGHLY RADIOACTIVE GLASS BLOCKS BURIED BELOW THE WATER TABLE
Fifteen years of results

W.F. MERRITT
Biology and Health Physics Division,
Atomic Energy of Canada Limited,
Chalk River Nuclear Laboratories,
Chalk River, Ontario,
Canada

Abstract

THE LEACHING OF RADIOACTIVITY FROM HIGHLY RADIOACTIVE GLASS BLOCKS BURIED BELOW THE WATER TABLE: FIFTEEN YEARS OF RESULTS.

The results from two test burials of high-level fission products incorporated into nepheline syenite glass indicate that the nuclear wastes from fuel processing for a 30 000 MW(e) nuclear power industry could be incorporated into such glass and stored beneath the water table in the waste management area of Chalk River Nuclear Laboratories (CRNL) without harm to the environment.

1. INTRODUCTION

One of the problems associated with the nuclear power industry is the ultimate storage of the radioactive wastes produced. The processing of spent fuels results in the accumulation of highly radioactive solutions which must be stored in such a manner that the chance of release to the environment is minimal. As yet, this is not a serious problem in Canada, since the CANDU[1] heavy water reactor system makes such efficient use of the uranium fuel that processing is not as yet economic. At present, spent fuel is stored under water at the reactor sites. Future storage methods were discussed at this symposium by Mayman, and co-workers[2]. How ever, it is envisioned that in the future it will become economically feasible, if not mandatory, to process spent fuel to recover the fertile and fissionable isotopes from it. The need is more acute if one is operating or intends to operate reactor systems using enriched fuel.

At present, fuel processing wastes are still generally stored as concentrated solutions in special tanks in a controlled area. Elaborate precautions must be taken to prevent or contain leakage, and monitoring plus maintenance must

---

[1] Canada Deuterium Uranium.

[2] Paper IAEA-SM-207/91, these Proceedings, Vol. 1.

be provided. Because of the mobility of liquids there is general agreement that eventual solidification of these wastes will be required, either in the tanks or as highly resistant glasses or ceramics [1], with storage in specially engineered facilities such as vaults or salt mines.

Many processes have been studied for converting highly radioactive solutions into solids [2]. The ultimate in such an approach to the problem would be to develop a material, the leaching rate of which would be so low that it would be possible to store it by burial in the ground without containment. The development of such a material at Chalk River Nuclear Laboratories (CRNL) has been described by Watson, Aikin and Bancroft [3]. The material is a glass based on nepheline syenite, a naturally occurring alumino-silicate mineral. To prepare the glass, a mixture of 85% nepheline syenite and 15% lime was combined with a nitric acid solution of fission products in ceramic crucibles. The resulting gel was dried, denitrated at 900°C and then melted at 1350°C. Volatile components, mainly ruthenium and cesium were adsorbed on a heated bed of fire brick and iron oxide.

By 1960, laboratory operations had produced glass blocks sufficient for two test burials. The first of these has been described by Bancroft and Gamble [4]. Twenty-five blocks containing about 300 Ci of aged fission products and made to a highly resistant formula were buried in a 5 x 5 block grid below the water table in the CRNL waste management area in August 1958. When early results from this test showed leaching rates too low to be readily measurable, a second test was carried out and has been described by Bancroft [5]. Twenty-five blocks containing about 1100 Ci of aged fission products were buried in a tighter 5 x 5 block grid in a similar location in May 1960. Laboratory leaching tests of this glass, which contained a higher proportion of uranium and corrosion products showed a leaching rate about ten times higher than that found for glass used in the first test. Considering this higher dissolution rate, larger quantity of fission products and tighter configuration, it was calculated that the amount of radioactivity leaving the blocks should be higher by a factor of 150 than the first disposal.

A description of the methods used in monitoring the second disposal has been given by Merritt and Parsons [6]. The ground water velocity past the blocks was measured using tritium as a tracer. The ground water downstream from the blocks has been sampled regularly and analyzed radiochemically. In 1963, 1966 [7] and 1971, soil samples were taken from downstream of the blocks and the patterns and concentrations of radioactivity which had left the blocks were determined. Results from this test have now been obtained over a period of 15 years.

TABLE I. STRONTIUM-90 LEAVING SECOND TEST DISPOSAL

Year	^{90}Sr Concentration in Ground Water (μCi/ℓ)	Total ^{90}Sr Leaving Blocks Per Year (μCi) (corr. to 1960)	% of 15 a Total
1960	1.3×10^{-1}	1060	70
1961	2.4×10^{-2}	350	23
1962	1.0×10^{-3}	15	1.0
1963	1.4×10^{-3}	21	1.4
1964	1.0×10^{-3}	15	1.0
1965	7.0×10^{-4}	11	0.73
1966	2.6×10^{-4}	4.2	0.28
1967	2.1×10^{-4}	3.5	0.23
1968	1.7×10^{-4}	2.9	0.19
1969	1.9×10^{-4}	3.3	0.22
1970	1.7×10^{-4}	3.0	0.20
1971	1.3×10^{-4}	2.3	0.15
1972	2.0×10^{-4}	3.8	0.25
1973	1.2×10^{-4}	2.3	0.15
1974	1.2×10^{-4}	2.5	0.17
		1500 μCi in 15 a	

2. RESULTS

Three ground water samplers positioned 0.6 m apart and 1 m downstream of the glass blocks have been sampled monthly from April to November each year since 1960. The water was analyzed for ^{90}Sr and the average content for each year is shown in Table I. The ground water velocity was measured to be 18 cm/d, the flow cross-section of the blocks was 5600 cm^2 and the soil porosity was 38%. Therefore 39 ℓ/d of water flows past the blocks. Using this figure the amount of ^{90}Sr leaving the blocks per year was calculated and is shown in the same Table.

The amount of ^{90}Sr leached from the blocks has remained essentially constant for the past seven years. The amount of ^{90}Sr that has left the blocks in 15 years is about 1.5 mCi, 90% of which was in the first two years.

Sampling in 1971 showed that the ^{90}Sr front had reached 33 m from the glass blocks. This is in good agreement with predictions from the surveys in 1963 and 1966, and indicates that ground water conditions have not altered significantly since the start of the experiment.

Laboratory studies of the glass used in the first test indicated that it would have a leaching rate about 1/10 of that used in the second test. To date, it has not been possible to detect any ^{90}Sr in the water samples taken downstream of the first test disposal. The sensitivity of our sampling and counting methods was such that it should have been possible to detect ^{90}Sr in the ground water if the leaching rate was 1/5 that of the glass used in the second test. Therefore, we can assume that it is possible to make glass with a leaching rate five times lower than that observed for the second test.

3. DISCUSSION

Thirty-nine ℓ/d of water flows by the blocks. The concentration of ^{90}Sr in the glass blocks was 1.0×10^{-2} Ci/g. The total surface area of the blocks is 1.1×10^{4} cm^2. Therefore, using the values for ^{90}Sr in the ground water given in Table I, it is possible to calculate the amount of glass being leached daily. For example in 1974,

$$\frac{1.2 \times 10^{-4} \times 39}{1.0 \times 10^{-2} \times 10^{6} \times 1.1 \times 10^{4}} = 4 \times 10^{-11}\ g \cdot cm^{-2} \cdot d^{-1}.$$

For the past seven years the ^{90}Sr leaving the blocks has remained essentially constant and averaged 1.6×10^{-4} $\mu Ci/\ell$. Table II shows the leaching rate based on the ^{90}Sr, up to the end of 1974.

In a study made in 1958, Watson, et al. [8] calculated the parameters needed to evaluate methods of permanent storage. Using their data as applied to a 1200 MWe CANDU heavy water reactor, and assuming that all the fuel is processed and the fission products are incorporated in nepheline syenite glass

TABLE II. LEACHING RATE OF GLASS BLOCKS

Year	Leaching Rate ($g \cdot cm^{-2} \cdot d^{-1}$)
1960	4×10^{-8}
1961	7×10^{-9}
1962	4×10^{-10}
1963	3×10^{-10}
1964	3×10^{-10}
1965	2×10^{-10}
1966	8×10^{-11}
1967	6×10^{-11}
1968-74	5×10^{-11}

and buried below the water table in the CRNL waste management area, then it is possible to use the leaching rates shown in Table II and calculate the cumulative release of ^{90}Sr from the blocks. The results are shown in Figure 1. After 200 years of continual operation of a 1200 MWe reactor, about 200 Ci of ^{90}Sr would be in the plume downstream of the blocks. An examination of the curve shows that about 4 Ci/a of ^{90}Sr leaves the blocks. After about 30 years the decay of ^{90}Sr begins to become apparent and equilibrium is reached in about 200 years. The lower curve shows the result of stopping operations after ten years when a maximum amount of 40 Ci would be leached.

Suppose such a waste storage facility was set up at CRNL and the glass blocks buried in the Perch Lake basin. The measured velocity of ^{90}Sr from the second test disposal has been 33 m in 12 a or 2.75 m/a. The distance from a suitable storage site to Perch Lake is 500 m. Therefore it would take 180 years, or six half-lives for the ^{90}Sr to reach the lake. By this time, the 4 Ci/a released from the blocks would have decayed to

$$\frac{4 \times 10^3}{64} = 62.5 \text{ mCi/a} \quad \text{or } 170 \ \mu\text{Ci/d}.$$

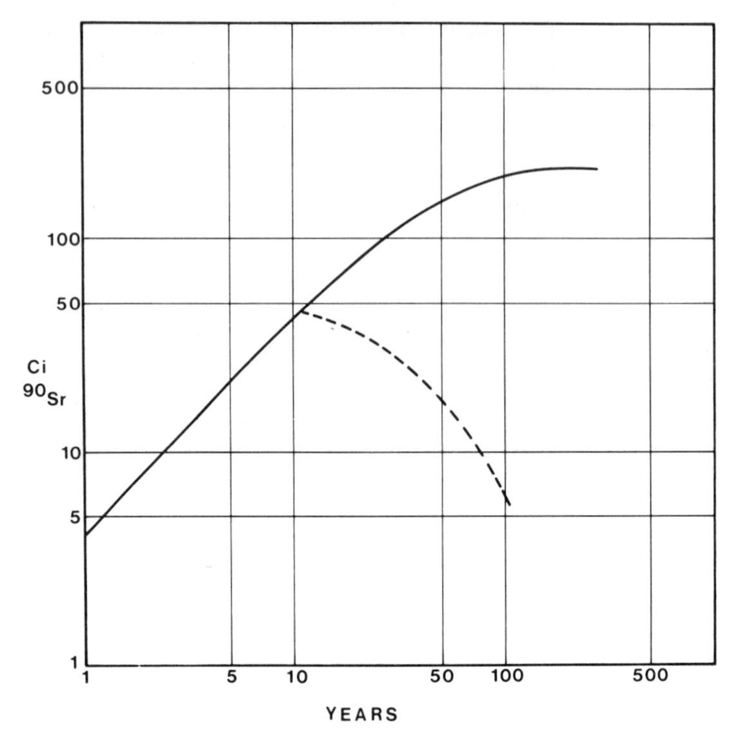

FIG.1. Total ^{90}Sr leached from blocks for 1200 MW(e) CANDU reactor.

High-level wastes from a fuel processing plant servicing a 30 000 MWe nuclear industry would, under these conditions, release only 5 mCi/d to Perch Lake.

Methods of improving the performance of the glass blocks are available. For example, the first test demonstrated it was feasible to manufacture glass with a leaching rate at least five times lower than the glass used in the second test. Also, the data in Table I shows that over 90% of the ^{90}Sr leaving the blocks did so in the first two years. Laboratory tests [3] have shown that pre-leaching of the blocks would prevent this original large release. The pre-leaching might have to be done in situ, and would generate a secondary liquid waste stream but would seem to be quite feasible. Again, making larger blocks with a lower surface to volume ratio than the blocks used would reduce the leaching rate.

It should be pointed out that these figures only hold if the glass is actually buried beneath the water table and in a configuration such that the ground water temperature is not significantly increased. Laboratory results [3] have shown that high water temperatures increase the rate of leaching considerably. Also it has been shown that devitrification of the glass will also increase the leaching rate. On the other hand, tests have shown that the irradiation received by the glass has no effect on the leaching rate.

When the glass was buried for the second test, three of the blocks lodged above the grid and have not been considered in the calculations. It should be possible to recover at least one of these blocks without disturbing the experiment. It would then be feasible to examine and test the block in a hot cell. The author would be interested to get the opinion and advice of others working in this field on the usefulness of such a procedure and on the type of examination and measurements required.

4. CONCLUSION

The results indicate that the incorporation of high-level nuclear wastes in nepheline syenite glass and burial of the glass directly in the soil in the CRNL controlled area would be a suitable method for permanent storage of such wastes from a 30 000 MWe nuclear power industry.

REFERENCES

[1] Proceedings of a Symposium on the Management of Radioactive Wastes from Fuel Reprocessing, OECD, Paris (1972) 181-446.

[2] Ibid 448-788.

[3] WATSON, L.C., AIKIN, A.M., BANCROFT, A.R. "The permanent disposal of highly radioactive wastes by incorporation into glass", Disposal of Radioactive Wastes (Proc. Conf. Monaco, 1959) 1, IAEA, Vienna (1960) 375-399.

[4] BANCROFT, A.R., GAMBLE, J.D. "Initiation of a field burial test of the disposal of fission products incorporated into glass", Atomic Energy of Canada Limited, Report AECL-718 (1958).

[5] BANCROFT, A.R. "A proposal for a second test of ground burial of fission products in glass", Atomic Energy of Canada Limited, Unpublished Report (1960).

[6] MERRITT, W.F., PARSONS, P.J. "The safe burial of high-level fission product solutions incorporated into glass", Health Phys. 10 (1964) 655-664.

[7] MERRITT, W.F. "Permanent disposal by burial of highly radioactive wastes incorporated into glass", Disposal of Radioactive Wastes into the Ground (Proc. Symp. Vienna, 1967), IAEA, Vienna (1967) 403-408.

[8] WATSON, L.C., RAE, H.K., DURHAM, R.W., EVANS, E.D., CHARLESWORTH, D.H. "Methods of storage of solids containing fission products", Atomic Energy of Canada Limited, Report AECL-649 (1958).

DISCUSSION

F. GERA *(Chairman)*: The leach rates that you have reported are several orders of magnitude lower than those usually given. Do you think this might be due, at least partially, to differences in the testing conditions?

W.F. MERRITT: We feel our leach rates are correct. We checked the amount of radioactivity leaving the blocks by both groundwater and soil sampling, and the results obtained by the two methods tallied. The whole history of the leaching is in the plume downstream from the blocks.

K. KÜHN: I would agree that the figures for the leach rates you have found in your long-term experiments are much lower than those normally reported from laboratory tests. Since the normal leaching medium in the laboratory is distilled water, can you give some details on the groundwater chemistry and also the groundwater flow?

W.F. MERRITT: The soil in the experimental area is a fine water-deposited sand from glacial melt. The groundwater is soft, with a low total solid content of about 40 ppm, with silica representing about 12 ppm, iron 3 ppm, aluminium 5 ppm, calcium 5 ppm and smaller amounts of sodium and magnesium, and low potassium. The pH value is about 6.5 and the mean temperature about 7°C. Soil bacteria maintain reducing conditions in the soil.

J.A.C. MARPLES: I do not think that the initial leach rates observed in your experiments – 4×10^{-8} g/cm²·d in 1960 – are very different from those observed during laboratory tests, if the difference in temperature is taken into account. A change from 5 to 100°C will make a difference of a factor of nearly 10^3. The decrease in leach rate with time under your realistic test conditions is very encouraging.

Nina V. KRYLOVA: Mr. Merritt, I would like to commend your very well organized experiments on the long-term behaviour of glasses when buried in the ground. But do you think it would be possible to bury actual vitrified waste in the ground without additional protection?

W.F. MERRITT: Our results show that it is technically feasible from the point of view of radioactivity release. There are, of course, engineering conditions to be considered, the volume of glass and so on. Heat production should not be a problem, since for natural uranium fuelled reactors there is no pressure for early processing and we can afford to wait for several years before processing our fuel.

S.A. MAYMAN: It should be clarified, I think, to avoid any misunderstanding, that in replying to the last question you expressed a personal opinion about the technical feasibility of such an operation. The present Canadian policy on the disposal of reprocessing wastes is that they will be placed, in some solidified form, in mined cavities 300 to 1000 m below the surface, in either granitic or salt formations. The paper I presented earlier this week gives some details of our development programme (Paper IAEA-SM-207/91, these Proceedings, Vol. 1).

W.F. MERRITT: Thank you, Mr. Mayman.

A. DONATO: Have you also determined the leaching of ^{137}Cs under the conditions that would prevail in a real disposal operation, as you did with ^{90}Sr?

W.F. MERRITT: ^{137}Cs migrates much more slowly through the soil than ^{90}Sr. It has moved only about 2 m from the blocks. We determine it in our soil sampling programme, but most of it is retained close to the blocks, where we do not do any sampling for fear of disrupting the experiments.

J.B. MORRIS: I am not quite certain whether the buried glass blocks are entirely submerged and surrounded by water or merely in contact with damp or saturated soil. Can you clarify?

W.F. MERRITT: They are in the saturated zone. The groundwater velocity is about 15 cm/d and about 39 l/d passes through the grid.

H. KRAUSE: The great merit of your experiments is that they have been carried out under the most favourable conditions that can be found in nature and that they have been pursued over a very long period of time. Their value would be increased still more if they could provide data on the fate of plutonium. Have you by any chance made any measurements of the plutonium escaping from the glass?

W.F. MERRITT: Our experience with plutonium is that it is firmly retained by the Canadian soils. Hence any plutonium leached from the blocks is adsorbed on the soil immediately adjacent to the blocks, where we are unwilling to sample it for the reason I gave.

CONFINEMENT DE LA RADIOACTIVITE DANS LES VERRES

F. LAUDE, R. BONNIAUD, C. SOMBRET, G. RABOT
CEA, Centre de Marcoule,
Bagnols-sur-Cèze, France

Abstract–Résumé

CONFINEMENT OF RADIOACTIVITY IN GLASS.

It is now universally recognized that the concentrated solutions of fission products currently stored in stainless steel tanks will have to be converted into a solid exhibiting maximum long-term stability. In France the Commissariat à l'énergie atomique has developed industrial vitrification processes. As a material that is mineral in composition, homogeneous and non-porous, glass has a certain number of advantages; for example, it can dissolve most oxides in the hot state and hence almost all fission products are easily incorporated into it. Since the various types of fission product solutions have different chemical compositions, they require the use of a relatively broad range of glass formulae – these are the subject of systematic studies. The selected glasses should possess the property of adequate nuclear, thermal and chemical stability. The study of these characteristics is carried out individually, according to the amount of the alpha emitters that the glasses contain. Investigations include hot tests with glass blocks of approximately 700 cm^3 volume, manufactured by processes similar to those employed in industry. It can be taken that the effect of the beta and gamma emitter fission products on the behaviour of the borosilicate glasses selected, given a leaching rate between 10^{-8} and 10^{-6} $g \cdot cm^{-2} \cdot d^{-1}$, remains small. The effect of the alpha emitters is yet to be clarified. These glasses were studied by incorporating increasing quantities of alpha emitters in the form of ^{238}Pu, ^{241}Am and ^{244}Cm into blocks approximately 700 cm^3 in volume. The results obtained will govern the policy that is to be pursued with regard to possible separation of actinides.

CONFINEMENT DE LA RADIOACTIVITE DANS LES VERRES.

Il est maintenant unanimement reconnu que les solutions concentrées de produits de fission actuellement stockées dans des cuves en acier inoxydable devront être transformées en un solide présentant le maximum de stabilité à long terme. En France, le Commissariat à l'énergie atomique a développé des procédés industriels de vitrification. Le verre, matière minérale, homogène et non poreux, présente un certain nombre d'avantages. Il peut dissoudre à chaud la plupart des oxydes; presque tous les produits de fission sont donc facilement incorporables. Les différents types de solutions de produits de fission, ayant des compositions chimiques différentes, conduisent à l'utilisation de formules de verre relativement variées faisant l'objet d'études systématiques. Les verres sélectionnés doivent posséder des qualités de stabilité nucléaire, thermique et chimique suffisantes. L'étude de ces propriétés est menée séparément suivant l'importance des émetteurs α qu'ils contiennent. Ces études comprennent des tests en actif effectués avec des blocs de verre d'environ 700 cm^3, réalisés par des procédés similaires aux procédés de fabrication industriels. On peut considérer que l'influence des produits de fission émetteurs $\beta\gamma$ sur la tenue des verres borosilicatés sélectionnés, à taux de lixiviation compris entre 10^{-8} et 10^{-6} $g \cdot cm^{-2} \cdot d^{-1}$, reste faible. Par contre l'influence des émetteurs α est encore à préciser. Une étude de ces verres a été entreprise en incorporant dans des blocs d'environ 700 cm^3 des quantités croissantes d'émetteurs α sous forme de ^{238}Pu, ^{241}Am et ^{244}Cm. Des résultats obtenus dépend la politique qui sera suivie concernant la séparation éventuelle des actinides.

1. INTRODUCTION

Parmi les déchets radioactifs engendrés par l'industrie nucléaire, les solutions concentrées de produits de fission, résidus du traitement des combustibles irradiés dans les réacteurs, méritent une attention particulière. Ces déchets liquides sont actuellement stockés dans des cuves en acier inoxydable sur les sites de production dans des conditions de sûreté satisfaisantes.

TABLEAU I. DIFFERENTS TYPES DE SOLUTIONS DE PRODUITS DE FISSION ETUDIES EN VITRIFICATION

Type de combustible	Taux de combustion ($MW \cdot d \cdot t^{-1} \cdot 10^3$)	Taux de concentration des produits de fission ($litres \cdot t^{-1}$)	H^+ (N)	Al[a]	Na	Mg	Fe	Ni	Cr	U	F	Gd	Mo hors fission	Produits de fission
Sicral G, uranium naturel	1	30	2	30–35	19–23	4–5	15–17	1–2	1–2	3	8			25
U,Mo à 1,1% en Mo, uranium naturel	2–3	110	2	2,5	5	2,4	2	0,5	0,5				150	28
Sicral EDF, uranium naturel	4	100	1,5	15	8		1,8			2	4			45
U,Al uranium enrichi	220 à 700	12 000	-1,8	81	2		2				12			10
Oxyde, réacteurs à eau légère, uranium enrichi	33	500	1,5	0,1	5		1,2	0,2	0,4	9,5				78
Oxyde, réacteurs à neutrons rapides U + Pu	60	2 000			20		15	1	2			25		24

[a] Toutes ces valeurs sont données en $g \cdot l^{-1}$

TABLEAU II. EXEMPLE DE COMPOSITIONS DE VERRE

Type de combustible	Volume de verre par tonne de combustible $(1 \cdot t^{-1})$	Oxydes apportés par la solution (%)	SiO_2 [a]	Na_2O	B_2O_3	Oxydes de PF	Al_2O_3	Fe_2O_3	MgO	Divers	P_2O_5	MoO_3	Gd_2O_3
Sicral G	8	30	48,8	15	14,2	4,8	8,4	2,6	6,1	0,2			
U, Mo	29	25	19	9,5	16,4	5	17,4	2,6	0,7		8,2	21,2	
U, Al	3200	26	34,8	19,1	19,1	2	22,8	0,2		1,7	0,3		
Sicral EDF	14	27,5	42,7	14,2	17,8	12,7	8,6	1,6	1	1,4			
Oxydes, eau légère	67	26	49	8,2	13,3	24,5	5						
Oxydes, neutrons rapides	330	20	41	19,7	13,6	6,9	5,8	4,9		1,5			6,6

[a] Toutes ces valeurs sont données en % pondéral.

Mais du fait de leur forte radioactivité et de la longue période de certains émetteurs, la solution du stockage sous forme liquide ne peut être que provisoire. Ces effluents devront être transformés en un solide offrant le maximum de garanties de stabilité pour pouvoir confiner à long terme la radioactivité.

En France, le Commissariat à l'énergie atomique s'est orienté dans cette voie, en développant des procédés industriels de vitrification et en poursuivant des études sur le comportement des produits obtenus [1, 2].

Le verre présente sur beaucoup d'autres matériaux utilisés pour le confinement de la radioactivité un certain nombre d'avantages.

Le produit obtenu est homogène, isotrope et exempt de porosité.

Le verre ne présente pas un caractère de spécificité chimique très marqué. La nature désordonnée de l'état vitreux fait que le solide peut dissoudre à chaud presque tous les oxydes et qu'en conséquence la plupart des produits de fission sont facilement incorporables. Cette propriété est importante étant donné la variété des éléments chimiques constituant les solutions de produits de fission.

Les différents types de solutions de produits de fission qui existent ou existeront en France selon le type de combustible utilisé sont rappelés dans le tableau I. Leur nature chimique différente conduit à l'utilisation de formules de verre relativement variées dont un exemple est donné dans le tableau II, pour chaque type de combustible.

Les verres utilisés sont en général des compositions siliceuses ayant une structure plus stable que celle des compositions phosphatées. Ils contiennent de l'anhydride borique ayant pour but d'abaisser la viscosité du verre à une valeur compatible avec les impératifs technologiques de coulée et de favoriser la digestion de l'oxyde molybdique.

Nos techniques de production aboutissent toujours à la coulée du verre; ainsi pour les coulées en pot métallique à 1150°C, la viscosité du verre doit être comprise entre 1000 et 100 poises à 1100°C.

Les volumes de verre produits par tonne de combustible sont, pour les combustibles à uranium naturel: 8 litres pour un taux de combustion de 1000 $MW \cdot d \cdot t^{-1}$, 14 litres pour 4000 $MW \cdot d \cdot t^{-1}$, 29 litres pour les combustibles à uranium allié au molybdène. Pour les combustibles à uranium enrichi, ces volumes sont de 3200 litres pour les combustibles des réacteurs de recherche MTR, 67 litres pour les combustibles oxydes des réacteurs à eau légère, 330 litres pour les combustibles oxydes des réacteurs à neutrons rapides.

Le pourcentage d'oxydes apportés par la solution représente 20% à 30% du poids du verre produit.

Les différentes sortes de verre font l'objet de nombreux tests ayant pour but de sélectionner les meilleures compositions.

Les verres sélectionnés doivent posséder les propriétés fondamentales suivantes:

- la stabilité nucléaire, autrement dit, une sensibilité faible à leur propre irradiation;
- la stabilité thermique: l'élévation de température du verre inévitable pendant les premières années de stockage ne devra pas trop altérer sa structure;
- la stabilité chimique et particulièrement la résistance à l'action de l'eau.

L'étude de ces propriétés a jusqu'ici été menée séparément suivant que le verre contenait ou non des actinides, ceci pour faciliter les expériences et cerner les problèmes.

2. COMPORTEMENT DES VERRES CONTENANT DES EMETTEURS $\beta\gamma$

Actuellement, après une dizaine d'années d'expérience, nous avons accumulé de nombreuses données dans ce domaine. Les essais inactifs sont toujours complétés par des tests actifs effectués sur des blocs de verre d'un volume de 700 cm^3 préparés par des méthodes identiques aux procédés industriels et dans lesquels la radioactivité incorporée provient d'authentiques solutions de produits de fission.

2.1. Stabilité sous rayonnement

Pour simuler l'auto-irradiation, des doses de 10^{11} rad ont été intégrées au moyen d'un irradiateur à électrons dans une vingtaine d'échantillons de verre représentant la gamme des compositions sélectionnées [3].

Pour atteindre cette dose, les échantillons, sous forme de pastilles de 20 mm de diamètre et de 4 mm d'épaisseur, ont été soumis à un flux d'électrons de $1,75 \cdot 10^{13}$ $cm^{-2} \cdot s^{-1}$, de 3 MeV d'énergie, pendant 288 heures, avec une puissance d'irradiation de 1,35 kW.

L'énergie de choc inélastique ainsi développée est 60 fois plus forte que celle des chocs dus à l'auto-irradiation.

Pendant l'irradiation, la température des échantillons était maintenue à 445°C. Les échantillons irradiés ont été comparés à des échantillons témoins et soumis à des examens pour déceler une éventuelle accumulation d'énergie, une variation du taux de lixiviation ou une modification de structure.

2.1.1. Accumulation d'énergie

Les mesures faites sur les échantillons maintenus à 50°C par analyse thermique et calorimétrie différentielles n'ont révélé, pour les conditions de l'expérience, aucune accumulation d'énergie mesurable.

2.1.2. Taux de lixiviation

La comparaison des taux de lixiviation de pastilles actives avant et après irradiation a montré que cette irradiation n'avait pas eu d'influence sur la lixiviation.

2.1.3. Modification de structure

Ces examens ont été effectués par diffraction X et par spectrométrie infrarouge. Les résultats de ces mesures ont permis de constater qu'il n'y avait aucune modification de structure notable, et aucune cristallisation inquiétante.

La verre, produit minéral, résiste donc suffisamment bien à l'action des rayonnements $\beta\gamma$.

2.2. Stabilité thermique

Les blocs de verre stockés seront le siège d'un échauffement dû à l'absorption d'une partie du rayonnement. L'élévation de la température au cœur du verre, dépendant de l'activité spécifique et de la géométrie des blocs, place le verre dans des conditions plus propices à sa cristallisation. Si cette cristallisation permet d'incorporer dans les cristaux la totalité des produits de fission à période longue, elle ne devrait pas avoir d'influence défavorable sur la lixiviation. Par contre, si la forme cristalline formée au détriment du verre confine insuffisamment les produits de fission, le verre restant assurant la cohésion des cristaux sera plus fragile.

Dans l'incertitude actuelle sur la répartition des produits de fission entre la phase vitreuse et la phase cristalline, nous jugeons préférable d'éviter au maximum sa cristallisation.

L'étude de la stabilité thermique du verre commence par la détermination de la tendance du verre à cristalliser sous l'action de la température, par la mesure des températures minimales de cristallisation et des vitesses de croissance des cristaux en fonction de la température.

Sur la plupart des verres étudiés et notamment ceux qui ont été utilisés par Piver[1] et ceux qui le seront dans l'atelier industriel de Marcoule, les températures inférieures de cristallisation

---

[1] Premier atelier pilote de vitrification à Marcoule, utilisant le procédé en pot.

se situent au-dessus de 750°C et la vitesse maximale de croissance des cristaux est de 0,01 à 0,03 $\mu m \cdot min^{-1}$ (verre sodocalcique industriel: 1 à 2 $\mu m \cdot min^{-1}$, verre Pyrex: 0,1 $\mu m \cdot min^{-1}$).

Les verres plus riches en oxydes de produits de fission tels que ceux prévus pour les combustibles des réacteurs à eau légère ont une température inférieure de cristallisation d'environ 600°C à 640°C et une vitesse de formation de cristaux plus grande.

Les formes cristallines sont souvent des composés chimiques du molybdène.

La présence du fluor dans les verres augmente la tendance à la dévitrification sans présenter un caractère de gravité.

Pour compléter ces mesures, une trentaine de verres dont les compositions peuvent être utilisées pour la gamme complète des combustibles ont été soumis à un séjour d'un an à des températures de 500°C et 600°C proches des températures maximales de stockage.

Tous les trois mois, des échantillons étaient prélevés pour suivre leur évolution par des examens macro et microscopiques.

D'une façon générale, les verres silicoboratés à haute teneur en silice se comportent bien (teneur en SiO_2 supérieure à 40%). Néanmoins cette condition n'est pas suffisante.

Si la teneur en oxydes de produits de fission est plus importante (de l'ordre de 25%), on observe une tendance à la cristallisation. La présence de molybdène favorise nettement la cristallisation. Par contre, la présence de bore (teneur en B_2O_3 de l'ordre de 15%) est bénéfique pour la stabilité de ces verres.

Le même traitement thermique est appliqué à des verres radioactifs, pour juger de l'incidence éventuelle de la cristallisation sur la lixiviation.

2.3. Stabilité chimique

La qualité du confinement de la radioactivité dans les verres est essentiellement appréciée par la lixiviation. Cette étude, comportant la fabrication de blocs de verre pesant environ 2 kg et dont l'activité peut dépasser 100 Ci, s'effectue dans une chaîne de cellules blindées implantée à Marcoule. Cette installation, dont la description détaillée a déjà été faite [4], est constituée d'une suite de cinq cellules reliées les unes aux autres. Les plus importantes de ces cellules sont:

– celle de vitrification, équipée d'un pot de 12 cm de diamètre et de 50 cm de hauteur chauffé par induction et d'un système complet d'épuration des gaz;
– celle de lixiviation, équipée pour tester simulanément huit blocs de verre.

2.3.1. Conditions normales de lixiviation

Les échantillons, sous forme de cylindres de 10 cm de diamètre et d'environ 10 cm de hauteur, sont placés chacun dans un récipient en acier inoxydable pour être soumis à la lixiviation à l'eau.

Les blocs sont arrosés avec 700 cm^3 d'eau ordinaire à la température ambiante et selon le cycle suivant:

– arrosage des pastilles pendant 30 secondes sans évacuation;
– immersion des pastilles dans le volume total pendant 30 secondes;
– évacuation de l'eau pendant 30 secondes.

L'eau est renouvelée chaque jour à la même heure et le test est poursuivi jusqu'à ce que l'équilibre soit atteint, c'est-à-dire lorsque la fraction d'activité entraînée devient sensiblement constante. L'ensemble de ces opérations est entièrement automatisé. Les mesures d'activité sont faites par comptage d'un échantillon de lixiviat.

Les taux de lixiviation sont exprimés en fraction d'activité passée dans l'eau par surface spécifique et par jour.

Nous avons choisi ce mode de lixiviation car une étude systématique des conditions de la lixiviation [4] avait démontré que la géométrie des blocs pour une même surface et le volume de l'eau étaient pratiquement sans influence.

Le renouvellement d'eau journalier augmente les taux de lixiviation d'environ 15%. L'influence de la circulation de l'eau (lixiviation dynamique) est faible, néanmoins nous préférons la maintenir pour avoir une meilleure homogénéité du lixiviat.

2.3.2. *Résultats obtenus*

Pour une même catégorie de verre, les taux de lixiviation restent compris dans une fourchette de valeurs relativement faibles.

Pour l'ensemble des verres sélectionnés, les taux de lixiviation se situent entre 10^{-8} et 10^{-6} $g \cdot cm^{-2}$.

Au-delà d'une teneur de 6%, le fluor dégrade légèrement le taux de lixiviation. Le remplacement de LiO_2 par Na_2O a une influence notable dès que l'on substitue la moitié des moles de Na_2O par LiO_2.

Les coefficients de diffusion de certains radioéléments dans le verre ont pu être déterminés et donnent les valeurs suivantes à 25°C:

- pour le strontium et le cérium: de 10^{-18} à 10^{-16} $cm^2 \cdot s^{-1}$;
- pour le ruthénium et le cérium: de $3 \cdot 10^{-21}$ à 10^{-20} $cm^2 \cdot s^{-1}$.

Ces coefficients varient assez fortement avec la température; par exemple pour le ruthénium, un coefficient de $3 \cdot 10^{-21}$ mesuré à 25°C atteint $6{,}7 \cdot 10^{-20}$ à 50°C et $5 \cdot 10^{-19}$ à 70°C. D'autres tests ont été faits en changeant la nature de la solution. Ainsi, avec une solution chargée en sels comme l'eau de mer, les résultats sont pratiquement les mêmes. Avec une solution acide à pH = 3, le taux de lixiviation est multiplié par 10.

Des tests de lixiviation par l'air circulant autour des blocs de verre à 4 $cm \cdot s^{-1}$ ont révélé qu'il n'y avait aucune radioactivité entraînée.

Pour ces essais on avait utilisé de l'air sec, de l'air à 40% et 80% d'humidité, et de l'air à 40% d'humidité et 0,1% de NO_2.

2.3.3. *Evolution de la lixiviation dans le temps*

Les taux de lixiviation contrôlés avec des verres actifs ayant été stockés 7 ans n'ont subi aucune modification.

Des prélèvements de blocs de verre produits par le pilote Piver seront effectués dans quelques années pour contrôler l'évolution de la lixiviation.

3. COMPORTEMENT DES VERRES CONTENANT DES EMETTEURS α

Les actinides que l'on risque de rencontrer dans les futures solutions de produits de fission sont caractérisés par leur énergie plus élevée et leur période beaucoup plus longue. D'autres phénomènes peuvent alors intervenir tels que la formation et le dégagement d'hélium, les réactions (α, n) et (α, p), l'accumulation de l'énergie pouvant éventuellement entraîner des changements de structure du verre et finalement affecter son taux de lixiviation et ses propriétés.

Pour simuler le vieillissement de tels verres, une étude a été entreprise en «dopant» des blocs de verre avec du plutonium, de l'américium et du curium.

L'énergie libérée cumulée en 1000 ans de stockage dans 1 litre de verre atteint 850 $kW \cdot h^{-1}$, si on admet un facteur de réduction de volume égal à 7,5 et en considérant que la solidification est effectuée après 1 an de refroidissement. Nous avons cherché à augmenter progressivement dans les verres l'énergie développée jusqu'à atteindre une énergie cumulée en 1 an supérieure au chiffre précité.

Le tableau III donne les principales caractéristiques des verres déjà réalisés.

Nous continuerons la réalisation de blocs avec 50 g et plus d'un mélange de Pu à 30% de ^{238}Pu et avec 1 g de ^{242}Cm.

TABLEAU III. CARACTERISTIQUES DES VERRES CONTENANT DES α

Actinides	Poids d'actinides par bloc (g)	Teneur en PuO_2-AmO_2 dans le verre (%)	Activité alpha par bloc (Ci)	Poids du bloc (g)	Energie cumulée par litre de verre en 1 an ($kW \cdot h^{-1}$)	Nombre de blocs réalisés
Mélange de Pu à 0,17% de ^{238}Pu	60–100	3,4–5,6	10–20	2000	5,5	15
Mélange de Pu à 30% de ^{238}Pu	50	2,2	245	2000	96	2
^{241}Am	50	2,3	160	2000	66	1
^{244}Cm	0,6		48	50	800	1

3.1. Comportement chimique des émetteurs α

Du point de vue chimique, on a constaté que le plutonium s'incorporait dans le verre plus difficilement que l'uranium. Des séparations de phases ont été notées lorsque les concentrations en PuO_2 étaient supérieures à 4% du verre. Ces séparations ont eu une influence sur la viscosité du verre et ont entraîné pour certains essais des difficultés de coulée. Ces inconvénients ne doivent cependant pas se présenter pour les teneurs réelles en Pu d'un verre non dopé. On n'a pas observé les mêmes difficultés avec l'américium.

3.2. Lixiviation

Pour les verres contenant du plutonium, les taux de lixiviation en α se situent entre 10^{-8} et 10^{-7} $g \cdot cm^{-2}$ selon les compositions. Les compositions contenant 25% d'oxydes de produits de fission sont légèrement plus lixiviables. Ces résultats de lixiviation sont en général plus dispersés que les résultats de lixiviation des produits de fission. La dispersion est due à une hétérogénéité du liquide de lixiviation, qui laisse supposer la présence de particules en suspension de composés de plutonium (fig. 1).

Le taux de lixiviation de ^{241}Am se situe au voisinage de $5 \cdot 10^{-9}$ après 40 jours.

3.3. Fluence neutronique et réactions (α, n)

Les flux neutroniques émis par ces pastilles révèlent des spectres ayant plus de 80% d'émission de neutrons d'énergie comprise entre 0,2 et 10 MeV (fig. 2).

L'importance des réactions (α, n) a pu être précisée: on constate une émission de 1 à 2 neutrons provenant de ces réactions pour 10^6 α dans les verres dopés au ^{238}Pu ou ^{241}Am et 4 neutrons pour 10^6 α dans les verres dopés au ^{244}Cm.

La composition des verres n'a pratiquement pas d'influence sur les réactions (α, n).

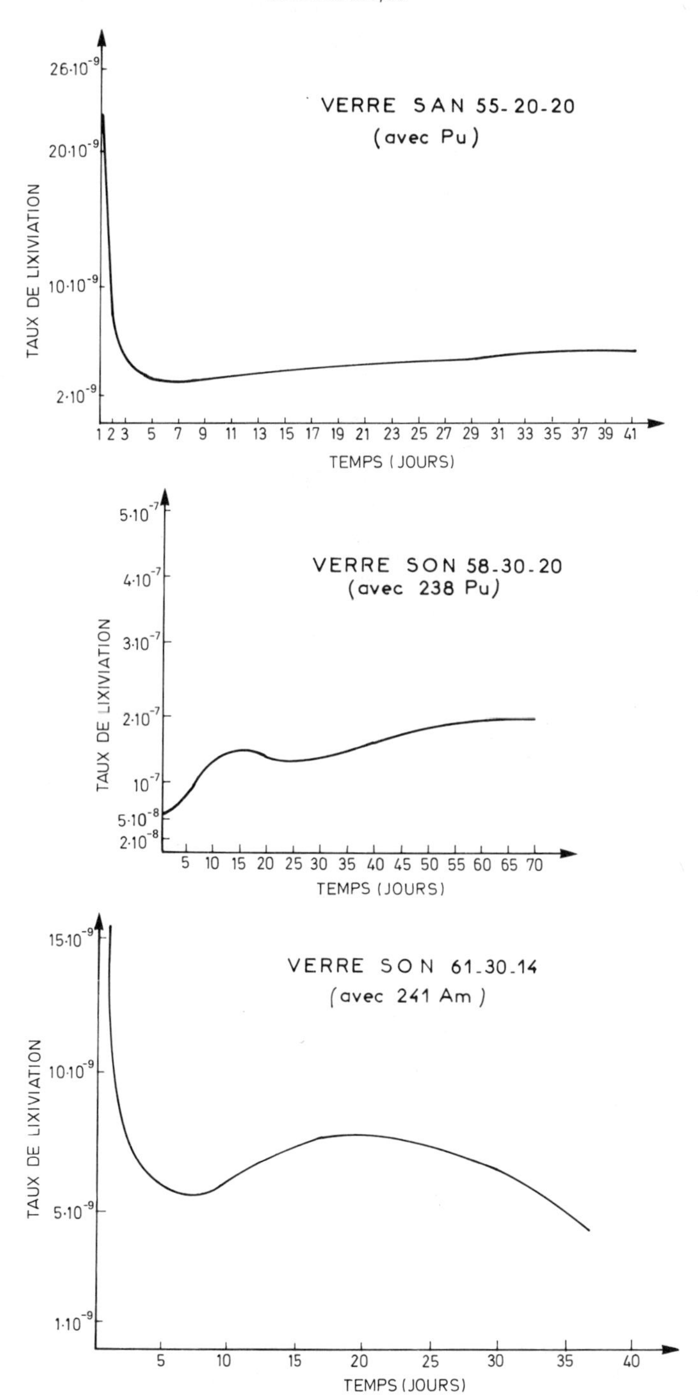

FIG.1. Courbes de lixiviation des blocs de verre contenant ^{238}Pu et ^{241}Am (SAN: silice, alumine, sodium; SON: silice, oxydes, sodium).

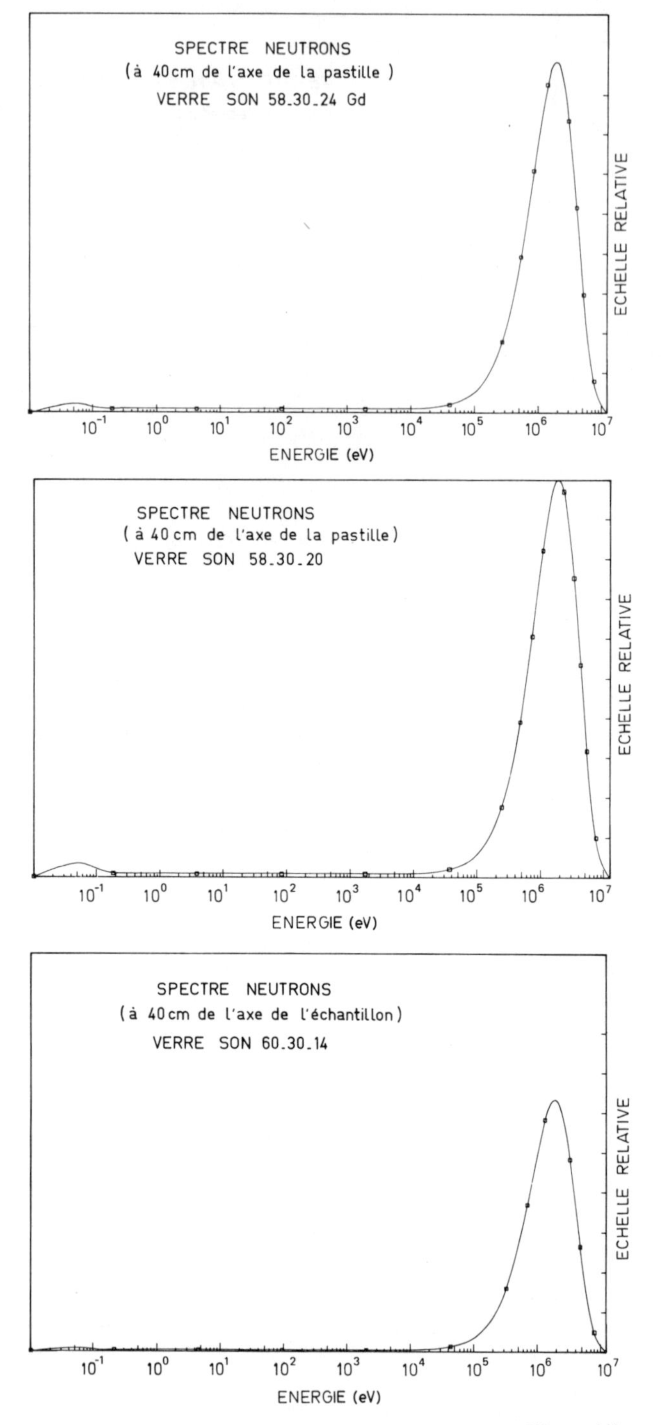

FIG.2. Spectres neutroniques émis par les blocs de verre contenant ^{238}Pu et ^{241}Am.

3.4. Dégagement d'hélium

Des tests sont en cours pour mesurer l'hélium occlus dans le verre.

Les premières mesures réalisées sur des bâtonnets de verre contenant de l'américium-241 et du plutonium stockés à la température ambiante pendant près d'un an ont révélé que les quantités d'hélium recueillies dans le verre correspondaient à environ 75% des quantités théoriques, calculées en supposant tous les α émis transformés en hélium.

3.5. Mesures de la dureté du verre

Les microduretés KNOOP effectuées sous 205 g varient de 500 à 600 $kg \cdot mm^{-2}$ sur 14 mesures faites le long du rayon d'une tranche du bâtonnet au ^{241}Am stocké pendant 6 mois. Ces mesures seront poursuivies dans le temps.

3.6. Examens complémentaires

Un certain nombre de tests seront réalisés lorsque les verres auront une ou deux années de stockage: parmi ceux-ci on peut citer:

- examen de verre en microscopie optique et électronique pour juger de l'influence éventuelle des émetteurs α sur la cristallisation des verres;
- mesures de stabilité thermique;
- mesure de l'énergie accumulée;
- tests mécaniques: dureté et résistance à la compression;
- mesure de l'hélium diffusé;
- mesure des coefficients de dilatation et des déformation éventuelles;
- essai de mesure des défauts provoqués par les émetteurs α.

4. CONCLUSIONS

En ce qui concerne les seuls produits de fission émetteurs de rayonnements $\beta\gamma$, le confinement assuré par les verres silicoboratés sélectionnés semble présenter une garantie satisfaisante.

En effet c'est surtout pendant les 10 premières années de stockage que se produiront les transformations dues aux effets thermiques et nucléaires, puisque environ la moitié de l'énergie totale est libérée pendant cette durée.

Les verres choisis résistent bien à l'action des rayonnements; quant aux effets thermiques, il suffit de prendre soin de stocker les verres en dessous d'une température de cristallisation, ce qui implique une contrainte de refroidissement pendant les premières années. Nous aurons bientôt l'expérience complète de la tenue de ces verres sur une période de 10 ans.

Après cette première période, leur comportement se rapproche beaucoup du vieillissement des verres classiques; or un acquis existe dans ce domaine, avec les verres anciens, et dans la plupart des cas cet acquis est positif.

En ce qui concerne l'influence des émetteurs α nous devons compléter les informations obtenues par les verres dopés.

Les taux de lixiviation de ces verres se sont révélés être du même ordre que ceux obtenus avec les émetteurs $\beta\gamma$, bien que les fluctuations trouvées nécessitent un complément d'études.

Les autres tests en cours ou programmés permettront de mieux apprécier l'influence des émetteurs α. Il est donc actuellement prématuré de se prononcer.

Si les garanties de stabilité à long terme de ces verres s'avéraient insuffisantes, la séparation des actinides réalisée de préférence au cours du traitement des combustibles irradiés permettrait soit leur transmutation dans les réacteurs, soit leur incorporation dans des verres plus riches en SiO_2, donc mieux structurés et plus stables. Ces compositions seraient semblables aux compositions des verres naturels telles les obsidiennes, qui ont parfaitement résisté à l'épreuve du temps.

REFERENCES

[1] BONNIAUD, R., LAUDE, F., SOMBRET, C., «Expériences acquises en France dans le traitement par vitrification des solutions concentrées de produits de fission», C.R. Coll. sur la gestion des déchets radioactifs résultant du traitement du combustible irradié (Paris, 1972), OECD/AEN, Paris (1973).

[2] BONNIAUD, R., «La solidification des solutions de produits de fission», Bull. Inf. Sci. Tech. (Paris) n° 188 (janv. 1974).

[3] BONNIAUD, R., PACAUD, F., SOMBRET, C., «Quelques aspects du comportement de radioéléments confinés à long terme sous forme de verre ou de produits à forte phase vitreuse», Le confinement de la radioactivité dans l'utilisation de l'énergie nucléaire (Actes VIIe Congrès Int., Versailles, 1974), Soc. Fr. Radioprotection, Montrouge (1974).

[4] IMBERT, J.C., PACAUD, F., «Contribution à l'étude de la diffusion en relation avec la lixiviation des verres», Rapport CEA-R-4550 (août 1974).

DISCUSSION

Y.P. MARTYNOV: What do you think of the idea of de-activating the filled containers by a wet method, i.e. by immersion in a de-activating solution? This technique was mentioned by Mr. Jouan yesterday (see paper IAEA-SM-207/27, these Proceedings, Vo.1). I should have thought there might be a problem of thermal stresses arising in the glass which could affect its properties. Would you agree?

F. LAUDE: For glasses with high specific activity corresponding, say, to a power of 100 W/l or more, the temperature of the cylindrical containers might indeed rise to 450°C. In these circumstances, decontamination of the container or storage under water is in fact likely to produce a thermal shock which would adversely affect the glass. This is a problem that should be studied. If appropriate shapes are chosen for the containers – annular canisters might be helpful, for example – the temperature and accordingly the thermal shock could be considerably reduced. There are other methods which aim at increasing the apparent thermal conductivity of the glass block.

A. DONATO: What equation did you use to calculate the diffusion coefficients in glass? Did you leach the samples over the entire surface?

F. LAUDE: We used Fick's law, solved with respect to the x-axis, and leached the whole of the surface of the glass block. This procedure is justified because with water leaching transfer is governed by a very thin layer close to the glass surface. We obtain the expression: $D = \pi l^2/4d^2t$, where l is the sum of the daily leaching rates at the end of time t in days, and d is the density of the glass in g/cm^3.

THERMAL AND RADIATION EFFECTS ON BOROSILICATE WASTE GLASSES

J.E. MENDEL, W.A. ROSS, F.P. ROBERTS,
R.P. TURCOTTE, Y.B. KATAYAMA, J.H. WESTSIK, Jr.
Battelle Pacific Northwest Laboratories,
Richland, Washington,
United States of America

Presented by A.M. Platt

Abstract

THERMAL AND RADIATION EFFECTS ON BOROSILICATE WASTE GLASSES.

Glasses for the immobilization of high-level radioactive waste must preserve their integrity for long time periods in spite of severe thermal and radiation conditions. The major thermal effect is devitrification. A detailed characterization of the phase formed upon devitrification of a typical HLW glass was carried out. Maximum devitrification effects occur at 700 to 750°C and include an increase in porosity and an approximately tenfold increase in leachability. The major radiation effects will be due to alpha decay and the associated recoil nuclei. Accelerated alpha effect studies using curium-244-doped glass specimens have simulated 2000 years storage of a typical waste glass with the only effects being an insignificant accumulation of stored energy (26 cal/g) and an approximately 0.5% increase in density.

INTRODUCTION

Glasses in which high-level radioactive wastes are incorporated differ from glasses of common experience in two important ways:

(a) Their compositions are much more complex, and they contain less silica than ordinary glass.
(b) During storage, portions of the waste glass bodies will be subjected to prolonged elevated temperatures, and all of the glass will receive high doses of self-irradiation.

Because both the compositions and storage conditions are different from known experience these glasses are receiving much study in many countries. At the Pacific Northwest Laboratories (PNL) this study dates back over ten years, beginning with the Waste Solidification Engineering Prototypes (WSEP) programme [1,2]. This paper will give a survey of recent in-depth investigations of thermal and radiation effects on a specific waste glass composition.

WASTE GLASS COMPOSITION

The waste glass composition being used is a zinc-borosilicate formulation which has been studied extensively at PNL. The composition is shown in Table I. Thermal effects are being characterized on glass specimens containing non-radioactive isotopes of all of the fission products except technetium. Uranium is the only actinide present. Radiation effects are being characterized using specimens in which curium-244 is substituted for uranium. Glass specimens have also been prepared using waste produced from spent power reactor fuel elements but no data are yet available from these specimens.

TABLE I. COMPOSITION OF WASTE GLASS USED FOR CHARACTERIZATION OF THERMAL AND RADIATION EFFECTS

SiO_2	27.31 wt%	NiO	0.66 wt%	TeO_2	0.46 wt%
B_2O_3	11.15	P_2O_5	0.42	Cs_2O	1.82
Na_2O	4.06	Rb_2O	0.22	La_2O_3	0.93
K_2O	4.06	Y_2O_3	0.38	CeO_2	2.11
ZnO	21.34	ZrO_2	3.13	Pr_6O_{11}	0.94
CaO	1.47	MoO_3	4.03	Nd_2O_3	2.95
MgO	1.47	RuO_2	1.88	Sm_2O_3	0.58
SrO	2.15	Rh_2O_3	0.30	Eu_2O_3	0.13
BaO	2.47	PdO	0.94	Gd_2O_3	0.09
Fe_2O_3	0.95	Ag_2O	0.06	UO_2	1.26
Cr_2O_3	0.22	CdO	0.06		

TABLE II. THERMAL TREATMENTS FOR THERMAL EFFECT STUDIES

Temperature	Time					
(°C)	2 hours	1 day	1 week	2 months	1 year	5 years
1000	A					
900		A	A	A		
800		A	A	A		
750		A	A	A		
700			A	A	A	
600			A	A	A	
500				SS	SS	SS
400				SS	SS	SS
300				SS	SS	SS

Note: 'A' indicates alumina crucible; 'SS' indicates stainless steel crucible.

THERMAL EFFECTS

Thermal effects which could adversely affect the long-term integrity of waste glass include:

(i) Devitrification
(ii) Phase separation
(iii) Gravity settling
(iv) Deleterious interactions with canister.

In order to investigate these effects separate specimens of the glass are being held for various lengths of time at various temperatures, as shown in Table II. The specimens were initially melted at 1000°C and cooled in the furnace to the temperatures shown.

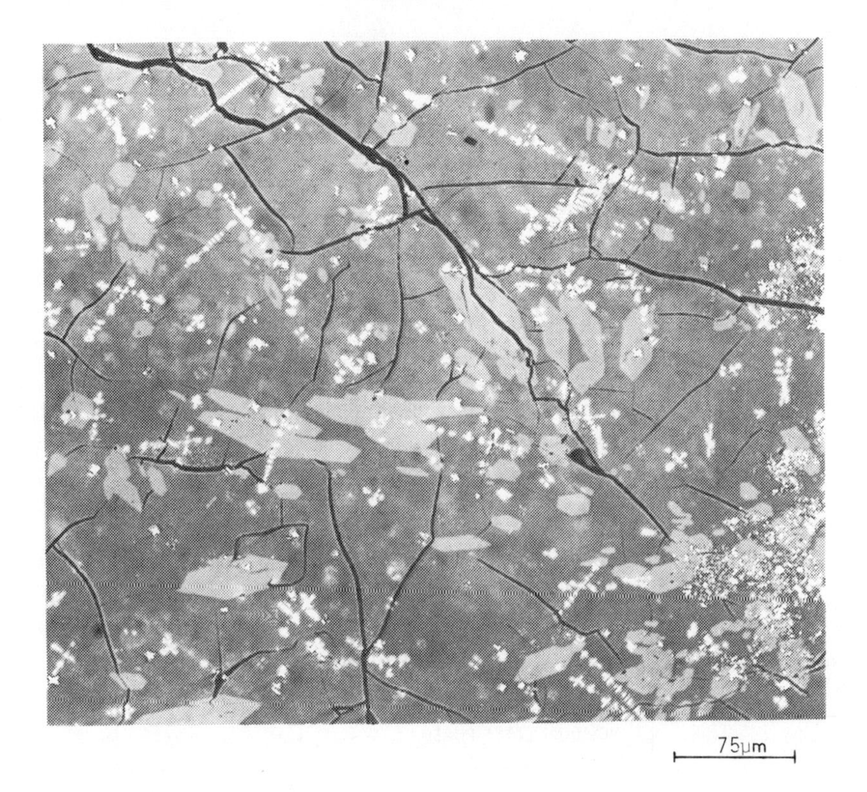

FIG.1. Zinc borosilicate waste glass devitrified by holding 1 week at 750°C.

TABLE III. OBSERVED PHASES IN ZINC BOROSILICATE WASTE GLASS 72-68

Phase	Conditions of formation	Maximum concentration (wt%)	Observed concentrations (% of maximum)	Size range (μm)
Pd	Melt	0.82	~100	<10-spherical
RuO_2	Melt	1.88	~100	<1-powder
CeO_2	Melt ... Devitrification	2.10	0–50	<1-powder <10-crystals
$SrMoO_4$	Devitrification 600–900°C	5.11	0–50	<100 × 2-dendrites
Zn_2SiO_4	Devitrification 600–1000°C	29.2	0–50	<500 × 100-columnar
$(Zn,Ni)(Fe,Cr)_2O_4$	Melt ... Devitrification	~1.3	0–10	<50-cubic
Rh-rich	1000°C melt	~0.3	0–30	<100-irregular

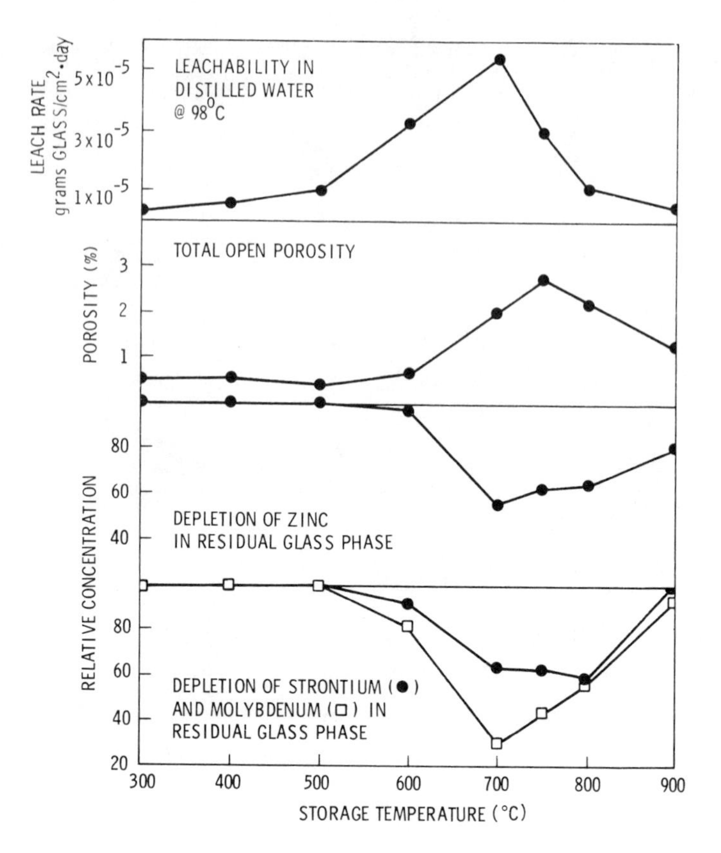

FIG.2. Physical and chemical changes due to devitrification (2 months storage at temperature).

Devitrification

Phase characterization of the thermally treated samples is being done using optical and scanning electron microscopy, microprobe analysis, and X-ray diffraction. A micrograph of a typical devitrified specimen is shown in Fig.1. Elemental distributions, obtained by X-ray fluorescence, have established that uniform solution in the glass matrix is obtained for most elements (including the actinides, U, Pu and Cm, as discussed later). At least seven crystalline phases have been identified however, some of which are residual in the as-formed glass and some of which are devitrification products. The observed phases, their concentration range, and approximate crystallite size are given in Table III. With the exception of a few very weak lines, the X-ray diffraction patterns of fully devitrified glasses are completely described as arising from these crystalline phases.

In the devitrified glass the major crystalline phases formed are Zn_2SiO_4 (to approximately 15 wt%) and $SrMoO_4$ (to approximately 2.5 wt%). The $SrMoO_4$ dendrites do not vary much in size but the Zn_2SiO_4 crystals range from approximately 1 μm at 600°C to several hundred μm's at 900°C and cause extensive cracking due to a large thermal expansion mismatch.

Using microprobe X-ray fluorescence analysis, we have obtained glass matrix concentrations as shown in Fig.2 for Zn, Sr and Mo, as a function of storage temperature. The results shown are for two months storage. Quite similar results were obtained for the specimens stored for 1 day and

1 week indicating most of the devitrification occurs rapidly. The results suggest that much more Mo leaves the glass phase than is necessary to form stoichiometric $SrMoO_4$. Recently we have established that the compound actually formed is $(Sr,Ba)MoO_4$.

Leaching behaviour and porosity

The leaching behaviour of the specimens is being determined at 98°C using a Soxhlet leach test [3] and at 25°C using a variation of the IAEA test [4]. As shown in Fig.2. Soxhlet leach tests show an increase in leachability of the thermally-treated samples, for the samples stored between 500 and 800°C, with the maximum observed at 700°C. Total open porosity, determined by mercury infusion, peaked 50°C higher than the leach rate. The increased porosity is a result of microcracking caused by the difference in thermal expansion of the glass and Zn_2SiO_4. Since the variations in leaching behaviour do not coincide precisely with either microcracking or devitrification it is believed that some, as yet not defined, phase separation may also be occurring.

Settling

As discussed above, there will be insoluble phases present in typical waste glasses, especially RuO_2 and Pd, but also CeO_2, Rh and several spinel type phases which may occur, depending on melt temperatures and the glass composition. It was anticipated that these materials or others, such as PuO_2 might settle to the bottom of the central hottest region of the canister resulting in an unfavourable, anisotropic distribution. The insoluble species (except the spinel phases) have densities near either 7 or 12 g/cm^3 and individual particle sizes range from less than 1 μm to about 10 μm. Often, however, the fine material tends to form conglomerates with diameters which range up to 500 μm. Based on these densities and diameters, approximate settling rates have been calculated to be 40 cm/h at 1050°C, 0.3 cm/h at 800°C and 3×10^{-5} cm/h at 600°C.

The calculations indicate that, at the 1050°C melting temperature, it is very likely that conglomerated particles will settle. This has been observed in large scale melts held at temperature overnight. The calculations also indicate that significant settling would be expected at 800°C, or lower, in a longer time frame. Experimentally the settling has not been observed in zinc borosilicate glasses at temperatures below 900°C. The predominant reason for this is that devitrification occurs at temperatures below 930°C which greatly increases viscosity [5]. A quasi-continuous network of precipitate regions is formed which greatly reduces the settling rate.

RADIATION EFFECTS

High-level waste glasses will be subjected to intense alpha, beta and gamma radiation as well as lesser amounts of neutron radiation. The radiation spectrums and intensities are dependent on reactor exposure conditions and cooling time. The alpha activity concentration in the glasses is also dependent on the efficiency of plutonium, uranium and neptunium separation achieved during fuel reprocessing.

The cumulative alpha dose to a glass made from the HLW, assuming 170 kg glass per tonne of heavy metal, is shown in Fig.3 in terms of alpha decays per gram of glass. Also shown in this figure is the alpha dose to a glass prepared from the waste of the plutonium fraction of mixed oxide plutonium recycle fuel.

Experimental approach

A number of experiments are in process to evaluate the long-term radiation effects on HLW glass. Because most of the radiation damage is expected to be related to the alpha dose which

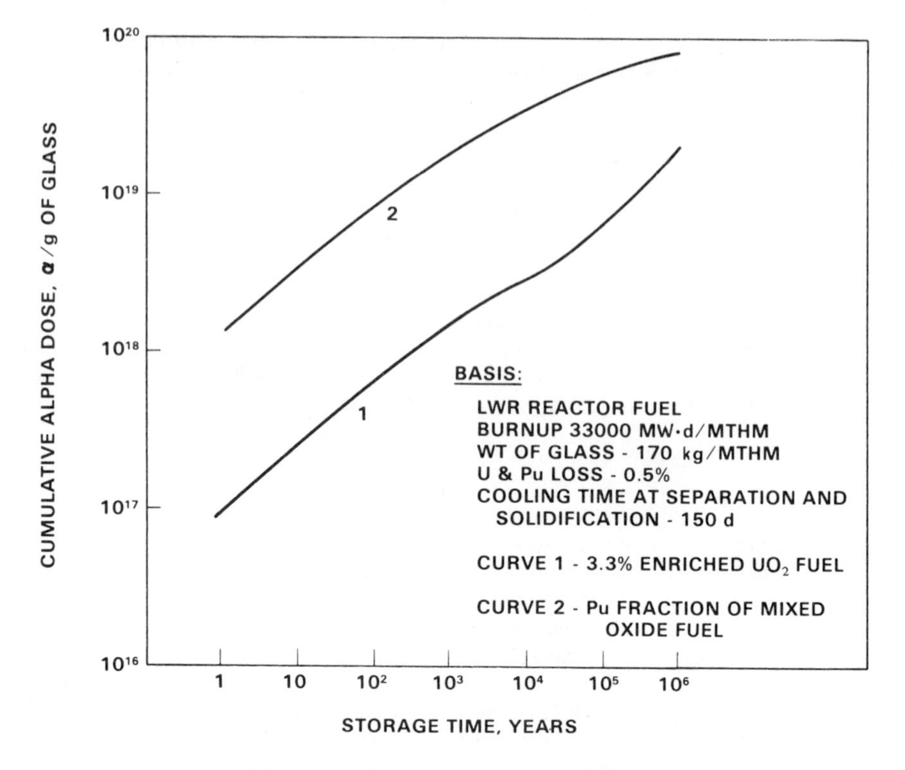

FIG.3. Cumulative alpha dose for typical HLW glass.

accumulates over very long time periods it is necessary to accelerate the effects in order to be able to observe them in reasonable time periods. By incorporating large amounts of ^{244}Cm in synthetic waste glasses it is possible to obtain alpha doses in five years or less equivalent to storage periods for actual HLW glass of up to 10^4 years.

Much of the alpha effect studies completed to date has been done with a zinc borosilicate glass having a composition similar to the glass used for the thermal effect studies but with curium substituted for uranium. Specimens of this glass were prepared in the form of buttons poured from a 1200°C melt. The buttons were about 3 cm in diameter and 0.6 cm thick and weighed about 10 g. The buttons were annealed at 400°C for two hours and slowly cooled to room temperature. Some specimens were devitrified by holding at 700°C for seven days and slowly cooling to room temperature.

The curium oxide used in the preparations was a mixture containing 84 wt% $^{244}Cm_2O_3$ and other curium isotopes as well as some ^{240}Pu, ^{243}Am and Fe. The specific alpha activity was 1.34×10^{14} $\alpha \cdot min^{-1} \cdot g^{-1}$ at the beginning of the study. The individual specimens contained nominally 1 wt% Cm_2O_3 giving them a specific activity of 1.34×10^{12} $\alpha \cdot min^{-1} \cdot g^{-1}$. (The actual activities ranged from 1.2 to 2.2×10^{12} $\alpha \cdot min^{-1} \cdot g^{-1}$.)

Phase characterization

The vitreous product contained only RuO_2 and Pd as second phases, as determined by X-ray diffraction and microprobe analysis. After one week at 750°C, the only additional phase observed in

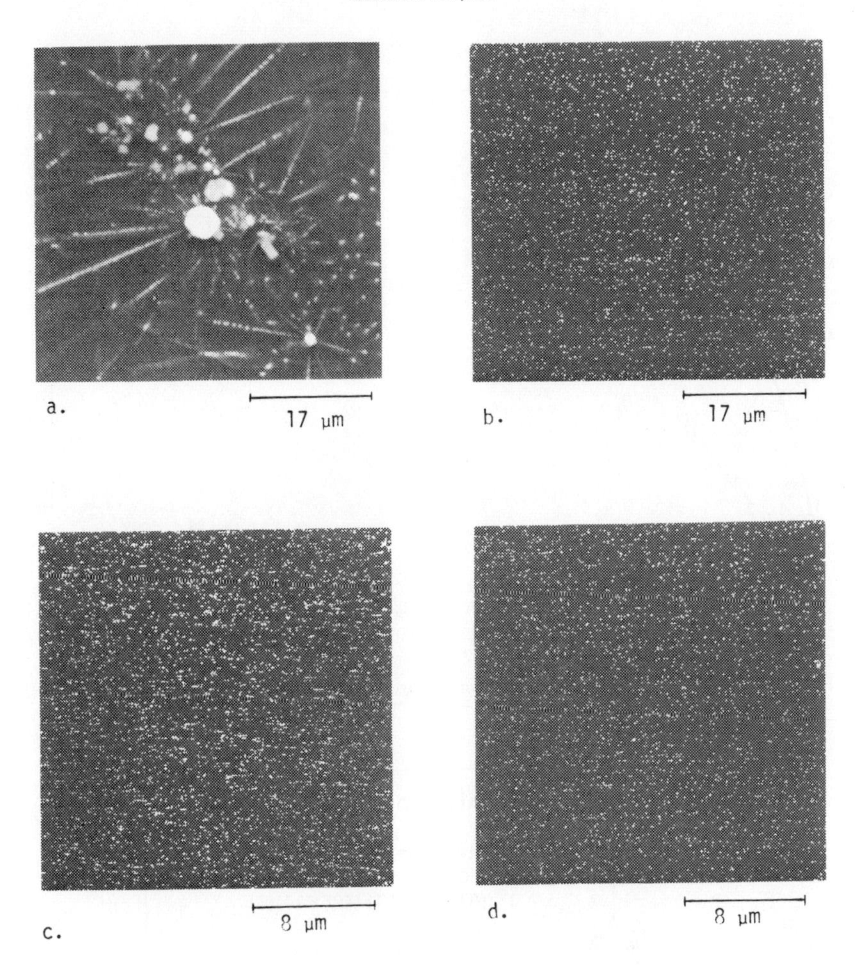

FIG.4. Microstructure of zinc borosilicate waste glass.
(a) Palladium and RuO_2 act as nucleating sites for $SrMoO_4$.
(b,c,d) Uranium, plutonium and curium are uniformly distributed in devitrified glass.

the diffraction pattern was $SrMoO_4$. Scanning electron micrographs, as well as some X-ray fluorescence maps, are shown in Fig.4 for the devitrified glass. The spherical, bright contrast phase is Pd metal. These particles, as well as RuO_2, clearly serve as nucleating sites for growth of the $SrMoO_4$ needles. The lack of CeO_2 and perhaps other low concentration 'insolubles' in the Cm doped glass, which was melted at 1200°C, delays formation of Zn_2SiO_4 to higher temperatures. After one day at 900°C Zn_2SiO_4 was observed in the X-ray diffraction patterns. A more detailed examination of the influence of 'insolubles' on devitrification kinetics is in progress.

Of some importance to the radiation damage question is location of the actinide elements, since if they were present as concentrated phases, the α-decay damage could be sufficiently anisotropic to crack the glass. As shown by the X-ray maps (Fig.4), the evidence for U (1.2 wt%), Pu (0.2 wt%), and Cm (0.8 wt%) is favourable, with no segregation apparent even in devitrified glass. This question will continue to need reassessment as new glass compositions are developed.

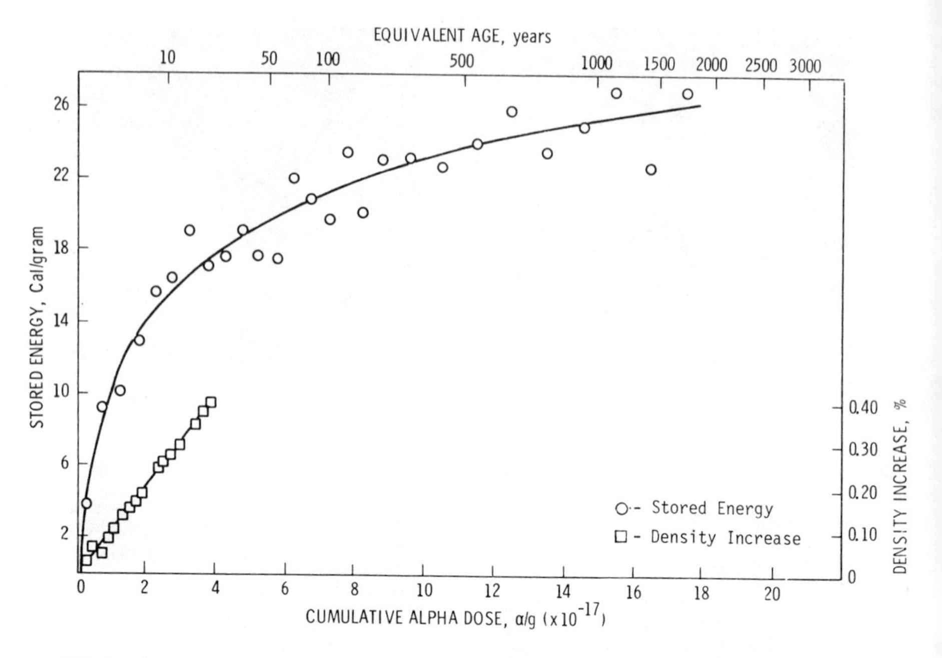

FIG.5. Alpha-induced stored energy and density changes in zinc borosilicate waste glass stored at 25°C.

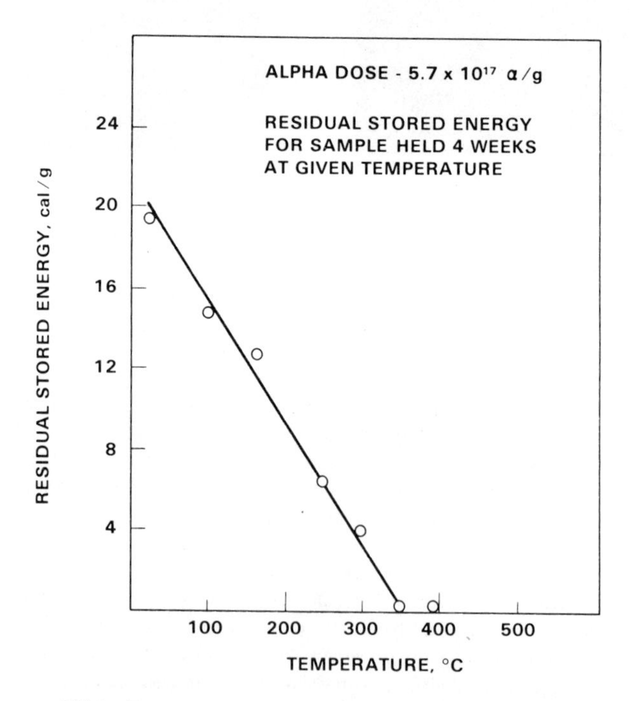

FIG.6. Temperature dependence of stored energy accumulation.

Stored energy

One of the observable effects of radiation in solidified HLW is the build-up of stored energy as the result of displacements of atoms from their normal lattice positions. These arise predominantly from alpha recoil collisions [6,7] and therefore the amount of stored energy should be related to the accumulated alpha dose in the solidified waste.

Stored energy results for the curium-doped glass are shown in Fig.5, where the amount of energy released in cal/g is plotted as a function of alpha dose and equivalent storage time. Doses in some specimens have reached 1.75×10^{18} α/g, equivalent to nearly 2000 years storage. The amount of stored energy may still be increasing but the rate of increase is definitely levelling off and extrapolation indicates the stored energy should not exceed 35 cal/g even at 10^6 years.

During the initial period that the wastes are self heating the build-up of stored energy will be decreased because of thermal annealing. The results given in Fig.6 show that residual stored energy is inversely proportional to storage temperature and is completely absent in glass stored at 350°C and above.

The results of this study show that large temperature excursions (>150°C) resulting from the release of stored energy are not possible, corroborating the results of previous theoretical analyses [8].

Density and mechanical properties

It is well known that radiation can cause structural modifications which result in changes in density. For waste glasses the direction of density change is apparently not readily predictable. Precise density measurements made on the curium-doped zinc borosilicate waste glasses by the water displacement, or buoyancy, method show that the density is increasing (Fig.5). On the other hand we have initial results which indicate that the density of curium-doped borosilicate and lead borosilicate waste glasses may be decreasing. Kelly has also reported a decrease in borosilicate waste glass density with alpha radiation [9]. The fact that both volume decreases and increases have been observed suggests that it may be possible to develop waste glass formulations without radiation-induced volume changes.

Helium behaviour

Helium produced from alpha decay in waste glass must either be accommodated interstitially, or diffuse to internal voids or to a plenum in the case of a sealed canister. Helium release from curium-doped specimens is being measured as a function of temperature with a mass spectrometer to determine helium diffusion coefficients in waste glass. Typical release data obtained at 350°C are shown in Fig.7. Comparison of the curves obtained with the curium-doped specimen and a similar non-radioactive glass which was equilibrated with 1 atmosphere of helium show a different release pattern for the irradiated glass. Trapping by radiation-induced defects giving rise to decreased release rates at longer times has been previously observed for inert gas diffusion in other ceramic and metallic systems [10]. Although some anomalies appear at higher temperatures, a simple approach is to consider the measured release as arising from untrapped gas and trapped gas. In both cases, we find a constant activation energy for diffusion (~15 kcal/mol He), and only a small dependence on release fraction. The diffusivity can therefore be expressed [$150 < T(°C) < 350$] as

$$D_U = 2.1 \times 10^{-3} \exp(-7500/T) \text{ cm}^2/\text{s, and } D_T = 1.7 \times 10^{-4} \exp(-7500/T) \text{ cm}^2/\text{s}$$

for untrapped (D_U) and trapped gas (D_T), respectively. Thus, the influence from radiation induced trapping is to reduce apparent diffusivity by about one order of magnitude. The diffusion coefficients increase one order of magnitude with a temperature rise ~160°C. At 25°C, extrapolation gives $D_U = 2.5 \times 10^{-4}$ cm²/s and $D_T = 2.0 \times 10^{-5}$ cm²/s.

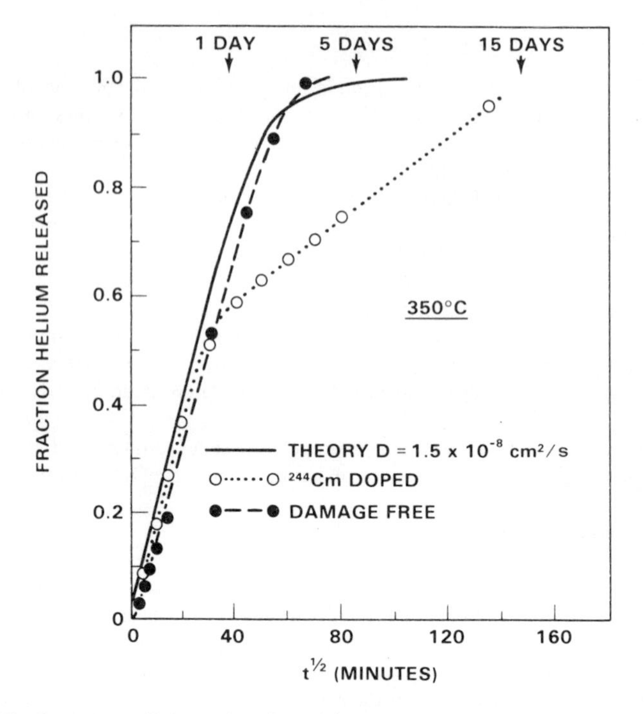

FIG. 7. Comparison of helium release from alpha-containing and non-irradiated waste glass.

If the glass canisters are sealed they may have to be treated as pressure vessels if the helium pressure can potentially exceed 1 atmosphere. Calculations made using the diffusion coefficients above indicate that the helium pressure in canisters containing waste from UO_2 fuel reprocessing will never exceed 1 atmosphere but that the helium pressure in sealed canisters containing waste from plutonium recycle fuel reprocessing may exceed 1 atmosphere in 10 to 50 years.

Leaching

Seven curium glass specimens were leached in distilled water at 25°C to determine the effect of alpha radiation on the chemical durability of HLW glasses. The button specimens were suspended in 500 ml distilled water which was sampled and changed on a weekly basis for eight weeks and on a monthly basis thereafter.

The measured leach rates are presented in Fig.8. The curium leach rates are one hundredfold lower than potassium and do not show a significant correlation with alpha dose. However, the potassium leach rate of the non-radioactive specimen was slightly lower than that of the curium-doped specimens. This may indicate that the radiation damage affecting leachability occurs at lower levels than covered by this set of measurements. In any case, the leachability of the zinc borosilicate waste glass is not grossly affected by alpha doses up to 8×10^{17} α/g, equivalent to waste storage of 100 years.

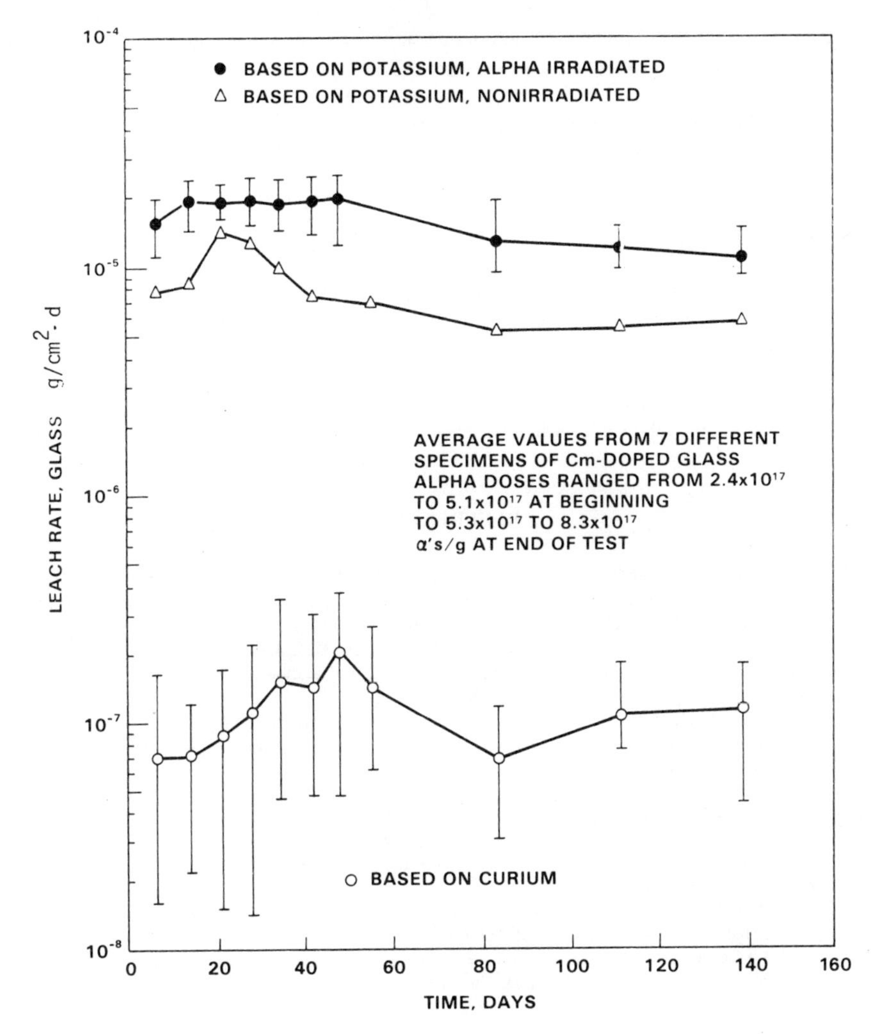

FIG.8. Leachability of ^{244}Cm-doped zinc borosilicate waste glass.

Fully radioactive glass

Glass specimens containing calcined HLW prepared from commercial LWR fuel are being fabricated and characterized for comparison with non-radioactive glasses of similar chemical composition as a function of storage conditions:

The first fully radioactive specimens were prepared using wastes prepared from 2764 g of UO_2 fuel discharged in October 1974 after an average burn-up of 54 000 MW·d per tonne heavy metal. Filtration of the nitric acid dissolver solution resulted in the recovery of 29.4 g of undissolved solids. Most of the HNO_3 – insoluble solids were dissolved by a series of further treatments and found to contain ^{106}Ru, ^{110}Ag^{m}, ^{95}Nb and ^{125}Sb as the major constituents. About 3.5 g of solids could not be dissolved even after HNO_3-HF leaching and $NaHSO_4$ and Na_2O_2-NaOH fusions.

After removal of uranium and plutonium by solvent extraction in a counter-current centrifugal contactor the HLW was calcined in a stainless steel beaker. Glass specimens were fabricated in batch type in-can melting equipment. The first 100 g specimen of fully radioactive waste glass was prepared in January of this year and more specimens are being made. The first results of their characterization will be available in a short time.

CONCLUSION

Our approach has been to use one glass composition to develop experimental techniques and to obtain detailed data on thermal and radiation effects. These data can now begin to serve as a baseline to evaluate new waste compositions and glass formulations. Although more information is needed the experience to date indicates that glass exhibits sufficient radiation stability to serve as an excellent medium for the immobilization of high-level waste. Optimum glass compositions are process dependent and will evolve with experience.

REFERENCES

[1] McELROY, J.L., SCHNEIDER, K.J., HARTLEY, J.N., MENDEL, J.E., RICHARDSON, G.L., BLASEWITZ, A.G., McKEE, R.W., Waste Solidification Program Summary Report, Volume II, Evaluation and Application Studies of WSEP High Level Waste Solidification Processes, BNWL-1667 (1972).
[2] BLASEWITZ, A.G., RICHARDSON, G.L., McELROY, J.L., MENDEL, J.E., SCHNEIDER, K.J., "The high level waste solidification demonstration program", Management of Radioactive Waste from Fuel Reprocessing (Proc. IAEA/NEA Symp. Paris, 1972), OECD/NEA, Paris (1973).
[3] MENDEL, J.E., WARNER, I.M., Waste Glass Leaching Measurements, BNWL-1741 (1973) 6.
[4] HESPE, E.D., "Leach testing of immobilized radioactive waste solids, A proposal for a standard method", Atomic Energy Review **9** 1 (1971) 195.
[5] ROSS, W.A., Viscosity of Molten Glasses, BNWL-1841 (1974) 31.
[6] JENKS, G.H., BOPP, C.D., Energy Storage in High Level Radioactive Waste and Simulation and Measurement of Stored Energy with Synthetic Wastes, ORNL-TM-3781 (1973).
[7] ROBERTS, F.P., JENKS, G.H., BOPP, C.D., Radiation Effects on Solidified High Level Wastes – Part I. Stored Energy, BNWL-1944 (1976).
[8] LASER, M., MERZ, E., The Question of the Energy Accumulation in Solids Through Nuclear Irradiation in the Storage of Highly Radioactive Wastes, JUL-766-CT (1971).
[9] KELLY, J.A., Evaluation of glass as a matrix for solidification of Savannah River Plant waste, DP-1382 (1975).
[10] ELLEMAN, T.S., FOX, C.H., MEARS, L.D., The influence of defects on rate gas diffusion in solids, J. Nucl. Mater. **30** (1969) 89.

DISCUSSION

S.O. NIELSEN: Is it possible to eliminate the many different phases in the glass that you have described by increasing the temperature before cooling the glass? I am aware, of course, that an increase in temperature may introduce additional problems.

A.M. PLATT: Table II shows the phases that appear in the original glass and those which are formed on devitrification. I do not know whether by simply increasing the melt temperature you could eliminate the former problem.

W. LUTZE: I am surprised that the first 60% of the helium is released without any particular influence from the radiation damage present in the solid. It seems that only the remaining 40% diffuses at a decreased rate, if I understand the plot in Fig.7 correctly. Has this behaviour been interpreted? Could the portions be related to the contents in the vitreous and non-vitreous phases, respectively?

A.M. PLATT: Although it may not have been clear from my oral presentation, we have concluded that there is a constant activation energy for helium diffusion (~ 15 kcal/mol), and only a slight dependence on the fraction released. Diffusivities for the trapped and untrapped gases are given in the paper.

A. SCHNEIDER: What is the composition of the 1% fraction of the high burn-up LWR fuel that is not dissolved by HNO_3? More particularly, is there a significant amout of plutonium associated with this residue?

A.M. PLATT: Unfortunately, the characterization of these solids is not yet complete, but we should have the data you request some time in the near future.

FIXATION OF FISSION PRODUCTS IN GLASS CERAMICS

A.K. DE, B. LUCKSCHEITER, W. LUTZE,
G. MALOW, E. SCHIEWER, S. TYMOCHOWICZ
Nuclear Chemistry and Reactor Division,
Hahn-Meitner Institut für Kernforschung Berlin GmbH,
Berlin,
Federal Republic of Germany

Abstract

FIXATION OF FISSION PRODUCTS IN GLASS CERAMICS.

Lab-scale and semi-technical scale experiments were performed to transform borosilicate glasses into glass ceramics. The preparation and characterization of the products are described. The results show that appropriate parent glasses can be developed meeting technologically relevant conditions such as melting temperature, viscosity and fission product content. The subsequent preparation of glass ceramics may be the tail-end step of a vitrification process. The final product exhibits considerable increase in thermodynamical stability and impact resistance as well as a fixation of long-lived fission products in chemically resistant host-phases. Hot-cell experiments are under way to demonstrate the feasibility of these new products further.

INTRODUCTION

High-level radioactive waste solutions must be transformed into a stable solid form that is suitable for interim and final disposal. Vitrification is at present the most thoroughly investigated method and technology. Two types of glass are used: borosilicate and phosphate glass. The work presented here is based on borosilicate glasses. It is unlikely that phosphate glasses can be transformed into useful glass ceramics.

The possible steps in making an appropriate stable solid product are illustrated in Fig. 1. The figure shows schematically the free energies for some solid waste forms. Calcination of the waste products after evaporating the solvent is the first step and yields an amorphous and fairly water soluble product, the calcine. This is the step of solidification. The next comprises the addition of a specially developed and prefabricated base glass to the calcine and subsequent melting. The resulting product is a solid glass. This is the step of immobilization. The final is the conversion of the glass into a glass ceramic right after melting. This leads to the lower energy level of Fig. 1. The conversion can be combined with a fixation of long-lived fission products in crystalline phases.

The last step is desirable because a glass is thermodynamically unstable and thus tends to stabilize by crystallization. There is experimental evidence for considering the spontaneous crystallization of a waste glass an undesirable process, which may decrease the degree of fission product immobilization already achieved [1, 2].

The first section of this paper gives a brief outline of the glass ceramics development programme. The preparation of appropriate glass ceramics on a lab and a semi-technical scale will be described in the second section. Their properties are discussed in the third section.

1. DEVELOPMENT OF GLASS CERAMIC

A glass ceramic is a ceramic product derived from a homogeneous glass by means of a special heat treatment that consists of two processes, as shown in Fig. 2. The first one is annealing at the

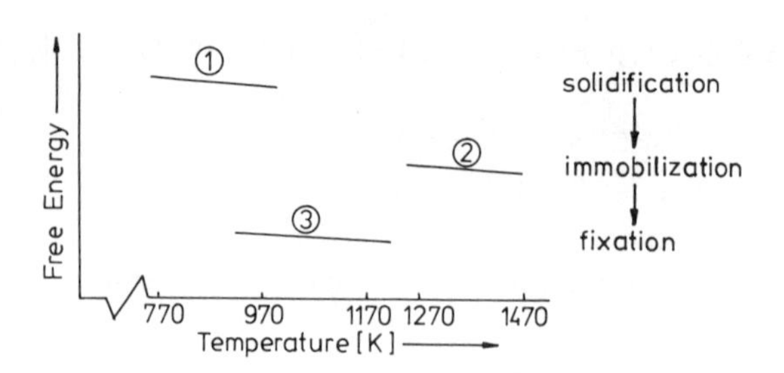

FIG.1. Schematic view of free energy of anticipated wastes forms: (1) calcine, (2) glass, (3) glass ceramic.

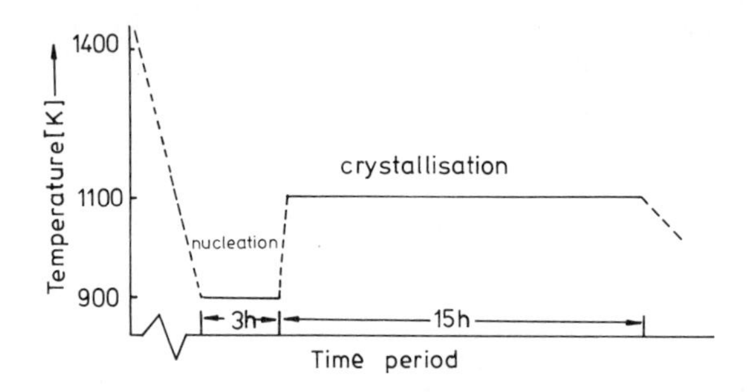

FIG.2. Typical annealing programme to transform borosilicate 'parent' glass containing fission products into glass ceramic.

temperature of maximum nucleation rate, the second is annealing at a sufficiently high temperature to yield the optimum crystal growth rate. By the proper choice of the glass composition crystalline species may come into existence that are resistant host phases for long-lived radionuclides.

The current programme for the development of fission product-containing glass ceramics comprises the following investigations:

(a) Search for suitable base glass compositions to meet the technologically relevant conditions (such as melting temperature and viscosity) for the preparation of the fission product-containing glass and to obtain the desired crystalline host phases for the dangerous fission products when converting the waste glass into a glass ceramic.

(b) Preparation and characterization of lab-scale (100 cm^3) and of semi-technical scale (up to 1000 cm^3) borosilicate glasses and glass ceramics that contain 20 wt% simulated fission product oxides of the WAK-type, see Table I.

(c) Hot cell demonstration of glass ceramic preparation in the Vulcain cell of the CEA at Marcoule, France.

The investigations under (a) are mainly basic research. The results have been published elsewhere [3–5] and will not be reviewed here. Attention is focussed on the technique of preparation and on characterization. Some preliminary remarks will be made concerning hot cell experiments.

TABLE I. COMPOSITION OF ORIGINAL AND SIMULATED FISSION PRODUCT OXIDE MIXTURES

(Original: Reprocessing plant at Karlsruhe – WAK)

Constituents	Fission product waste	
	From reprocessing plant (wt%)	Simulated for non-radioactive glass ceramics (wt%)
Cs_2O	8.46	8.73
Rb_2O	0.78	0.81
SrO	2.95	3.04
BaO	3.43	3.55
Y_2O_3	1.68	1.73
La_2O_3	3.30	3.40
CeO_2	6.51	6.72
Pr_2O_3	2.97	3.07
Nd_2O_3	10.31	10.64
Pm_2O_3	0.54	
Sm_2O_3	1.03	
Eu_2O_3	0.09	Nd_2O_3 1.73
Gd_2O_3	0.005	
ZrO_2	12.50	12.90
MoO_3	11.72	12.10
TcO_2	2.58	MnO_2 1.77
RuO_2	4.58	TiO_2 2.82
Rh_2O_3	1.20	Co_2O_3 0.81
NiO	1.15	1.19
PdO	0.57	0.36
Ag_2O	0.009	0.009
CdO	0.02	0.02
SnO_2	0.08	0.08
Sb_2O_3	0.02	0.02
TeO_2	1.28	1.32
U_3O_8	2.03	2.09
PuO_2	0.02	
NpO_2	0.007	U_3O_8 0.26
AmO_2	0.23	
Fe_2O_3	7.29	7.53
Cr_2O_3	2.08	2.14
Na_2O	10.79	11.14

TABLE II. CHEMICAL COMPOSITION RANGES FOR BASE GLASSES FREE OF FISSION PRODUCTS (wt%)

(C, P, D, E refer to Celsian, Perovskite, Diopside, and Eucryptite, respectively – see Table IV).

Constituents	C-type	P-type	D-type	E-type
SiO_2	32–37	32–50	40–52	41–60
Al_2O_3	10–17	12–15	10–18	14–24
B_2O_3	5–8	5–10	5–10	4–11
CaO	5–10	12–14	10–15	2–9
BaO	16–23			0–5
Na_2O	2–6	5–8	0–8	0–3
Li_2O	0–3	0–3	K_2O 0–3	5–10
TiO_2	5–11	10–14	3–5	4–8
ZrO_2	0–2		Fe_2O_3 2–5	ZrO_2 0–2
ZnO	4–6	0–3	0–3	1–10
MgO	0–3		5–10	0–2
Cs_2O	0–2	0–2		
As_2O_3	0–0.5			

2. PREPARATION

2.1. Base glass compositions

Four chemical composition ranges have been found from which base glasses can be prepared in order to meet the conditions mentioned under (a) in section 1. The composition ranges are given in Table II. The existence of different base glass types and rather broad ranges within each group reflects a desirable flexibility in view of the technological realization of glass ceramics.

2.2. Inactive fission product containing glasses

2.2.1. Lab-scale experiments

Four selected base glasses – one of each type given in Table II – have been prepared in the form of a frit.

The dry powder of each frit was mixed with a powder of fission product oxides. The mixture also contained some corrosion products from the reprocessing plant, as can be seen from Table I. The material was melted in platinum crucibles and yielded homogeneous glass melts. After rapid cooling non-crystallized bubble-free glasses of a black colour were obtained. These glasses are called 'parent glasses' as they constitute the desired starting material for suitable glass ceramics, as defined in section 1(a).

With regard to the digestion of molybdenum in the glass, there was no evidence for phase separation in the melt and in the solid. Molybdenum digestion was, however, a crucial point for the development of a proper base glass composition. An incomplete dissolution of molybdenum is

TABLE III. SUMMARY OF DATA FOR GLASS CERAMICS PREPARATION

	C-type	P-type	D-type	E-type
Frit melting temp. (K)	1520	1620	1520	1520
Glass melting temp. (K)	1470–1570	1520–1620	1470–1570	1370–1670
time (h)	3–4	3–4	3–4	3–4
Viscosity 100 poise temp. (K)[a]	1420	1420	1400	1430
Transform temp. (K)	830–860	860–890	880–930	740–830
Nucleation temp. (K)	–	900–940	925–970	800–850
time (h)	–	3–5	5	3–5
Crystallization temp. (K)	1040–1130	1070–1140	1130–1230	900–1020
time (h)	6–15	12–24	10	10–24

[a] For the low melting glasses.

often observed in borosilicate glasses and leads to the formation of a water-soluble molybdenum-rich phase, consisting of alkaline molybdates and substantial fractions of the fission products caesium and strontium [3].

The conditions of the appearance of this phase are not yet fully understood, though a lot of experimental data are available [3, 6]. A comprehensive understanding could further simplify the conditioning of base glass compositions. Some further experimental work is certainly necessary in this field.

The conditions of preparation and the properties of the parent glasses are listed in Table III. As can be seen from the table, all glasses could be prepared at temperatures below 1200°C (1470 K) and the viscosities at 1150°C (1430 K) are about 100 P. These data meet the currently envisaged requirements for vitrification in hot cells.

The nucleation and crystallization temperatures were determined from differential thermal analysis and experiments in a temperature gradient oven. Annealing times were determined empirically and are sufficiently long to yield homogeneously crystallized glass ceramics. The tendency of crystallization was found to be the maximum in the case of C-type glass. Here the nucleation step could be omitted. For the other three types nucleation was necessary before crystallization in order to get high crystal number densities in the whole sample. It is important to note that the thermal treatment of the glass can be completed within one day. Thereafter the crystal growth rate is very low.

2.2.2. *Semi-technical scale experiments*

The aforementioned C, P, D and E-type products were prepared in the laboratory of the Schott Company in the form of cylindrical blocks of up to 1000 cm^3 in volume (some 3 kg). Substantially the same conditions were applied as determined by the lab-scale experiments. The frits were used, however, as coarse products and the melts were stirred to obtain similar homogeneity as in the lab-scale experiments. Figure 3 shows, as an example, the viscosity of a parent glass melt as a function of temperature. The melts can easily be poured at about 1100°C (1370 K).

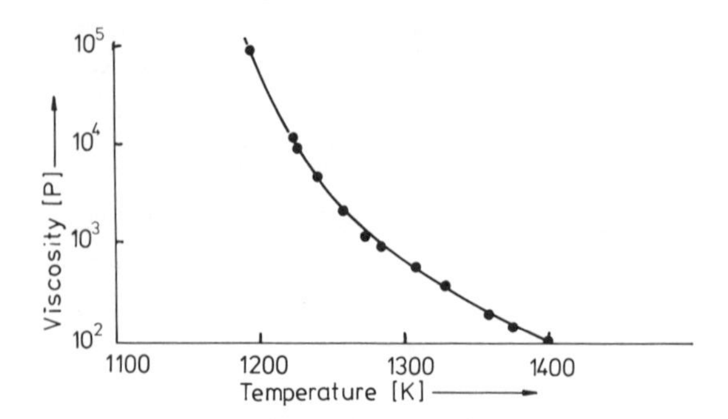

FIG.3. Viscosity of a parent glass melt (C-type) as a function of temperature, after Schott.

The solid C, P and D-type glasses showed no visible quantities of a phase-separated molybdenum-rich product. The E-type showed a thin film of this phase on the surface. The subsequent heat treatment led to bulk crystallized products with no cracks and no bubbles.

The C, D and E-type ceramics were homogeneously crystallized products. The P-type appeared inhomogeneous (large areas of dark brown and light brown colour became visible). In the E-type product some molybdenum-rich phase was found in the bulk.

2.2.3. Hot cell experiments at CEA, Marcoule

In consequence of the encouraging results of the scale-up experiments a campaign was started in collaboration with the CEA at Marcoule to prepare radioactive glass ceramics in the Vulcain cell. In the frame of this work the celsian ceramic was prepared recently by using a real waste solution. The melting temperature was 1150°C. A homogeneous parent glass was obtained and after thermal treatment the ceramic was formed having 200 Ci/l. Further investigations are in progress.

3. PROPERTIES OF GLASS CERAMICS

In this section the glass ceramics will be characterized and compared with their own parent glasses. A description of experimental methods can be found in previous work [3].

3.1. Lab-scale products

Table IV contains a summary of the properties of the glass ceramics and their parent glasses (all containing 20 wt% simulated fission product oxides).

3.1.1. Crystalline host phases for fission products

As can be seen from Table IV, various crystalline phases have been detected in the glass ceramics. Most of them are host phases for long-lived fission products, e.g. pollucite for Cs, h-celsian, perovskite and calcium lanthanum silicate for Sr, (Ce, Zr)O_2, perovskite and calcium lanthanum silicate are suspected to be potential host phases for transuranium elements. The crystals were found to be embedded in a residual boron-rich glass phase.

TABLE IV. PROPERTIES OF GLASS CERAMICS AND PARENT GLASSES

	Glass ceramics			
Crystal phases in the glass ceramics	*h-celsian* [$BaAl_2Si_2O_8$]	*Perovskite* [$CaTiO_3$]	*Diopside* [$CaMg(SiO_3)_2$]	*Eucryptite* [$LiAlSiO_4$]
	pollucite [$(Cs, Na)AlSiO_2O_6$]	Mo-nosean [$Na_8(AlSiO_4)_6 \cdot MoO_4$]	$(Ce, Zr)O_2$	$CaMoO_4$
		$(Ce, Zr)O_2$		
	$(Ce, Zr)O_2$	$SrMoO_4$	$CaMoO_4$	$Ca_4La_6(SiO_4)_6(OH)_2$
	$BaMoO_4$	pollucite [$(Cs, Na)AlSiO_2O_6$]		
Leaching rates $g \cdot cm^{-2} \cdot d^{-1} \times 10^6$				
parent glass				
grain-test	9	7	9	5
Soxhlet-test	9	11	11	14
ceramics				
grain-test	7	8.5	14	4
Soxhlet-test	10	7.1	19	21
Thermal expansion $\alpha_{290-570\,K} \times 10^7$				
glass	110	110	100	100
ceramics	97	89	88	53

The yields of the various phases have not yet been determined quantitatively. There are, however, considerable differences in the crystal yields. Generally, one phase dominates in each glass ceramic. This phase is underlined in Table IV. It is important to note that both low and high number densities of host phases may be efficient for the fixation of fission products. This depends upon whether the fission product in question is an essential constituent of its host phase or whether it will be accommodated in a substitutional way. In the first case a considerable enrichment of a substantial fraction of the respective ion can be expected even with low crystal yields. This is typical of caesium being accommodated in pollucite. A substantial accommodation with low enrichment is expected for strontium, which may substitute calcium and barium. The minerals celsian and perovskite should be mentioned here. These minerals are dominating crystals in the glass ceramic and are, therefore, suitable materials for efficient strontium fixation.

Figure 4(a) shows an SEM picture of the celsian ceramic and Fig. 4(b) gives evidence for strong caesium enrichment in pollucite. Caesium is depleted in all other crystalline phases and in the residual glass phase.

Figure 5(a) is an SEM picture of the perovskite ceramic. The cerium enrichment in perovskite can be seen in Fig. 5(b). This microprobe pattern is also typical of other rare earth elements like La, Pr and Nd.

In view of these figures one may suggest that the optimum glass ceramic is found when all long-lived hazardous fission products are highly enriched in leach-resistant crystalline phases. The residual glass phase, which is known to be less durable, would then be strongly depleted with respect to these ions. As yet a quantitative estimate of the distribution of the fission products has not been made. However, a remarkable enrichment of some relevant ions is already evident.

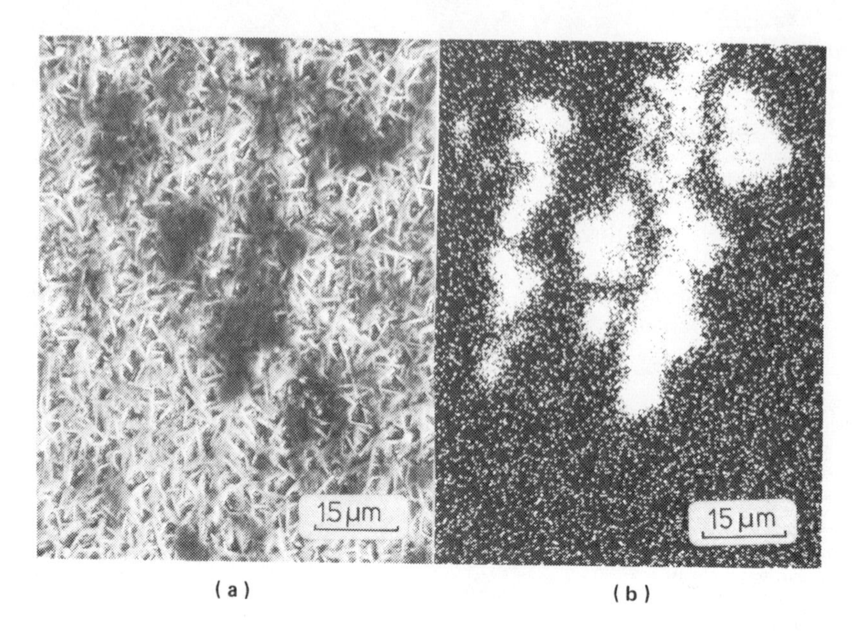

FIG.4. (a) Scanning electron micrograph of a celsian glass ceramic; (b) electron microprobe pattern for Cs of a celsian glass ceramic.

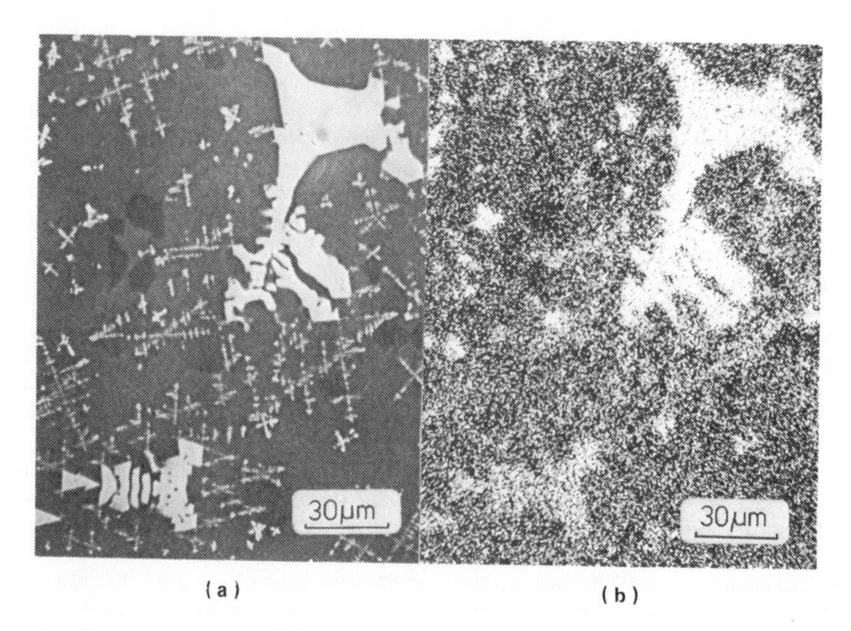

FIG.5. (a) Scanning electron micrograph of a perovskite glass ceramic; (b) electron microprobe pattern for Ce of a perovskite glass ceramic.

In the case of radioactive glass ceramics containing about 20 wt% fission product oxides the nuclear transformation of Cs (into Ba) and Sr (via Y into Zr) in their host phases requires some attention.

When the ceramics are made six years after reprocessing 30 000 MW · d/tHM burned-up fuel the ratio of $^{137}Cs/Cs$ is 0.44. Furthermore, if the pollucite contained all the caesium, then $\leqslant 25\%$ of the total alkali content of pollucite ($\leqslant 50\%$ Cs, $\geqslant 50\%$ Na, K, Rb) would decay to Ba within 300 years, i.e. ten half-lives. A similar consideration for strontium leads to even much lower figures for the respective ratio of Zr/(Ca, Sr, Ba) after 300 years, because of the lower enrichment of strontium in its host phases. It is difficult to evaluate the macroscopic effects of nuclear transformation. However, a deleterious influence on the properties of the whole product seems unlikely. Nevertheless, this point needs further consideration.

3.1.2. Leach resistance

It can be seen from Table IV that the leach rates of the glass ceramics and of their parent glasses are very similar. The grain test values refer to the total alkali loss [7], whereas the Soxhlet values refer to the weight loss of the sample. The results are well in the range of what has been reported for other borosilicate glasses. Glass ceramics show a phase selective leachability. This can be seen from Fig. 4(a). The residual glass phase was leached preferentially, whereas the crystal phases remained intact.

3.1.3. Impact resistance

Impact tests have revealed that glasses tend to form a substantial fraction of powder, whereas glass ceramics do not [3, 6]. The impact resistance of parent glasses is considerably improved by crystallization. ZnO und TiO_2 favour this improvement.

3.1.4. Softening

The dilatometric softening temperatures for glass ceramics are in the range of 970 to 1070 K. The respective values for the parent glasses are only 800 to 870 K. Hence, a considerable increase of the softening temperature is obtained by crystallization.

3.1.5. Thermal expansion

The coefficients of thermal expansion for the glass ceramics were found to be $(50-100) \times 10^{-7}$, whereas the glasses have higher values $(110-115) \times 10^{-7}$. These coefficients determine the tensile strength, i.e. the probability for rupture in the presence of a temperature gradient within the glass. The lower values of the glass ceramics indicate a slightly increased thermal stability as compared to the glasses.

3.2. Semi-technical products

The investigation of various samples taken from the blocks have revealed the following results:

The *host-phases* in the celsian ceramics are the same as given in Table IV. An additional phase, (Ca, Sr, Ba)TiO_3, was observed. Microprobe analysis of the caesium enrichment, e.g., yields analogous patterns as shown in Fig. 4(b). In the perovskite ceramic no (Ce, Zr)O_2 was detected though it was found in lab-scale samples (Table IV). Perovskite can digest the cerium necessary to form (Ce, Zr)O_2 and, therefore, sometimes (Ce, Zr)O_2 does not appear. The cerium enrichment was detected in perovskite and is qualitatively represented by Fig. 5(b).

The diopside ceramic has qualitatively the same composition as given in Table IV. The crystals were, however, too small to detect any enrichment. The eucryptite ceramic showed some Mo-phase separation, as mentioned above. The base glass composition must be modified.

The *leach-resistance* of the four products – not including the molybdenum rich-phase in eucryptite – is within a factor of two the same as for the lab-scale ceramics.

The values of *thermal expansion* and *dilatometric softening temperature* are the same as quoted for the lab-scale samples.

Impact resistance is expected to be as good as measured for the lab-scale samples.

4. CONCLUSIONS

In the present state of development an evaluation of the semi-technical scale glass ceramics in terms of fabrication and properties yields:

Glass ceramic making could be the tail-end step of the currently developed waste vitrification technology.

The perovskite ceramics would be the most desirable product because perovskite is the only aluminium and silicon-free dominating host phase and therefore a durable residual glass phase is expected. However, the fabrication of a homogeneous ceramic as obtained for lab-scale samples must be demonstrated. The diopside ceramic is readily fabricated. There is, however, only one desirable host-phase.

Eucryptite ceramic development calls for minor changes in chemical composition in order to completely suppress the molybdenum separation. Crystallization of more host phases should be achieved.

Celsian is easy to fabricate. The ceramic is homogeneously crystallized and a variety of chemically resistant host phases are formed. The product can be prepared in a hot cell.

ACKNOWLEDGEMENTS

The authors are grateful to Dr. G. Müller, Schott u. Gen., Mainz, for the preparation of the semi-technical products. The authors also wish to thank CEA/CEN Marcoule for preparing the radioactive ceramic.

REFERENCES

[1] BLASEWITZ, A.G., et al., Management of Radioactive Wastes from Fuel Reprocessing (Proc. IAEA/NEA Symp. Paris, 1972), OECD/NEA, Paris (1973).
[2] HEIMERL, W., et al., Management of Radioactive Wastes from Fuel Reprocessing (Proc. IAEA/NEA Symp. Paris, 1972), OECD/NEA, Paris (1973).
[3] DE, A.K., et al., 1st European Nuclear Conf. Paris 1975, TANSAO **20** (1975) 666.
[4] DE, A.K., et al., Bull. Am. Ceram. Soc. (to be published).
[5] DE, A.K., et al., Atomwirtschaft **20** (1975) 359.
[6] DE, A.K., et al., HMI-Report (to be published).
[7] Normblatt 12111, Berlin und Köln: Beuth-Vertr. (1962).

DISCUSSION

R. BONNIAUD: To get good glass ceramics I believe you have to produce fine, controlled crystallization. The glasses used have a high crystallization capacity and for that reason have to be cooled down to the nucleation temperature rapidly. Do you not anticipate some difficulties in the case of actual glasses with a high specific activity? There would, after all, be a problem in controlling the cooling prior to nucleation.

A.K. DE: If you foresee difficulties in cooling down a high-activity sample to the nucleation temperature, let us say 600°C, you then try to make glass ceramics that do not need annealing at the nucleation temperature. In such a case cooling to about 800°C, the crystallization temperature, would be sufficient. Pots of annular shape could be used. Celsian ceramics would be the preferred material to use at the present time.

W. HEIMERL: I think one way of solving the problem just raised, namely avoiding unwanted devitrification during the cooling period, would be to produce smaller glass samples, for example glass beads instead of huge glass blocks. The beads could be rapidly quenched from the melt temperature to the nucleation temperature. Complete transformation into a glass ceramic material would then be a simple matter because a homogeneous temperature distribution in the sample could be obtained owing to its small size. The glass ceramic particles could then be embedded in a metal matrix.

A.K. DE: Yes, that is an interesting idea, but I think it would require some investigation.

J. SAIDL: Do you know the thermal conductivity of your product?

A.K. DE: No, we have not measured it.

ТЕРМИЧЕСКАЯ, ХИМИЧЕСКАЯ И РАДИАЦИОННАЯ УСТОЙЧИВОСТЬ ОСТЕКЛОВАННЫХ РАДИОАКТИВНЫХ ОТХОДОВ

В.В.КУЛИЧЕНКО, Н.В.КРЫЛОВА,
Н.Д.МУСАТОВ
Всесоюзный научно-исследовательский институт
неорганических материалов
Государственного комитета по использованию
атомной энергии СССР,
Москва,
Союз Советских Социалистических Республик

Abstract–Аннотация

THERMAL, CHEMICAL AND RADIATION STABILITY OF VITRIFIED RADIOACTIVE WASTE.

The authors consider delocalization of radionuclides from high-level silicate materials in the course of two storage periods. The first period is characterized by high-temperature conditions in which there is no contact with water, the second by a reduction in the radioactive decay temperature and possible contact between the radioactive materials and water. The paper contains a classification of silicate materials from the standpoint of their thermal, chemical and radiation stability.

ТЕРМИЧЕСКАЯ, ХИМИЧЕСКАЯ И РАДИАЦИОННАЯ УСТОЙЧИВОСТЬ ОСТЕКЛОВАННЫХ РАДИОАКТИВНЫХ ОТХОДОВ.

В докладе рассматривается делокализация радионуклидов из высокоактивных силикатных материалов в течение двух периодов хранения. Первый период – в условиях высоких температур при отсутствии контакта с водой и второй – при снижении температуры радиоактивного распада и возможном контакте радиоактивных материалов с водой. Приведена классификация силикатных материалов с точки зрения их термической, химической и радиационной устойчивости.

Разрабатываемые в настоящее время процессы отверждения высокоактивных отходов основаны на включении радионуклидов в материалы, пригодные для длительного хранения. Опытно-промышленную проверку проходит процесс остекловывания, в результате которого получается материал с наиболее прочной фиксацией радиоизотопов. В СССР в течение длительного времени изучались свойства и поведение при хранении различных типов силикатных материалов, получаемых в результате остекловывания отходов разного состава, в том числе с удельной активностью до 10^4 Ки/л. Основное внимание уделялось факторам, оказывающим влияние на степень закрепления радионуклидов.

Механизм выщелачивания радионуклидов водой из стеклоподобных материалов и зависимость скорости выщелачивания от различных факторов описаны в предыдущих работах [1-6]. Было показано, что для наиболее устойчивых плавленных материалов с базальтоподобной структурой, для которых скорость выщелачивания может достигать величины 10^{-8} г·см$^{-2}$·сут$^{-1}$, существует опасность заражения воды радионуклидами при активности ^{90}Sr в блоке более 10 Ки/л и ^{137}Cs – более 100 Ки/л. Это обстоятельство не позволяет рекомендовать для таких отходов создание могильников без герметизации (естественной или искусственной).

Однако, пренебрегать химической устойчивостью отходов не следует, так как обеспечение надежной гидроизоляции на сотни лет осуществить практически невозможно, а снижение активности при длительном хранении отвержденных отходов может привести к отсутствию опасности распространения радионуклидов за счет выщелачивания. Необходимый для этого промежуток времени будет тем меньше, чем ниже скорость выщелачивания радионуклидов из отвержденных отходов.

В процессе хранения высокоактивные материалы будут длительное время находиться при повышенной температуре [7] под воздействием ионизирующей радиации. Поэтому наиболее важным параметром является влияние этих факторов на изменение скорости выщелачивания радионуклидов. Кроме того, изучалась возможность делокализации радионуклидов в различные периоды хранения стеклоподобных материалов вне контакта с водой.

1. ВОЗМОЖНОСТЬ ДЕЛОКАЛИЗАЦИИ РАДИОНУКЛИДОВ ПРИ ХРАНЕНИИ ОСТЕКЛОВАННЫХ ОТХОДОВ В ОТСУТСТВИИ ВОДЫ

Оценивая одну из возможностей заражения окружающей среды за счет диффузии радионуклидов в пограничные твердые материалы на примере стронция-90, можно видеть (рис.1), что расстояние, на которое может распространиться стронций за время хранения, незначительно даже для материалов с коэффициентом диффузии $10^{-8} - 10^{-9}$ см^2/с.

Таким образом, следует предположить, что при хранении твердых радиоактивных материалов опасность распространения радионуклидов за счет диффузии исключается.

На рис.2 приведены экспериментальные данные по определению скорости делокализации цезия-137 в процессе хранения силикатного препарата при различных температурах[1].

Увеличение перехода в газовую фазу с увеличением температуры отмечается также для стронция-90 (табл.1). Испарение радионуклидов при повышенных температурах может быть вызвано увеличением упругости пара соединений, входящих в состав стекла. Однако, загрязнение через газовую фазу наблюдается и в процессе хранения стеклообразных препаратов при комнатной температуре. Так, стеклообразный препарат, приготовленный из реальных высокоактивных отходов на основе боросиликатного стекла и имеющего исходную удельную активность 10 000 Ки/л, хранился в течение восьми лет в закрытом стальном сосуде. В момент вскрытия сосуда, крышка его, не касавшаяся препарата, была сильно загрязнена радионуклидами. Расчет дал среднюю скорость делокализации β-излучающих радионуклидов 10^{-11} г/см^2·сут.

Исследование этого явления показало:

- Скорость делокализации радионуклидов значительно повышается в результате хранения материалов на воздухе при комнатной температуре.

- В определенных условиях наряду с делокализацией радионуклидов наблюдается и сублимация макрокомпонентов препаратов. Наличие радиационных возгонов наблюдается не только при хранении радиоактивных материалов, но и при облучении нерадиоактивных на источнике ^{60}Co.

---

[1] За единицу скорости улетучивания принята величина, имеющая размерность г/см^2·сут, которая может быть получена по формуле: $\frac{\text{мКи}\cdot\text{вес препарата в граммах}}{\text{мКи}\cdot\text{поверхность препарата в см}^2\cdot\text{сут}}$

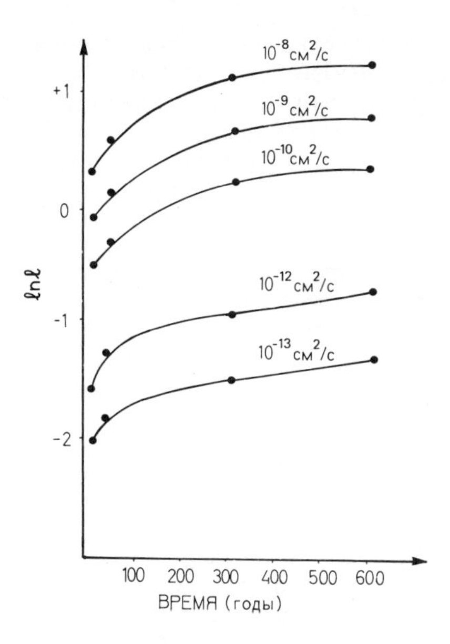

Рис.1. Расстояние ℓ в см, на которое может продиффундировать ^{90}Sr в различных средах с различными коэффициентами диффузии.

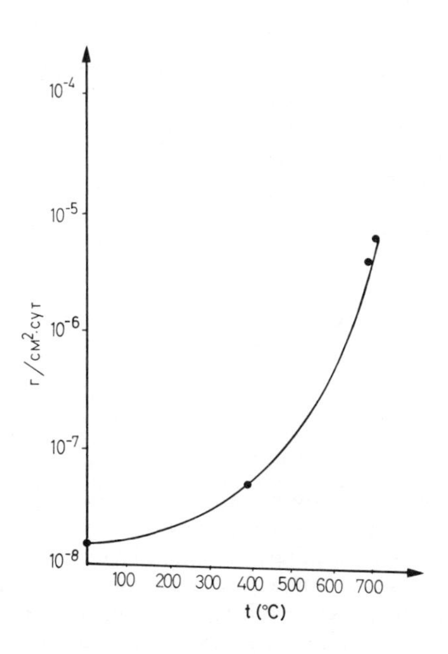

Рис.2. Зависимость скорости улетучивания ^{137}Cs от температуры хранения стекловидного материала состава: SiO_2 – 37%, Al_2O_3 – 7%, B_2O_3 – 15%, Na_2O – 5%, Cs_2O – 2%, CaO – 34%.

ТАБЛИЦА I. СКОРОСТЬ ПЕРЕХОДА СТРОНЦИЯ-90 В ГАЗОВУЮ ФАЗУ В ПРОЦЕССЕ ХРАНЕНИЯ РАДИОАКТИВНЫХ СТЕКЛООБРАЗНЫХ ПРЕПАРАТОВ В ТЕЧЕНИЕ 100 ЧАСОВ ПРИ 550°С

Удельная активность Sr-90 в препарате (Ки/л)	Скорость улетучивания Sr-90 (г/см²·сут)
595	$1,4\cdot10^{-9}$
600	$1,0\cdot10^{-9}$
560	$1,1\cdot10^{-9}$
125	$7,0\cdot10^{-8}$
128	$3,2\cdot10^{-9}$
7	$3,6\cdot10^{-9}$

ТАБЛИЦА II. ДЕЛОКАЛИЗАЦИЯ СТРОНЦИЯ-90 ИЗ ПРЕПАРАТОВ С УДЕЛЬНОЙ АКТИВНОСТЬЮ 270 Ки/л В РЕЗУЛЬТАТЕ ХРАНЕНИЯ В ТЕЧЕНИЕ 3,5 МЕСЯЦЕВ

Состав препарата	Сорбированный стронций-90			
	алюминий		стекло	
	мКи	мКи/см² Al	мКи	мКи/см² стекло
Na_2O - 10% SiO_2 - 40% B_2O_3 - 30% SrO - 20%	$1\cdot10^{-4}$	$8\cdot10^{-6}$	$6\cdot10^{-5}$	$4\cdot10^{-7}$
Na_2O - 11% CaO - 26% Fe_2O_3 - 1% Mn_3O_4 - 12% SiO_2 - 7% B_2O_3 - 25% CaF_2 - 18%	$2\cdot10^{-5}$	$2\cdot10^{-6}$	$2\cdot10^{-5}$	$1\cdot10^{-7}$

При дозах поглощенной энергии около 10^9 рад на внутренней стороне крышки из алюминиевой фольги, закрывающей сосуды с образцами, появляется кристаллический налет, видимый невооруженным глазом. По данным рентгенофазового анализа в сублиматах обнаруживается борная кислота и бораты натрия, α-кварц, нитрат натрия. Как было показано ранее [3,8], эти соединения образуются на поверхности стеклообразных материалов, находящихся в контакте с воздухом за счет радиационно-химических реакций.

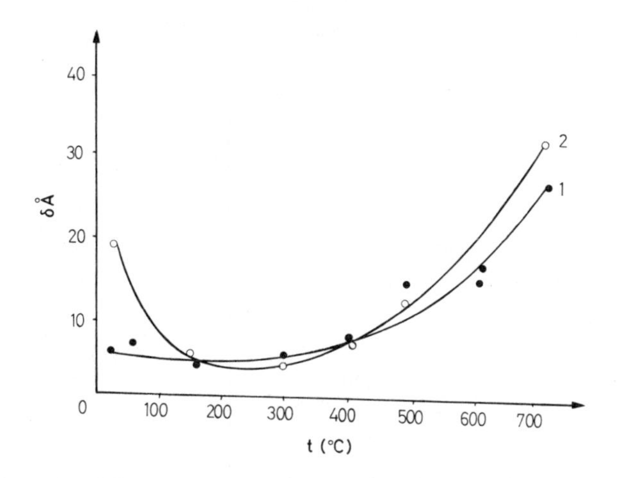

Рис.3. Влияние отжига и облучения на выщелачивание водой в течение 1 часа материала состава: SiO_2 – 35%, B_2O_3 – 10%, Na_2O – 10%, CaO – 16%, ΣFe_2O_3, Cr_2O_3, Mn_3O_4 – 25%. 1 – образец после отжига без облучения, 2 – образец после одновременного отжига и облучения при дозе $6 \cdot 10^8$ рад.

- Удаляющиеся из препарата радионуклиды сорбируются преимущественно на незаземленных металлических поверхностях (табл.II). Причиной этих явлений может быть кулоновское отталкивание высокодисперсных частиц продуктов гетерогенных радиационно-химических реакций от поверхности, которая приобретает заряд в результате эмиссии σ-электронов. Этот процесс является определяющим при температуре ниже 100°C.

При температурах выше 100°C основную роль в делокализации радионуклидов играет упругость пара их соединений. Высокие температуры, в зоне которых продолжительное время будут находиться высокоактивные препараты в процессе хранения, могут оказывать влияние на поведение радиоизотопов, как в процессе хранения, так и при последующем контакте препарата с водой.

2. ВЛИЯНИЕ ТЕМПЕРАТУРЫ ХРАНЕНИЯ НА ПОВЕДЕНИЕ РАДИОНУКЛИДОВ ПРИ ПОСЛЕДУЮЩЕМ КОНТАКТЕ МАТЕРИАЛОВ С ВОДОЙ

В результате хранения материалов разных составов при повышенных температурах химическая устойчивость их меняется неоднозначно: независимо от наличия или отсутствия кристаллизации для разных препаратов можно наблюдать увеличение, уменьшение или отсутствие изменения химической устойчивости (рис.3).

По аналогии с промышленными стеклами имеются три причины изменения скорости выщелачивания в результате отжига [9-12].

1. Миграция щелочей к поверхности для уменьшения поверхностного натяжения, а также за счет освобождения натрия при разрыве связей $-\overset{|}{\underset{|}{Si}}-O-Na$ с образованием связей $-\overset{|}{\underset{|}{Si}}-O-\overset{|}{\underset{|}{Si}}-$

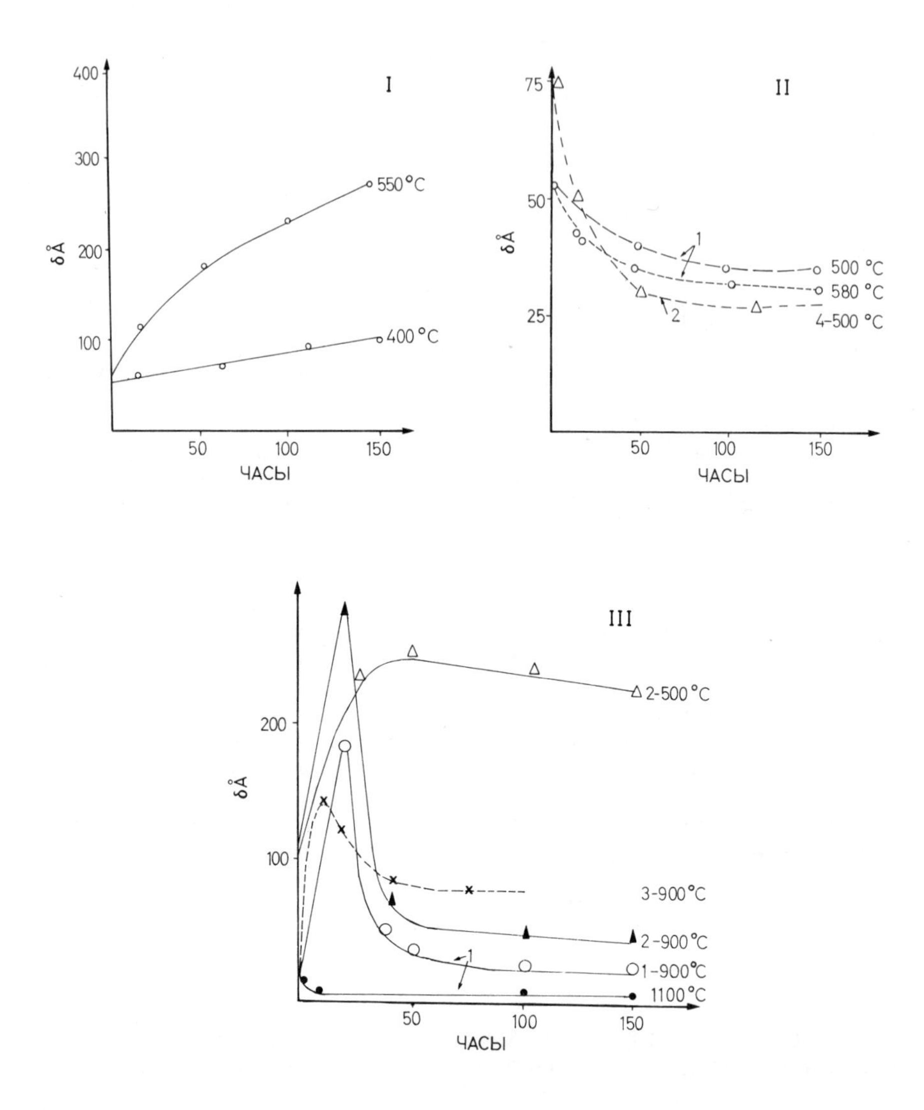

Рис.4. Зависимость глубины разрушения (δÅ) водой от времени и температуры отжига плавленых препаратов: I – боросиликатного состава: SiO_2– 35%, B_2O_3– 10%, Na_2O – 10%, CaO – 16%, ΣFe_2O_3, Cr_2O_3, Mn_3O_4– 25%. II – алюмоборосиликатного состава: 1 – SiO_2 – 35%, B_2O_3– 20%, Al_2O_3– 15%, Na_2O – 16%, CaO – 15%, ΣFe_2O_3, Cr_2O_3, Mn_3O_4– 4%. 2 – SiO_2 – 35%, B_2O_3– 15%, Al_2O_3– 7%, CaO – 34%, ΣFe_2O_3, Cr_2O_3, Mn_3O_4–1%, прочие – 8%. III – базальтоподобные: 1 – SiO_2– 35%, Al_2O_3– 9%, CaO – 16%, MgO – 13%, ΣFe_2O_3, Cr_2O_3, Mn_3O_4– 20%, прочие – 8%. 2 – SiO_2 – 33%, Al_2O_3 – 37%, Na_2O – 12%, CaO – 18%. 3 – SiO_2 – 26%, Al_2O_3 – 3%, CaO – 18%, Cs_2O – 7%, Na_2O – 10%, MgO – 1,5%, ΣFe_2O_3, Cr_2O_3, Mn_3O_4 – 18%, прочие – 16,5%.

2. Обеднение поверхности стекла щелочами за счет улетучивания.

3. Глубокие химические превращения в стекле, приводящие к закреплению ионов щелочных металлов в решетке.

Первая причина должна привести к увеличению скорости выщелачивания, две вторые – к ее уменьшению. Так как в отсутствие кристаллизации скорость выщелачивания для большинства исследованных препаратов с течением времени уменьшается, миграция щелочей к поверхности не играет существенной роли.

Оценка изменения концентрации щелочных металлов на поверхности проводилась с помощью газохроматографического метода по изменению абсолютных величин удерживаемых объемов специфических сорбатов (метиловым спиртом, в частности). Результаты исследования показали, что концентрация ионов натрия в результате отжига при температуре 500°C и 550°C практически не меняется вне зависимости от ухудшения или улучшения химической устойчивости образцов. Следовательно, увеличение химической устойчивости нельзя объяснить обеднением поверхности щелочными компонентами за счет улетучивания и, очевидно, основной причиной являются структурные изменения препаратов.

По своей структуре исследованные материалы могут быть разделены на три структурные группы.

1. Препараты, содержащие не более 35% стеклообразующих добавок, не менее 10% окислов щелочных металлов и не менее 50-10% борного ангидрида, и значительного количества многовалентных металлов. Препараты обычно представляют собой стеклокристаллическую структуру уже в результате его приготовления (рис.4).

В процессе отжига кристаллизация усиливается с образованием новых кристаллических соединений, сравнительно легко разрушаемых водой.

Ионизирующая радиация при высоких температурах не приводит к изменению состава кристаллической фазы, но увеличивает скорость кристаллизации, в связи с чем при одновременном воздействии температуры и радиации эффект изменения химической устойчивости усиливается (рис.3).

Полная замена окиси натрия в препаратах на борный ангидрид, который способствует растворению окислов железа, хрома и марганца, приводит к отсутствию кристаллизации даже в результате длительного отжига и к улучшению химической устойчивости стекла (рис.5). При замене B_2O_3 на окись натрия в результате отжига препараты кристаллизуются не только с поверхности, но и в объеме. Кристаллизация приводит к очень незначительному изменению поверхности и не увеличивает количество щелочных ионов на поверхности. Состав кристаллической фазы существенно отличается от исходного препарата и в данном случае растворимость препаратов с увеличением температуры отжига уменьшается (рис.5).

К этой структурной группе следует отнести материалы, полученные при отверждении отходов, содержащих значительные количества натрия и алюминия, с применением в качестве флюсующих добавок природного минерала датолита $Ca_2B_2[SiO_4]_2(OH)_2$. Включение в датолит до 40 вес% кальцинированных отходов позволяет получить материалы, химическая устойчивость которых характеризуется величиной выщелачивания цезия и стронция 10^{-5} - 10^{-6} г/см^2 · сут и 10^{-6} г/см^2, соответственно. Несмотря на значительное содержание окиси натрия в препаратах (до 22%) в результате

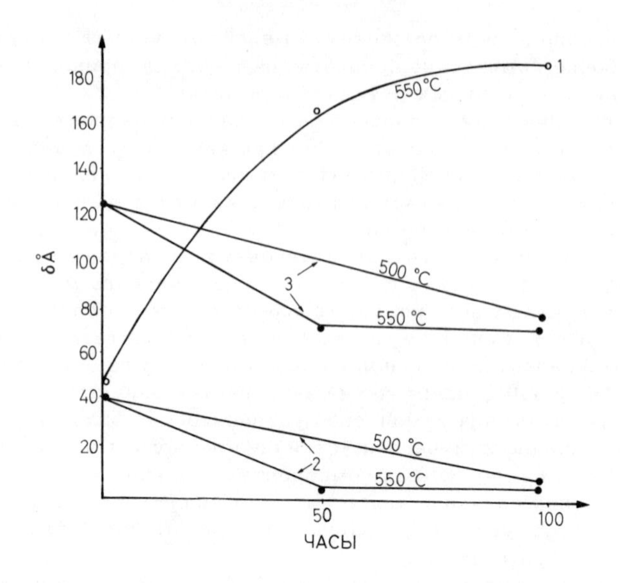

Рис.5. Зависимость глубины разрушения водой от продолжительности отжига боросиликатного материала состава: (1) SiO_2 – 35%, B_2O_3 – 10%, Na_2O – 10%, CaO – 16%, ΣFe_2O_3, Cr_2O_3, Mn_3O_4 – 25 и при полной замене в нем Na_2O на B_2O_3 (2) и B_2O_3 на Na_2O (3).

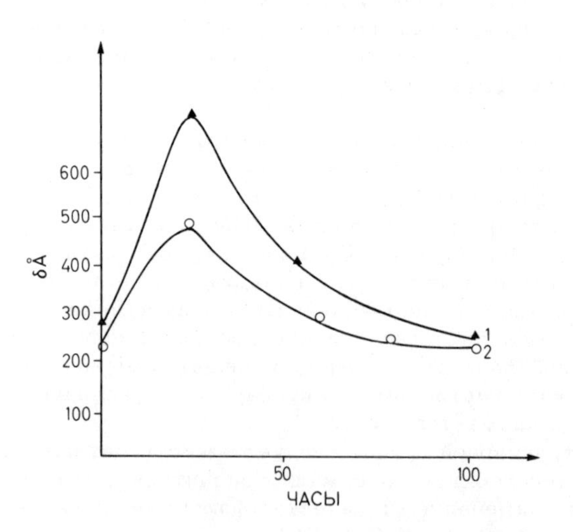

Рис.6. Зависимость глубины разрушения водой материалов, полученных при сплавлении кальцинированных отходов с датолитом. 1 – состав материала: Na_2O – 22%, CaO – 22%, B_2O_3 – 13 SiO_2 – 24%, Al_2O_3 – 15%, ΣFe_2O_3, Cr_2O_3, Mn_3O_4 – 3%. 2 – состав материала: Na_2O – 16,5%, CaO – 26%, B_2O_3 – 16%, SiO_2 – 28%, Al_2O_3 – 11,3%, ΣFe_2O_3, Cr_2O_3, Mn_3O_4 – 2%.

отжига после первого периода ухудшения химической устойчивости материалов к воде происходит перестройка их структуры с постепенным улучшением химической устойчивости (рис.6).

II. Препараты, содержащие не менее 50% стеклообразующих компонентов (SiO_2, Al_2O_3), не более 10% окислов щелочных металлов, 10% борного ангидрида, а также не более 10% окислов железа, хрома и марганца. Эти препараты не содержат кристаллической фазы, не кристаллизуются вплоть до температур на 100-200°C ниже температуры их размягчения. Их структурный каркас значительно укреплен наличием окиси алюминия и большим общим количеством стеклообразующих добавок. Химическая устойчивость их в результате отжига при указанных температурах улучшается (рис.4-II).

III. Материалы, отличающиеся высокой концентрацией кристаллической фазы в объеме материала уже в результате их приготовления. Их структура приближается к структуре базальтов. Они отличаются высокой температурой приготовления и высокой вязкостью расплава, имеют высокую химическую устойчивость образцов (рис.4-III).

Химическая устойчивость существенно не изменяется от времени при температурах 550°C, имеет сложную зависимость при 900°C и увеличивается при 1100°C.

Рентгенофазовый анализ показал, что структура препаратов не претерпевает существенных изменений при температуре 550°C и значительно изменяется от времени отжига при 900°C и 1100°C.

Аномальный характер изменения химической устойчивости при 900°C может быть объяснен высокой скоростью кристаллизации при нагреве, что делает возможным образование микротрещин и сопровождается увеличением скорости разрушения препарата водой. Увеличение времени отжига приводит к упорядочению структуры и образованию термодинамически устойчивых мелких кристаллов, подобных ситаллам, приводящих к резкому увеличению химической устойчивости.

Такие материалы были получены в процессах с использованием тепла химических реакций [13].

ЗАКЛЮЧЕНИЕ

Таким образом, материалы всех представленных выше групп могут быть использованы для целей длительного хранения или захоронения радионуклидов. В процессе остекловывания отходов возможно получение материалов с заранее заданной структурой. Это позволяет получить высокоактивные материалы, степень закрепления радионуклидов в которых в результате длительного хранения не будет существенно ухудшаться, а в ряде случаев и улучшаться. Наиболее стабильно радионуклиды фиксируются в материалах с базальтоподобной структурой, для получения которых необходимы температуры 1200°C и выше.

Очевидно, такие материалы будут в будущем наиболее приемлемы при переработке отходов ТВЭЛ на быстрых нейтронах.

В связи с этим следующей стадией после освоения процесса с получением плавленыххстеклоподобных материалов при температуре 900-1100°C необходима разработка аппаратурно-технологических схем для осуществления процесса при 1200-1400°C.

ЛИТЕРАТУРА

[1] ЗИМАКОВ, П.В., КУЛИЧЕНКО, В.В., Ат.Энерг.10(1961) 59.
[2] ЗИМАКОВ, П.В., КУЛИЧЕНКО, В.В. и др., Treatment and Storage of High-Level Radioactive Wastes (Proc. Symp., Vienna, 1962) IAEA, Vienna (1963) 397.
[3] ДУХОВИЧ, Ф.С., КУЛИЧЕНКО, В.В., Ат. Энерг.18(1965) 361.
[4] КУЛИЧЕНКО, В.В., ДУХОВИЧ, Ф.С. и др., Изучение физико-химических свойств различных типов твердых материалов с целью определения возможности и целесообразности их использования для закрепления радиоизотопов при обработке радиоактивных отходов, Отчет по контракту с МАГАТЭ № 340/RB/RL за 1969 г.
[5] КУЛИЧЕНКО, В.В., Ат. Энерг. 29(1970) 320.
[6] At. Energy Rev. 9 1 (1971) 195-207.
[7] ВЕРИГИН, Н.Н., МАРТЫНОВ, Ю.П. и др., Disposal of Radioactive Wastes into the Ground (Proc. Symp., Vienna, 1967) IAEA, Vienna (1967) 441.
[8] ДУХОВИЧ, Ф.С., КУЛИЧЕНКО, В.В., Сборник докладов конференции специалистов стран-членов СЭВ по проблеме обезвреживания радиоактивных отходов, Брно (1965) 255.
[9] ЭЙТЕЛЬ, В., Физическая Химия Силикатов, ИЛ, Москва, 1962.
[10] ГРЕБЕНЩИКОВ, И.В., ФАВОРСКАЯ, Т.А., Труды Госуд. Оптич. Инст., 5 45(1929).
[11] WILLIAMS, K., WEYL, W., Glass Industry (1945) 275.
[12] ДУБРОВО, С.К., Журн. Прикл. Хим., 20(1947) 714.
[13] ВЕРЕСКУНОВ, В.Г., ЗАХАРОВА, К.П. и др., Disposal of Radioactive Wastes into the Ground (Proc. Symp., Vienna, 1967) IAEA, Vienna (1967) 455.

DISCUSSION

D.W. CLELLAND: Have I understood correctly that the volatilization of fission products into the air from vitrified waste surfaces is extremely low and represents a negligible problem?

Nina V. KRYLOVA: The volatilization of the fission products, especially caesium, at temperatures above 500°C is quite considerable and special conditions for storing the vitrified materials are therefore necessary.

A.M. PLATT: What technology did you apply and what equipment was used to make the radioactive samples described in your paper? Those were silicates, as opposed to the phosphates described by Mr. Dolgov.

Nina V. KRYLOVA: The silicate sample containing actual radioactive waste was produced in a pilot plant. The equipment described by Mr. Dolgov is designed to produce silicates as well as phosphates.

СВОЙСТВА ФОСФАТНЫХ И СИЛИКАТНЫХ СТЕКОЛ ДЛЯ ОТВЕРЖДЕНИЯ РАДИОАКТИВНЫХ ОТХОДОВ

Н.Е.БРЕЖНЕВА, С.Н.ОЗИРАНЕР,
А.А.МИНАЕВ, Д.Г.КУЗНЕЦОВ
Институт физической химии Академии наук СССР,
Москва,
Союз Советских Социалистических Республик

Доклад представлен Н.В.Крыловой

Abstract–Аннотация

PROPERTIES OF PHOSPHATE AND SILICATE GLASSES USED FOR THE SOLIDIFICATION OF RADIOACTIVE WASTE.

The authors study the chemical and thermal stability of phosphate and borosilicate glasses intended for fixing radioactive waste containing sodium and aluminium salts. Vitrification regions in the systems $Na_2O - Al_2O_3 - P_2O_5$ and $Na_2O - Al_2O_3 - CaO - B_2O_3 - SiO_2$ are investigated together with some of the properties of the glasses themselves, such as chemical stability, softening point and the temperature at which crystallization begins. It is demonstrated that these systems can yield chemically and thermally stable glasses capable of fixing up to 50% radioactive waste (as calcined residue).

СВОЙСТВА ФОСФАТНЫХ И СИЛИКАТНЫХ СТЕКОЛ ДЛЯ ОТВЕРЖДЕНИЯ РАДИОАКТИВНЫХ ОТХОДОВ.

Работа посвящена исследованию химической и термической устойчивости фосфатных и боросиликатных стекол, предназначенных для фиксации радиоактивных отходов, содержащих соли натрия и алюминия. Исследованы области стеклообразования в системах $Na_2O - Al_2O_3 - P_2O_5$; $Na_2O - Al_2O_3 - CaO - B_2O_3 - SiO_2$, а также некоторые свойства стекол в этих системах: химическая устойчивость, температура размягчения, температура начала кристаллизации. Показано, что в этих системах есть химически и термически устойчивые стекла, позволяющие фиксировать до 50% (по кальцинированному остатку) радиоактивных отходов.

Широкое развитие атомной энергетики вызывает в свою очередь рост объемов радиоактивных отходов, в том числе и высокоактивных.

Безопасное их захоронение поэтому является сейчас одной из наиболее важных проблем. В настоящее время, по мнению большинства ученых всего мира, наиболее безопасным методом хранения высокоактивных отходов является их фиксация в устойчивых стеклах с последующим захоронением стеклянных блоков в могильниках. Однако, до сих пор не решен вопрос о том, какие стекла применять в технологии отверждения. Хотя данная работа и не дает окончательного ответа, в ней сделана попытка сопоставления некоторых свойств фосфатных и силикатных систем, применяемых для остекловывания алюминий и натрий содержащих отходов. Так как соотношения компонентов отходов могут меняться в зависимости от способа переработки твэлов, мы не привязывали свои исследования к определенному составу отходов, а ставили вопрос несколько шире, т.е. изучали системы типа:

$Na_2O - Al_2O_3 - P_2O_5$ $\qquad$ $Na_2O - Al_2O_3 - SiO_2$ $\qquad$ и т.д.

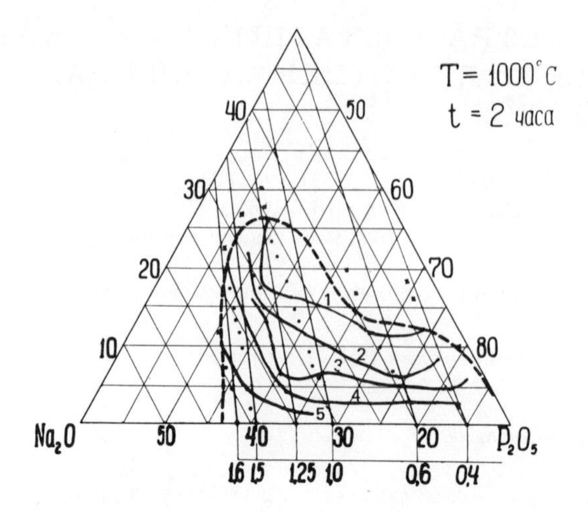

Рис.1. Область стеклообразования в системе $Na_2O - Al_2O_3 - P_2O_5$ (температура варки 1000°С). 1) $1 \div 3 \cdot 10^{-5}$ г/см^2, 2) $3 \div 10 \cdot 10^{-5}$ г/см^2, 3) $1 \div 4 \cdot 10^{-4}$ г/см^2, 4) $4 \div 10 \cdot 10^{-4}$ г/см^2, 5) $1 \div 9 \cdot 10^{-3}$ г/см^2

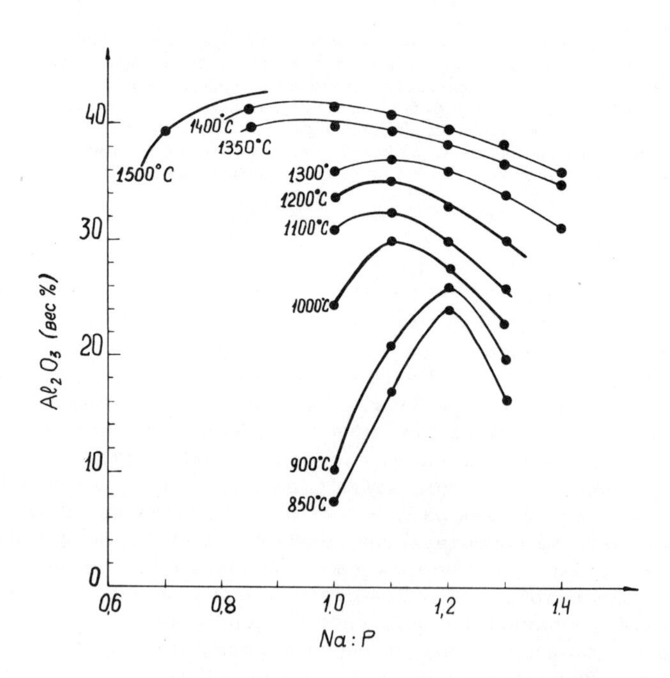

Рис.2. Зависимость предельного содержания окиси алюминия в стекле в системе $Na_2O - Al_2O_3 - P_2O_5$ в зависимости от температуры варки и мольного отношения Na : P.

При сравнении этих систем прежде всего возникает вопрос об области стеклообразования в них, т.е. какое количество отходов может быть включено в стекло и возможность выплавки подобных стекол в промышленных условиях.

Область стеклообразования, изученная при температуре выплавки 1000°С в системе $Na_2O-Al_2O_3-P_2O_5$, представлена на рис.1. Такая температура выбрана потому, что она довольно легко достижима в промышленных печах с большим сроком их службы и надежностью в работе. Как видно, уже при температуре 1000°С суммарное содержание окиси натрия и окиси алюминия достигает величины 50%. Варку таких стекол значительно легче осуществить в производстве, чем варку силикатных стекол, так как флюс вводится в отходы в виде фосфорной кислоты в жидком виде и при сушке и кальцинации образуются соединения типа $xNa_2O \cdot yAl_2O_3 \cdot zP_2O_5$, которые плавятся в области температур 700-750°С.

Процесс варки силикатных стекол, при прочих равных условиях, протекает либо при высокой температуре, либо медленнее, так как имеют место твердофазные процессы.

Температура начала размягчения фосфатных стекол, выплавленных при Т = 1000°С, не очень высокая (600-700°С), что не очень хорошо с точки зрения хранения их в могильниках. Повысить точку размягчения и следовательно температуру варки этих стекол можно, увеличив в них содержание окиси алюминия или уменьшая мольное отношение натрия к фосфору (М).

Зависимость предельного содержания окиси алюминия от температуры варки и отношения натрия к фосфору приведены на рис.2.

Как и следовало ожидать, при увеличении содержания окиси алюминия резко возрастает температура варки стекла и при содержании ее около 40% достигает величины 1500°С.

Температура варки возрастает также при уменьшении М. Поэтому, несмотря на то, что эти стекла обладают высокой химической стойкостью и высокой температурой начала размягчения, стекла, содержащие $>35\%$ окиси алюминия или с мольным отношением $M<0,9$, можно будет использовать только при наличии надежных дистанционно управляемых высокотемпературных (1300-1500°С) стекловаренных печей. Были изучены также случаи стеклообразования в некоторых боросиликатных системах (так как чисто силикатные системы очень тугоплавки и их применение вряд ли будет возможно).

Были изучены разрезы 4-х компонентной системы $Na_2O-Al_2O_3-B_2O_3-SiO_2$ с постоянным содержанием борного ангидрида – 5,10,15 и 20 вес%. На рис.3,4 приведены области стеклообразования в этой системе при содержании борного ангидрида 10 и 20% при различных температурах.

При всех значениях концентрации борного ангидрида область стеклообразования ограничена содержанием окиси алюминия ~30% и окиси натрия – 20%. При этом, если количество окиси натрия становится меньше 20%, стекла не провариваются даже при Т = 1400°С. Следует отметить, что при содержании окиси натрия больше 25%, стекла становятся химически не очень устойчивыми. Следовательно, область оптимальных стекол лежит в довольно узком интервале по содержанию окиси натрия. В отходах содержание ее может колебаться в широких пределах. Поэтому стекла системы $Na_2O-Al_2O_3-B_2O_3-SiO_2$, на наш взгляд, не

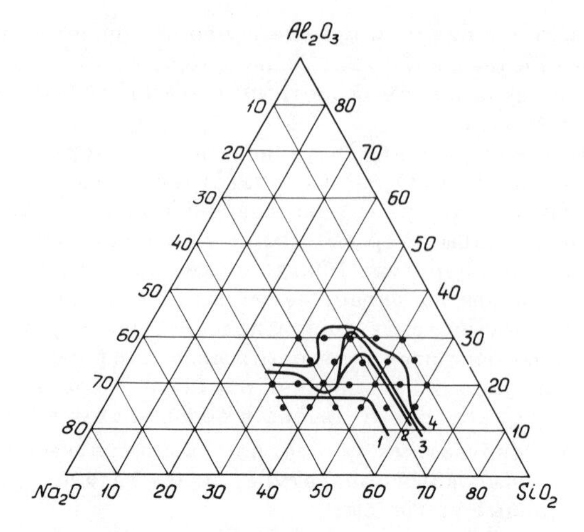

Рис.3. Область стеклообразования в системе $Na_2O - Al_2O_3 - B_2O_3 - SiO_2$; B_2O_3= 10% (вес). Кривые: 1-Т= 1100°С, 2-Т = 1200°С, 3-Т = 1300°С, 4-Т = 1400°С. На рис. 4-8 обозначения те же.

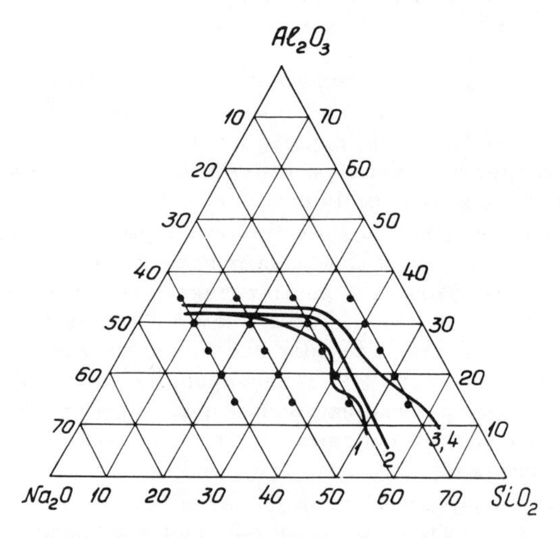

Рис.4. Область стеклообразования в системе $Na_2O - Al_2O_3 - B_2O_3 - SiO_2$: B_2O_3 = 20%.

могут применяться для целей отверждения натрий и алюминий содержащих отходов.

Как известно из практики стекловарения, окись кальция является хорошим плавнем для силикатных стекол, поэтому были изучены разрезы системы $Na_2O - Al_2O_3 - CaO - B_2O_3 - SiO_2$ с постоянным содержанием окиси кальция и борного ангидрида – 10,10%; 10,15%; 15,10% и 15,15%, соответственно. Области стеклообразования их приведены на рис.5,6,7,8.

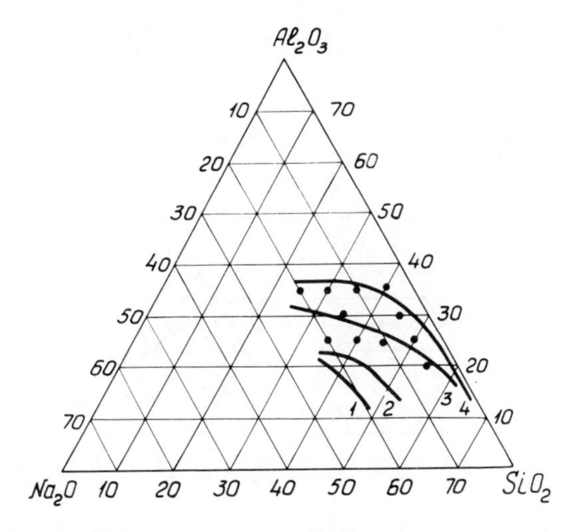

Рис.5. Область стеклообразования в системе $Na_2O - Al_2O_3 - B_2O_3 - SiO_2 - CaO$; $CaO = 10\%$; $B_2O_3 = 10\%$.

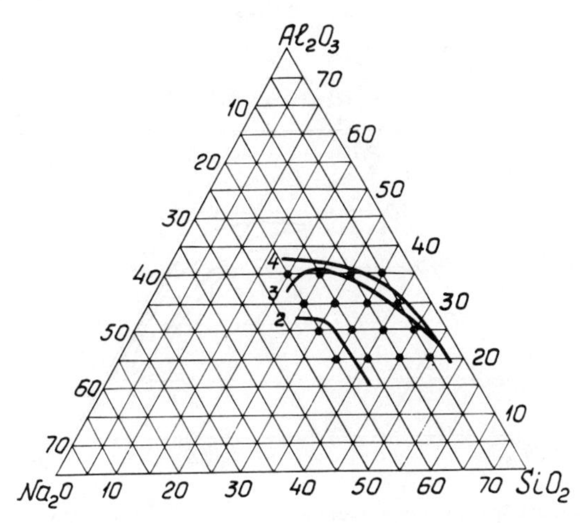

Рис.6. Область стеклообразования в системе $Na_2O - Al_2O_3 - CaO - B_2O_3 - SiO_2$; $CaO = 15\%$; $B_2O_3 = 10\%$.

Видно, что введение CaO значительно расширяет область стеклообразования и в сторону увеличения содержания окиси алюминия (до 35-40%) и в сторону уменьшения содержания окиси натрия (до 5%). Суммарное содержание окислов отходов может доходить до 50% и более. Такие стекла могут быть использованы для целей отверждения отходов.

Оптимальные составы стекол лежат в системе Na_2O-Al_2O_3-CaO-B_2O_3-SiO_2 с содержанием CaO = 15% и B_2O_3 = 15%.

Следует отметить, что при увеличении содержания окиси алюминия в стекле более 30% температура варки повышается до 1300-1400°C. То же самое происходит при уменьшении содержания окиси натрия меньше 10%.

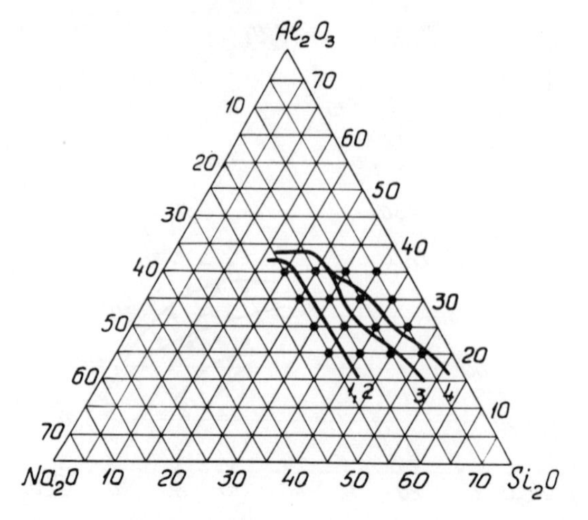

Рис.7. Область стеклообразования в системе $Na_2O - Al_2O_3 - CaO - B_2O_3 - SiO_2$; CaO = 10% ; B_2O_3 = 15%.

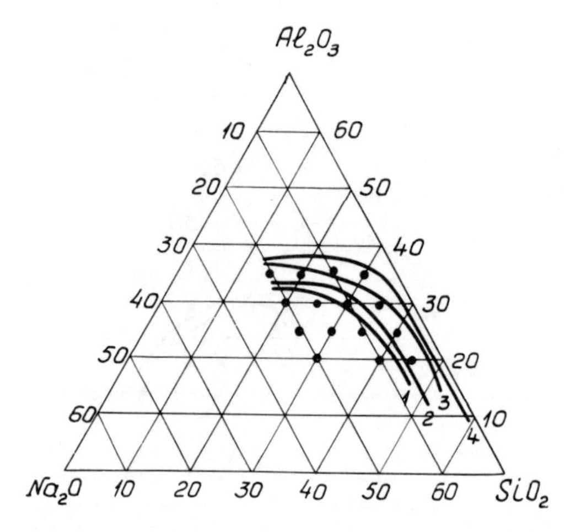

Рис.8. Область стеклообразования в системе $Na_2O - Al_2O_3 - CaO - B_2O_3 - SiO_2$; CaO = 15% ; B_2O_3 = 15%.

Химическая устойчивость этих стекол сопоставима с химической устойчивостью фосфатных стекол, что видно из таблиц I и II.

Их химическую устойчивость можно оценить величиной 10^{-6} г/см^2·сут (по вымываемости натрия в воду при температуре 20-25°C).

Изучалась также кристаллизационная способность исследуемых стекол и влияние кристаллизации на их химическую устойчивость. Эти зависимости играют существенную роль при выборе условий хранения отходов в могильниках и, в конечном счете, влияют на экономику всего процесса остеклования.

ТАБЛИЦА I. ХИМИЧЕСКАЯ УСТОЙЧИВОСТЬ НЕКОТОРЫХ ФОСФАТНЫХ СТЕКОЛ

№ стекла	Состав (вес%)			Химическая устойчивость (г/см^2·сут·10^{-6})
	Na_2O	Al_2O_3	P_2O_5	
5	25,4	26	48,6	0,2
15	30,1	31	38,9	0,29
16	28,2	34	37,2	0,26
17	28,4	35	36,6	0,84
18	27,5	37	35,5	0,23
19	26,6	39	34,4	1,18
20	25,8	41	33,2	1,1

ТАБЛИЦА II. ХИМИЧЕСКАЯ УСТОЙЧИВОСТЬ НЕКОТОРЫХ СИЛИКАТНЫХ СТЕКОЛ

№ стекла	Состав (вес%)					Химическая устойчивость (г/см^2·сут·10^{-6})
	Na_2O	Al_2O_3	CaO	B_2O_3	SiO_2	
3	20	20	0	10	50	9,3
4	25	20	0	10	45	11,7
33	20	15	0	10	55	5,7
52	20	15	0	20	45	0,9
53	30	15	0	20	35	7,1
64	40	20	0	5	35	47,0
95	20	30	10	15	25	1,7
97	20	20	10	15	35	5,0
106	10	30	15	15	30	0,2
108	10	20	15	15	40	0,1
109	20	30	15	15	20	13,0
111	20	20	15	15	30	18,0

Кристаллизационная способность силикатных и фосфатных стекол изучалась в одинаковых условиях по экспресс методике. Она заключалась в выдержке кусочков стекла в печи при заданной температуре в течение 1 часа.

Начало кристаллизации определялось по появлению опалесценции, а начало плавления по исчезновению острых сколов.

Хотя эта методика завышает температуру начала кристаллизации, она дает хорошие сравнительные данные при сравнении разных стекол и позволяет определить зависимость химической устойчивости от кристаллизации.

ТАБЛИЦА III. КРИСТАЛЛИЗАЦИОННАЯ СПОСОБНОСТЬ И ТЕМПЕРАТУРА НАЧАЛА ОПЛАВЛЕНИЯ НЕКОТОРЫХ ФОСФАТНЫХ СТЕКОЛ

№№	Состав (вес %)			Мольное отношение Na : P	T(°C)	
	Na_2O	Al_2O_3	P_2O_5		Начало кристаллизации	Начало оплавления
1	25,3	22,0	52,7	1,1	450	800
2	24,6	26,0	49,4	1,1	500	650
3	22,7	30,0	47,3	1,1	650	600
4	26,8	22,0	51,2	1,2	450	700
5	25,4	26,0	48,6	1,2	600	500
6	24,0	30,0	46,0	1,2	650	550
10	28,3	22,0	49,7	1,3	500	750
11	26,8	26,0	47,2	1,3	600	500
12	25,4	30,0	44,6	1,3	600	500
16	28,8	34,0	37,2	1,0		700
18	27,5	37,0	35,5	1,0		800
20	25,8	41,0	33,2	1,0		900
22	20,6	41,0	38,4	0,8		1300
23	18,0	41,0	41,0	0,7		1350

ТАБЛИЦА IV. КРИСТАЛЛИЗАЦИОННАЯ СПОСОБНОСТЬ И ТЕМПЕРАТУРА НАЧАЛА ОПЛАВЛЕНИЯ НЕКОТОРЫХ СИЛИКАТНЫХ СТЕКОЛ

№	Состав (вес %)					T(°C)	
	SiO_2	Al_2O_3	CaO	B_2O_3	Na_2O	Начало кристаллизации	Начало оплавления
3	50	20	0	10	20	850	800
4	45	20		10	25	750	800
33	55	15		10	20	900	750
52	45	15		20	20	800	750
53	35	15		20	30	650	600
64	35	20		5	4	500	850
95	25	30	10	15	20	650	850
97	35	20	10	15	20	650	800
106	30	30	15	15	10	850	800
108	40	20	15	15	10	850	750
111	30	20	15	15	20	650	950

Зависимость температуры начала кристаллизации и оплавления от состава стекла приведены в табл.III и IV. Для фосфатных стекол температура начала оплавления (если брать незакристаллизованные стекла) мало зависит от состава стекла в интервале содержания окиси алюминия от 20 до 30% и мольного отношения M = 1,1-1,3 и резко возрастает при увеличении содержания окиси алюминия и уменьшении M.

Температура начала кристаллизации несколько возрастает при увеличении содержания окиси алюминия и в интервале 26-30% и M =1,1-1,3 равна 600-650°C. При длительной выдержке эта величина будет ниже на 100-150°C (это справедливо и для силикатных стекол).

О силикатных стеклах, судя по приведенным данным, можно кратко сказать следующее: увеличение содержания окисей натрия и кальция, а также борного ангидрида несколько снижает температуру размягчения (при этом снижается и температура варки).

В основном температура размягчения не полностью закристаллизованных стекол колеблется в интервале 700-800°C. Это выше, чем у фосфатных стекол.

Температура начала кристаллизации сильно уменьшается при повышении концентрации окиси натрия в стекле; при содержании ее около 20% при наличии даже небольших количеств B_2O_3 и CaO температура начала кристаллизации снижается до 650°C.

Те же стекла, у которых температура начала кристаллизации около 800°C, имеют высокую вязкость даже при температуре 1100-1200°C.

Следовательно, температура начала кристаллизации приемлемых силикатных стекол лежит вблизи 650°C, что не намного выше этой температуры у фосфатных стекол. Можно также отметить одно общее явление как для фосфатных, так и для силикатных стекол: температура начала плавления закристаллизованных образцов существенно выше, чем у незакристаллизованных.

Следует отметить, что эти результаты носят сравнительный характер, так как проведенные исследования по длительному отжигу показали, что температура начала кристаллизации на 100-150°C ниже, чем полученная по ускоренной методике.

Исследования зависимости химической устойчивости от кристаллизации показали, что химическая устойчивость фосфатных стекол уменьшается в 5-10 раз, а силикатных в 20-50 раз. При этом неизвестно, как поведут себя закристаллизованные стекла с точки зрения механической прочности. Поэтому на данном этапе, видимо, следует рекомендовать хранение стекол в могильнике при температуре ниже начала кристаллизации.

ВЫВОДЫ

Проведенное исследование натрийалюмофосфатных и натрийалюмокальцийборосиликатных стекол показало, что стекла в обеих системах с температурой выплавки 1000-1100°C обладают примерно одинаковой химической и термической устойчивостью. Они могут быть применены для прочной фиксации радиоактивных отходов, если температура их хранения не превышает 400-450°C.

В обеих системах имеются стекла с бо́льшей термической устойчивостью, однако, температура их выплавки существенно выше – вплоть до 1500°С. Их применение возможно лишь при создании надлежащей аппаратуры.

DISCUSSION

H. KRAUSE: Investigations carried out by some laboratories have shown that glasses with the same chemical resistance need not necessarily have the same radiation resistance. I would therefore like to know whether, in addition to your tests on chemical and thermal stability, you have also carried out irradiation tests?

Nina V. KRYLOVA: No, we did not carry out any such tests with our system.

CONDITIONING
MEDIUM-LEVEL WASTE
(Session VII)

Chairman:

E. DETILLEUX, OECD/NEA

INCORPORATION OF RADIOACTIVE WASTES FROM NUCLEAR POWER PLANTS INTO CONCRETE AND BITUMEN*

E.K. PELTONEN, J.U. HEINONEN
Technical Research Centre of Finland,
Reactor Laboratory, Espoo

J. KUUSI
Oy Finnatom Ab,
Helsinki,
Finland

Abstract

INCORPORATION OF RADIOACTIVE WASTES FROM NUCLEAR POWER PLANTS INTO CONCRETE AND BITUMEN.

On the basis of published data and mathematical calculations the study aims to critically review and assess in general the suitability of cementation and bituminization methods for the solidification of radioactive liquid wastes originating from nuclear power plant operation (i.e. excluding fuel reprocessing wastes). Particular endeavours were made on the grounds of the information gained to study and assess the applicability of these two techniques to wastes originating from the first Finnish power reactor VVER-440 in Loviisa. In evaluating the data obtained it turned out that solidification technology as a whole is still under development. There is an urgent need for the standardization of conceptions, terms, units and tests. The relevant properties of solidification products and their criteria should be clearly specified simultaneously considering further waste management steps, particularly final storage or disposal. In assessing the suitability of the two solidification techniques under Finnish circumstances it was stated that both cementation and bituminization would be applicable for the solidification of relevant wastes from the Loviisa plant. On the basis of data available one cannot definitively give either of the two techniques precedence. As regards the process, both techniques would be applicable with certain conditions. The solidified products do not show remarkable differences with respect to handling and transport. Under the conditions prevailing in Finland the space available for temporary storage on the plant site might play a comparatively important role, which will give precedence to the smaller amount of bituminized products, this being about one fifth of the equivalent concrete products.

1. INTRODUCTION

Nuclear power generation in Finland will be started by a Soviet VVER-440 unit (440 MW(e), PWR) at the end of the year 1976. This unit will be followed by another VVER-440 unit at Loviisa and two Swedish 660 MW(e) BWR units at Olkiluoto. In addition, the 1980's will bring yet two units to Loviisa.

Several decisions connected with the management of radioactive wastes at the plant site and at the place of final storage are to be made before and shortly after the start of nuclear power production. In the preparatory studies experience obtained elsewhere will be compiled and the effects of local circumstances evaluated.

As an example of preparatory studies of this type work carried out in 1974 to 1975 at The Technical Research Centre of Finland for evaluating the suitability of two solidification processes – cementation and bituminization – for the management of radioactive wastes from the first Finnish nuclear power plant (excluding fuel reprocessing wastes) is described. In addition to problems

* Work performed within a joint Nordic research programme on radioactive waste.

connected with the solidification process at the plant site the problems connected with the transportation and final storage of the wastes and manufacture of the solidification process units and handling equipment are to be taken into account when choosing the process to be used. The course of the work and the conclusions made indicate the problems and potential solutions faced in this special area by countries starting nuclear power production.

2. WASTES OF INTEREST

The present study deals with the wastes originating from nuclear power plant operation, i.e. excluding wastes from the reprocessing of spent fuel. Wastes of this type are usually treated on site, while the spent fuel is transported to reprocessing plants, which do not arouse current interest in Finland.

Those wastes with importance from the solidification point of view originate from the treatment of contaminated waters. In power plants ion exchange and evaporation are used for the purification of liquids from the primary circuit, decontamination, drainage, regeneration, etc. Under normal circumstances the most problematic component of radioactive wastes consists of fission products, although more than 99.9% of them remain in the fuel elements and will be released only in the reprocessing stage.

There are several different types of ion-exchange resins; organic and inorganic, synthetic and natural, anion and cation active. Furthermore, resin materials can also be classified into the granular and powder types.

In the case of evaporator concentrates the radioactivity is usually not seen as the critical factor, but rather the content of salts and chemicals.

3. AN ASSESSMENT OF THE SOLIDIFICATION TECHNIQUES

In assessing the solidification techniques the relevant factors should be estimated simultaneousl taking into account the different stages, viz:

(i) Solidification process
(ii) Interim storage of products
(iii) Transportation
(iv) Long-term storage or even final disposal.

The criteria to be applied in the assessment of the different steps mentioned are unfortunately not commonly specified or accepted, neither the properties according to which these criteria would be evaluated. The need to eliminate these flaws is urgent in order to render possible conclusive comparisons of different studies and the assessment of the suitability of a particular solidification technique. In Table I a number of the most relevant properties are specified on the basis of which one can estimate the suitability of the method with regard to the different steps of treatment mentioned.

3.1. Characteristics of the process

According to most specialists, the cementation process tends to be safe and technically feasible. Perhaps the most important advantage from the safety point of view is the absence of the fire and explosion risk. The most notable disadvantage when viewing safety and the technical feasibility are constituted by dust and the fact that the equipment can be stopped up. A drawback of great weight in the steps after the process is the increase in volume and mass. It must be emphasized that the cementation process is not simple in every respect, e.g. it is quite complicated

TABLE I. IMPORTANT CHARACTERISTICS OF SOLIDIFICATION AGENT, MIXTURE OR PRODUCT

Characteristic	Cementation phase				Bituminization phase			
	P	I	T	L	P	I	T	L
1 Density	+		+		+		+	
2 Penetration						+	⊕	+
3 Viscosity	⊕				⊕			
4 Softening point					⊕	⊕	⊕	⊕
5 Plasticity							⊕	
6 Homogeneity	+			⊕	+			⊕
7 Porosity			+	⊕		+	+	⊕
8 Compressive strength		⊕	⊕	⊕				
9 Shock resistance		+	⊕	+		+	⊕	+
10 Cold resistance		⊕	⊕	⊕		⊕	⊕	⊕
11 Water/cement ratio	⊕	⊕	⊕	⊕				
12 Content of solids	⊕	+	+	⊕	⊕	+	+	⊕
13 Thermal conductivity		+	⊕	+	⊕	+	⊕	⊕
14 Thermal expansion			⊕		⊕	+	⊕	⊕
15 Ignition point					⊕	+	⊕	+
16 Flash point					⊕	+	⊕	+
17 Burning point					⊕	+	⊕	+
18 Burning rate					⊕	+	⊕	+
19 Phase separ. during burning		+	⊕	+	⊕	+	⊕	+
20 Radiation attenuation	+	⊕	⊕	+	+	⊕	⊕	+
21 Dose rate	⊕	⊕	⊕	⊕	⊕	⊕	⊕	⊕
22 Gas generation				⊕		+		⊕
23 Swelling (during gas gener.)				⊕		+		⊕
24 Swelling (due to water)		+	+	⊕		+	+	⊕
25 Water absorption		+	+	⊕		+	+	⊕
26 Water content	+	+	+	⊕	+	+	+	⊕
27 Solubility in water		+	+	⊕		+	+	⊕
28 Leaching		+	+	⊕		+	+	⊕
29 Corrosion				⊕				⊕
30 Effect of micro-organisms				⊕				⊕
31 Ageing				⊕				⊕

P Solidification process
I Interim storage
T Transportation
L Long-term storage
\+ Characteristic is important for this stage.
⊕ Characteristic is very important for this stage.

to run the process if certain limits have been set for the water/cement and salt/cement ratio. In general it can be said that these ratios, above all their influence on the final product, are not particularly well known. There are quite a few data available treating the resin/cement ratio and the maximum resin content in the product.

The structure of bitumen has commonly been explained as colloidal, but there is not full certainty on the point. The composition of bitumen varies depending on the origin of the crude oil and on the further treatment. Therefore, it is rather difficult to predict the behaviour of bitumen during the solidification process.

The great advantage of bituminization, i.e. the reduction of volume and mass, is above all its economic profitability. This fact is, however, a disadvantage from the safety and technical point of view. The necessary elimination of water entails a need for heating, from which the most serious problems posed by the bituminization process originate, viz. the danger of fire and explosion. The risk of fire or explosion will be extremely low in bituminization under the following conditions:

(i) Temperature $\leqslant 200°C$
(ii) Short mixing and heating time (minutes)
(iii) Small mixture portion/heating surface ratio
(iv) Waste/bitumen ratio exactly adjustable
(v) Equipment remotely controlled and sealed in a ventilated space.

3.2. Relevant features of solidification products and their testing

Setting the criteria of feasibility for the solidification product and placing the properties concerned in determining those criteria in order of their relative importance is a complicated task. In addition to the task being of a complicated nature, the criteria always depend upon the local circumstances and the particular methods used. Table I gives properties of importance in feasibility estimates. By far the most important criterion of judging the product is how well the radioactive materials remain in the final product.

3.2.1. Leaching properties

The release of radioactive nuclides from the product has been examined almost exclusively by studying the leaching of active materials out of the product submerged in water. Naturally it is a primary question how well the tests correspond to actual conditions and, furthermore, what sort of conclusions can be drawn from the results obtained.

If leaching process is accepted as a diffusion phenomen – as is in most cases – then the amount of activity leached out in a certain time can be estimated [1]. Sometimes, particularly in the case of a bitumen product containing soluble salts, the experimental data fit a diffusion model quite well – but only over rather a limited period of time. There is no evidence of the behaviour of leaching over long periods of time, i.e. decades or centuries. Furthermore, there is no theory that takes into account the product's special internal properties and the circumstances outside the product. In conclusion it can be said that the information value of the estimates of leached activity is relatively poor at present.

On the other hand, some preliminary estimate of the degree of leaching is necessary in order to determine the extent of the consequences of the product's possible immersion. The determination of leaching is equally important for cementation and for bituminization products.

As leaching of concrete products is quite high in many cases, it is considered to be a critical factor that can restrict the amount of activity acceptable in the product.

Leaching from bituminized products is slight enough in general, 1 to 3 decades lower than in concrete products. In some cases, however, leaching can be of the same order of magnitude as in concrete, e.g. a high content of Na_2CO_3 and $Ca(NO_3)_2$ or water in the product can cause such a situation [2].

3.2.2. *Relevant mechanical properties*

Radioactive materials can also be released in connection with the breaking down of the product. In the case of bituminization products shock, cold resistance and plasticity are the most important properties, whereas in the case of concrete products shockproof and compression strength are important.

Among the critical factors for the hardening and strength of conrete are the salt content and composition of the waste. Although it is profitable in most cases to neutralize the waste, one has to be careful to avoid salts that weaken the properties of the final product, e.g. $Ca(NO_3)_2$, $Fe(OH)_3$ [2, 3]. The dependence of the concrete's strength on the salt/cement ratio has not been thoroughly investigated, although the influence can be critical. However, according to Soviet studies, which are supported by Czech results, a concrete product may not contain more than 130 g of salts/kg cement to produce a product of the necessary strength [4].

The influence of the water/cement ratio upon the final solidified waste products is also not completely clear. The strength of normal concrete increases when this ratio decreases down to 0.3. The situation with values lower than 0.3 has not been properly investigated. However, it is known that the increase in strength ceases and that with very low ratios (<0.2) the strength goes down rapidly [5]. Moreover, when the water/cement ratio is too large, for instance more than 1, even normal concrete is relatively weak, the compression strength then being about one tenth of the maximum.

Where waste is used instead of water and sand the situation is more complicated. Due to the lack of research data one can only estimate an acceptable water/cement ratio to be roughly 0.2–1, though it may be even more restrictive.

The additives (e.g. vermiculite) used in cementation have two main purposes viz. first to reduce the leaching and secondly to eliminate the excess water. On the other hand, the additives may reduce the mechanical strength as well. This particular field of additives is still under development and may involve future possibilities of the concrete incorporation technique. No appropriate publications dealing with the ageing of concrete solidification products have been met with by the author.

Pure bitumen is quite durable, stiff and plastic. Moreover, it can be stated in general that the maximum content of solid components determined by the leaching characteristics is not crucial to the mechanical properties of the bituminization product. Actually, the only uncertainty is the durability of the product at low temperatures below 0°C. This problem seems to have been universally neglected. Under stress bitumen may even break suddenly at a higher temperature than the essential break point. This temperature, called the 'glass point', depends on the type of bitumen and is some degrees below zero. Some very preliminary studies on the cold resistance of pure bitumen have been carried out, e.g. in Finland.

3.2.3. *Radiation stability*

The changes caused by radiation in the solidification product have an effect on practically all the other properties. Consequently, the product's radiation resistance is a vital property.

Radiation causes different processes in the bituminization product, viz:

(a) Radiolysis of hydrocarbons resulting in radicals, some of them gaseous
(b) Reactions between the radicals and the constituents of bitumen and waste
(c) Oxidation of bitumen.

Radiolytic gas generation and the swelling caused by it have been measured primarily as a function of the absorbed dose. In most studies rather little attention has been paid to such factors as the dose rate, the manner of irradiation, type of radiation, type of waste etc., though they might have an essential influence on the phenomen. It seems that the results acquired by high dose rate external gamma irradiation are too positive and can lead to erroneous and too optimistic conclusions.

The effects of radiation are not adequately known. Therefore it is rather difficult to predict in general what will happen and to what extent in a solidification product in a certain time with a certain specific activity and a certain composition.

4. SUITABILITY OF THE CEMENTATION AND BITUMINIZATION TECHNIQUES FOR WASTES FROM LOVIISA NUCLEAR POWER PLANT

The study attempts to give an example of an evaluation task of current interest in a country that is starting nuclear power production.

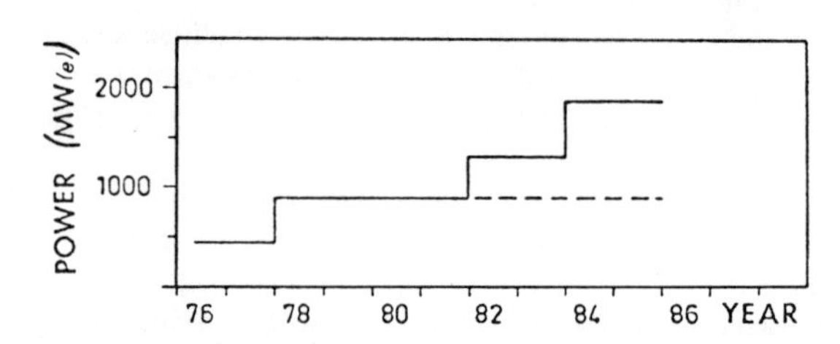

FIG.1. The capacity estimated at Loviisa nuclear power plant.

4.1. Wastes to be solidified

The wastes dealt with in the present study will be produced at the Loviisa nuclear power plant [6]. The planned capacity of four power units is shown in Fig.1. There is, however, uncertainty regarding the time schedule of the last two units, indicated by the dotted line. The wastes expected to undergo treatment by solidification are ion-exchange resins and evaporator concentrates.

The spent ion-exchange resins and the solutions used for their regeneration will be divided into two groups while stored at the plant site, viz.:

(1) Lower activity wastes containing spent resins from steam generator blow-down purification and from the purification of condensate.

(2) Higher activity wastes including resins from the purification of the primary circuit, boron concentrate drain water, decontamination water.

The total amounts of resins and corrosion sludge predicted for the two power plant units at Loviisa (LO 1 and LO 2) are: 17 m^3/a in 'the lower activity tank', and 54 m^3/a in 'the higher activity tank', where 5 m^3/a originate from the primary circuit purification. The density of resins is about 1–1.2 g/cm^3 and the water content is expected to be ca. 80–90%.

The estimates of the salinity of evaporator concentrates vary from 250 to 400 g/1, the saline concentration thus being 22–33 wt%, respectively. The estimated annual amount of evaporator concentrate for the two units together, LO 1 and LO 2, depends on the salinity as well, viz. 250 m^3/a with a salinity of 400 g/1, and 400 m^3/a with a salinity of 250 g/1.

4.2. The solidification process

The necessary capacity was estimated to be ca. 160 l waste/h, or ca. 130 l water/h in the case of bituminization. Both bituminization and cementation equipment with an adequate process capacity were taken to be available. The evaporator concentrate's salinity of ca. 400 g/l is apparently no hindrance to the cementation or bituminization process.

The composition of the evaporator concentrate – where the solution is not acid – does not present any obstacle to the cementation process. In bituminization $NaNO_3$ is the critical component. Its content in a dry salt-bitumen mixture easily rises over 20 wt%, which already has a considerable effect upon the physical and chemical properties of the mixture. For instance, the softening point and the viscosity can increase and the penetration diminish by a factor of 2. However, there is no instant danger of fire or explosion if the temperature is kept sufficiently low, e.g. at approximately 200°C. Sodium tetraborate, $Na_2B_4O_7$, has been observed to act as a strongly hardening agent on the mixture but the data available do not indicate whether the $Na_2B_4O_7$ content of 10 wt% is definitely harmful or not. The concentrate cannot be seen to contain any other salts that would be particularly injurious for the process. Some data obtained indicate that the same conditions of bituminization equipment would not be suitable for both ion-exchange resins and evaporator concentrates [7]. This question is quite essential in the case of Loviisa.

4.3. Temporary storage

The storage of wastes at the plant site before solidification is reasonable because of the consequent decrease in total activity and particularly because of the decrease of the volatile iodine ^{131}I, since storage for one month already reduces the amount of ^{131}I by over a decade. However, a storage of several months cannot reduce the radiation hazards of the waste to an essential degree, since the long-lived γ-emitters ^{137}Cs and ^{60}Co begin to become dominant in that time.

As to the safety aspects of temporary storage after solidification no distinction can be made between the two solidification techniques with regard to the storage container, i.e. the container will hardly weaken during temporary storage. However, because of the rather long setting time of concrete, temporary storage of at least six months after solidification is deemed necessary in order to achieve an acceptable strength for safe transport.

Since the question of centralized long-term storage in Finland will still be under consideration over the next few years, the annual and accumulating amounts of solidification products will play an important role when comparing the feasilibity of the two solidification techniques.

4.4. Transport

The product's permissible activities according to IAEA regulations on LLS materials do not prevent transport of the solidification products from Loviisa in a 'strong industrial package' [8].

It is obvious that the radiation of the bituminization product is more critical owing to its greater specific activity and poorer capability to absorb radiation. An over-large dose rate outside a container does not necessarily entail a reduction in the waste content of the product or fixed radiation shielding in each container. Temporary shields on the outside can be used where necessary.

Both bituminization and cementation products of resins from 'the higher activity tank' in Loviisa plant will need radiation shields during transportation.

4.5. Long-term storage

The assessment of the properties of solidification products was made on rather a general basis because the whole task of long-term storage is just under study in Finland.

4.5.1. Mechanical strength

According to Czech studies, the salt content should not exceed certain limits when evaporator concentrates are solidified into cement [4]. The $NaNO_3$ concentration of 180 g/l and the total salt content of 400 g/l in the Loviisa concentrate would thus be about 20 wt% and 70 wt% too high, respectively. Moreover, Soviet experts think it unwise to cement salt concentrates of over 200 g/l [4] since the product should not contain more than 130 g salts/kg cement. There are two alternatives possible where water is not added, namely an increase in the salt content given or too strong a decrease in the water/cement ratio will weaken the product.

The composition and content of salt in the Loviisa concentrate do not appear to be harmful for the durability of the evaporator concentrate's bituminization product provided the content of salts in the product does not exceed 60 wt%.

Nothing definite can be said about the effect ion-exchange resins have on the mechanical durability of a cementation or bituminization product, there being no research results to be had. Recent tests in Norway have indicated the bituminization product of ion-exchange resins to be poor, when the resin content exceeds 60 wt% and sometimes even before this content is attained. These studies are reviewed by Bonnevie-Svendsen et al. [9]. No data on the durability of the products over tens and hundreds of years are available and so no comparisons of this respect can be made.

4.5.2. Radiation stability

The dose from the most active waste at Loviisa, i.e. the higher activity resins in a 60 wt% bituminization product, has been calculated to be about 3×10^7 rads at the most. The water content of the resin waste has an influence upon the bituminization product's dose, but the dose will be less than 10^8 rads in all cases.

The dose and gas generation induced by the higher activity resins may be too great in a bituminized product to allow the use of hermetic containers. Under no circumstances can resins be bituminized and stored as a normal routine without making tests on the Finnish bitumens. Evaporator concentrates are not expected to produce gas generation problems because of the insignificance of the dose.

Gas generation has proved to be no problem in the cementation products.

It has been calculated that Loviisa primary circuit resins induce a ca. 0.2°C temperature rise in the centre of a 208 litre-barrel of a bituminization product of 60 wt%, when the temperature on the outer surface is 20°C. The temperature rise occasioned by the evaporator concentrate is negligible. The activity of the cementation products is of no consequence at all as regards an increase in temperature.

4.5.3. Leaching

According to some foreign estimates available the maximum specific total activity for a cementation product would be 10^{-5} to 10^{-2} Ci/l, whereas the maximum specific activity for isotopes ^{90}Sr and ^{137}Cs would be 10^{-4} Ci/l on burial in the ground, when leaching of a normal concentrate is taken as the criterion. In the case of bituminization products the limit is taken to be ca. 1 Ci/l for ^{90}Sr and ^{137}Cs [4].

The highest specific activity for the isotopes ^{90}Sr and ^{137}Cs in the Loviisa wastes would be 3×10^{-2} Ci/m^3 for ^{90}Sr and 9 Ci/m^3 for ^{137}Cs, in ion-exchange resins from the higher activity tank. This implies that the specific activity of no waste is too great with regard to leaching of the bituminization product, even though the specific activity might increase fivefold on bituminization. This conclusion, however, holds true only if leaching from the bituminization product of resins is nearly equal to that from concentrates. No results on this are available and thus the applicability for resins is still questionable.

However according to the limits mentioned, the ^{137}Cs specific activity of 'the higher activity tank' resins is approximately one decade too high for cementation, although it might decrease to one third in the process. The activities of the 'lower-level resins' and evaporator concentrates are, however, no hindrance to cementation.

With regard to leaching the evaporator concentrate's solid content in the bituminization product can be approximately 60 wt% and considerably smaller, approximately 10 wt%, in the cementation product. Data concerning the permissible content of resin in the product are not available.

4.6. Changes in volume and mass of wastes

There is a great difference between the two techniques as regards the changes in volume and mass during the solidification process. The amounts likely to be produced were estimated in the following way.

4.6.1. Coefficients of the change

The coefficient of the volumetric reduction K_V is defined as a ratio

$$K_V = \frac{V_w}{V_p}$$

where V_w is the volume of wet waste, and V_p is the volume of the solidified product. The coefficient of mass reduction K_M is defined analogously as:

$$K_M = \frac{M_w}{M_p}$$

where M_w is the mass of wet waste, and M_p is the mass of the solidified product. The relation between the two coefficients is

$$K_M = \frac{\rho_w}{\rho_p} K_v$$

where ρ_w is the density of wet waste, and ρ_p is the density of the solidified product.

In the cementation of evaporator concentrates one cannot keep the salt content in the product C_{sp} as a free variable because it depends on the content of cement in the product in the following way

$$\frac{C_{sp}}{C_{cp}} \lesssim a$$

where C_{sp} is the salt content in product, C_{cp} is the cement content in product, and a is taken to be ca. 0.13.

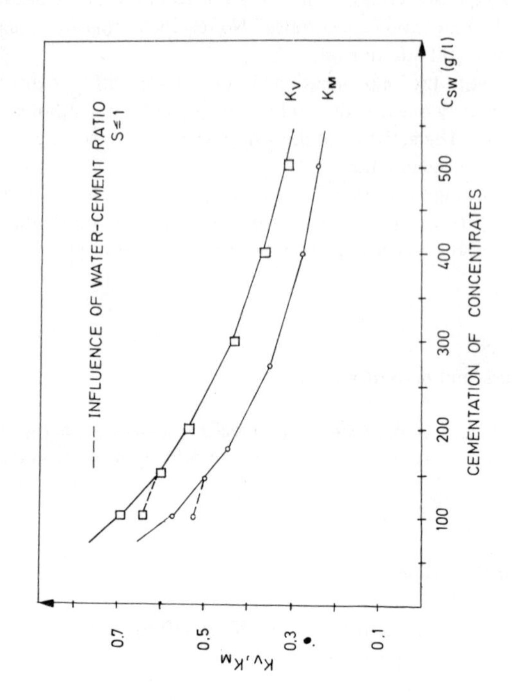
INFLUENCE OF WATER-CEMENT RATIO
S=1
K_V
K_M
K_V,K_M
0.7
0.5
0.3
0.1
100
200
300
400
500
C_SW (g/l)
CEMENTATION OF CONCENTRATES

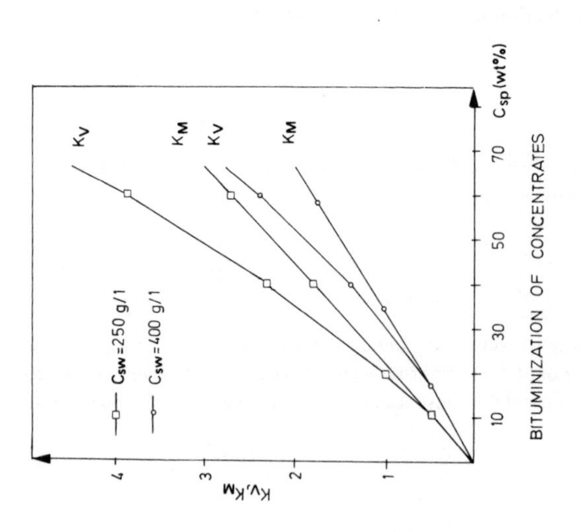
C_SW=250 g/1
C_SW=400 g/1
K_V
K_M
K_V
K_M
K_V,K_M
4
3
2
1
10
30
50
70
C_SP (wt%)
BITUMINIZATION OF CONCENTRATES

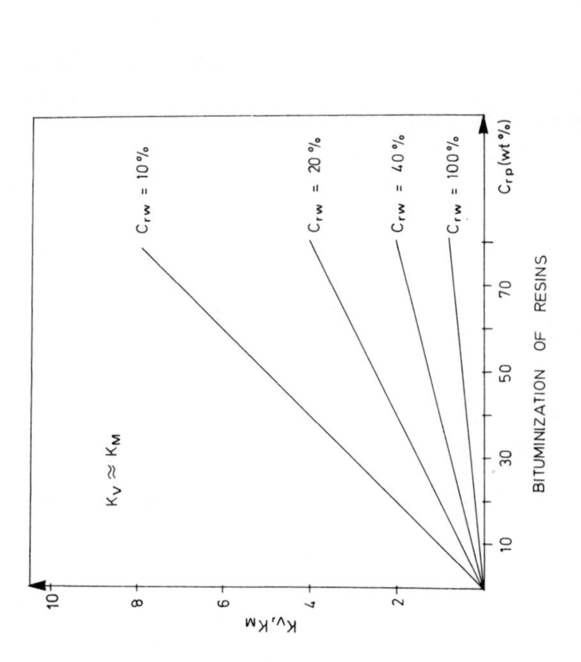

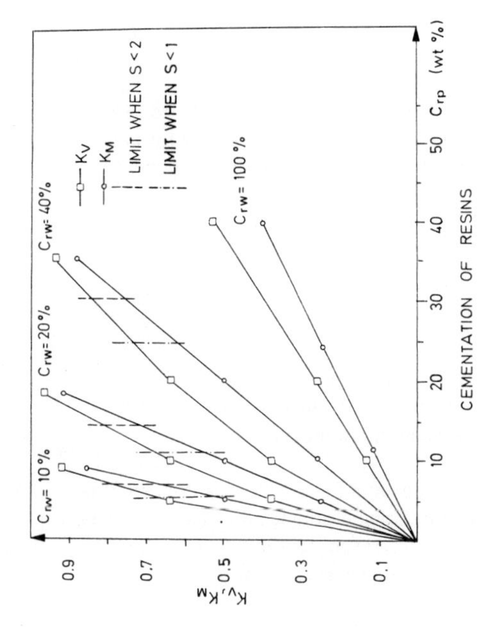

FIG.2. Coefficient of volumetric reduction K_V *and mass reduction* K_M *in solidification of evaporator concentrates and ion-exchange resins*

c_{sw}: *content of salts in waste*
c_{sp}: *content of salts in product*
c_{rw}: *content of resin in waste*
c_{rp}: *content of resin in product*

Besides there is the following restricting ratio between water and cement:

$$S_l \lesssim \frac{C_{wp}}{C_{cp}} \lesssim S_u$$

where C_{wp} is the water content in product, S_l is assumed to be ca. 0.2, and S_u is ~1.

There are no data on the ratio between resins and cement for ion-exchange resins. However, the water/cement ratio restricts the resin content in the product C_{rp} as follows:

$$C_{rp} \lesssim \frac{1}{1 + \left(\frac{1 - C_{rw}}{C_{rw}}\right)\left[1 + \frac{1}{S}\left(1 + \frac{C_{ap}}{C_{cp}}\right)\right]}$$

where S is the water/cement ratio
C_{rw} is the content of resin in resin waste (wt%)
C_{ap} is the content of additives in product (wt%)
C_{cp} is the content of cement in product (wt%).

The maximum values of C_{rp} are shown in Fig.2 with water/cement ratio of 1 and 2. The coefficients of volumetric and mass reduction during bituminization and cementation are also shown in Fig.2.

Accepting the maximum salt and resin content in the bituminization product to be 60 wt%, the coefficients considered realistic are shown in Table II.

4.6.2. *Amount of solidification products*

The solidification was assumed to take place in 208-l metal barrels. The amount of product contained in a barrel was 170 l in the case of bituminization and 200 l for cementation. The number of barrels needed for each unit of 440 MW(e) for the different solidification alternatives is shown in Fig.3. A prognosis of the annual amount of barrels, considering the increase in plant units given in Fig.1., can be seen in Fig.4 and the accumulating number of barrels in Fig.5.

TABLE II. COEFFICIENTS OF VOLUMETRIC AND MASS REDUCTION K_V AND K_M IN SOLIDIFICATION

Solidification technique		Evaporator concentrates		Ion-exchange resins		
		Salt content in waste		Resin content in waste		
		400 g/l	250 g/l	40 wt%	20 wt%	10 wt%
Bituminization	K_V	1.4–2.4	2.2–3.9	1.0–1.5	2.0–3.0	4.0–6.0
	K_M	1.2–1.8	1.9–2.8	1.0–1.5	2.0–3.0	4.0–6.0
Cementation	K_V	0.35–0.40	0.50–0.55	0.6–0.7	0.55–0.65	0.5–0.6
	K_M	0.25–0.30	0.33–0.38	0.5–0.6	0.45–0.55	0.4–0.5

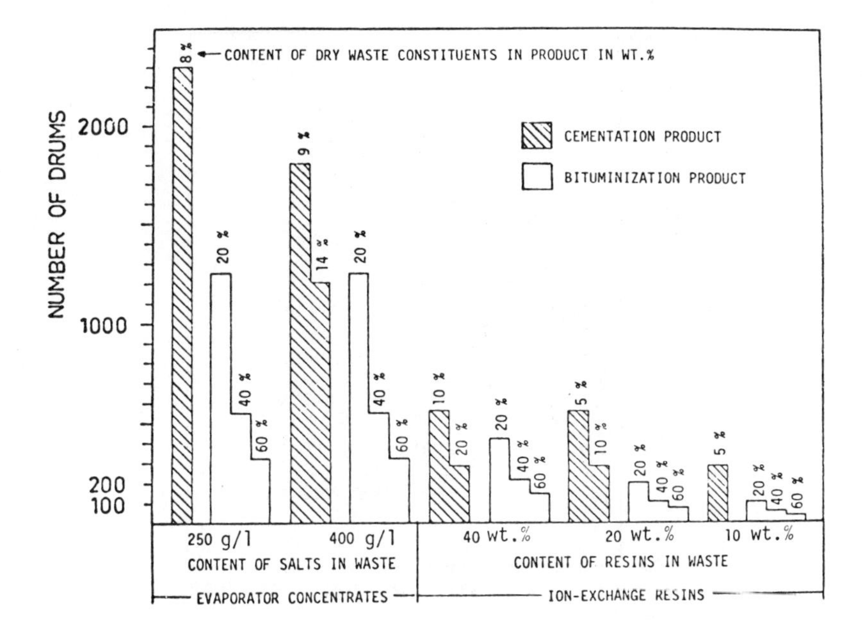

FIG.3. The annual amount of solidification products at Loviisa per one 440 MW(e) unit.

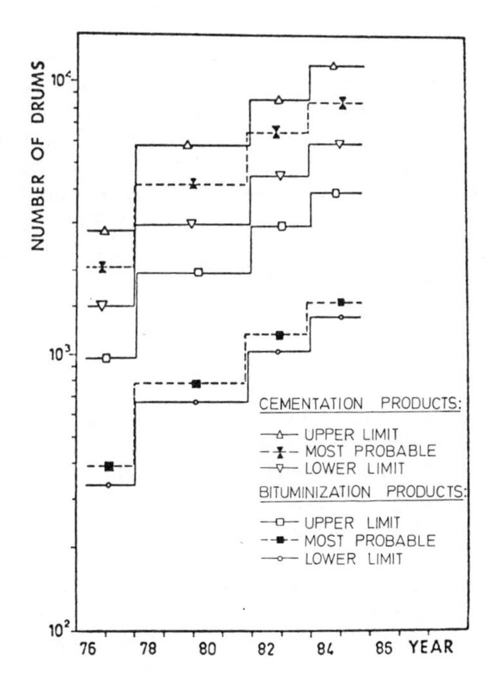

FIG.4. The estimated annual amount of solidification products at Loviisa. Each step caused by one 440 MW(e) unit.

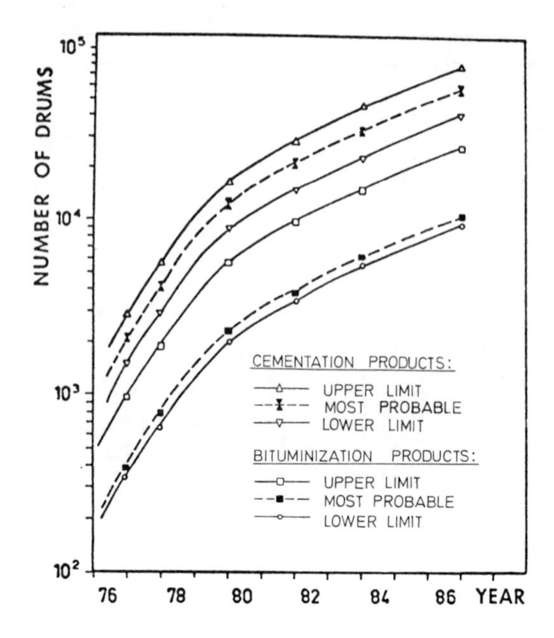

FIG.5. The estimated cumulative amount of solidification products at Loviisa.

4.7. Comparison of cost

In calculating the costs one should take into consideration all the factors associated with the operation and investments including solidification process, transportation and interim and long-term storage. However, a detailed cost calculation is not included in this study.

When handling large amounts (>100 m^3/a) of waste the ratio of the total costs for sludges and concentrates between bituminization and cementation is, according to most foreign studies, ca. 0.3–1 and the range of input costs ca. US$ 100--1000/m^3. Thus it has been estimated that the total costs per year would be about US$ 10^5 more for cementation than for bituminization when considering the waste products of Loviisa 1 and 2.

5. CONCLUSION

As a general conclusion, it is justified to state that both techniques, namely cementation and bituminization, would be applicable to the solidification of wastes from the Loviisa plants. A definitive preference cannot be given to either of the two methods on the basis of the data on hand.

The absence of fire and explosion hazard makes the cementation process safe in this respect. The risk will also be extremely small in bituminization if equipment answering certain qualifications is used.

The solidified products do not manifest notable differences with respect to handling and transportation. The higher dose rate from the bituminized products can be eliminated by means of shielding. Furthermore, the smaller bulk of bituminization products, this being about one fifth of the cementation products, will also compensate for the higher dose rate. Since the bituminized products are less liable to be injured, they are more suitable for handling and transport.

Decisive criteria and limitation of the properties of solidification products might be set up by the way of long-term storage, which is still under consideration in Finland. Thus there are no criteria that would definitively eliminate either of these two products. However, cementation of spent resin from 'the higher activity tank' might lead to concrete products of too high a total activity and a specific activity of ^{137}Cs, should there be a chance of the groundwater becoming polluted.

Under the present circumstances in Finland the space allotted for temporary storage at the plant site will play a comparatively important role. This fact will favour bituminized products.

In Finland one must endeavour to achieve such a level of knowledge and such a readiness in testing as to make possible a reliable estimation of solidification techniques. In this instance the first task is to develop criteria estimating:

(a) A particular solidification method's applicability for treating radioactive waste in general and, on the other hand,
(b) Comparability between different methods or between the same methods employed in different places.

In relation to the field of solidification of radioactive wastes on a world-wide basis the topics actual and calling for further research and development work are as follows:

(i) Investigation and development of the process in solidifying ion-exchange resins
(ii) The cold resistance of solidified, above all bituminized, products
(iii) The effects of the water/cement ratio
(iv) The effects of dry waste constituents/cement/additives ratio.

REFERENCES

[1] BELL, M.J., An Analysis of the Diffusion of Radioactivity from Encapsulated Wastes, ORNL-TM-3232 (1971).
[2] BRODERSEN, K., Erfaringer med solidificering af radioaktivt affald (Experience on solidification of radioactive wastes) in Danish, Nordic Seminar on Radioactive Wastes, Lidingö, Sweden (1974).
[3] SMITH, A.J., Leaching of Radioactive Sludges incorporated in Cement and Bitumen, AERE-M-2223 (1969).
[4] IAEA, Waste Management Techniques and Programmes in Czechoslovakia, Poland and in the Soviet Union, Study Tour Reports, IAEA, Vienna (1971) 154 pp.
[5] PIHLAJAVAARA, S.E., Technical Research Centre of Finland, private communication (1975).
[6] KALLONEN, I., Imatran Voima Oy, private communication and design data concerning Loviisa nuclear power plant (1975), unpublished.
[7] MEIER, G., BÄHR, W., Kernforschungszentrum Karlsruhe, Rap. KFK-2104 (1975).
[8] IAEA, Regulations for the Safe Transport of Radioactive Materials, 1973 Revised Edition, Safety Series No.6, IAEA, Vienna (1973) 148 pp.
[9] BONNEVIE-SVENDSEN, M., TALLBERG, K., AITTOLA, B., TOLLBÄCK, H., these Proceedings, Vol.2.

DISCUSSION

G. STOTT: Have any limits been set for the discharge of liquid waste to the environment from the Loviisa station since this may depend on the number of drums of solid waste which is expected to be produced. Furthermore, there are many kinds of cement. How many of the different types have been used in your tests on maximum salt content and radiation damage?

J.U. HEINONEN: The estimates I gave regarding the number of drums of solidified material were made without regard for the limitations on the release of liquid effluents. Generally speaking, it can be said that the IAEA recommendations will be followed in Finland.

As regards the cement, we did not carry out the experiments ourselves but collected and evaluated the available data. The information on the maximum salt content stems from Ref.[4], cited here, which refers to cementation in general. The same reference, however, recommends Portland cement on the basis of a comparative study. The radioactivity of the wastes included in our study will be so low that radiation is unlikely to cause any great damage in the concrete product.

W. HILD: Your paper mentions a figure of 20 wt% for $NaNO_3$ and also refers to the hardening effects of the bitumen due to the presence of the nitrate. Where does the nitrate in the LWR effluents come from? Second, with regard to hardening, we have made extensive tests in this area and have not been able to detect any particular hardening when the concentrates are evaporated at pH 8 to 9.

J.U. HEINONEN: Your experimental conditions may have been different from our own. I might mention that Ref. [4] restricts the $NaNO_3$ content to 40% in bituminized products.

I. KALLONEN: I think the $NaNO_3$ originates mainly from the use of nitric acid for the regeneration of ion-exchange resins, and in smaller amounts from decontamination.

J. SAIDL: Could you give more detail regarding the specific activities of your solutions, wastes and products treated?

J.U. HEINONEN: The specific activity in the tanks with 'lower activity' is approximately 10^{-5} Ci/m^3, and in the tanks with 'higher activity' about 10^2 Ci/m^3.

SEDIMENTATION TECHNIQUE OF WASTE BITUMINIZATION AND THERMOGRAVIMETRIC CHARACTERISTICS OF THE FINAL PRODUCTS

J. ZEGER, K. KNOTIK
Institut für Chemie,
Österreichische Studiengesellschaft für Atomenergie GmbH,
Seibersdorf

H. JAKUSCH
Vereinigte Edelstahlwerke AG,
Ternitz,
Austria

Abstract

SEDIMENTATION TECHNIQUE OF WASTE BITUMINIZATION AND THERMOGRAVIMETRIC CHARACTERISTICS OF THE FINAL PRODUCTS.

In the research centre of the Österreichische Studiengesellschaft für Atomenergie GmbH a semi-technical plant has been installed for waste bituminization, which has been tested inactively since 1973. This plant uses a new technological process for embedding. One of the important features of this new process is that the solution water, which is normally inactive, is distilled off before embedding, resulting in dry and powdery salts. The second important feature is that these dry salts are mixed with the thin fluid bitumen by sedimentation. A special feature is that there is no mechanical aid used for mixing. Thermogravimetric analysis of samples which simulated the final products of this pilot plant was carried out to verify the best working parameters and to study the possible chemical damage to the bitumen. Analysis was done by heating the samples, consisting of various mixtures of bitumen and inorganic salts, in a Mettler-thermoanalyser up to 500°C using different atmospheres (air, nitrogen). It was shown that only nitrate and nitrite, especially in combination with Fe(III)-ions, are of negative influence on the thermostability of bitumen. They lead to a sudden and quick weight loss of the samples between 370 and 410°C (above the melting point of both $NaNO_2$ and $NaNO_3$). The Fe-ions have a catalytic influence, as it could be shown that a 1% addition of $Fe(NO_3)_3$ to $NaNO_3$ leads to a considerable acceleration of the incineration. This influence of the Fe(III)-ion can be suppressed to some extent by a hydrolysis before the embedding. There is, however, no danger to the embedding process from these effects since the process temperature of maximum 200°C is well below the ignition temperatures. In the preparation of further studies on the behaviour of radiation-damaged bitumen a method of measuring the dose rate of an unknown radioactive salt mixture at any point of this mixture has been developed. This is done by making two measurements with glass dose-meters, one with a beta-absorber to get a pure gamma dose and the other without it to get the combined beta and gamma dose. During the first measurement the dose-meters were protected against contamination by a thin layer of rubber.

1. INTRODUCTION

Bitumen seems to be the medium best suited for the final embedding of low to intermediate radioactive wastes. It has great advantages because of a good resistance against water leaching, sufficient plasticity even at lower temperatures, and a fair stability against irradiation. Bitumen is also commercially attractive because of its fairly low price and easy attainability.

Its disadvantages may be the low conductibility for heat and the possibility of chemical reactions with oxygen containing salts at higher temperatures.

Most of the low or intermediate radioactive waste in the near future will be the operational waters of power reactors.

The embedding of radioactive waste, especially from evaporator concentrates, into bitumen using currently known techniques leads to some difficulties:

(1) Because of the low thermoconductivity of bitumen it is very difficult and uneconomical to introduce heat to evaporate the water.

(2) The system components needed are contaminated by two different substances: bitumen and radioactive waste. A decontamination for service and repair jobs is therefore very complicated.

(3) The conditions during the mixing of the radioactive solutions with the bitumen are optimal for chemical reactions of the substances: high temperatures, changing pressures and highly reactive solutions.

(4) These processes are accompanied by an unpleasant steam distillation of the bitumen resulting in a considerable contamination of the filters.

According to the new Austrian technique these disadvantages can be overcome or at least decreased [1]. This is done by two important steps during the process. Primarily the normally inactive solution water is distilled off before the embedding process, resulting in dry and powdery salts. Afterwards these salts are mixed with the thin fluid bitumen using only sedimentation under normal gravity. A special feature is that there is no mechanical aid used for mixing the salt and bitumen.

In the Seibersdorf research centre of the Österreichische Studiengesellschaft für Atomenergie a semi-technical pilot plant has been installed by the Vereinigte Edelstahlwerke AG for waste bituminization, which has been tested inactively since 1973. The testing of the system components has been almost completed. The pilot plant is capable of treating up to 25 kg/h of concentrate solutions and precipitation sludge. This process makes it possible to operate almost without environmental hazards, as even the condensate can be recycled into the evaporator.

2. TECHNICAL DESCRIPTION OF THE BITUMINIZATION PROCESS

The process, which is schematically depicted in Fig.1, starts with a concentrate from an evaporizer amounting to less than one percent of the original volume. This concentrate contains approximately 10 to 40% salts and may also contain insoluble particles.

The first step is a further concentration in a thin film evaporator. The residue from this evaporation, which is powdery and contains a very small amount of moisture, falls, by its own weight, into the bitumen beneath it.

The bitumen is kept at a suitable temperature to guarantee a small viscosity. Through the force of its own weight the residues form a highly packed sediment which gradually fills the mixing vessel. The mixing vessel when filled is then emptied into a standard barrel. The whole system, from the thin film evaporator to the standard barrel is closed and kept under reduced pressure. It is thereby possible to operate without any contamination of the environment.

The vapours of the drying process of the evaporizer residue are recondensed in a cooled water bath. The condensate is tested for its activity and either dispensed into a waste water system or recycled into the process. Because of the small amounts of condensate this is of almost no influence on the economy of the whole process.

There are practically no distillation products of bitumen in the condensate. This is due to two important facts:

(1) There is no steam distillation of bitumen because all water is removed before the radioactive salts come into contact with the hot bitumen.

(2) Eventually escaping distillation products have to pass the thin film evaporator. There they are adsorbed and transported back into the bitumen mixing vessel with the down-falling dried salts.

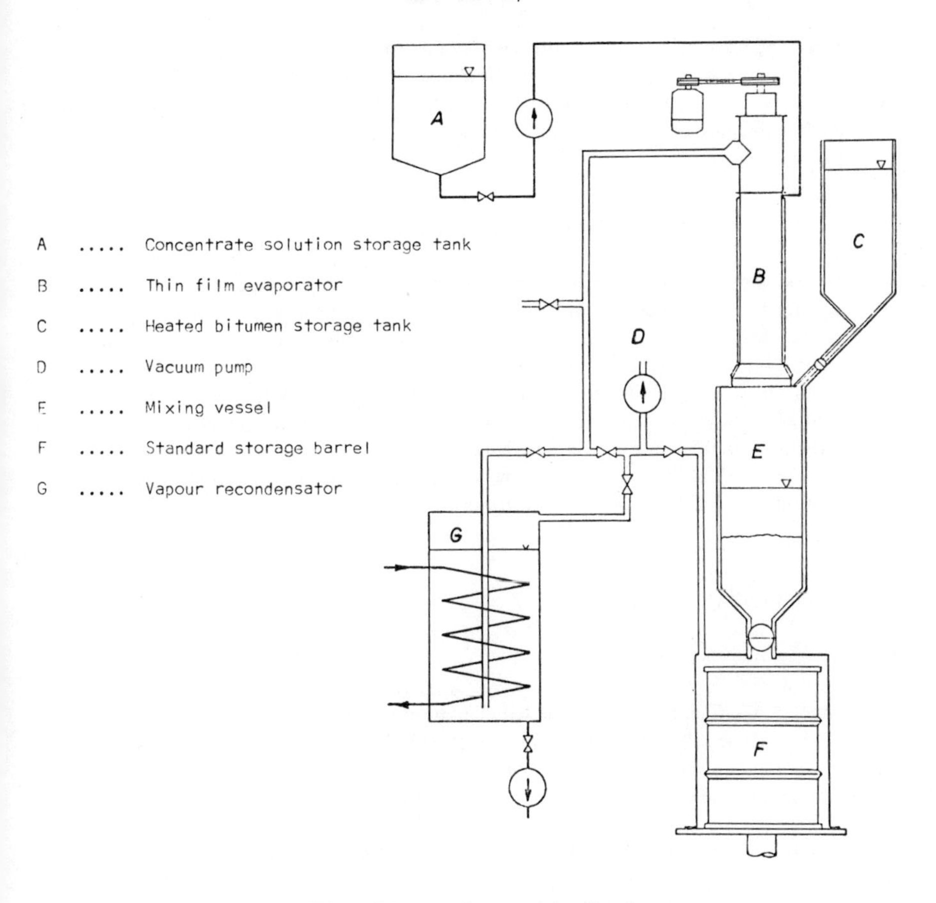

FIG.1. Schematic diagram of the pilot plant.

The pilot plant is built for a batchwise operation, but a continuously operating plant, using a barrel changer, is already under construction [2].

The decontamination of the complete plant is very easy. The evaporator is self-cleaning and can be rinsed additionally with water or any other solution agent. Any residues in the mixing vessel are normally removed by the recharge with fresh bitumen. However, there is also the possibility of an easy cleaning with solution agents even at higher temperatures because of the smooth inner surface of the mixing vessel and the absence of any stirring apparatus.

3. DESCRIPTION OF THE FINAL PRODUCT

The final product, which is stored in standard barrels, reaches a packing density of 50 to 70 wt% salts in bitumen. It can be shown that a complete coating of the single salt particles remains even after a forced breakage of the block. There is a very distinct boundary between the sediment and any remaining surplus pure bitumen above it. The packing density is constant over the whole bitumen-salt body.

The leaching resistance of the mixtures is very important for their storage ability. A series of analyses have been carried out to study the leaching rate in distilled water. A comparison shows that the results are in good accord with published values. For example, a sample with 62 wt% NaCl had a leaching rate of $3 \times 10^{-4}\ g \cdot cm^{-2} \cdot d^{-1}$ after 120 days of leaching [3].

4. INTRODUCTION TO THE THERMOANALYTICAL STUDIES

The working conditions during the embedding process – high temperatures, changing pressures, oxidizing atmosphere – and the constitution of the final product – a mixture between an organic substance and oxygen-containing salts – may lead to the assumption that these final products may react as explosives.

It has therefore been the intention of this thermoanalytical study to investigate the reactions between bitumen and embedded salts that occur from the embedding temperature up to 500°C. Although the temperatures of the new Austrian embedding process will at no stage surpass 200°C there is still the possibility that during the final storage of the embedded products higher temperatures may be reached owing to external influences such as fire.

5. SAMPLE PREPARATION

5.1. Salts

To simulate the final products of the sedimentation process samples were prepared using p.a. grade salts. These salts were ground and a fraction $\leqslant$ 100 μm was used for sample preparation. Before embedding into the bitumen these salts were dried for one day at 120°C.

A small part of the salts thus prepared was used to measure the reference thermogravimetric curves.

The study was performed using the following salts and salt-mixtures:

$NaNO_2$, $NaNO_3$, K_2HPO_4, $Fe(NO_3)_3$
$KH_2PO_4 + Na_2B_4O_7$
$KCL + NaNO_3 + Na_2SO_4 + K_2HPO_4$
$CuSO_4 + PbCl_2 + KH_2PO_4 + Na_2B_4O_7$
$NaNO_2 + NaNO_3$
$NaNO_3 + Fe(NO_3)_3$
$NaNO_2 + Fe(NO_3)_3$
$NaNO_2 + NaNO_3 + KH_2PO_4 + Na_2B_4O_7$

All salt mixtures consisted of equal weight percents of their constituents. The only exceptions were the mixtures with $Fe(NO_3)_3$, where two additional mixtures containing each 99% of $NaNO_3$ and 1% of $Fe(NO_3)_3$ were prepared.

5.2. Bitumen

The bitumens used for this study had been selected in a previous study [4]. Three types were used: Two of the Österreichische Mineralöl Verwaltungs AG, type B40 and B85/25, and the Austrian type Shell Mexphalt 10/20. Their characteristics will be found in Table I. The bitumen type B85/25 was only used in a one-by-one mixture with B40 to get a product with a softening point more like that of Mexphalt M 10/20.

TABLE I. CHARACTERISTICS OF BITUMEN

	B 40	B 85/25	M 10/20
Density 25/25 ($g \cdot cm^{-3}$)	1.015	1.015	1.02–1.07
Penetration 25°C (mm/10)	43.5	28.5	10–20
Softening point (°C)	54.2	84.5	65–75
Breaking point (°C) (Fraass)	- 14	- 16	+ 3
Flashpoint (°C)	324	287	250

5.3. Mixing of bitumen and salts

The bitumen-salt mixtures were prepared by placing the weighted bitumen portions in small stainless-steel containers, which were heated to 180°C in an oil bath. As the bitumen became fluid the weighted amounts of salt were added and dispersed by vigorous hand-stirring. After the salt-addition had been completed, the samples were cooled first by water then using liquid nitrogen.

There were different weight percentages used in preparing the bitumen-salt mixtures. The samples contained:

(1) Two parts bitumen and one part salt for $NaNO_3$, K_2HPO_4, KCL + $NaNO_3$ + Na_2SO_4 + K_2HPO_4

(2) Equal amounts of bitumen and salt for $NaNO_2$, KH_2PO_4 + $Na_2B_4O_7$, $CuSO_4$ + $PbCl_2$ + KH_2PO_4 + $Na_2B_4O_7$

(3) One part of bitumen and two parts of salt for $NaNO_2$ and the remaining salt mixtures.

All these salts and salt mixtures were selected according to possible compositions of waste waters that will be processed in the Austrian embedding plant.

6. MEASUREMENT CONDITIONS

A Mettler-Thermoanalyser was used for all the thermogravimetric analysis in this work.
The parameters of each single analysis were kept constant at:
sample mass approximately 50 mg
temperature gradient 4 degC/min
maximum temperature 500°C.

Each sample was analysed twice in air and twice in nitrogen as previous tests had shown that the reproducibility of the measurements was satisfactory.

There were several conditions used for the comparison of the results: the ignition temperature of the bitumen salt mixture, the weight loss after the complete temperature programme and the types of exothermic reactions during the heating process.

7. DISCUSSION OF THE MEASUREMENTS

As expected it was found at the evaluation of the measurements that of all tested salts only those that will release oxygen already at lower temperatures are of great influence on the thermo-

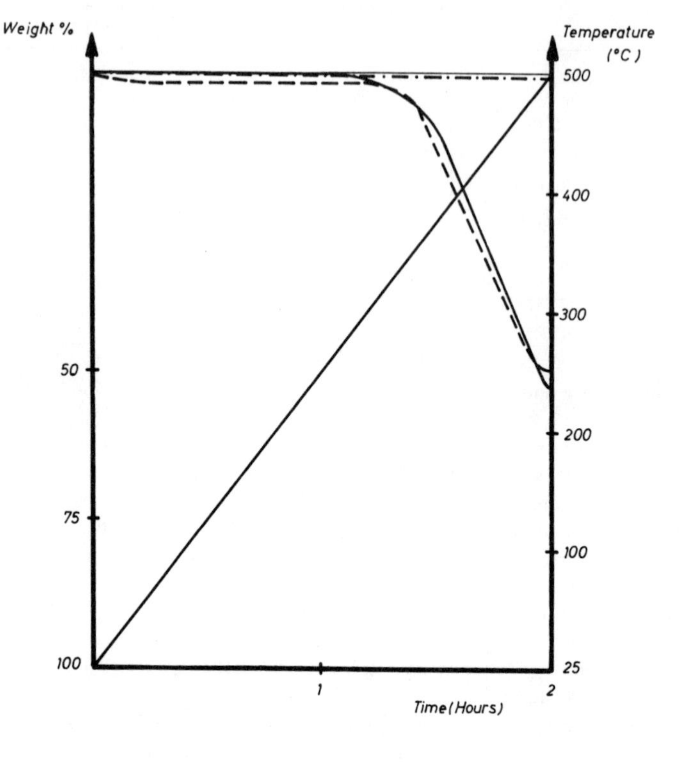

FIG. 2. Thermogram of bitumen-sodium nitrate mixture.

stability of the bitumen. Very important for these reactions are the velocity of the oxygen release and the oxidation potential.

It could be found that thermodynamically unstable salts such as KH_2PO_4 and $Na_2B_4O_7$ are of almost no influence on the bitumen although they easily release water. They do not lead to decomposition of the bitumen or to formation of gas bubbles.

On the other hand salts that release highly oxidizing gases in large amounts, such as the nitrates and nitrites, are of great negative influence on the bitumen. With these salts it is necessary to strictly control and observe the temperature at every step of the embedding process and also during storage afterwards. If there is at any time an overheating of the bitumen salt mixture above 200°C there is the risk of damage to the integrity of the salt coating.

The decomposition of nitrates and nitrites under production of highly oxidizing nitrogen oxides leads to an intensive formation of gas bubbles in a thin fluid bitumen at temperatures above 200°C. If these ions are present in larger amounts, as, for example, in the samples with

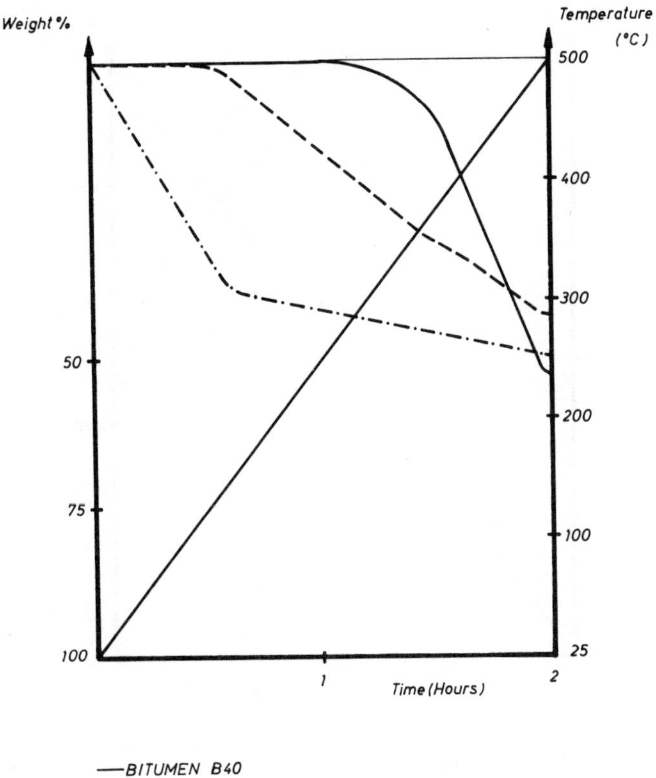

FIG.3. Thermogram of bitumen-sodium nitrate mixture with 50 wt% iron nitrate added.

~ 66 wt% $NaNO_3$, this sudden increase in the bitumen surface by bubble formation leads to an intensive oxidizing attack at the reactive newly formed surface. The oxidation of the bitumen is almost instantaneous and leads to the formation of a slaglike product which is very porous and lacks any mechanical stability. There is no remaining coating of the incorporated salt particles.

This reaction is furthermore accelerated by the addition of Fe(III) ions. During the experiments no other heavy metal ion has been found with the same accelerating force. This is especially important, as it has been found that even a 1% addition of Fe(III) ions to $NaNO_3$ has an accelerating effect. It seems that this is due to catalytic influences of the Fe(III) ion and as long as slightly acid solutions are pumped through steel tubes there is always the chance of some iron ions going into solution.

It is therefore very important in the technical embedding of nitrate, nitrite and iron-containing salts for either the oxidizing ability of the nitrates and nitrites to be destroyed before the embedding into the bitumen or the catalytic influence of the Fe(III) ions to be inhibited.

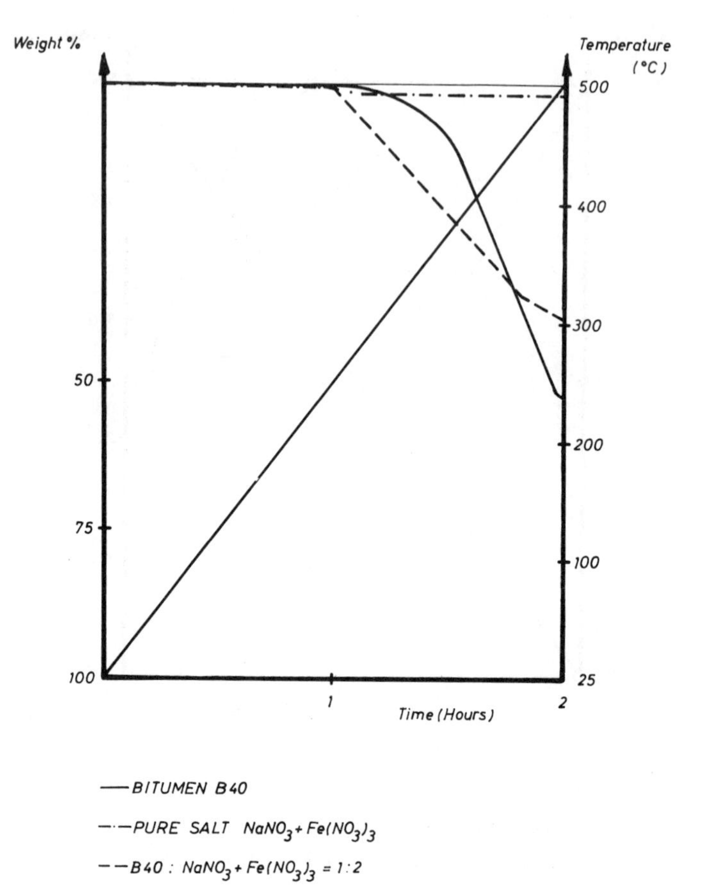

FIG.4. Thermogram of bitumen-sodium nitrate mixture with 50 wt% iron nitrate added and pH adjusted to 11.

Denitrification could be done by treating the solutions with formaldehyde or similar reducing agents. The inhibition of the activity of the Fe(III) ions was tested during this study by a simple hydrolysis using NaOH.

It could be shown that by adjusting the pH of the solutions containing $NaNO_3$ and $Fe(NO_3)_3$ to a value between 9 and 11 it was possible to shift the 'ignition' range of the mixture almost 100 degC.

There was, however, no change in the reaction sequence and almost no difference in the total weight loss and the structure of the final reaction product (Figs 2–4). As has also been found in a similar study [5], no exothermic reactions were found below 300°C when using nitrate or nitrite mixtures (i.e. between the melting point of $NaNO_2$ = 271°C and $NaNO_3$ = 311°C). The 'ignition' range was always situated between 390 and 420°C.

It should be explicitly stated that a very sudden weight loss of the samples and an intensive formation of gas bubbles was always found after the ignition but there has never been an explosion of the bitumen-salt mixture [6].

As mentioned above, all the measurements were also done using a nitrogen atmosphere. By using an inert atmosphere it was found that the distillation of lower bitumen components was somewhat more intensive than in air.

The reaction of bitumen with nitrates, nitrites and Fe(III) ions was in principle the same. Owing to the absence of an oxidizing atmosphere there was a delayed start to the reaction but the reaction itself was faster than in air. There was almost no difference in the final products after the heat treatment.

8. DOSIMETRY

In preparing future studies on the behaviour of radiation-damaged bitumen during a similar heat-treatment, a method of measuring the dose rate of unknown radioactive salt mixtures at any point of the mixtures has been developed [7]. This is done by two measurements with glass dose-meters. To determine the gamma dose only the dose-meter is used with its beta shielding and protected against contamination by a thin rubber layer. To determine the sum of beta and gamma dose only the glass body without any protection was used in direct contact with the radioactive material. Afterwards the dose-meters had to be decontaminated using hydrochloric acid, water and alcohol.

Before evaluation the dose-meters were always conditioned for one day. The reproducibility of the measurements was better than 5%.

9. CONCLUSIONS

According to the results of this study there are no chemical difficulties to expect at embedding low to intermediate radioactive solutions in bitumen after preconcentration and drying.

As most of these waste solutions will come from power reactor operational waters, from solutions for decontamination and from regeneration waters for ion-exchange resins, there exist no reasons for very high concentrations of nitrates and nitrites.

These two ions especially in combination with Fe(III) ions are as far as we know the only source of any troubles concerning the embedding process. But also these difficulties can be overcome by some form of chemical conditioning of the waste solutions.

ACKNOWLEDGEMENTS

The authors wish to express their gratitude to the Jubiläumsfonds der Österreichischen Nationalbank and the Forschungsförderungsfonds der gewerblichen Wirtschaft for the generous financial support of this work.

REFERENCES

[1] JAKUSCH, H., Oesterr. Z. Elektrizitaetswirtsch. **28** 6, Beilage "Das Kernkraftwerk" Nr. 1/2 (1975).
[2] JAKUSCH, H., KNOTIK, K., Nuklex 75, Spezialkolloquium D 3, Basel (Oct. 1975).
[3] SORANTIN, H., HUBER, I., JAKUSCH, H., KNOTIK, K., Nucl. Eng. Des. **34** (1975).
[4] ZEGER, J., KNOTIK, K., SCHOPPER, G., Berichte der Österreichischen Studiengesellschaft für Atomenergie Ges.m.b.H. CH-119/73 (December 1973).

[5] Eurochemic IDL-Report No. 67 (Feb. 1973).
[6] BACKOF, E., DIEPOLD, W., Institut für Chemie der Treib- und Explosivstoffe, Technische Mitteilung Nr. 9/70 (1970).
[7] PATEK, P., private communication.

DISCUSSION

I. LEVIN: In your thermogravimetric analysis did you try a rate of temperature increase greater than 4 deg/min? We have found that this has a marked effect on the ignition point. At 40 deg/min, for example, the bitumen may ignite at 280 to 300°C.

J. ZEGER: Yes, we have tried higher rates of temperature rise in previous tests (up to 10°C/min), but have found that the thermoanalytical curves obtained with the higher rates are not suitable for evaluation. We therefore selected a temperature rise rate of 4 deg/min for our investigations.

G. LEFILLATRE: What is the time of contact between the hot bitumen and the salts in the vessel? Secondly, since you do not have any form of stirring, the mixture is produced by sedimentation. Have you encountered any problem of binding of the dry residues, or non-coating of the salts, and, more generally, non-homogeneity of the final product?

J. ZEGER: Since the complete sedimentation process takes place in the ‘mixing vessel’, the hot bitumen is in contact with the dry salt for the time taken to accumulate 200 litres of bitumen-salt mixture at a rate of roughly 20 l/h, i.e. about 10 hours.

As regards your second point, we did indeed have some forms of agglomeration of the salt from the thin-film evaporator during the early stages of development of our sedimentation process. These difficulties were overcome by changing the type of bitumen used for embedding and also by altering the operational parameters of the thin-film evaporator in such a way that it is now impossible for the slurry zone to ‘leave’ the evaporator; with the result that the downfalling salts are always dry and powdery.

P.M. STIPANITS: If there is a fault in the evaporation process you will get slurries instead of dry solids. In such a case, how would the sedimentation process work?

J. ZEGER: By pretreating the solutions and signalling the powder moisture to the feed rate of the thin film evaporator, we have been able to ensure that no slurries are introduced into the hot bitumen. Unfortunately, I cannot give you the details as the process is being patented. Further information will be made available later.

PROGRES DANS LES TECHNIQUES DE BITUMAGE DES EFFLUENTS LIQUIDES DES CENTRALES NUCLEAIRES A EAU PRESSURISEE

G. LEFILLATRE
CEA, Centre d'études nucléaires de Cadarache,
Saint-Paul-lez-Durance,
France

Abstract–Résumé

ADVANCES IN TECHNIQUES FOR BITUMINIZING LIQUID WASTE FROM PRESSURIZED WATER NUCLEAR POWER REACTORS.

The author gives an account of recent progress in the technique of direct bituminization (without pre-concentration) of spent effluents from a nuclear power station with a light-water reactor (PWR). Direct bituminization involves complete dehydration in a vertical evaporator with a thin layer of dilute emulsion formed by the on-line mixing of bituminous emulsion and effluents at pH 9.6. The bituminized material containing 40 to 50% salts and less than 1% water flows by gravity into a drum, where it solidifies. There is complete decontamination of the distillates by passage over activated charcoal and ion-exchange resin columns. The process was developed at the pilot plant of the Cadarache Nuclear Research Centre, which is equipped with an LUWA L150 evaporator that can treat 60 l/h of effluents with an activity of the order of 1 Ci/m^3. The decontamination factors for the equipment range from 10^3 to 10^5, and the actual concentration factor attains 290 with effluents containing 2 g/l. The characteristics of the solidified waste, i.e. density, bitumen content, softening point, leaching rate, and thermal stability, are given, Finally, from the example of an industrial direct bituminization plant with a capacity of 3 m^3/h, the author estimates the annual cost of treating spent effluents from two 900 MW(e) power plants (PWR) at 1.5 million francs. A comparison of a number of drums treated with bitumen and with cement suggests that, in terms of the solidified waste to be stored, there is a gain in volume of 2.7 in favour of bitumen.

PROGRES DANS LES TECHNIQUES DE BITUMAGE DES EFFLUENTS LIQUIDES DES CENTRALES NUCLEAIRES A EAU PRESSURISEE.

Le mémoire rend compte des récents progrès réalisés dans la technique de bitumage direct, sans pré-concentration, des effluents usés issus d'une centrale nucléaire à eau légère du type PWR. Le bitumage direct consiste à déshydrater complètement dans un évaporateur vertical à couche mince l'émulsion diluée formée par le mélange en ligne d'émulsion bitumineuse et d'effluents à pH 9,6. L'enrobé bitumineux renfermant 40 à 50% de sels et moins de 1% d'eau s'écoule par gravité dans un fût où il se solidifie. La décontamination des distillats est parfaite par passage sur des colonnes de charbon actif et de résines échangeuses d'ions. La mise au point du procédé a été réalisée dans l'installation pilote du Centre d'études nucléaires de Cadarache, équipée d'un évaporateur LUWA L150 qui permet de traiter 60 litres par heure d'effluents d'une activité de l'ordre de 1 Ci/m^3. Les facteurs de décontamination des appareillages sont compris entre 10^3 et 10^5, et le facteur de concentration réel atteint 290 avec des effluents à 2 g par litre. Les caractéristiques du déchet solidifié sont mentionnées: densité, teneur en bitume, point de ramollissement, vitesse de lixiviation, stabilité thermique. Enfin, sur la base d'une installation industrielle de bitumage direct d'une capacité de 3 m^3 par heure, le coût annuel du traitement des effluents usés de deux centrales PWR de 900 MW(e) est estimé à un million et demi de francs. Une comparaison du nombre de fûts conditionnés par le bitume et le ciment fait ressortir un gain en volume pour les déchets solidifiés à stocker de 2,7 en faveur du bitume.

INTRODUCTION

Les centrales nucléaires à eau légère pressurisée rejettent des effluents liquides usés qui ont la particularité, pour ceux provenant des circuits primaires, de renfermer du bore soluble sous forme d'acide borique et de borates alcalins.

Ces effluents usés subissent un traitement de décontamination par distillation dans des évaporateurs primaires. Les distillats, après contrôle, sont dilués et rejetés dans le fleuve ou la mer proches du site des réacteurs. Les concentrats sont stockés avant d'être solidifiés. En raison de la faible solubilité de l'acide borique dans l'eau, ces concentrats ont une salinité limite admissible de 300 g par litre (valeur normale 40 000 ppm de bore).

Dans ces conditions, avant l'opération de solidification, les concentrats d'évaporateur doivent être maintenus en température (80–90°C) dans le réservoir de stockage et brassés par un agitateur. Actuellement le procédé de solidification le plus utilisé pour ce type de concentrats est la cimentation. Cependant d'autres procédés de solidification sont ou vont être prochainement mis en œuvre, employant, soit des résines polymérisables à la température ambiante du type urée-formaldéhyde (en milieu humide) ou du type polyester non saturé (à sec), soit des bitumes fondus.

Quel que soit le procédé de solidification appliqué, de nombreuses difficultés apparaissent avec ces concentrats borés chauds: cristallisations fréquentes dans les cuves de stockage et obturation des tuyauteries de transfert, nécessité d'assurer un chauffage permanent des appareillages contenant les concentrats, risques de prise en masse dans le réseau de ventilation du ciment avec la vapeur d'eau des concentrats chauds lors des opérations de malaxage.

Dans le but de s'affranchir des contraintes apportées par la concentration préalable des effluents borés, nous avons développé un nouveau procédé de bitumage direct des effluents sans préconcentration. Ce procédé présente l'avantage de réaliser en une seule opération continue l'évaporation de la totalité de l'eau des effluents contaminés et l'enrobage par le bitume des sels neutralisés que renferment ces effluents. Cette technique est applicable également au bitumage des suspensions de résines usées et des boues d'adjuvants de filtration (diatomées, cellulose, charbon actif, amiante).

Compte tenu de l'expérience acquise avec les évaporateurs à couche mince dans le domaine du bitumage des concentrats d'évaporateur, des suspensions de diatomées et de résines usées [1, 2], nous avons mis au point cette technique des bitumage direct des effluents en utilisant un évaporateur LUWA-SMS du type L150. Cet évaporateur équipe notre installation pilote au Centre d'études nucléaires de Cadarache et permet de traiter 60 litres par heure d'effluents radioactifs d'une activité de l'ordre de 1 Ci/m^3.

1. DESCRIPTION DU PROCEDE

Le procédé de bitumage direct des effluents (fig. 1) comprend les étapes suivantes:

– Neutralisation des effluents et ajustement de leur pH à 9,6 par de la soude, dans le cas des effluents borés, pour transformer l'acide borique en tétraborate de sodium.

– Mélange à froid en continu des effluents traités (pH = 9,6) et d'une émulsion aqueuse à 50% de bitume (M 40/50).

– Déshydratation complète de l'émulsion diluée dans un évaporateur à couche mince vertical: l'enrobé bitumineux renfermant 40 à 50% de sels et moins de 1% d'eau s'écoule par gravité dans un fût à la base de l'appareil, la vapeur d'eau chargée d'huiles et de goudrons est condensée à la sortie supérieure de l'évaporateur. Après refroidissement le déchet se trouve solidifié et conditionné pour un stockage définitif.

– Décontamination des distillats par passage, éventuellement sur des préfiltres spéciaux à cartouche, et/ou directement sur des colonnes de charbon actif suivies de colonnes de résines

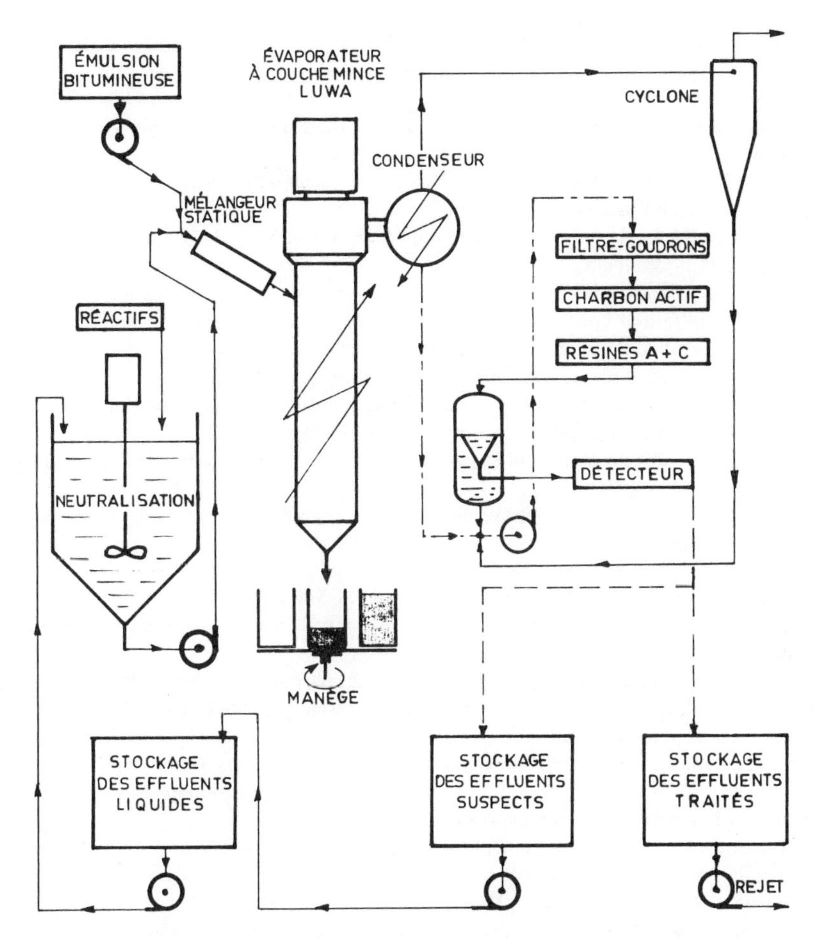

FIG.1. Bitumage direct des effluents radioactifs sans préconcentration.

échangeuses d'ions cationiques et anioniques. Le charbon actif et les résines sont saturés après traitement respectivement de 2000 à 7000 volumes de lit et de 5000 à 10 000 volumes de lit. Ces déchets solides sont alors transférés par hydro-éjecteurs et les suspensions obtenues sont bitumées séparément suivant la même technique que pour les effluents.

– Rejet des effluents dans l'environnement, après contrôle, et recyclage éventuel des distillats ne satisfaisant pas aux normes de rejet.

2. DESCRIPTION DE L'INSTALLATION PILOTE DE BITUMAGE DIRECT

L'installation comprend:

– Un ensemble d'alimentation des effluents entièrement en acier inoxydable composé d'une cuve agitée de 1 m^3 pour la neutralisation des effluents avec prise d'échantillons, d'un poste de dosage des effluents avec une pompe centrifuge de 1 m^3/h , une roue doseuse à débit variable (20 à 100 l/h) et un mélangeur statique.

– Un poste de distribution d'émulsion de bitume constitué par un bac de 200 litres et une pompe doseuse à piston (0,3 à 30 l/h) qui injecte en continu l'émulsion en ligne à la sortie de la roue doseuse.

– Un évaporateur à couche mince LUWA L150 entièrement en acier inoxydable comprenant:

- un corps d'évaporation vertical de 0,5 m^2 disposant d'une double enveloppe chauffée à 215–225°C par un fluide caloporteur Gilotherm ALD,
- un rotor type L à 3 pales tournant à 1100 tours/min avec entrefer de 1,5 mm et palier inférieur lubrifié par un polyglycol,
- un dévésiculeur précédant la tubulure de sortie des vapeurs en partie supérieure,
- une vanne chauffante à boisseau sphérique télécommandée à la base de l'appareil.

– Une centrale thermofluide à chaudière électrique de 96 kW comportant quatre allures de marche et disposant d'une pompe de circulation de 20 m^3/h.

– Un poste de mise en fûts constitué par une cellule ventilée équipée d'un chemin de roulement non mécanisé et de portes avec protections blindées amovibles.

– Un ensemble de traitement des vapeurs et des distillats à la sortie de l'évaporateur comportant:

- un condenseur vertical en acier inoxydable à faisceaux tubulaires démontables d'une surface d'échange de 2 m^2,
- un pot de collecte de condensats de 50 litres,
- une pompe centrifuge de circulation des condensats de 500 l/h,
- une batterie de deux filtres spéciaux à cartouches perdues Cuno, l'un en service, l'autre en réserve,
- une batterie de deux colonnes en série de 10 litres de charbon actif Filtrasorb 400,
- une batterie de deux colonnes en série de 5 litres de résines Duolite C20 et A101D avec débitmètres, manomètres et prises d'échantillons,
- une cuve de 1 m^3 en acier inoxydable pour le contrôle des distillats décontaminés avant rejet.

– Un groupe d'extraction et de filtration des incondensables (2000 m^3/h) raccordé au réseau de ventilation du bâtiment comprenant:

- un cyclone avec bac de refroidissement raccordé à l'avant du condenseur pour piéger les traces d'eau et de goudrons dans les vapeurs,
- un caisson de préfiltres métalliques Mioval pour arrêter les traces d'huiles et de goudrons aspirées dans la cellule de mise en fûts. La dépression dans l'évaporateur est réglée entre −5 et −15 mmCE au moyen d'un volet de réglage.

3. EXPERIENCE ACQUISE AVEC DES EFFLUENTS USES DE LA CENTRALE DE LA SENA

Indépendamment des essais d'enrobage effectués sur d'autres types d'effluents, nous avons solidifié par le bitume un lot d'environ 6 m^3 d'effluents usés provenant du réacteur PWR de Chooz (300 MW(e)) et des résines Amberlite saturées.

3.1. Caractéristiques des effluents traités

Composition chimique

Bore	320 mg/l
Sodium	400 mg/l
Ammonium	4,5 mg/l
Lithium	0,6 mg/l
Carbonate	160 mg/l (CO_3^{2-})

Nitrate	19 mg/l (NO_3^-)
Détergents anioniques	4,5 mg/l (exprimés en dodécylbenzène sulfonate de sodium)
Extrait sec à 105°C	2,02 g/l
Matières solides en suspension	27 mg/l
pH	9,2

Composition radiochimique

^{3}H	1,25 Ci/m^3
– Activité γ globale	$1{,}3 \cdot 10^{-2}$ Ci/m^3 dont:
^{137}Cs	$6 \cdot 10^{-3}$ Ci/m^3
^{54}Mn	$3 \cdot 10^{-3}$ Ci/m^3
^{134}Cs	$2{,}4 \cdot 10^{-3}$ Ci/m^3
^{106}Ru	$7{,}8 \cdot 10^{-4}$ Ci/m^3
^{58}Co	$3{,}7 \cdot 10^{-4}$ Ci/m^3
^{60}Co	$3{,}1 \cdot 10^{-4}$ Ci/m^3
^{131}I	$2 \cdot 10^{-4}$ Ci/m^3
^{124}Sb	$8{,}7 \cdot 10^{-5}$ Ci/m^3
^{140}La	$1{,}4 \cdot 10^{-5}$ Ci/m^3
– Activité β globale hormis ^{3}H	$8{,}9 \cdot 10^{-3}$ Ci/m^3 dont:
^{90}Sr	$3{,}1 \cdot 10^{-4}$ Ci/m^3
– Activité α globale	$8{,}5 \cdot 10^{-5}$ Ci/m^3 dont:
^{239}Pu	$7{,}2 \cdot 10^{-5}$ Ci/m^3
^{241}Am	$1 \cdot 10^{-5}$ Ci/m^3

3.2. Caractéristiques des suspensions de résines

Résines IRA 400 et IR 120	
Matières en suspension	10 g/l
pH	7,8

Composition radiochimique

– Activité γ globale	$2{,}25 \cdot 10^{-1}$ Ci/m^3
– Activité β globale	$5{,}5 \cdot 10^{-4}$ Ci/m^3
– Activité α globale	$4 \cdot 10^{-4}$ Ci/m^3

TABLEAU I. FACTEURS DE DECONTAMINATION A LA SUITE DES DIFFERENTS POSTES DE TRAITEMENT

FD	Total	^{137}Cs	^{54}Mn	^{134}Cs	^{106}Ru	^{58}Co	^{60}Co	^{131}I	^{90}Sr
Evaporateur à couche mince LUWA L150	860 à 1200	800 à 1500	350 à 1400	900 à 1900	170 à 330	150 à 360	160 à 390	100 à 160	3000 à 30 000
Filtration sur charbon actif	900 à 3000	1100 à 2500	800 à 1900	1300 à 2000	190 à 350	–	175 à 400	–	30 000 à 60 000
Echange ionique	1500 à 6000	1800 à 6000	1500 à 5500	2000 à 10000	370 à 1500	–	450 à 2000		100 000 à 300 000

3.3. Facteurs de décontamination de l'installation pilote

Les facteurs de décontamination obtenus à la suite des différents postes de traitement de l'installation sont mentionnés dans le tableau I.

Le facteur de décontamination de l'évaporateur à couche mince est compris, globalement pour les émetteurs γ, entre 860 et 1200 et pour le ^{90}Sr entre 3000 et 30 000.

Après traitement sur charbon actif et résines cationiques et anioniques, le facteur de décontamination se situe, globalement pour les émetteurs γ, entre 1500 et 6000, et pour le ^{90}Sr entre 100 000 et 300 000.

3.4. Caractéristiques des distillats

Avant traitement		
Conductivité	30 à 50 μS/cm	
pH	8,3 à 9,4	pour les effluents borés, et
	7,4 à 8,2	pour les résines
Teneur en bore	1 à 4 mg/l	
Matières organiques (huiles et goudrons)	6 à 50 mg/l	
Matières solides en suspension	1 à 3 mg/l	
Après traitement		
pH	6,8 à 7,5	
Matières organiques (huiles et goudrons)	1,5 à 4 mg/l	
Activité spécifique	toujours inférieure à la CMAP (hormis le tritium).	

3.5. Caractéristiques des enrobés

Densité réelle	1,38 à 1,57	pour les effluents borés
	1,23	pour les résines
Densité apparente	1,25 à 1,48	pour les effluents borés
	1,23	pour les résines
Facteur moyen de réduction de volume tenant compte du conditionnement du charbon actif saturé et des résines usées	290	
Teneur en eau	0 à 0,7%	
Teneur en bitume	60 à 50%	
Point de ramollissement	73–78°C	

Vitesse de lixiviation dans l'eau déminéralisée au bout de 123 jours d'immersion (normes AIEA)

- pour le ^{137}Cs: $3 \cdot 10^{-6}$ cm/j
- pour le ^{90}Sr: $2 \cdot 10^{-5}$ cm/j
- pour le ^{60}Co: $7 \cdot 10^{-5}$ cm/j
- pour le ^{106}Ru: $5 \cdot 10^{-5}$ cm/j.

L'analyse thermique gravimétrique et différentielle (fig. 2) de l'enrobé ne fait pas apparaître de pic exothermique avant 482°C, et le début de décomposition thermique se situe après 300°C (perte de poids à 293°C : 0,3%; à 324°C : 4,5%). Par ailleurs des échantillons d'enrobé maintenus pendant 24 h à des températures allant de 250°C à 400°C n'ont subi de dégradation thermique sensible qu'au-delà de 300°C, sans toutefois déclencher de réactions exothermiques significatives, même à 400°C.

La perte de poids après 24 h à 300°C est de 9,4%, par contre celle à 400°C atteint 46%; presque la moitié du déchet solidifié est donc décomposé après un stockage d'un jour à 400°C.

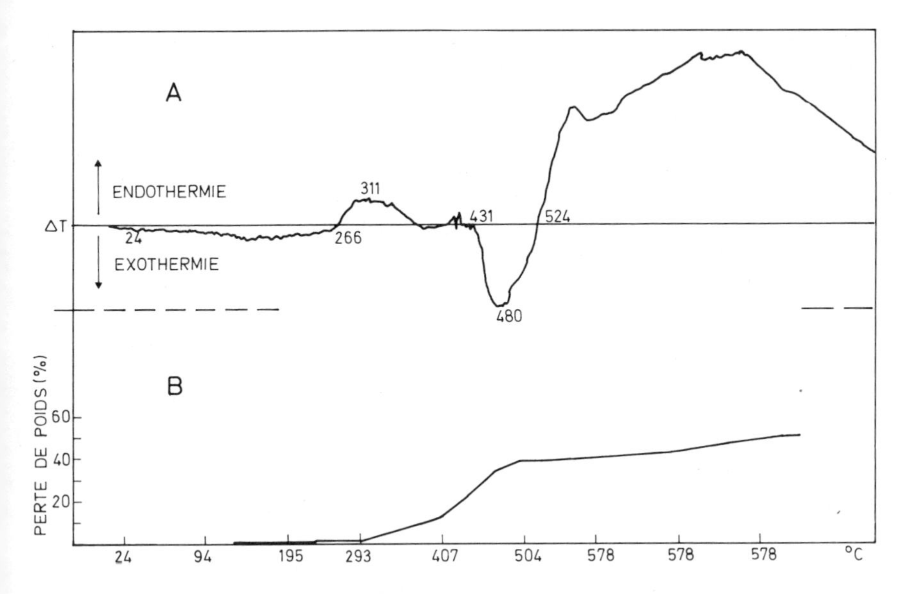

FIG.2. Enrobé bitumineux de Chooz.
Sels: 46,9%; Mexphalte 40/50 : 52,7%; eau : 0,4%.
A. Analyse thermique différentielle. Vitesse de montée en température : 6°C/min.
B. Analyse thermogravimétrique.

4. ESTIMATION DU COUT D'EXPLOITATION D'UNE STATION CAPABLE DE CONDITIONNER LES EFFLUENTS LIQUIDES USES DE DEUX CENTRALES PWR DE 900 MW(e)

Cette installation doit permettre de traiter

- 9000 m³ d'effluents liquides usés par an dont la minéralisation moyenne est supérieure à 1 g/l et se répartissant comme suit: 1000 m³ par an d'effluents primaires (drains résiduaires; effluents de servitude, laverie et douches); 4000 m³ par an d'effluents secondaires (purges de déconcentration des générateurs de vapeur); 4000 m³ par an d'effluents exceptionnels (purges des circuits primaires, purges des générateurs de vapeur, effluents de décontamination);
- 80 m³ de résines usées par an.

La station de traitement, basée sur la technique de bitumage direct sans préconcentration, serait équipée d'un évaporateur à couche mince LUWA LN 3200, d'une surface d'échange de 32 m², capable de traiter en continu 3 m³/h d'effluents, et chauffé par une centrale thermofluide de 3000 thermies.

Le stockage avant traitement serait constitué par quatre réservoirs de 35 m³, deux réservoirs de 30 m³ et une cuve agitée de 10 m³ pour les résines.

L'unité de traitement des distillats, après passage dans un condenseur refroidisseur de 72 m², comprendrait deux colonnes de 2 m³ pour le charbon actif et deux colonnes de 500 litres pour les résines à lits mélangés, toutes les quatre équipées d'hydro-éjecteurs.

Les distillats traités seraient stockés dans une cuve de contrôle de 10 m³ avant d'être envoyés dans quatre réservoirs de 30 m³.

La station disposerait en outre des cuves de réactifs, des pompes de transfert et de dosage, du matériel de contrôle et de régulation nécessaires à la bonne marche de l'installation.

Un poste de mise en fûts avec manège à fonctionnement automatique, contrôlé par caméra de télévision, et une salle de stockage temporaire, d'une capacité de 3000 fûts de 225 litres, desservie par un pont roulant pneumatique, compléteraient l'installation.

Le coût estimatif d'une telle station avec son bâtiment propre serait d'environ 25 millions de francs.

Sur la base d'un fonctionnement de l'installation 45 semaines par an, en continu cinq jours par semaine avec trois postes de deux opérateurs et un chef d'équipe de jour, l'évaporateur marchant 75% du temps, le coût annuel d'exploitation serait le suivant:

Matières consommables (fuel, émulsion de bitume, réactifs, fûts, etc.)	678 000 F
Energie (électricité, fluides)	86 000 F
Main d'œuvre	504 000 F
Frais d'entretien (matériel et bâtiment)	232 000 F
soit un montant global des frais de fonctionnement, hors amortissement, de	1 500 000 F par an.

5. COMPARAISON ENTRE CONDITIONNEMENT PAR LE BITUME ET CONDITIONNEMENT PAR LE CIMENT

Sur les bases d'un même volume de résidus liquides à conditionner, et en prenant l'hypothèse d'une teneur moyenne en sels de 150 g/l dans les concentrats de l'évaporateur des effluents usés, avant conditionnement par le ciment, nous avons établi le tableau II.

De ce tableau il ressort que deux réacteurs PWR de 900 MW(e) produiraient 700 fûts «ciment» et seulement 255 fûts «bitume», soit une réduction de volume d'un facteur 2,7 et une réduction de poids d'un facteur 3,2.

TABLEAU II. COMPARAISON ENTRE CONDITIONNEMENT PAR LE BITUME ET CONDITIONNEMENT PAR LE CIMENT

Résidus à solidifier	Quantité (m^3/an)	Activité moyenne (Ci/m^3)	Nombre de fûts de 225 litres par an	
			Ciment[a]	Bitume
Effluents usés à 1 g/l	9000	0,4		80[b]
Concentrats à 150 g/l	60		300[c]	
Résines saturées	80	360	400[c]	175[d]
Totaux			700	255

[a] Le nombre de fûts de résidus solidifiés par le ciment a été déterminé en partant des données de la référence [3].

[b] 12 fûts proviennent du bitumage du charbon actif saturé et 3 fûts du bitumage des résines usées, issues du traitement des distillats.

[c] 35% de résines ou concentrats + 65% d'eau, de ciment et d'agrégats.

[d] 55% de bitume + 45% de sels ou de résines sèches.

6. CONCLUSIONS

Le procédé de bitumage direct des effluents liquides des centrales nucléaires à eau pressurisée présente de nombreux avantages par rapport à la technique actuelle de concentration des effluents par évaporation thermique suivie d'un conditionnement des concentrats chauds par le ciment:

- importante réduction de volume et de poids;
- simplification de la station de traitement des effluents en supprimant l'évaporation de pré-concentration et le stockage réchauffé des concentrats, tout en respectant les normes de rejet (facteurs de décontamination compris entre 10^3 et 10^5);
- garanties d'homogénéité et bonne résistance à la lixiviation des résidus solidifiés obtenus qui satisfont, en outre, aux critères de sûreté [4] pour le stockage temporaire, le transport et le stockage à long terme des déchets radioactifs.

Par ailleurs, le coût annuel d'exploitation, hors amortissement, d'une station de traitement d'effluents usés pour deux centrales PWR de 900 MW(e), basée sur le procédé de bitumage direct, serait compétitif puisque de l'ordre de 1 500 000 francs.

REFERENCES

[1] SOUSSELIER, Y., et al., «Conditioning of wastes from Power Reactors», ANS Winter Meeting, San Francisco, novembre 1973.

[2] LEFILLATRE, G., LE BLAYE, G., Traitement des effluents radioactifs par le bitume, NUCLEX 75, Bâle, Colloque D , série 3, octobre 1975.

[3] USAEC, Rapport WASH-1258 (juillet 1973).

[4] LEFILLATRE, G., «Bitumage des résidus radioactifs – Problèmes de sûreté et domaines d'application», Conférence nucléaire européenne, Paris, avril 1975, sous presse.

DISCUSSION

H. KRAUSE: For some years we have had a programme of close collaboration in the field of bituminization. We have always found the same results, encountered the same problems and reached the same conclusions as you, except with regard to radiation resistance. I notice that in this respect you always get higher values than we do. Since the tests have been carried out on the laboratory scale, the results depend on the mode of operation. To throw further light on this problem, we have planned to collaborate with Eurochemic in the preparation of several 200 litre samples with fairly high activity levels. It is intended to take samples for quality control and also to take gas samples. I would like to ask whether you have already taken samples from drums that have been in storage for several years so as to check their quality.

G. LEFILLATRE: In 1967 we carried out some experiments at Marcoule in connection with a study of leaching from coated blocks from our plant (with a volume of about 150 litres each). Three blocks were used, one placed in ordinary water, one in sea-water and one in soil taken from the Centre and which was permanently watered.

In 1971 we discontinued these experiments. The blocks were left in the water (but not the sea-water) and in the ground. In 1975 we resumed tests with the blocks kept in ordinary water, and took samples of the coating at that time. Analyses were made showing the softening point (95°C), water content (2%), salt content and radioactivity, but we did not observe any sign of radiolytic gas release in the blocks (bubbles or cavities in the coating).

Unfortunately no check on the quantity or quality of radiolytic gases has been made in the storage facilities. Since degradation is very slow in the case of activity levels below 100 Ci/m^3, it is questionable whether it would be possible in any case to have valid results from such determinations. The experiments you have planned to undertake at Mol with activities of the order of 1 Ci/l will certainly be more interesting and I am personally hoping very much that the results will provide an exact idea of the radiation resistance of bitumens.

W. HILD: Thank you, Mr. Lefillatre, for an excellent paper. With regard to your tests on the bituminization of ion-exchange resins, at what temperature were you working and at what temperature did you find partial degradation of the resins, i.e. formation of amines?

G. LEFILLATRE: We worked with a thin layer evaporator at temperatures between 215 and 235°C. For some of the anion resins, for example non-saturated Microionex, we observed decomposition and the release of NH_3 in the distillates (pH ~9–10). Conversely, with saturated Amberlite and Diaprosim resins we did not observe this effect; the pH value of the condensates ranged from 7.9 to 8.2 (the pH value of the suspension was adjusted to 8 at the start).

J. SAIDL: Were you able to suppress the volatility of the orthoboric acid completely just by alkalization, i.e. with soda?

G. LEFILLATRE: By treating the boric acid effluents with soda and adjusting the pH to 9.6 so as to convert H_3BO_3 into $Na_2B_4O_7$, we obtained boron concentrations in the distillates ranging from 1 to 4 mg/l (6 to 24 mg/l of H_3BO_3).

V.V. DOLGOV: Can you give us more details of the way in which you store the bituminized products?

G. LEFILLATRE: In France all bituminized wastes are stored in metal drums holding 225 litres. Depending on the Centre concerned the drums are sent either to interim storage or to final storage: for example, at Marcoule they are stored in ventilated bunkers; at Cadarache on concrete pads inside sheds, and at Valduc in ventilated halls to await shipment to the La Hague storage site; from Saclay they are shipped directly to the La Hague storage site by road in shield transport.

Y.P. MARTYNOV: Have you studied the water-resistance of the bituminized materials containing ion-exchange resins? And are you troubled to any extent by the problem of such materials swelling?

G. LEFILLATRE: We have kept samples of bituminized resins in ordinary water and in demineralized water. We have noticed a certain amount of swelling in some of the samples – about a 20% increase in volume at the end of three months. Furthermore, we have tested radioactive samples in accordance with IAEA standards. These experiments are still under way, but the first results are similar to those obtained with evaporated concentrates.

The swelling of bituminized blocks immersed in water is not a new problem; we have also found it in some of the evaporated concentrates. We believe that if we store the waste under adequately safe conditions, the storage area will not be permanently in contact with water and the blocks will therefore not swell. Tests involving the burial of 100 litre blocks with evaporated concentrates have been under way at Cadarache for two years. The burial area is marshy and is flooded for about 5 months of the year. We have dug up blocks after two years and put them back in the ground. No swelling or bacterial proliferation on the surface (through the action of soil microorganisms) has been observed.

I. LEVIN: What is the residence time for the bitumen mixture in the evaporator?

G. LEFILLATRE: In the LUWA L150 evaporator the residence time of the bitumen-solution mixture does not exceed three minutes.

IAEA-SM-207/81

RECENT EXPERIMENTS ON THE TREATMENT OF MEDIUM LEVEL WASTES AND SPENT SOLVENT AND ON FIXATION INTO BITUMEN

W. BÄHR, W. HILD, S. DROBNIK, L. KAHL, M. KELM,
W. KLUGER, H. KRAUSE
Gesellschaft für Kernforschung mbH,
Karlsruhe

W. FINSTERWALDER
Gesellschaft zur Wiederaufarbeitung
von Kernbrennstoffen mbH,
Karlsruhe

W. RUTH
Nukem GmbH,
Wolfgang,
Federal Republic of Germany

Abstract

RECENT EXPERIMENTS ON THE TREATMENT OF MEDIUM LEVEL WASTES AND SPENT SOLVENT AND ON FIXATION INTO BITUMEN.

The paper gives a summary of the main research and development work performed at the Nuclear Research Centre Karlsruhe on the treatment of the low and medium level waste solutions that are expected to be produced in a typical large Purex-type reprocessing plant of 1500 t U/a capacity. Results are presented that have been obtained in investigations accompanying the operation of the industrial bituminization plant of the Research Centre, which till now has produced about 2000 drums of solidified waste. A description of lab- and pilot-scale work on the destruction of nitric acid in medium level waste (MLW) solutions and evaporator concentrates by means of formic acid is given. A new approach for the treatment of MLW solutions that aims at a splitting into a very small fraction of high level waste and a low level waste solution of less or equal volume than the original MLW is presented, together with the first encouraging decontamination factors, averaging 10^2 to 10^3. A discontinuous procedure for the treatment of spent solvent containing TBP was successfully demonstrated by processing more than 100 m^3 of original reprocessing waste. Results are described, together with first investigations on the scale up to a continuous process version for a large reprocessing plant.

1. INTRODUCTION

Reprocessing plants produce large volumes of low and medium level waste solutions. In the densely populated Federal Republic of Germany only minor amounts of radionuclides can be discharged to the environment. Extensive waste treatment is necessary to attain the low permissible levels. The radioactive residues have to be converted into a form that ensures a high degree of safety for final disposal. To meet these requirements, all radioactive effluents are evaporated and the residues incorporated into bitumen.

To reduce the amounts of radioactive residues, to improve their properties further and to treat spent solvent a special R & D programme has been started. The paper presents the first results from this programme.

Flowsheet studies show that in a 1500 t/a reprocessing plant about 50000 m^3 of low and medium active waste solutions of varying chemical compositions are produced per year at the various process areas and peripheral installations. The total activity amounts to some 10^6 Ci. Up to 3000 t of sodium nitrate have to be treated if the nitric acid-containing waste streams are neutralized. After evaporation this would lead to roughly 20 000 m^3, or 4000 m^3 of solidified waste if concreting or bituminization is applied.

2. BITUMINIZATION OF RADIOACTIVE WASTES

For almost four years the concentrates from the evaporation of low and intermediate level radioactive effluents produced in the Nuclear Research Centre Karlsruhe and the 40 t/a reprocessing plant WAK have been solidifed by incorporation into bitumen. Up to now roughly 700 m^3 of evaporator concentrates from 40 000 m^3 low and intermediate level effluents containing some 10^4 Ci have been solidified into bitumen. Approximately 2000 drums of 175 l each have been produced with an average salt content of 50 wt% and a specific activity of 100 Ci/m^3 and shipped to the ASSE salt mine for disposal. Operation experience of the plant is described in detail in Refs [1, 2].

Operation of the bituminization plant is accompanied by R & D activities aiming both at the definition of operational conditions for the bituminization of all types of radwaste and at the investigation of related safety aspects. To this end and for trouble-shooting experiments, a bench-scale unit with the same features as the operational plant is operated in Karlsruhe, too. The results of the R & D work are described in Refs [3–10].

In the course of four years of operation of the bituminization plant two incidents occurred in the shielded chamber where the hot bitumen is filled from the extruder into the drums. In both cases vapours caught fire but were brought under control within a short time and could be extinguished with CO_2. No radioactivity escaped from the plant. In the first incident the fire was caused by ignition of the vapours of organic solvents contained in the waste concentrate. In the other incident an evaporator concentrate was processed with a pH of 13.8 instead of normally 9 to 10. At the same time the agitator of the concentrate feed tank had failed so that separation of organic compounds contained in the waste (TBP, degradation products, antifoaming agents, polyethylene oxide adducts) took place. As demonstrated in experiments with the bench-scale unit, these products are decomposed into easily flammable volatile compounds during incorporation at high alkalinity. The resulting bitumen products had ignition points around 200°C, whereas at pH 8 to 10 no degradation occurs and the ignition points of the products are in the region of 400°C. After improvement of the control and safety measures about 700 drums have been produced without any further incident.

Tests on the radiation stability of bitumen products up to an integrated dose of several times 10^8 rads showed that the radiolytical H_2 production is in linear proportion to the total dose averaging 0.4 to 0.5 ml H_2/g product per 100 Mrad. This value is of importance for safety considerations in connection with disposal in salt cavities [9]. The radiation-induced porosity is generally in the order of some per cent, which does not pose particular problems. Internal-irradiation tests and time-lapse investigations are at present under way.

Leach tests with products coated with a layer of 5 mm pure bitumen showed an average leach rate of 3×10^{-8} $g \cdot cm^{-2} \cdot d^{-1}$ after five years, as compared with 10^{-4} to 10^{-5} $g \cdot cm^{-2} \cdot d^{-1}$ without coating.

Incorporation conditions for power reactor wastes, e.g. ion-exchange resins, boric acid etc. were recently determined and the resulting bitumen products were characterized [5]. From the experiments it can be concluded that these types of wastes can safely be incorporated into bitumen, too.

3. DENITRATION OF HNO_3-CONTAINING MLW STREAMS

In a reprocessing plant the amount of nitric acid in the waste solution is higher than the amount of alkaline solutions. Neutralization of excess nitric acid leads to an increase in the total salt content of more than 30%. For this reason in Karlsruhe a process was developed for the destruction of the nitric acid by reaction with formic acid [10]. According to the equation

$$2\ HNO_3 + 4\ HCOOH \rightarrow N_2O + 4\ CO_2 + 5H_2O$$

nitric acid and formic acid react to water, nitrous oxide and carbon dioxide.

Up to now the process has been carried out batchwise. The necessary amount of formic acid (85%) is heated to 100°C and the nitric medium level waste (MLW) solution is fed into the boiling formic acid. The reaction starts immediately. The feeding of the nitric acid solution takes about two to three hours. During this time more than 90% of the total gas is formed. After the dosage the solution has to be boiled another five hours for completion of the reaction. In a reflux condenser the water and formic acid vapours are condensed and fed back to the reaction vessel. The non-condensable gases CO_2 and N_2O are discharged to the atmosphere after passing a scrubbing column and a filter.

Investigations in lab and pilot-plant units with nitric acid solutions showed that the nitric acid can be destroyed to a residual concentration of 0.1 to 0.2 mol/l. The experiments were carried out with simulated medium level process waste, decontamination solutions and with MLW evaporator concentrates with a high salt content. High salt concentrations do not influence the results, showing that the denitration process is more effective after evaporation.

Experiments with 1–4 molar solutions of nitric acid showed that the presence of tensides, decontamination agents and other organic compounds such as TBP, kerosene and their degradation products in concentrations up to 1 g/l has no influence on the reaction. Only a small increase in foam formation is observed. In waste solutions containing small amounts of corrosion products the NO content in the gas is below 0.1 vol.%, whereas higher concentrations of these elements cause NO development up to 2 to 3 vol.%.

Table I shows some results of the denitration experiments.

TABLE I. DENITRATION OF NITRIC ACID IN THE PRESENCE OF $NaNO_3$ AND 0.01M $Fe(NO_3)_3$

HNO_3 conc. before denitration (M)	$NaNO_3$ conc. (M)	HNO_3 conc. after denitration (M)	Decomposition (%)	
0.6	0	0.12	80	Dosage time 1.5 h, reaction time 6 h
0.6	1	0.06	90	
0.6	2.5	0.05	94	
1.5	0	0.14	91	
1.5	1	0.1	94	
1.5	2.5	0.05	98	
4.0	0	0.15	96	Dosage time 4 h, reaction time 2 h
4.0	1	0.12	97	
4.0	2.5	0.05	99	

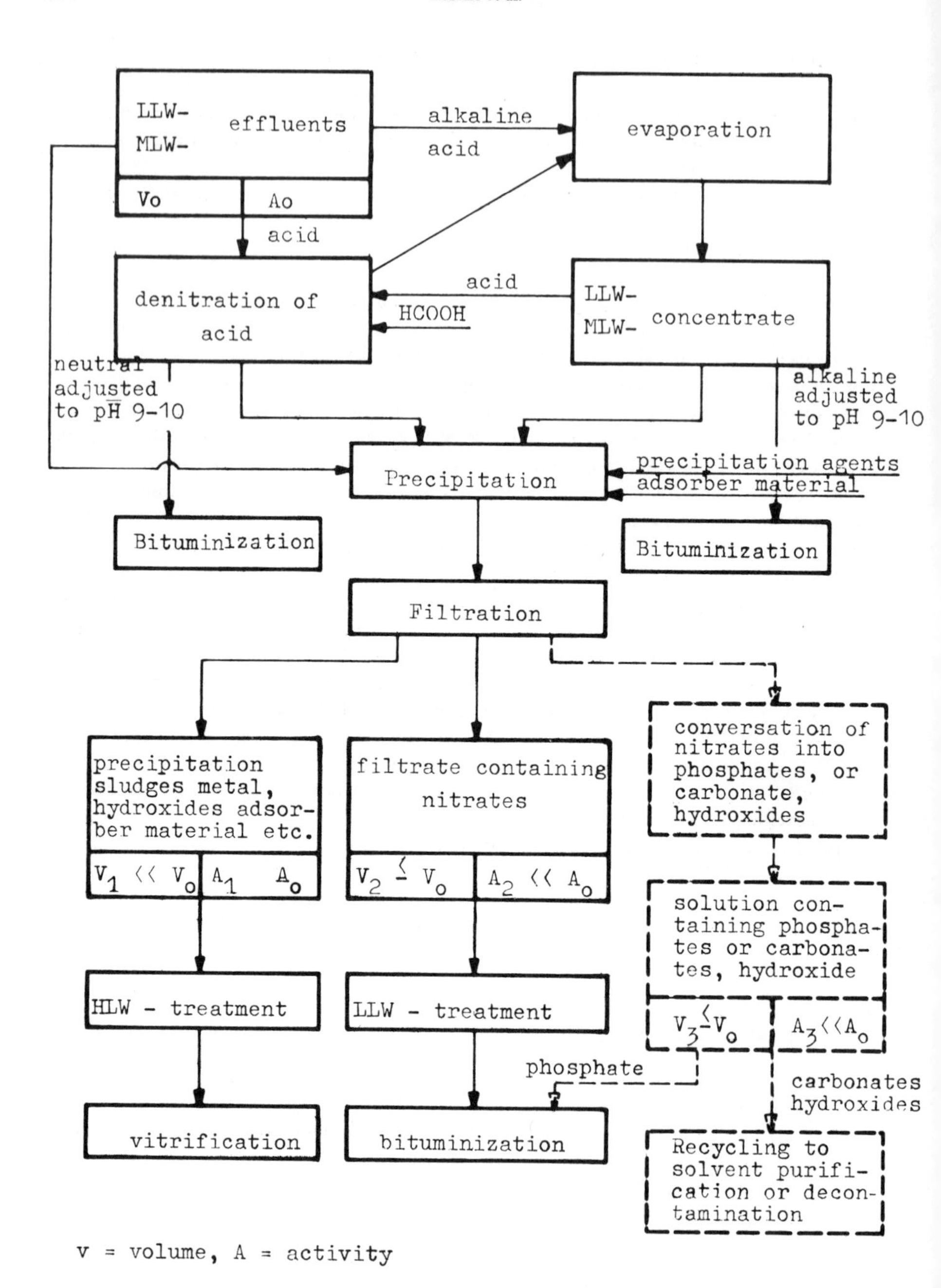

FIG.1. Alternatives for the management of LLW and MLW.

4. NEW TREATMENT FOR MEDIUM LEVEL LIQUID WASTES FROM REPROCESSING PLANTS

Due to the poor heat conductivity, the thermal behaviour and the radiation resistance of bitumen, the maximum activity concentration is limited, especially if stored in very large, non-ventilated caverns. A part of the MLW from the reprocessing plants could under some circumstances be above this activity limit. As concreting leads to relatively large waste volumes and to products with poor leach resistance, the incorporation into glass is too expensive, and other fixation materials are not available for the time being, the following procedure is under investigation at Karlsruhe (Fig.1).

After denitration, the fission products, corrosion products and actinides are separated from the MLW by precipitation and adsorption processes. The small volume of solids formed can be incorporated into glass together with the HLW. After filtration of the precipitates, the remaining solution has a maximum activity content of about 0.01 Ci/l but contains practically all the soluble salts (mainly $NaNO_3$). It can be incorporated into bitumen by the well-established and simple process without difficulty or limitations.

Numerous investigations have established the optimum conditions for separating the radionuclides from MLW by adsorption on surface active precipitates and anorganic ion exchangers.

The fission products Ce, Zr and Nb and the corrosion products Fe, Co, Cr and Ni and the actinide elements are precipitated almost quantitatively together with Al^{3+} and Si^{4+} by adjusting the pH to 9–10 with sodium hydroxide. The best results were obtained at a molar ratio of $Al_2O_3 : SiO_2 = 1:3$. However, the decontamination effect achieved by this precipitation procedure was not satisfactory for strontium, ruthenium and caesium. The addition of Sn^{2+} and Ti^{3+} leads to decontamination factors of $\leqslant 100$ for Ru and 300 for Sr.

The decontamination factor for caesium is increased to > 100 by addition of $K_4\{Fe(CN)_6\}$ and nickel salts, which form unsoluble precipitates. An alternative to the precipitation of caesium is separation by ion exchange. The best results have been obtained with the anorganic ion exchangers bentonite, vermiculite and filtrolite. The decontamination factor for caesium in a simulated MLW solution containing 50 g $NaNO_3$/1 was about 100 for filtrolite.

The best results in the decontamination of MLW solutions have been found by simultaneous addition of the following reagents per litre of waste: 0.5 g Al^{3+}, 1.8 g SiO_2, 0.5 g Sn^{2+} and 0.3 g Ti^{3+} and adjusting the pH to 10. Under these conditions the following decontamination factors have been obtained:

$DF_{Ce} > 2000$	$DF_{Sr} \leqslant 300$
$DF_{Nb} > 3000$	$DF_{Ru} \leqslant 100$
$DF_{Zr} > 1000$	

The DF for Cs is $\leqslant 700$ by precipitation with hexacyanoferrate and $\leqslant 100$ by the ion exchanger filtrolite.

To facilitate the incorporation of the MLW precipitates into glass, only reagents have been selected that are used in HLW glass fabrication or are compatible with glass. By this procedure the total amount of HLW glass is only slightly increased.

5. TREATMENT OF SPENT SOLVENT

Several hundred cubic metres of organic waste solutions are produced annually in large reprocessing plants. These waste solutions contain mainly kerosene with 3 to 5 vol.%. TBP and show activity concentrations between 10 and 100 Ci/m^3. Furthermore, solutions with up to 30 vol.% TBP and an activity concentration of up to 1000 Ci/m^3 are discharged.

TABLE II. TREATMENT OF TBP-KEROSENE MIXTURE

TBP content (%)	Before treatment		After treatment: With 5% carbonate sol.		After treatment: With phosphoric acid		After treatment: With SiO_2 column	
	α (Ci/m^3)	β (Ci/m^3)	α (Ci/m^3)	β (Ci/m^3)	α (Ci/m^3)	β (Ci/m^3)	α (Ci/m^3)	β (Ci/m^3)
9	21×10^{-1}	325×10^{-1}	4.6×10^{-3}	1.5×10^{-1}	2.7×10^{-5}	4.4×10^{-5}	1.2×10^{-7}	2.3×10^{-6}
9.2	22×10^{-1}	135×10^{-1}	6.3×10^{-2}	11.5×10^{-1}	5.2×10^{-6}	7.2×10^{-4}	1.8×10^{-7}	3.3×10^{-6}
9.0	20×10^{-1}	470×10^{-1}	3.4×10^{-2}	3.1×10^{-1}	2.4×10^{-5}	1.7×10^{-4}	1.3×10^{-7}	3.5×10^{-6}
7.5	3×10^{-1}	44×10^{-1}	1.1×10^{-2}	1.4×10^{-1}	2.8×10^{-6}	1.1×10^{-4}	1.4×10^{-7}	1.7×10^{-6}

FIG.2. View of the plant for treatment of spent solvent.

The aim of the treatment of organic solvents is to separate and purify kerosene and TBP for recycle. Such a process was developed in laboratory-scale investigations [10] and demonstrated in a hot plant.

The TBP-kerosene mixture is first treated with a sodium carbonate solution. In a second step the separation of TBP and kerosene is performed by the addition of concentrated phosphoric acid. In this way a compound containing TBP and phosphoric acid in a molar ratio of 1:2 is formed which separates from the kerosene due to its high density. After separation of the phases, kerosene is further purified by adsorption on a SiO_2 column to an activity content that allows reuse in the reprocessing plant. The TBP-phosphoric acid compound is decomposed with water into dilute phosphoric acid, which contains the bulk of the radioactivity and can be discharged to the intermediate level waste.

More than 100 m^3 of organic waste solution from the reprocessing plant in Karlsruhe (WAK) were processed batchwise. The purified kerosene with a residual activity of 10^{-6} Ci/m^3 was recovered and reused in the plant. The separated TBP, which contained a small fraction of the original activity, was fixed in PVC granules and disposed of. Typical results obtained by this procedure are shown in Table II. A view of the plant is shown in Fig.2.

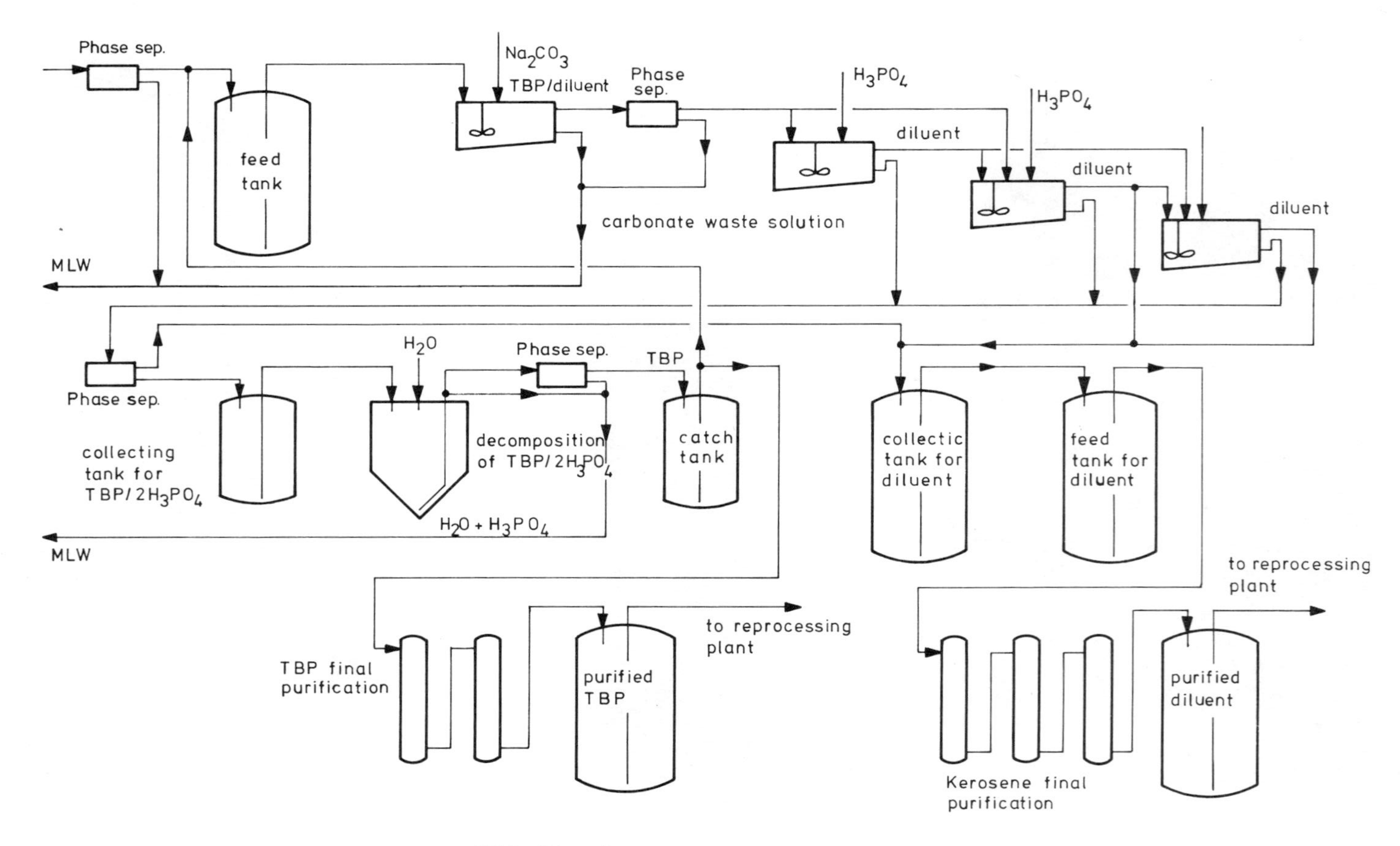

FIG.3. Scheme for a continuous solvent treatment process.

This process has already been developed for a continuous operation in a large reprocessing plant. The results gained with pot mixer-settlers look very promising.

In the first step the bulk of the α- and β-activity as well as the nitric acid are removed by washing with carbonate solution. The scrubbed solution is fed to a line of 3 pot mixer-settlers where the TBP is separated from the kerosene by the addition of phosphoric acid. The organic and aqueous phases are recycled through the mixing turbine of the mixer-settler to yield a high stage efficiency. The light kerosene phase is withdrawn over a weir, while the heavy phase ($TBP-H_3PO_4$) is discharged through a small airlift. The interface between the kerosene and the viscous adduct-emulsion can be sensed by a tuning-fork level indicator. The addition of the phosphoric acid is controlled by the density of the kerosene phase, which depends on the TBP content. The adduct-phosphoric-acid emulsions from the three pot mixer-settlers are fed to a separation vessel. The decomposition of the adduct can be done in similar pot mixer-settlers. This is achieved in countercurrent in two stages with water. Subsequently the TBP is scrubbed by sodium carbonate and water. After this treatment the TBP is ready for reuse.

A concept for technical realization in a large plant is shown in Fig.3.

REFERENCES

[1] BÄHR, W., HILD, W., KLUGER, W., KFK-2119 (1974).
[2] MEIER, G., BÄHR, W., KFK-2104 (1975).
[3] BACKOF, E., DIEPOLD, W., KFK-tr-450 (1975).
[4] KLUGER, W., NENTWICH, O., KFK-1037 (1969).
[5] HILD, W., et al., Nuclex 75 Basel (Oct. 1975) Colloq. D3, Paper 4.
[6] KRAUSE, H. (Ed.), KFK-2000 (1974).
[7] KRAUSE, H., RUDOLPH, G. (Eds), KFK 2126 (1975).
[8] KRAUSE, H. (Ed.), KFK-1830 (1973).
[9] KRAUSE, H., RUDOLPH, G. (Eds), KFK-2212 (1975).
[10] KRAUSE, H., RANDL, R., Management of Radioactive Wastes from Fuel Reprocessing (Proc. OECD/NEA Symp. Paris, 1972), OECD, Paris (1973) 199; KFK-1741 (1972).

DISCUSSION

J.B. LEWIS: Aluminium hydroxide precipitates are difficult to separate and to dewater. What technique do you intend to use to prepare the precipitates for the vitrification stage?

W. HILD: We have not yet moved on from the stage of laboratory-scale investigation in this particular area. As you will have noted, the precipitate we produce is formed from constituents that are at the same time glass-forming additives. On the other hand, we operate with evaporator concentrates, i.e. solutions with high salt concentrations and the resulting slurries are not so difficult to handle. We shall be studying the question of filtration during future investigations. On the basis of the laboratory experiments and previous experience gained from pilot-plant LLW treatment by chemical precipitation, we are confident that an appropriate solution will be found. The same goes for the dewatering. However, depending on the final step for vitrifying the HLW into glasses, the precipitates could even be dissolved in HNO_3 and stored with the liquid HLW, or mixed with it somewhere in the solidification process.

H. WITTE: You suggest separation of the medium- and low-level liquid waste streams in a reprocessing plant into a high-activity precipitate and a very low-activity filtrate by precipitation techniques. As justification you say that bitumen, as the proposed matrix for insolubilization, has only a limited radiation resistance. How much radioactivity can be incorporated into the bitumen then, if the assumed repository for final storage is a salt dome?

W. HILD: The treatment procedure under investigation is considered as one alternative for the pretreatment of MLW evaporator concentrates that is not dependent on the solidification procedure finally applied. The main advantage is that the MLW problem is eliminated by division into a LLW stream and a HLW stream, which you have anyway. Depending on the site selected for a large reprocessing plant, the procedure could involve chemical treatment followed by in situ solidification of the waste stream in cavities in a salt formation either after mixing the waste with a hydraulic binder such as cement, or by bituminization.

As I indicated, the threshold radioactivity for bitumen products is largely dependent on the final storage conditions. Thus, orderly disposal in a vented cavern would allow higher specific radioactivity than random disposal in an unvented cavern. Basically, the radioactivity limit is a question of the total absorbed dose from the decaying radionuclides, both long-lived and short-lived. On the basis of these considerations the radioactivity values lie within the range 0.1 to 1 Ci/l.

A. SCHNEIDER: I entirely agree with you regarding the advantages of eventually combining the radioactive components of medium-level liquid wastes with solidified high-level wastes. However, this task would be greatly facilitated by extreme care in reprocessing flow sheet development so as to avoid the introduction of potentially troublesome elements such as Na, Fe and Hg.

B. LOPEZ-PEREZ: With regard to the fires that you mentioned, have you installed a fire extinguishing system in the bituminization chamber? If so, does it operate automatically?

W. HILD: Yes, we have now installed such a system. It operates automatically and uses carbon dioxide as the fire-extinguishing medium.

B. LOPEZ-PEREZ: Did you not originally envisage the possibility of a fire?

W. HILD: We had carried out such a large number of thermal degradation and other fire tests that we were fairly confident there could not be an outbreak of fire in the filling chamber. So we did not install a system at the outset. The incidents that occurred, though troublesome, have been very helpful in terms of experience.

H.W. LAHR: I want to add a brief comment regarding waste disposal figures for the planned reprocessing plant in Germany. The new flow sheets for the 1400 t/a plant show that the amount treated will actually be 1000 m^3/a of concentrated solutions, with a salt content of roughly 600 t, since there is no excess of acid and no neutralization.

INCORPORATION OF RADIOACTIVE WASTES IN POLYMER IMPREGNATED CEMENT

A. DONATO
CNEN-CSN,
Casaccia,
Rome,
Italy

Abstract

INCORPORATION OF RADIOACTIVE WASTES IN POLYMER IMPREGNATED CEMENT.

Polymer-impregnated cement (PIC) has been examined from a technical viewpoint as a possible medium for the immobilization of low and intermediate level radioactive wastes. According to the process under investigation, the radioactive wastes are at first incorporated in cement or concrete, obtaining a solid product which, afterwards, is dehydrated and impregnated with an organic monomer (styrene or methylmethacrylate). Finally, the monomer contained inside the pores of the cement is thermally polymerized. The compressive strength of PIC containing radioactive wastes is considerably higher than plain concrete, in the best cases about 200% better. The flammability, even when nitrates are incorporated, is very low and the radiation resistance satisfactory at least up to 1×10^8 rads. The leachability of PIC is significantly lower than that of concrete and cement. The incremental leach rates R for ^{137}Cs, ^{58}Co and ^{85}Sr are 10, 8 and 312 times lower, respectively. A demonstration pilot plant for the incorporation of radioactive chemical sludges in PIC has been designed and is now under construction.

1. INTRODUCTION

Polymer impregnated cement or concrete (PIC) has been under investigation at the CSN Casaccia of CNEN since the end of 1974 as a possible medium for the immobilization of low and intermediate level radioactive wastes [1].

This new type of material has been developed in the USA [2], Italy and elsewhere with the aim of producing a new building material which, in view of its characteristics of high mechanical strength, extremely low porosity, resistance to chemical attack and weathering [2, 3], would be particularly suitable for use in conditions of very high corrosiveness. The same properties have been considered attractive for the incorporation of radioactive wastes, and therefore an experimental programme for estimating the advantages obtainable by using PIC for this purpose was started.

2. PREPARATION OF PIC INCORPORATING RADIOACTIVE WASTES

PIC-incorporated wastes are produced in the following steps:

(a) Waste incorporation in cement or concrete, and curing.

(b) Dehydration of consolidated cement or concrete at 165°C under vacuum. In this step the pore water is removed and a solid with a high percentage of empty pores is obtained. The amount of water removed depends on the waste/cement ratio and on the nature of incorporating material. In the case of cement it can be as high as 20 to 30% by weight, while for concrete it can vary up to 6 to 12% by weight.

(c) Impregnation of dehydrated cement with a catalysed organic monomer. This can be accomplished in several ways: for small samples soaking in the monomer for about two hours at normal pressure is sufficient.

(d) Heating of impregnated material for polymerization. Styrene catalysed with 2.5 wt% of diazoisobutyronitril needs a temperature of 85°C for about 40 hours, while methylmethacrylate polymerizes almost completely into the cement pores at 75°C for 19 hours. The polymerization of the monomer contained in the pores can be induced also by gamma irradiation.

Following this general procedure, samples of PIC incorporating two types of simulated wastes were prepared:

(A) Sludges generated in the chemical treatment of low level radioactive wastes by means of the calcium phosphate and ferric hydroxide coprecipitation process employed at the CSN Casaccia.

(B) Evaporation concentrates with high nitrate content. A solution containing 400 g/l of nitrates was considered representative.

The incorporation of simulated waste A was carried out by mixing Portland 425 cement, sand and waste according to the following ratios (by weight):

sand + gravel/cement = 3 sludge/cement = 0.8

Simulated waste B was incorporated in pozzolanic cement, without addition of inert material, using waste/cement weight ratio of 0.55 or lower. At higher ratios, in fact, the cement cannot harden satisfactorily and, moreover, solid nitrates separate. Cement and concrete cylindrical samples, 5 cm diameter and 10 cm height, were prepared in this manner and cured for 7 days before being treated in a small laboratory plant for PIC production.

The waste/cement ratios adopted are considerably higher than those commonly employed for the incorporation of wastes in cement or concrete at nuclear plants. Ratios of 0.4–0.5 are generally used there in order to obtain a good mechanical strength, a limited porosity [4, 5] and, therefore, for the lower surface area, also a better resistance to leaching. In the case of PIC, on the other hand, where the mechanical strength is given merely by the type and content of polymer, one can obtain at higher ratios a higher porosity and indeed a better permeability to monomers, at the same time increasing considerably the amount of incorporated waste. Moreover, the addition of inert material (sand and gravel) can be considered unnecessary. The advantages of using higher waste/cement ratios can be summarized in these figures: at a 0.45 ratio the Portland concrete can incorporate 10 wt% of sludges and Portland cement 32 wt%. At 0.75 ratio, on the other hand, the Portland cement (and therefore the PIC) can accommodate 48% (by weight) of the same waste.

The weight loss of cement after dehydration and therefore also the amount of monomer absorbed obviously increases at increasing waste/cement ratios. So the Portland cement prepared with a sludge/cement ratio of 0.45 loses 21% by weight and absorbs after impregnation 16% by weight of styrene, while at a 0.75 ratio it loses 33% and absorbs 26%.

It is important to note that the monomer does not completely replace the water removed from cement pores during the dehydration step: for example, the styrene absorbed in Portland cement containing chemical sludges and impregnated at normal pressure represents 77% by weight and 85% by volume of the evacuated water. The difference can be explained by the presence of very small pores, which cannot be filled in these impregnation conditions but could be at higher pressures.

Monomer polymerization can be carried out; besides heating, also by means of radiation: samples of PIC incorporating sludges and impregnated with styrene and methylmethacrylate have been prepared by irradiating them with a cobalt source at 23 and 7 Mrad, respectively. Nevertheless this technique does not appear promising to us for industrial application of the PIC for radioactive waste incorporation, unless the radiations emitted by the same nuclides contained in the wastes and trapped in the solid structure will be used for this purpose. Unfortunately, the doses obtainable in this way, unless high level radioactive wastes could be used, are not sufficient.

3. PROPERTIES OF PIC INCORPORATING RADIOACTIVE WASTES

The evaluation of the properties of PIC incorporating radioactive wastes has been limited to the following aspects: mechanical strength, flammability and heat resistance, radiation stability, and leachability.

3.1. Mechanical strength

In general the mechanical properties of polymer impregnated concrete are far better than those of plain concrete, as shown by other workers [2, 3]. This is true also for cement and concrete incorporating radioactive wastes, at least as far as compressive strength is concerned. Samples of them have been tested by means of a manual press. The results of these tests are shown in Table I as variations in respect to the properties of samples of plain Portland 425 concrete cured for 28 days. The compressive strength of PIC incorporating simulated radioactive wastes is considerably higher than plain concrete. In the best cases it is about 200% better.

3.2. Flammability and heat resistance

A knowledge of the flammability characteristics of the materials employed for radioactive waste incorporation is very important from a safety viewpoint. This is true mainly when the wastes to be incorporated are evaporation concentrates, with a high nitrate content. Everybody knows the problems arising when these wastes are incorporated into organic matrices, such as bitumen, and how unsatisfactory this conditioning can be considered in relation to the final product flammability [6, 7].

To ascertain the flammability of PIC, two types of tests have been employed:

(1) Direct heating of block cylindrical samples (volume 200 cm^3) by means of a Bunsen flame for five minutes.

(2) Flammability determination according to the ASTM D635–74 method [8]. The test specimens are in the form of strips 125 mm long, 12.5 mm wide and 8 mm ± 1 mm thick. They are submitted to the flame of a particular Bunsen burner for 30 s. The duration of the residual flame and the extension of burned strip are measured.

The materials tested for block flammability are of the type reported as samples under 1, 2, 3 and 4 in Table I. Cement impregnated with methylmethacrylate, exposed to the flame for five minutes, perfectly retains its initial form without breaking or cracking. After the test it appears blackened on the surface because of the presence of residual products from a merely superficial combustion. When the Bunsen flame is removed, it maintains a very light flame, which is self-extinguishing after one or two minutes. Cement impregnated with styrene shows a still better behaviour because it does not maintain any residual flame once the Bunsen burner is removed. For all samples no signs of violent reactions due to the presence of nitrates has ever been noticed. In Fig. 1 some samples of PIC incorporating radioactive wastes are shown. The black ones have been tested for block flammability.

The flammability and the burning rate of PIC are increased by the presence of nitrates, as the results obtained in the ASTM D635–74 method demonstrate. In fact, the ATB (Average Time of Burning) and the AEB (Average Extent of Burning) of pozzolanic cement impregnated with styrene (18% content) are 1 min 40 s and 10 mm, respectively, while the presence of nitrates raises these figures to 3 min 45 s and 20 mm, respectively. Nevertheless, these changes are not appreciable in block flammability tests, perhaps because of the low heat conductivity of the materials.

3.3. Radiation stability

The radiation stability of PIC incorporating simulated radioactive wastes has been evaluated by irradiating the samples to be tested by means of a cobalt source, at a dose rate of 0.27 Mrad/h.

TABLE I. INCREASE IN COMPRESSIVE STRENGTH OF PIC INCORPORATING SIMULATED RADIOACTIVE WASTES IN COMPARISON WITH PLAIN CONCRETE

Sample	Type[a]	Waste[b]	Waste / cement (wt/wt)	Waste (wt%)	Dehydration (%)	Monomer[c]	Polymerization	Polymer content (%)	Density (g/cm^3)	Compressive strength increase (%)
1	a	B	0.45	31.0	22.2	MMA	Heat	19.7	1.59	202
2	a	B	0.45	31.0	22.1	ST	Heat	18.1	1.61	100
3	a	B	0.55	35.5	23.4	MMA	Heat	19.8	1.45	174
4	a	B	0.55	35.5	23.3	ST	Heat	18.9	1.48	100
5	b	A	0.80	16.7	6.2	ST	Radiation	5.2	1.91	174
6	b	A	0.80	16.7	7.3	MMA	Radiation	5.5	2.01	119
7	b	A	0.80	16.7	9.9	ST	Heat	9.0	2.09	174
8	b	A	0.80	16.7	11.6	MMA	Heat	9.7	2.04	173

[a] a: pozzolanic cement, b: Portland concrete.
[b] A: chemical sludges, B: evaporation concentrates.
[c] MMA: methylmethacrylate, ST: styrene.

FIG.1. Samples of PIC.

The samples were placed in a Pyrex cell in order to trap and subsequently analyse the gases that might be generated by radiolysis. The post-irradiation examination has been carried out to consider the following factors:

(1) Weight variation,

(2) Quantitative and qualitative analysis of radiolysis gases,

(3) Macroscopic appreciation of sample integrity.

Two types of samples were examined: a Portland concrete sample containing chemical sludges impregnated with styrene (polymer content = 9.1 wt%), and a pozzolanic cement sample containing 35.5 wt% of evaporation concentrates impregnated with styrene (18.3 wt%).

Both samples were submitted to a total dose of 10^8 rads, without appreciable weight variation and gas generation. At the end of these tests the samples were unchanged from a macroscopic point of view. The dose absorbed by the samples would be equivalent to the infinite dose delivered to the PIC by an aged fission product mixture with a specific activity of about 5 Ci/l. This level of activity is far higher than those one could expect to be commonly incorporated in PIC.

3.4. Leachability

Among the properties of materials to be used for the solidification of liquid radioactive wastes, the leaching of radioactivity is undoubtedly the most important one. A knowledge of it can permit first of all the comparison of methods of waste insolubilization on an objective basis, and secondly estimates of the hazards arising when solidified radioactive wastes come in contact with water, unde given conditions, during either storage or shipping and even after disposal. Unfortunately, the methods developed in the past for the determination of leachability are quite different from each other, and in practice every laboratory working in this field has developed its own [9]. The only attempt to introduce a standard leaching test was made by the IAEA in 1970 [10], but it failed almost completely. Nevertheless, the IAEA method appears to be the best one available today because it is the only one with the benefit of a solid theoretical basis in the field of solid state diffusion [11, 12].

The leachability of PIC has been determined with radioactive tracers, using the following two methods:

(1) A rapid method, used in the past at our laboratory for assessing the leachability of ESTER glasses [13]. Cylindrical samples (diameter and height 50 mm) are leached by 500 cm^3 of distilled water at 95°C for 24 h. The fraction of initial activity removed by water is determined.

TABLE II. RAPID LEACHING TEST RESULTS OF PIC SAMPLES IN COMPARISON WITH NON-IMPREGNATED CEMENT

Sample	Waste[a]	Waste (wt%)	Dehydration (%)	Styrene (%)	Tracer	Sample activity (mCi)	Activity leached (mCi)	Activity leached (%)
Portland concrete	A	16.7	–	–	^{137}Cs	0.416	0.0550	13.2
PIC	A	16.7	12.8	9.1	^{137}Cs	0.418	0.0048	1.1
Pozzolanic cement	B	35.5	–	–	^{58}Co	0.815	0.0500	6.1
PIC	B	35.5	15	14.1	^{58}Co	0.815	0.0050	0.61

[a] A: chemical sludges; B: evaporation concentrates.

(2) A long-term leachability determination according to the IAEA method, with the exception that, for practical reasons, the whole surface of the samples (diameter and height 50 mm) was leached.

The results obtained in the rapid determination of PIC leachability are shown in Table II, together with those obtained for non-impregnated concrete and cement. The characteristics of the samples tested are also reported. The PIC leachability is significantly lower than that for concrete and cement. It decreases by a factor of 12 for ^{137}Cs, while in the case of ^{58}Co there is a 10-fold reduction.

From the viewpoint of long-term leaching, three pairs of samples have been examined; the results are shown in Table III. Each pair is formed by a sample of cement incorporating the simulated radioactive waste and a similar sample impregnated with styrene. The leaching curves for ^{137}Cs, ^{58}Co and ^{85}Sr are shown in Fig.2, where the incremental leach rate R, is plotted against a time function $\{t_n - \frac{1}{2}(t_n - t_{n-1})\}$. Here:

$$R = \frac{a/A_0}{(F/V)t_n}$$

where a is the activity in the distilled water of the isotope being considered, F the surface area (cm^2) and V the volume (cm^3) of the sample, A_0 the total activity of the isotope initially present in the sample, and t_n the time of the nth leaching water renewal. Leaching curves are also shown in Fig. 3, where the cumulative fraction of radioactivity leached from the specimens is plotted against the total time of leaching ($(\Sigma a_n/A_0)/(F/V)$ versus Σt_n). From the examination of these curves one can say that PIC can be considered as a very effective material for radioactive waste incorporation in comparison with cement and concrete, as far as its leachability is concerned. The ^{137}Cs incremental leaching rate is about 10 times lower than for concrete (1.2×10^{-5} cm/d versus 1.2×10^{-4} cm/d, both after 100 days). In the same manner the incremental leach rate of ^{58}Co decreases about 8 times (from 1.2×10^{-4} cm/d to 1.5×10^{-5} cm/d after 88 days), while in the case of ^{85}Sr the results are still better, as the leaching rate decreases from 1×10^{-4} cm/d for pozzolanic cement to 3.2×10^{-7} cm/d for PIC after 88 days, with a 312-fold decrease.

TABLE III. CHARACTERISTICS OF SAMPLES SUBMITTED TO LONG-TERM LEACHABILITY DETERMINATIONS

Sample	Waste[a]	Waste (wt%)	Dehydration (%)	Styrene (%)	Tracer	Sample activity (mCi)	R(cm/d)[b]	$\frac{\Sigma a_n}{A_0} / \frac{F}{V}$ [b]
Portland concrete	A	16.7	–	–	^{137}Cs	0.416	1.2×10^{-4}	0.052
PIC	A	16.7	12.8	9.7	^{137}Cs	0.418	1.2×10^{-5}	0.006
Pozzolanic cement	B	35.5	–	–	^{58}Co	0.815	1.2×10^{-4}	0.173
PIC	B	35.5	24.7	21.6	^{58}Co	0.815	1.5×10^{-5}	0.008
Pozzolanic cement	B	35.5	–	–	^{85}Sr	0.733	1.0×10^{-4}	0.207
PIC	B	35.5	23.4	20.6	^{85}Sr	0.733	3.2×10^{-7}	0.004

[a] A: chemical sludges; B: evaporation concentrates.
[b] ^{137}Cs after 100 days; ^{58}Co and ^{85}Sr after 88 days.

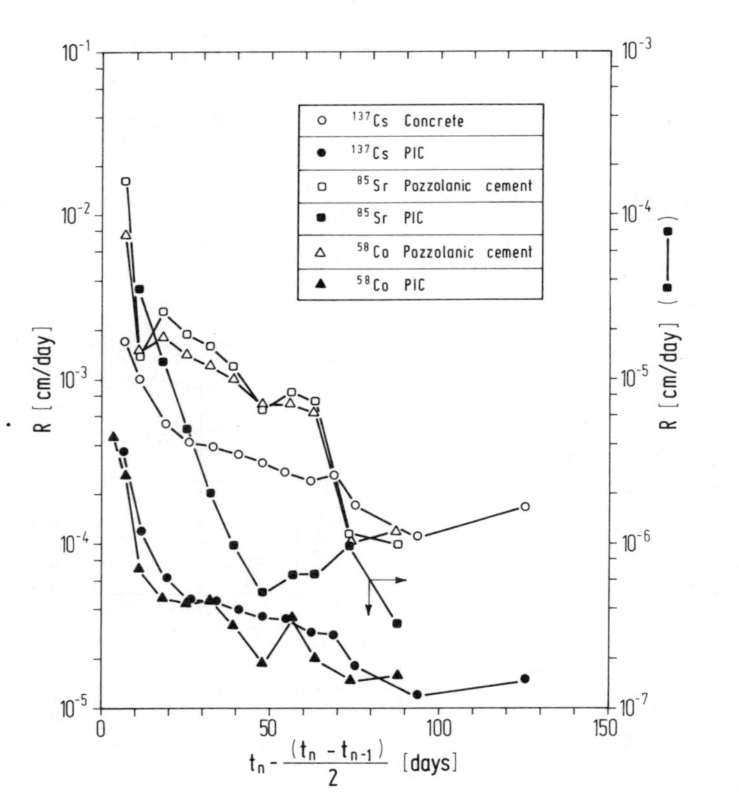

FIG.2. Long-term leachability of PIC in comparison with cement. Incremental leach rates of ^{137}Cs, ^{85}Sr and ^{58}Co.

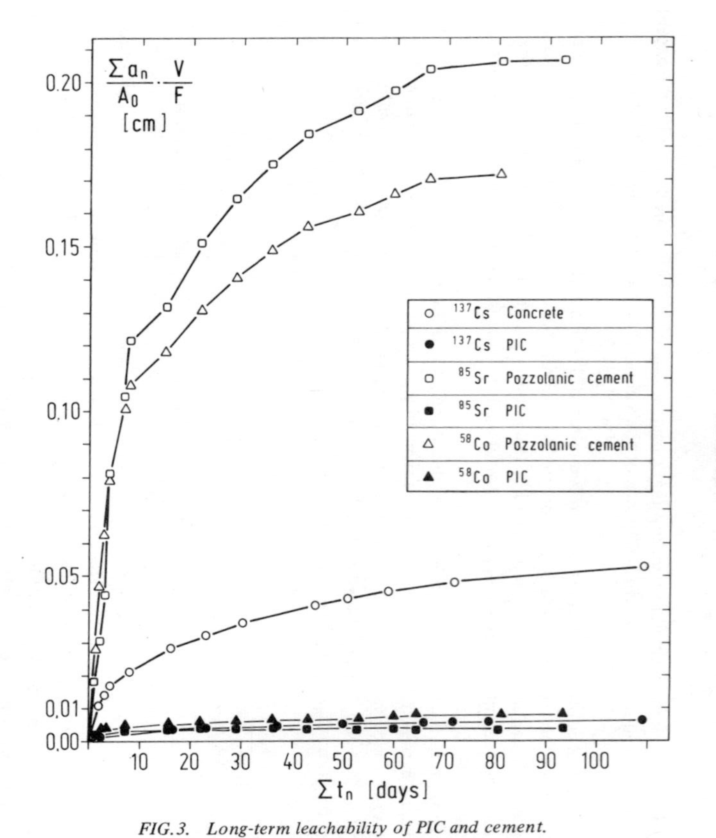

FIG.3. Long-term leachability of PIC and cement. Cumulative fractions leached of ^{137}Cs, ^{85}Sr and ^{58}Co.

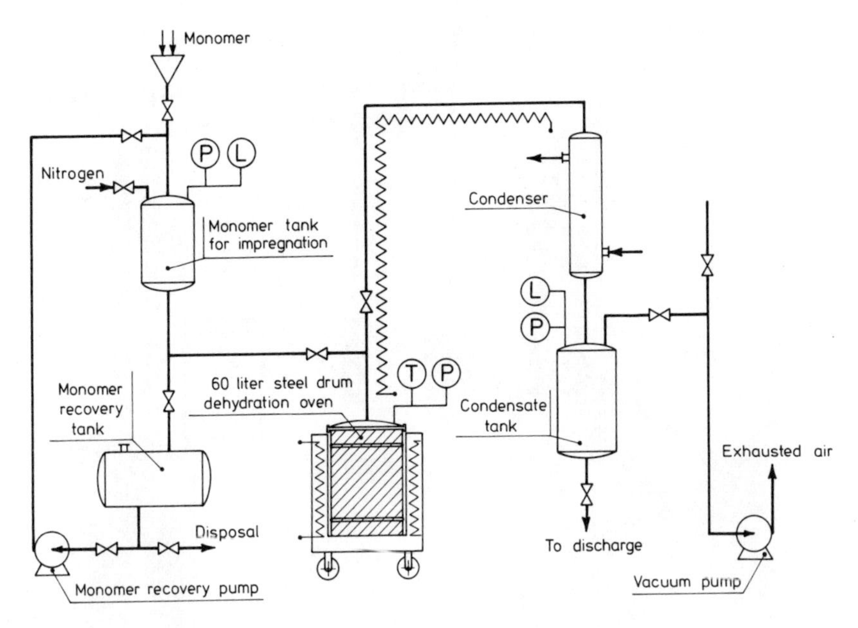

FIG.4. Pilot plant scheme for incorporation of chemical sludges in PIC.

Similar conclusions can be drawn from the examination of cumulative fraction curves (Fig.3). So the cumulative fraction of ^{137}Cs leached after 100 days is 5.2% X cm, while for PIC it is only 0.6% X cm. Analogously for ^{58}Co and ^{85}Sr the cumulative fractions after 88 days decrease from 17.3 and 20.7% X cm to only 0.77 and 0.40% X cm, respectively.

4. DESIGN AND BUILDING OF A DEMONSTRATION PILOT PLANT

The good qualities shown by PIC incorporating simulated radioactive wastes, together with the extreme simplicity of the process, justify in our opinion the design and building of a demonstration pilot plant. Thus a pilot plant capable of satisfying the needs of a nuclear centre such as the CSN Casaccia was conceived. The scheme of this plant is shown in Fig.4. It appears very simple, just as the process for the production of small samples. The plant is composed of a 6 kW oven for the dehydration of cement incorporating sludges contained in 60 l drums, and for polymerization after impregnation; a condenser and a tank for distillate collection; and two tanks and a pump for monomer feeding and recovery. A vacuum pump will keep the plant under a light vacuum to facilitate dehydration. The drums will be closed after impregnation, and the monomer polymerized. Each drum will incorporate about 45% (by weight) of chemical sludges. The choice of a good waste/cement ratio, the particular form of the consolidated cement and the operative conditions selected will, it is hoped, make the dehydration and the impregnation of the cement drums sufficiently fast for a good production rate. The pilot plant, which is at present being assembled, is shown in Fig.5.

FIG.5. The demonstration pilot plant for incorporation of sludges in PIC (during assembly).

5. CONCLUSIONS

The characteristics of PIC incorporating radioactive wastes appear very promising. The leachability is far better than plain cement and comparable with the best materials employed in this field. Furthermore, the mechanical strength of PIC is the best material at present obtainable, with the possible exception of polyester resins. But the most important aspects involved in the PIC application can be identified chiefly in the simplicity of the process and in the low initial investment cost. The experience that will be acquired with the demonstration pilot plant will make it possible to evaluate other important aspects, such as operational cost and safety on a more representative scale.

REFERENCES

[1] DONATO, A., Inglobamento dei rifiuti radioattivi in calcestruzzo impregnato con polimeri. Studio preliminare, CNEN – RT/PROT 75(9).

[2] STEINBERG, M., et al., Concrete polymer materials. First Topical Report, BNL 50134 (1968), and subsequent reports.

[3] RIO, A., CERNIA, E.M., Polyblends of cement concrete and organic polymers, J. Polym. Sci., Part D **9** (1974) 127.

[4] CHEKHOVSKY, J.V., et al. "Investigation into cement concrete porous structure and its relationship with technological factors and physical properties of concrete", Int. Symp. RILEM-IUPAC on Pore Structure and Properties of Materials, Prague 1973 (Proc. B51).

[5] YOSHIRO KOH, FIJI KAMADA, "The influence of pore structure of concrete made with absorptive aggregates on the frost durability of concrete", ibid (Proc. F15)

[6] IAEA, Bituminization of Radioactive Wastes, Technical Reports Series No. 116, IAEA, Vienna (1962) 64 pp.

[7] RODIER, J., et al. La solidification des déchets radioactifs par le bitume, CEA Rep. CEA-R-3982 (1970).

[8] ASTM, "Standard method of test for flammability of self supporting plastics", D635–74.

[9] MENDEL, J.E., A Review of Leaching Test Methods and the Leachability of Various Solid Media containing Radioactive Wastes, BNWL 1765 (1973).

[10] HESPE, E.D., "Leach testing of immobilized radioactive waste solids", At. Energy Rev. **9** (1971) 195.

[11] BELL, M.J., An Analysis of the Diffusion of Radioactivity from Encapsulated Wastes, ORNL-TM-3232 (1971).

[12] GODBEE, H.W., JOY, D.S., Assessment of the Loss of Radioactive Isotopes from Waste Solids to the Environment. Part 1: Background and Theory, ORNL-TM-4333 (1974).

[13] DONATO, A., BOCOLA, W., The ESTER program: high level radioactive waste solidification. Borosilicate glass experimental evaluation and bench-scale vitrification with simulated fission product solutions, Energ. Nucl. **19** 7 (1972) 459.

DISCUSSION

J. VAN GEEL: Have you studied the possibility of directly mixing the waste with the polymer-impregnated cement, thereby permitting some water in the mixture prior to polymerization?

A. DONATO: The cement is dehydrated beforehand and the polymerization is intended to fill the cement pores with polymer, but I do not think that the presence of water would necessarily inhibit the polymerization. I think experiments along those lines have been carried out in the United States of America.

H.W. LAHR: What are the times for dehydration, impregnation and polymerization? And what is the maximum size of the mixed concrete pieces for the three steps?

A. DONATO: The time required for dehydration and impregnation depends on many parameters, such as the waste/cement ratio, the type of waste, the operational pressure, and so forth. In the case of 60 litre drums, the figures we extrapolate are 12 to 16 hours for dehydration and 6 to 12 hours for impregnation.

As far as size is concerned, we feel at the moment that a 60 litre drum is the dimensionally optimum size. Polymerization is carried out at 85°C for about 28 to 40 hours, depending on the dimensions of the sample.

STUDIES ON THE INCORPORATION OF SPENT ION EXCHANGE RESINS FROM NUCLEAR POWER PLANTS INTO BITUMEN AND CEMENT

Moj BONNEVIE-SVENDSEN, K. TALLBERG
Institutt for Atomenergi,
Kjeller,
Norway

P. AITTOLA
Technical Research Centre of Finland,
Finland

H. TOLLBÄCK
AB Atomenergi,
Studsvik,
Sweden

Abstract

STUDIES ON THE INCORPORATION OF SPENT ION EXCHANGE RESINS FROM NUCLEAR POWER PLANTS INTO BITUMEN AND CEMENT.

The joint Nordic incorporation experiments should provide technical data needed for the assessment of solidification techniques for wastes from nuclear reactors in the Nordic countries. Spent ion exchange resins are a main fraction of such wastes, and more knowledge about their incorporation is wanted. The effects of simulated and real ion exchange wastes on the quality of bitumen and cement incorporation products were studied. Blown and distilled bitumen and three Portland cement qualities were used. Product characterizations were based on properties relevant for safe waste management, storage, transport and disposal. The applicability and relevance of established and suggested tests is discussed.

Up to 40–60% dry resin could be incorporated into bitumen without impairing product qualities. Products with higher resin contents were found to swell in contact with water. The products had a high leach resistance. Their form stability was improved by incorporated resins. Product qualities appeared to be less affected by physico-chemical variables than by mechanical process parameters. Pure resin-cement products tend to decompose in water. Product qualities were strongly affected by a variety of physico-chemical process parameters, and integer products were only obtained within narrow tolerance limits. Caesium was rapidly leached out. To attain integer products and improved leach resistance within technically acceptable tolerance limits it was necessary to utilize stabilizing and caesium-retaining additives such as Silix and vermiculite. Under the present conditions the water content of the resins limited the amounts that could be incroporated in 40–50 wt% or about 70 vol.% water-saturated (containing 20–40% dry) resin.

1. INTRODUCTION

The present incorporation studies are part of an integrated Nordic research programme (NIPA) on the management of low and medium level radioactive wastes from nuclear power reactors, emphasizing the coupling between the different stages of the waste cycle. Together with the basic review by Peltonen et al. [1], experiments – performed by Finnish, Swedish and Norwegian research workers at the Norwegian Institutt for Atomenergi – are expected to furnish the technical information needed for safety assessments, choice and optimizations of solidification techniques in the Nordic countries. Here both bitumen and cement incorporation is practised (Table I). Though bitumen is gaining, the planning of new plants still involves parallel evaluations of both techniques.

TABLE I. SOLIDIFICATION OF RADIOACTIVE WASTE IN THE NORDIC COUNTRIES

Country	Institute	Power reactor	Solidification technique		Start-up	References
			Bitumen	Cement		
Denmark	AEK-Risø	–	X		1970	[2]
Finland	–	Loviisa	?	?	1979	[1]
		Olkiluoto	X		1978/79	[1]
Norway	IFA-Kjeller	–		X	1962	[3]
Sweden	AE-Studsvik			X	1960	
		Oscarshamn		X	1972	[4]
		Ringhals		X	1975	[4]
		Barsebäck	X		1974	[4]
		Forsmark	X		1978	[4]

Spent ion exchange resins constitute a major fraction of radioactive wastes from nuclear power reactors. It appears desirable to develop special procedures for the separate handling of these resins without admixture of other waste categories. This renders relatively well defined systems and favours process optimizations. Where incorporation of these wastes is foreseen, the feasibility for ion exchange resins will affect the choice of incorporation technique. Though there is an abundant literature about the incorporation of various more or less specified waste concentrates, little has been published about the adaptation of these techniques to ion exchange resins.

As emphasized by Peltonen [1], assessments of solidification techniques should be based on criteria for safe waste management, and thus require comparable data for product qualities relevant for safe processing, interim and long-term storage, transport and final disposal. Another important factor is how far reasonably homogeneous product qualities can be maintained. In particular, cement products are easily impaired by small irregularities in process conditions. Where cement techniques are considered acceptable tolerance ranges must be ascertained.

The experimental work initiated to provide such required information concentrates upon the following closely coupled items

(a) Evaluations and development of standardizable tests for properties relevant to safe storage, transport and final disposal. Analyses and cross-checks of representative laboratory and plant samples to provide consistent comparable data for safety and process assessments;

(b) Studies on the impact of physico-chemical process variables, identification of critical parameters, trends and tolerance ranges. Investigations of means to improve product qualities and increase tolerance ranges for critical parameters.

The paper describes the establishment of simple test routines, screening experiments and more systematic studies on simulated and on a few real ion exchange wastes. Critical parameters and trends have been identified, means to improve cement product qualities are outlined. The results furnish a basis for further laboratory experiments with spent radioactive resins from nuclear power reactors and for cross-checks with plant samples.

2. EXPERIMENTAL CONDITIONS

Within the limits imposed on laboratory-scale experiments, conditions were adapted to those prevailing at existing and planned waste facilities in the Nordic countries. Some arbitrary

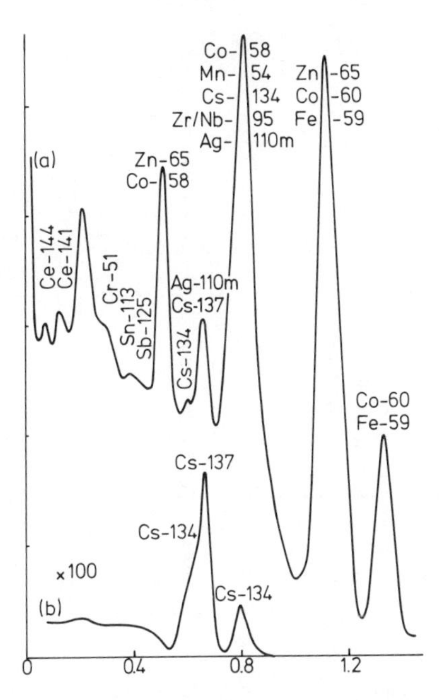

FIG.1. Gamma-spectra: (a) Spent resin, Oscarshamn BWR; (b) Leach-solution from its cement product (Osc-0-1).

simplifications as well as extended ranges have been introduced to enable systematic studies on the effects of physico-chemical process variables and to allow for realistic variations in other stages of the waste cycle.

2.1. Simulated (and real) ion exchange wastes

Types and state of the resins should be representative for those used or foreseen at nuclear power reactors in the Nordic countries. Strong basic, strong acid powdered (representing those from BWR condensate purification) and granular resins (as from PWR and from BWR primary circuit) were assayed.

Ratios (equivalents) of cation to anion exchanger (3 : 2 and 1 : 1 for granular, 2 : 1, 3 : 1, 4 : 1 for powdered) resins were adapted to specifications from the reactor industry, but the two types were also studied separately.

New resins, Dowex 50 W-21 K, 20-50 mesh and Microionex, 200-500 mesh, conditioned and equilibrated with water to simulate real water-saturated ion exchange wastes, were used in most experiments. To study varying degrees of saturation and effects of absorbed ions, resins in H^+ - H^- as well as Na^+, Ca^{2+}, Cl^-, SiO_3^{2-} form were assayed. Finally, a granular resin from a conventional power plant (Borås) and a slightly radioactive (Fig. 1) powdered resin from the condensate purification of the Swedish BWR, Oscarshamn, were incorporated.

2.2. Incorporation media

Three bitumen qualities — blown bitumen R 85/40 from Norwegian Shell A/S and distilled bitumen BIT 45 and BIT 15 from the Finnish OY Neste — and three qualities of Portland cement —

TABLE II. TESTS USED FOR QUALITY CONTROL OF RESIN-BITUMEN (B) AND RESIN-CEMENT (C) PRODUCTS. RELEVANCE FOR STORAGE (S), TRANSPORT (T), DISPOSAL (D), PROCESS STEERING (P).

Property	Relevance				Test	Matrix	Requirement	Remarks
	S	T	D	P				
1. Water resistance	+	+	++	+	Immersion in distilled water	B C	Not swell Not decompose	Screening test
2. Leach resistance	+	+	++	+	Leaching of Cs, Co in distilled water	B C		IAEA standard
3. Mechanical strength	+	++	+	+	Compressive strength, tensile strength	C	> 50 kgf/cm^2	Cement standard
4. Fall resistance	+	++			Fall from 9–14 m height	C (B)	Not break 9 m	IAEA transport accident
5. Form stability (viscosity)	++	+	+	+	Ring-ball, penetration, cylinder bending, hole migration	B		Asphalt standard suggested new tests
6. Heat resistance	+	++			Controlled heating	C	800°C $\frac{1}{2}$ h	IAEA transport accident
7. Frost resistance	+		+		Storage in freezer, cycling −40–20°C	B C		
8. Structure homogeneity				++	Microscopy, X-ray diffraction, auto-radiography, chemical analysis	C B		
9. Long term stability against radiation, chemical and bacterial attack	+	++			Structure changes, radiolytical gas evolution			

ordinary Portland cement, type 300 (NS 3050) (OC), rapid hardening (RHC) and low heat (LHC) Portland cement – were studied.

2.3. Additives

Silix GP, a Ca-soap forming commerical cement additive based on fatty acids (1.5% of cement weight), Barra 62 LV, a lignosulphate-based commercial cement additive, and vermiculite (ball-milled – 2.5, 5 and 10% of cement weight) were examined for their ability to improve the quality of resin-cement products.

2.4. Procedures

2.4.1. Bitumen incorporation

A laboratory mixing and kneading machine from Werner & Pfleiderer for work at controlled temperature (up to 180°C) and atmosphere with a working capacity of about 1.5 l was used for the experiments.

The water-saturated resins are mixed with bitumen and dewatered at reduced pressure and slowly increasing temperatures (50-110°C). After evaporation of water the temperature is raised to working temperature – 140, 160, 180°C, respectively – and kept there for $\frac{1}{2}$ hour. Finally the mix is fed by a screw drive to the test beakers and moulds and allowed to harden for at least 48 hours.

2.4.2. Cement incorporation

A standard mixing machine, test outfit and procedures specified for cement laboratories [5] have been adapted. The ratios of water-saturated resins and cement are adjusted to give a mix of optimal consistency or a specified water content. Only for special investigations (5.1) has excess water been added. The mix is thoroughly stirred to ascertain that the exchange of water between resin and cement is completed and is finally poured into the test moulds and allowed to harden in a water-saturated atmosphere for at least 7, 28, 70 days according to specifications for RHC, OC and LHC, respectively.

3. PRODUCT CHARACTERIZATIONS

Peltonen et al. [1] emphasize that adequate product characterizations must be based on properties relevant to safe storage, transport and final disposal, and that suitable standard tests for such properties are required. It is an aim of the Nordic solidification programme to establish simple test routines for such product characterizations. The coupling with studies of process parameters renders a broad spectrum of product qualities, well suited for checking the relevance and applicability of established and suggested tests.

Quantitative criteria and even the relative impact of various product qualities cannot be established without specifications about radioactivities, containment, mode and place of storage and disposal. Thus product characterizations had to be based on unweighed tests for integrity, leach resistance, mechanical strength and temperature resistance.

To rationalize the test programme we start with simple screening tests to verify that the products meet with some minimal requirements, i.e. are self-supporting, maintain their form and some mechanical strength in ambient atmosphere and water. Parallel with the establishment of representative products the test programme (Table II) is gradually extended, but will be restricted to a practicable minimum. A few structure tests have been included as a guidance for process optimalizations.

3.1. Screening tests for water resistance

Bitumen and cement product samples are inspected for changes during storage and are then immersed in water for 48 hours and 7 days, respectively, and classified according to their water-resistance. As an extension of this test the uptake of water is determined.

3.2. Leach tests

To control the immobilization of incorporated radioactivities the leaching of ^{137}Cs and ^{60}Co by distilled water is measured. The suggested IAEA standard procedure worked out by Hespe [6] is followed. Results are presented as accumulated leach rates

$$R_n = \frac{\Sigma a_n \cdot V}{A_0 \cdot F} \text{ cm}$$

plotted against time (d), or as average leach coefficients

$$L = \frac{\pi(\Sigma a_n)^2 \cdot V^2}{4 A_0^2 \cdot t \cdot F^2} \text{ cm}^2 \cdot \text{d}^{-1}$$

where V = volume (80 cm^3)
F = surface (20 cm^2, i.e. $V/F = 4$)
A_0 = original activity (~10 μCi ^{137}Cs, 20 μCi ^{60}Co for cement samples, about 5 times higher for bitumen) of the product sample
Σa_n = sum of leached activities.

Though the procedure is specified in details the general validity of the results and even their relevance for comparing cement with bitumen products can be questioned.

Effects of surface properties are illustrated by parallel tests on bitumen samples with cut (marked K) and original surfaces (Fig. 2).

Leach coefficients for different tracer elements may differ by several decades. They were in all cases highest for caesium. Even the chemical state and composition of the tracers have some influence on the results.

Tests with ^{90}Sr tracer have been initiated. Tests with other leachants and with fully exposed test cylinders are foreseen.

3.3. Mechanical strength of cement products

Compressive and tensile strength are determined with an Amsler press according to cement standards [5]. It seems to be accepted that the compressive strength of cement waste products should be at least 50 kg/cm^2 [7]. With few exceptions (extreme water/cement ratios) even the present cement-resin products meet this requirement. Most of them have a compressive strength of 100 to 200 kg/cm^2 and a tensile strength of 20 to 40 kg/cm^2. Correlations between these typical cement material tests and the more relevant fall tests can be established.

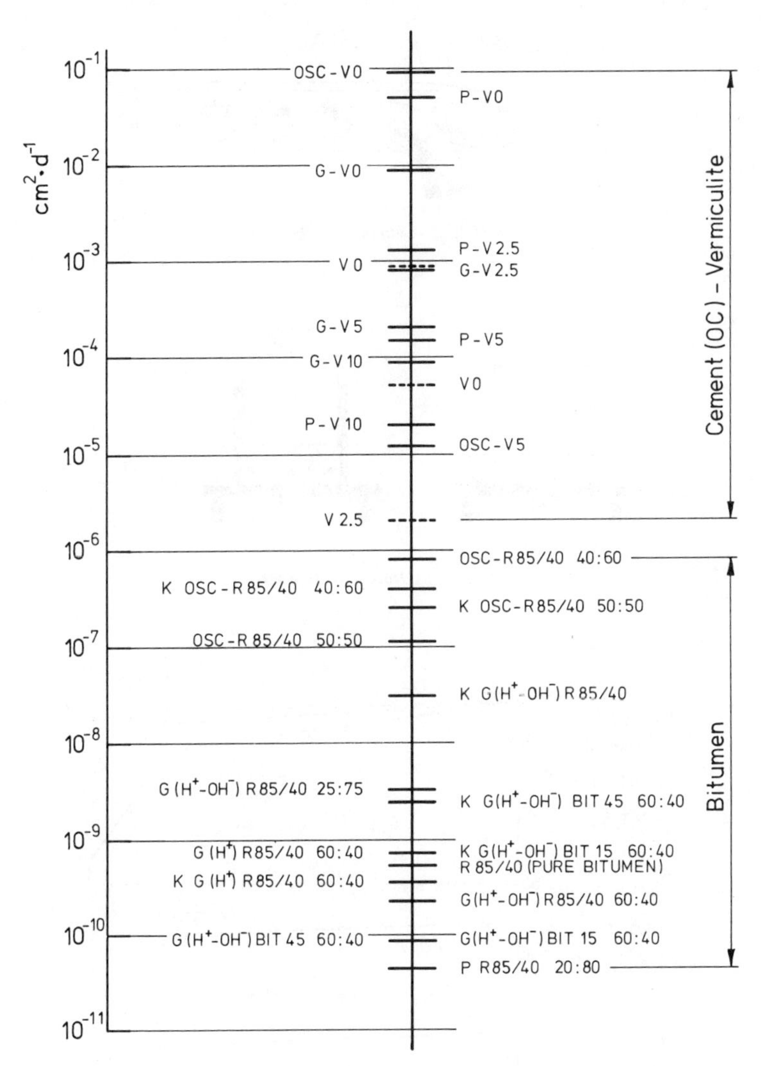

FIG.2. Cs leach coefficient ($cm^2 \cdot d^{-1}$) *for resin (P = powder, G = granular) cement (OC)-vermiculite (V) and resin-bitumen products.*

3.4. Fall test

To simulate a transport accident as specified in the IAEA's transport regulations cement cylinders (100 ml) were exposed to a free fall from 9 m height. Tested samples with a compressive strength of 100 kg/cm^2 or more withstood this test. To differentiate between samples the fall height was further increased and the height where breakage occurs was recorded. Similar tests will be carried out with bitumen products.

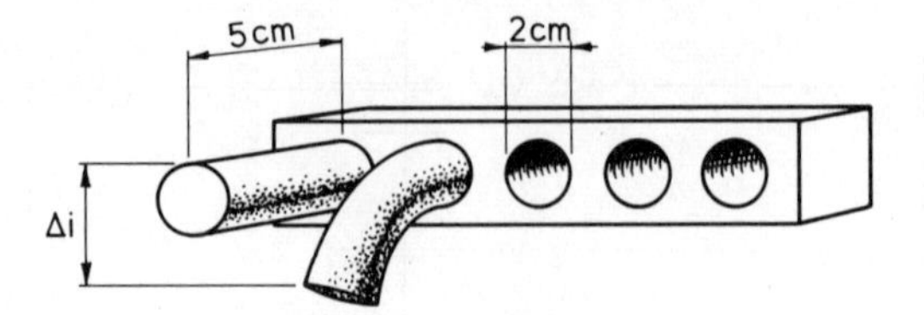

FIG.3(a). Cylinder bending test.

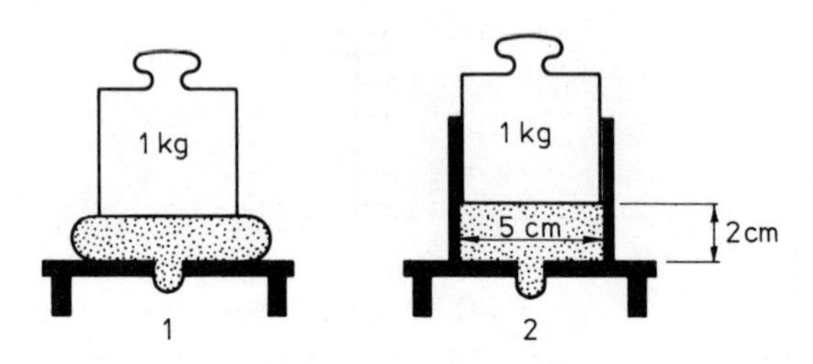

FIG.3(b). Hole migration test.

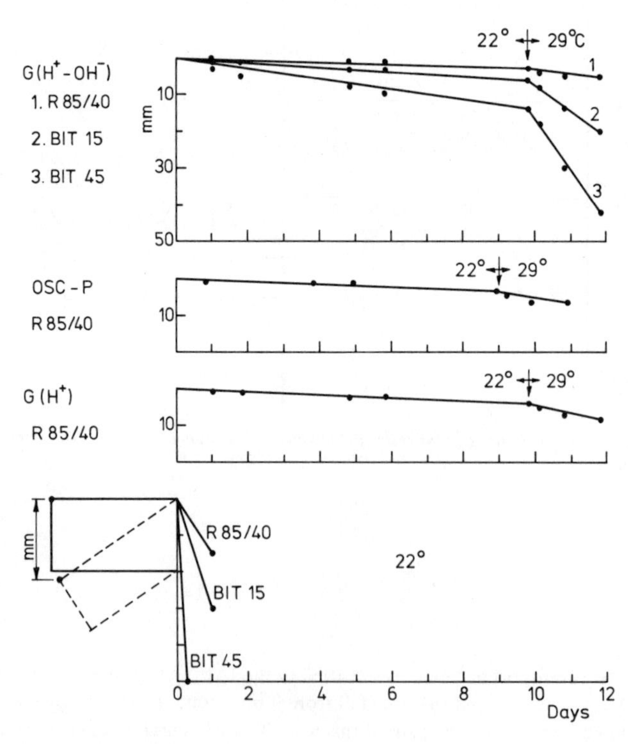

FIG.3(c). Cylinder bending test of resin (40%) – bitumen products and pure bitumen. (G = granular, P = powdered resin.)

3.5. Viscosity and form stability of bitumen resin products

The standard 'Penetration' and 'Ring and Ball' methods [8] – adopted from the petrochemical industry – did not give consistent results for bitumen resin products, due to interference from resin particles.

In co-operation with Brodersen from the Danish Atomic Energy Research Institute, Risø, [9] some simple tests for the form stability of such heterogeneous bitumen waste products are developed.

3.5.1. The 'cylinder bending' test (Fig. 3 a)

The bending of four to six cylinders held in a brass block is compared and measured. The deviation (Δi) of the free end from its original position at normal (20 ± 2°C) and elevated (30 ± 2°C) room temperature is recorded as a function of time. This simple device enabled significant differentiations between the tested products (Fig. 3 c).

3.5.2. The 'hole migration' test (Fig. 3 b)

The test is meant to simulate a case of damage to a container and provide a measure for the tendency of the bitumen to migrate through a hypothetical hole in the container bottom. Two versions are tried out. A bitumen cake is placed on a disc with a hole and pressed down by a weight. In the first case the increase in diameter is recorded simultaneously with the migration through the hole. In the second version deformations other than through the hole are prevented by surrounding the cake with a cylinder. Migration through the hole and deformation is recorded as a function of time and temperature (20, 30, 40°C). With the present bitumen-resin products migration was only observed at elevated temperatures.

3.6. High temperature test

To control whether the products meet with the IAEA's transport regulations (fire accident) cement products provided with a thermocouple were exposed to elevated temperatures for $\frac{1}{2}$ h. Pure cement lime was found to maintain its form and some compressive strength (ca. 80 kg/cm^2) after $\frac{1}{2}$ h at 800°C. Products with 8 to 10% (dry) powdered resin maintained their form (as a skeleton) but no mechanical strength under the same conditions, while products with 15 to 19% granular resin collapsed after $\frac{1}{2}$ h at 500°C.

3.7. Freezing tests

To control whether the products can withstand outdoor storage during a cold Nordic winter, effects of continued exposure to −40°C and of cycling between −40° and +20°C on mechanical properties and leachability are examined. So far no adverse effects of this temperature treatment have been identified. Tests for increased brittleness of bitumen (specially the BIT-qualities) have not yet been performed.

3.8. Additional tests to guide process optimizations

Visual inspection, optical and electron microscopy, X-ray diffraction and chemical analyses are used for further product characterizations and to gain a better knowledge of material structures and reaction mechanisms. The water content of bitumen products is controlled by ASTM standard procedures [8] and has been found to be below 0.5% in all tested products. In some cases the resin content was determined simultaneously to control the homogeneity of the sample. In future homogeneities will also be controlled by autoradiography.

FIG.4. The effect of water on the form stability of bitumen-ion exchange products. Ratio dry resin: bitumen 80 : 20, 70 : 30, 60 : 40 and 50 : 50.

3.9. Extended test programme

Tests for long-term effects of internal ionizing radiation (both gas evolution and structural effects), of corrosive water and soil components and of bacterial attack are foreseen. Thermal conductivities and heats of reactions (cement) should also be measured.

4. BITUMEN EXPERIMENTS

One aim of the bitumen experiments was to study process parameters of interest for the scheduled Finnish waste plant at Loviisa [1]. Starting with new resins to simulate ion exchange wastes, effects of resin concentrations and characteristics of process temperatures and bitumen qualities were examined. The rather elaborate programme was discontinued after it had been established that effects on product qualities were marginal and suitable tests for their quantitative measurement were not available. Some new tests – discussed in the previous section – are to be worked out before studies on marginal effects for the Finnish project can be continued. Finally a few experiments with spent resin from the Swedish Oscarshamn reactor were performed.

4.1. Quantities incorporated

Screening experiments with 20 to 90% (dry) resin showed that up to 40-60% could be incorporated without impairing the water resistance of the products. For products with more than 60% resin swelling in contact with water, for the highest concentrations even in air, was observed (Fig. 4). Conditions for granular and powdered resins were much the same except for differing consistences of the swelled products. Minor effects of resin concentrations on the consistence of products with 40 to 60% resin were sometimes detected. A few samples with 50 to 60% resin even showed some tendency to swell in contact with water. It was concluded that the maximal resin should be 40 to 60% and that further studies should concentrate upon this range.

4.2. Effects of process temperatures, bitumen qualities and resin characteristics on product qualities

In experiments with new resins only marginal effects on product qualities were observed.

An increase in working temperature from 140 to 160°C was found to reduce or eliminate swelling (3.1) of products with 50 to 60% resin.

The leachability of all examined products was low. Leach coefficients (L) for cobalt were in the order of 10^{-13} to 10^{-10}, for caesium 10^{-10} to 10^{-8} $cm^2 \cdot d^{-1}$. The analytical accuracy is poor for these low L-values, and differences of less than a decade are hardly significant. Leach coefficients for BIT 45 and BIT 15 products were essentially the same, and somewhat lower than for a corresponding product with blown (R 85/40) bitumen (see Fig. 2). Even a statistical treatment of co-ordinations according to visual surface classifications seems to indicate that the distilled bitumen qualities may be somewhat more compatible with ion exchange resins than the blown bitumen.

Cylinder bending tests (Fig.3(a) and 3(c)) revealed that the incorporated resins strongly increase the viscosity and form stability of bitumen products. Similar observations were made by Brodersen [9] who studied Mexphalt 40/50-resin products. The test showed the R 85/40 quality (with and without resins) to be far more form stable than the BIT-bitumen products. The low form stability of BIT 45 may prohibit its use for waste concentrates which further reduce product viscosities, but is hardly critical for the highly reinforced resin-bitumen products.

The effects will be further studied by means of the migration test. Even indications of increased brittleness in some BIT products are to be examined further.

Definite effects of resin characteristics (anion-cation, granular-powdered, H^+-Na^+, OH^--Cl^-) could not be identified in products with new resins.

The spent powdered resin from the Oscarshamn reactor behaved differently from the new resins. Attempts to make a product with 50% resin-R 85/40 bitumen failed because the mix got hard and brittle and could not be fed out by screw drive. A product with 40% of this resin was readily obtained. Leach rates from these products were somewhat higher than for corresponding products with new resins (see L-values, Fig. 2). The adverse behaviour of the spent resin may be due to absorbed ions (2.4 g Fe, 0.9 g Zn, 0.3 g Ca, 0.1 g Cr per 100 g dry resin) or to mechanical exhaustion.

4.3. Effects of process parameters on the process performance

Most difficulties experienced during the incorporation process were typical for the apparatus used and could be overcome by adjustment of process conditions.

The dewatering was found to be the most critical step. Blowout and a marked electrical charging of a powdered resin occurred due to inadequate preliminary mixing with bitumen, while too rapid increase of process temperatures in an experiment with granular resin-BIT 15 bitumen resulted in swelling and foaming in the mixing chamber.

Carry-over of organic decomposition products was most pronounced for the BIT 15 bitumen. With this quality even carry-over of activities was observed in experiments with new resins. During incorporation of the spent powdered resin from Oscarshamn with R 85/40 bitumen some carry-over of activities was also detected.

5. CEMENT EXPERIMENTS

Earlier experiments [10] had shown that resin-cement products tend to disintegrate in water. Thus the first task was to determine whether, and under what conditions, water-resistant products could be obtained. After identification of critical parameters their effect on further product qualities was examined. Means to extend tolerance ranges and to maintain stable products of

TABLE III. RESIN-CEMENT PRODUCTS. INDEPENDENT VARIATION OF RESIN AND WATER. GRANULAR RESIN Na^+–OH^-, RHC, NO ADDITIVES

% dry resin	w/c	Compressive strength (kgf/cm^2)	Water-resist.	% dry resin	w/c	Compressive strength (kgf/cm^2)	Water-resist.
4.3	0.56	204	good	8.6	0.67	18	no
4.3	0.45	272	good	8.6	0.60	181	poor
5.8	0.62	143	poor[a]	10	0.70	75	no
5.8	0.51	240	good	10	0.62	154	poor
7.3	0.64	124	poor	16	0.65	140	no
7.3	0.52	222	good	16	0.52	203	good

[a] Not decomposed, but loss of mechanical strength.

TABLE IV. EFFECTS OF RESIN CHARACTERISTICS, CEMENT QUALITIES, SILIX GP (Si)

Resin	Cement-additive	w/c	Water-resist.	Compressive strength (kgf/cm^2)	Remarks
Powdered	RHC	0.55	no	212	Powdered resin
H^+-OH^-	LHC	0.55	good	350	LHC > OC > RHC
8–10%	OC	0.55	poor	108	
	OC-Si	0.55	good	270	Stabilized by Silix
	OC	0.85	poor	30	
	OC-Si	0.85	good	97	W-resist. up to 0.85
	OC-Si	0.95	no	(d)	(d) decomp. in air
G H^+-OH^-	OC	0.55	no	(d)	Granular: H^+-OH^- less compatible
	OC-Si	0.55	poor	183	than Na^+-OH^- (Table III)
14%	OC-Si	0.65	no	63	
G H^+ -	OC	0.65	good	127	Cation alone: good
OH^-	OC	0.65	no	120	Anion alone: not
OH^-	OC-Si	0.65	no	140	water-resistant
G Na^+-OH^-	RHC	0.65	no	140	Granular
	RHC-Si	0.65	good	147	RHC > OC > LHC
	LHC-Si	0.65	no^x	153	
14–17%	OC	0.65	no	130	
	OC-Si	0.65	poor	107	After 90 days
	OC-Si	0.65	good^x	138	^{x}After 180 days
Borås	OC	0.65	no	130	
spent	OC-Si	0.65	good	150	After 28 days, already

homogeneous quality and low leachability within expected variations in technical process conditions were tested with simulated (water-saturated new resins) and eventually with real ion exchange waste.

5.1. Quantities incorporated (Table III)

In the first screening experiments the effects of resin and water content were studied separately. The content of water-saturated granular resin was varied between 15 and 50 wt% (i.e. 4 – 17% dry resin), the total water content between 30 and 50% to give water/cement ratios of 0.4 – 0.8. Within this range the measured product qualities were less affected by the resin content than by water/cement ratios. Consequently the resin contents were adjusted according to their water content to give acceptable water/cement ratios or a mix of optimal consistence. In this way products with 60 to 70 vol.% or 40 to 50 wt% water-saturated resins, corresponding to 12 to 20 wt% dry granular and 6 to 10% powdered resin, were attained.

B-1

A-1

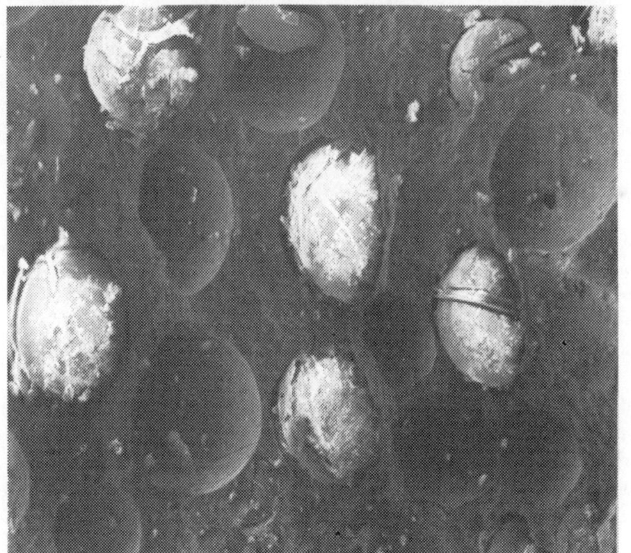
A-2

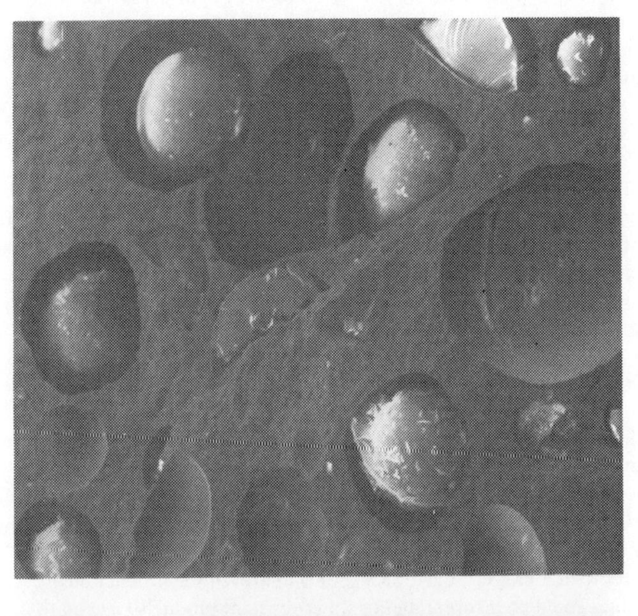
B-2

FIG.5. Cement products. A. Cation exchanger, 1. Effect of water; B. Anion exchanger, 2. Electron micrographs (30 x).

5.2. Water/cement ratios

Water-resistant products were only obtained within a narrow range of water/cement (w/c) ratios. The lower limit ($W_{min.}$) is governed by the water required to give a workable mix, and raises with the resin content. In these experiments it was within 0.45 to 0.5 for 4 to 20% (dry) resin. The upper limit ($W_{max.}$) has varied with resin characteristics, their grade of swelling, emulsifying and water-retaining properties. With powdered resins the products were water-resistant, though of low compressive strength up to water/cement ratios of 0.85 (Table IV), while for new granual resins, $W_{max.}$ could be so near to $W_{min.}$ that it was difficult to attain water-resistant products.

To be technically acceptable the tolerance ranges should at least be so wide that product qualities are not impaired by the varying amounts of water introduced with the water-saturated resins. It was attempted to extend the lower limit by addition of Barra to achieve a workable mix with a minimum of water. But this decreased the leach resistance and was therefore abandoned. The upper limit could be extended by addition of Silix (Table IV). Prolonged hardening times (e.g. 6 months) were needed to get the full benefit of its stabilizing effect.

5.3. Effects of resin characteristics

There is a marked difference in the compatibility of cation and anion exchange particles (Fig. 5). The former (both in H^+ and Na^+ form) were readily incorporated into water-resistant products (w/c ratios of 0.55 and 0.65), while such products were not obtained with anion exchange particles (OH^- form). This may be due to extreme swelling-shrinking, and formation of cavities, as illustrated in Fig. 5 (B).

In both cases absorbed ions (Na^+, Ca^{2+}, SiO_3^{2-}, Cl^-) improved the product qualities. Even anion exchange granules could be incorporated to water-resistant products (w/c up to 0.65) when their OH^--ions had been substituted by other ions (SiO_3^{2-}, Cl^-).

Cavities like those shown in Fig. 5 (B-2) were not observed in Microionex-cement products. Beside the much wider tolerance range for water these products even had a better temperature (3.6) and fall resistance (3.4) than corresponding products with granular resin. Attempts to modify the latter to similar conditions failed due to electrical charging of the resin during the (wet) grinding procedure.

The spent granular resin from the Borås power plant was more readily incorporated (w/c 0.55 and 0.65) than new resins, probably due to mechanical exhaustion (reduced swelling) and the stabilizing effects of absorbed ions (2.2% Na). The spent resin from Oscarshamn differed from new micro-resins mainly by prolonged setting times, apparently caused by retarding effects [11] of absorbed zinc (0.9%).

5.4. Cement qualities

The compatibility of the three tested Portland cement qualities (RHC, OC, LHC) showed opposite trends for powdered and granular resins. Granular resins were most readily incorporated to water-resistant products with rapid hardening cement, while the low heat quality was best suited for powdered resins (Table IV). In both cases ordinary cement was somewhat inferior to the most compatible quality. Corresponding trends were even observed for the leach rates (Fig. 6).

5.5. Means to reduce the leaching of caesium

High leaching of caesium is one main objection against cement incorporation of fission product dominated wastes. Cs leaching from the present resin-cement products was even higher than from pure cement products (Fig. 2).

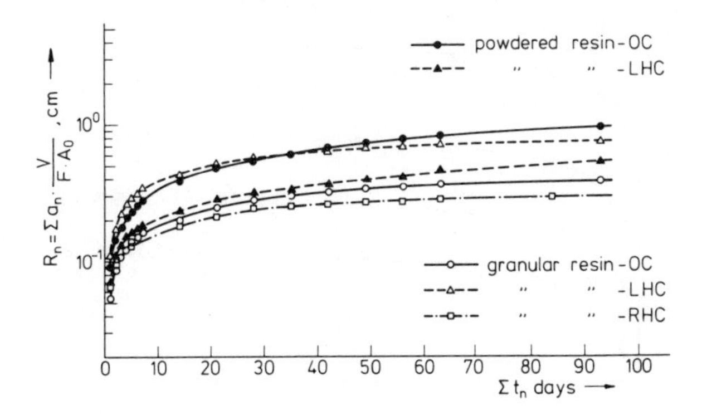

FIG.6. Effect of cement qualities on the Cs leach rates of resin-cement products.

The leaching of other fission products (perhaps except Sr) and the main activation products is much lower. Cobalt leach coefficients (L_{Co}) were about 10^{-7} $cm^2 \cdot d^{-1}$ in all experiments, while corresponding values for Cs varied between 10^{-5} and 10^{-2} $cm^2 \cdot d^{-1}$. The essentially selective leaching of Cs is also illustrated by γ-spectra of Oscarshamn resin (^{137}Cs too low for quantitative determination) and of a leach solution from its cement product (Fig.1). The Cs leach rates were significantly affected by resin characteristics, water/cement ratios and cement qualities.

Earlier tracer experiments [10] with 10 potential leach-reducing and Cs-retaining additives showed vermiculite to be by far the most efficient. With 2.5% vermiculite of optimal particle size (30-70 mesh) Cs leach coefficients were reduced from 10^{-3} - 10^{-4} to 10^{-6} $cm^2 \cdot d^{-1}$ (Fig. 2, VO and V 2.5) without impairing the mechanical properties.

Similar effects achieved for cement products with about 18% new granular resins (H^+ - SiO_3^{2-}), 10% new powdered resin (H^+ - OH^-) and 9% spent powdered resin are illustrated in Figs 2 and 7 to 9. The reduction of Cs leach rates with increasing vermiculite content (2.5, 5 and 10% of cement weight) is significant. For products with spent Oscarshamn resin the leach coefficient (L) was reduced from 10^{-2} (without vermiculite) to 10^{-5} $cm^2 \cdot d^{-1}$ with 5% vermiculite, somewhat less for the products with new resin (Fig. 2). Adverse effects on other properties were not observed. Vermiculite even buffers the sensitivity to variation in the water content. When it was used together with Silix, stable water-resistant products were readily attained even with new granular resins in H^+-OH^- form.

6. CONCLUSIONS

Bitumen is obviously more compatible with ion exchange resins than cement. Up to 40-60% dry resin can be incorporated without essentially impairing the product qualities. The resins improve the form stability of bitumen products and only slightly reduce their high leach resistance. Product qualities are less affected by physico-chemical variables than by purely mechanical process parameters. Conditions may, however, be somewhat less favourable for real than for simulated ion exchange wastes.

Without special precautions the incorporation of resins into cement involves several problems and renders products with a tendency to decompose in water and with poor resistance towards the leaching of caesium. Product qualities are strongly affected by resin characteristics, water/cement ratios, cement qualities and additives. The sensitivity towards physico-chemical variables could make cement unsuited for the technical incorporation of ion exchange resins, but this sensitivity also provides ample possibility of affecting the system in a positive way.

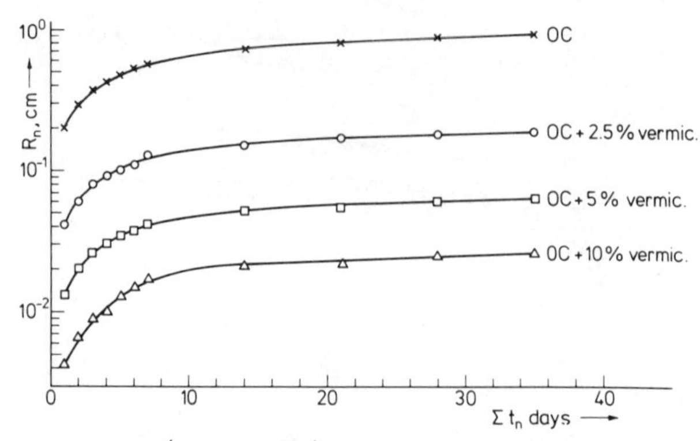

FIG.7. Accumulated Cs leach rates $\left(R_n = \Sigma a_n \cdot \frac{V}{F \cdot A_0}\right)$ *for powdered resin-cement-vermiculite products.*

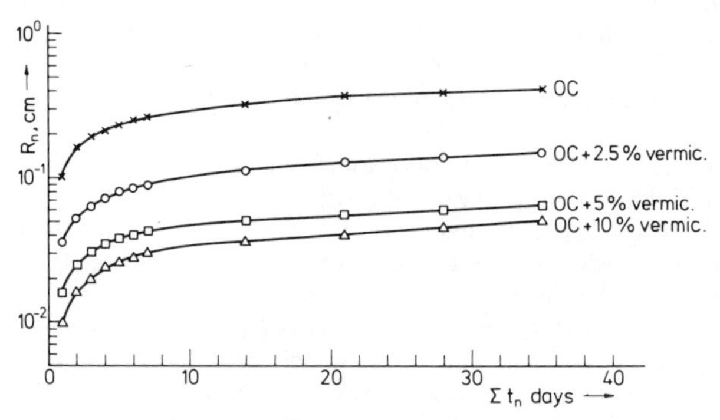

FIG.8. Accumulated Cs leach rates $\left(R_n = \Sigma a_n \cdot \frac{V}{F \cdot A_0}\right)$ *for granular resin-cement-vermiculite products.*

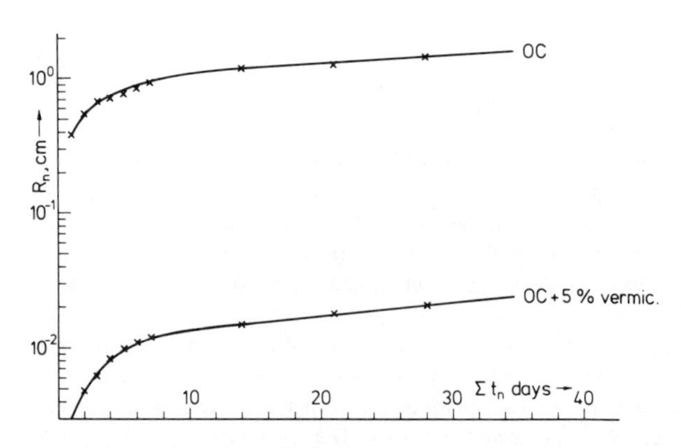

FIG.9. Accumulated Cs leach rates $\left(R_n = \Sigma a_n \cdot \frac{V}{F \cdot A_0}\right)$ *for Oscarshamn spent resin-cement-vermiculite products. (Cs-tracer added.)*

Product qualities can be significantly improved and tolerance ranges extended by means of stabilizing additives such as Silix GP and vermiculite. It seems possible, by further optimization of this or similar systems, to maintain stable products of homogeneous quality and relatively low leachability within technically acceptable tolerance ranges, provided the mechanical problems can be solved satisfactorily.

Another main disadvantage, the increase in weight and volume, cannot be circumvented. To keep these as low as possible the introduction of excess water should be avoided. With just water-saturated resins the lowest weight and volume increase factors (product/water-saturated resins) were 2–2.2 and 1.4, respectively.

Unlike bitumen, cement appears to be more compatible with spent than with new resins. Thus comparative feasibility assessments of the two techniques require further experience with representative real ion exchange wastes.

REFERENCES

[1] PELTONEN, E.K., HEINONEN, J.U., KUUSI, J., these Proceedings, Vol.2.
[2] LARSEN, I., Danish Atomic Energy Research Establishment, Risø 276 (1972).
[3] NESET, K.M., LUNDBY, J.E., NIELSEN, P.O., Management of Low- and Intermediate-Level Radioactive Wastes (Proc. Symp. Aix-en-Provence, 1970), IAEA, Vienna (1970) 45.
[4] Aka-utredningen, Radioaktivt avfall, DS I (1975) 8.
[5] Norwegian Standards, NS 3049 (1970).
[6] HESPE, E.D., At. Energy Rev. 9 (1971) 195.
[7] BURNS, R.H., At. Energy Rev. 9 (1971) 547.
[8] ASTM Designation: D36–66T, D5–65, D95–62.
[9] BRODERSEN, K., NIPA (75) DK-2, NIPA (76) DK-3.
[10] LUNDBY, J.E., NIPA (75) N-1 to N-7.
[11] LIEBER, W., Zement-Kalk-Gips 20 (1967) 91.

DISCUSSION

K.A. GABLIN: Granular and bead ion exchange resin wastes, when stored in operating reactors in the United States of America, develop excessive bacterial populations in the phase separators and the storage tanks. In view of this fact, do you consider that bacterial growth inhibitors should be added to the waste prior to solidification?

Moj BONNEVIE-SVENDSEN: We have not thought of using bacterial growth inhibitors, but studies on the effect of bacterial attack are planned and we shall certainly have to consider ways of preventing such effects.

N.J. KEEN: I find your study very interesting, but although you have described the conditions for incorporating resins into bitumen so as to produce a product with an acceptable leach rate, you have not given any information on the changes that may take place in the composite material due to radiation decomposition of the resin. Have you any plans to study the leach rates of 'aged' material?

Moj BONNEVIE-SVENDSEN: Yes, long-term tests on the effects of internal radiation, corrosive atmospheres, water and soil components, together with bacterial attack, are planned. We shall be studying leach rates as well as mechanical properties of such 'aged' material.

W. HILD: I am interested to see that your results tie in very well with our own. I should like to ask you whether you found thermal degradation of the ion exchange resins to the same extent that we reported in our paper (see paper IAEA-SM-207/81, these Proceedings, Vol. 2).

Moj BONNEVIE-SVENDSEN: Yes, we have certainly observed some degradation of the anion-exchange resins and we intend to ascertain whether this is more pronounced during dewatering or at the actual bituminization stage. It would, of course, affect the choice of temperature profile.

J.B. LEWIS: First, let me make an observation, which also relates to the paper that your Finnish colleague, Mr. Heinonen, presented. I think that in comparing the relative costs of the concreting and bituminizing processes, the additional shielding costs needed for bitumen should be included.

The question I should like to ask is whether you have considered combining the system proposed in the United States for curing concrete with polymers, with the addition of compatible ion-exchange materials.

Moj BONNEVIE-SVENDSEN: Yes, we have considered studying this possibility. Up till now we have given priority to inorganic additives so as to keep the advantages of using cement, but I agree that for the incorporation of ion-exchange resins, organic polymers would be an attractive alternative.

I agree with your comment that the additional shielding costs for bitumen should be taken into account. The impact of this factor will, of course, depend upon the mode of handling, storage and disposal as well as on the amount and radioactivity of the incorporated wastes.

ENROBAGE PAR RESINES THERMODURCISSABLES DE DECHETS RADIOACTIFS

A. BAER, Anne-Marie TRAXLER
Groupement pour les activités atomiques avancées,
Le Plessis-Robinson

A. LIMONGI, D. THIERY
CEA, Centre d'études nucléaires de Grenoble,
Service de protection et des études d'environnement,
Grenoble,
France

Abstract–Résumé

INCORPORATION OF RADIOACTIVE WASTES INTO HEAT-HARDENED RESINS.

The paper discusses the process developed and used at the Grenoble Nuclear Research Centre for incorporating radioactive wastes into heat-hardened resins and the application of this process to the treatment of radioactive waste originating from nuclear power stations with light-water reactors (PWR and BWR). The various types of waste are listed and assessed in terms of their activity and quantity, for example, evaporator concentrates, ion-exchange resins, filtration sludges, filters, various solid wastes, and so forth. The paper describes the lines along which research has been conducted and indicates, for each type of waste under consideration, the processing cycle – ranging for example, from the insolubilization of radionuclides, through the drying of the concentrates, to final incorporation. A study is made of the safety aspects of the process while in operation and of the shipping and storage of the incorporated waste; the authors outline the main technical characteristics relating to the safety of the plant and the final product obtained. More especially, they give the results of fire, irradiation, leaching and other types of endurance tests – characteristics which can be regarded in effect as safety criteria. Lastly, the authors touch on the economic aspect of the procedure by describing the contributions made in that respect by a reduction in volume and weight of the waste to be stored, the simplicity of the equipment and the cost of raw materials. They conclude that the sphere of application of heat-hardened resins appears, at the present stage of research and operational experience, to meet the requirements for the treatment of low- and medium-level waste.

ENROBAGE PAR RESINES THERMODURCISSABLES DE DECHETS RADIOACTIFS.

Le mémoire présente le procédé d'enrobage de déchets radioactifs en résines thermodurcissables mis au point et utilisé au Centre d'études nucléaires de Grenoble, et son application au traitement des déchets radioactifs des centrales nucléaires à eau légère (eau sous pression et eau bouillante). Les différents types de ces déchets sont énumérés et estimés en activité et quantité: concentrats d'évaporateurs, résines échangeuses d'ions, boues de filtration, filtres, déchets solides divers, etc. Les auteurs rappellent les orientations des études effectuées et indiquent, pour chaque type de déchets considéré, le cycle des opérations de traitement, depuis, par exemple, l'insolubilisation des radioéléments, la mise à l'état sec des concentrats jusqu'à l'enrobage final. La sûreté du procédé en exploitation et la sûreté du transport et du stockage des déchets enrobés sont étudiées; le mémoire dégage les caractéristiques techniques essentielles relatives à la sécurité de l'installation et du produit final obtenu. En particulier, sont donnés des résultats d'essais de tenue au feu, à l'irradiation, à la lixiviation, etc., caractéristiques qui constituent autant de critères de sûreté. Enfin, les auteurs abordent l'aspect économique du procédé en présentant les incidences qu'ont, à cet égard, réduction de volume et de poids de déchets à stocker, simplicité des installations et coût de matière première. En conclusion, le domaine d'application du procédé par résines thermodurcissables paraît, dans l'état actuel des études et de l'expérience d'exploitation déjà acquise, correspondre aux exigences de traitement des déchets de faible et moyenne activité.

1. PREAMBULE

Le développement de l'énergie nucléaire a entraîné de nombreuses recherches dans le domaine du traitement des déchets radioactifs en vue de leur stockage.

Ce traitement a pour but de convertir les déchets radioactifs, qui se présentent à l'état liquide (concentrats d'évaporation), pulvérulent, voire granuleux (résines échangeuses d'ions) ou solide (filtres, déchets métalliques, etc.), en blocs solides, homogènes, dont les caractéristiques physiques et chimiques permettent de conserver l'intégrité dans le temps.

Le but principal du conditionnement est d'éviter la dispersion de l'activité dans l'environnement

Les traitements des effluents liquides amènent finalement à concentrer l'activité sous un volume le plus faible possible.

La méthode de conditionnement la plus ancienne et encore la plus couramment utilisée consiste à enrober les déchets de ciment ou de béton. Ce mode d'enrobage présente certains inconvénients: volume et poids importants, mauvaise tenue aux intempéries, taux de lixiviation important, etc.

Les exigences de sûreté et les perspectives d'accroissement de volume des déchets radioactifs ont incité au développement d'autres modes de conditionnement: bitume, urée-formol, polyéthylène, etc.

De son côté, le Centre d'études nucléaires de Grenoble (CENG) du Commissariat à l'énergie atomique a mis au point le procédé d'enrobage par résines thermodurcissables. Ce procédé, en exploitation au CENG depuis 1971, est en cours de développement industriel par le Groupement pour les activités atomiques avancées (GAAA).

L'installation opérationnelle pour les concentrats d'évaporateur du CENG a permis d'expérimenter le procédé pour les diverses catégories de déchets produits, notamment dans les réacteurs à eau bouillante (BWR) et à eau sous pression (PWR).

Dans son principe [1,2], le procédé consiste à disperser les résidus radioactifs dans une résine thermodurcissable, par exemple dans une résine de polyester en solution dans le styrène, et à provoquer la polymérisation après avoir obtenu un mélange homogène. Cette polymérisation s'effectue *sans apport extérieur de chaleur.* D'autres résines thermodurcissables peuvent être utilisées

Le procédé, de mise en œuvre simple, permet d'enrober des résidus secs, des résines échangeuses d'ions ou des boues de filtration avec leur eau absorbée, des déchets solides secs ou humides, etc.

L'étude de ce procédé comporte:

- des essais de compatibilité entre les matériaux d'enrobage et les déchets de propriétés physico-chimiques diverses;
- des essais de qualité et de sûreté portant sur les caractéristiques du produit final obtenu après enrobage et sur les conditions de sûreté dans la mise en œuvre du procédé;
- des essais d'orientation destinés à fournir les données nécessaires à la conception et au dimensionnement de projets d'installation; ils sont complétés par des essais de qualification des équipements.

Cette communication porte essentiellement sur l'application du procédé d'enrobage par résines thermodurcissables, qui a été décrit ailleurs [3,4], aux réacteurs à eau légère (PWR et BWR).

2. APPLICATION DU PROCEDE AUX REACTEURS A EAU LEGERE

Ces réacteurs produisent des déchets de faible et moyenne activité, contaminés en produits de fission ou en produits d'activation. Du point de vue de l'exploitation de la station de traitement, ces déchets peuvent être classés en deux catégories:

- déchets humides transférables par canalisation: concentrats d'évaporation, à base de sulfates (BWR) ou de borates (PWR), de détergents; résines échangeuses d'ions en grains ou en poudre; boues de filtration; cette catégorie est la plus importante en volume et en activité (voir tableaux I et II);

TABLEAU I. DECHETS PRODUITS PAR UN REACTEUR BWR[a]

Déchets	Volume ($m^3 \cdot an^{-1}$)	Activité ($Ci \cdot m^{-3}$)
Résines échangeuses d'ions:		
– en grains	10	1,25 à 0,125
– en poudre (circuit primaire)	10	80
Concentrats d'évaporation	400	1
Boues de filtration	60	50

[a] D'après GESSAR (General Electric Standard Safety Analysis Report), Docket 50 447.

TABLEAU II. DECHETS PRODUITS PAR UN REACTEUR PWR[a]

Déchets	Volume ($m^3 \cdot an^{-1}$)	Activité ($Ci \cdot m^{-3}$)
Résines échangeuses d'ions:		
– circuit primaire	15	10^2 à $5 \cdot 10^3$
– circuits auxiliaires	25	environ 5
Concentrats d'effluents divers	30	1,5
Concentrats d'effluents à base de bore	3	15
Concentrats d'effluents à base de détergents	1	$3,5 \cdot 10^{-3}$
Déchets solides:		
– filtres	7	10^2 à $5 \cdot 10^3$
– divers		négligeable

[a] D'après SWESSAR (Stone & Webster Standard Safety Analysis Report), Docket STN 50 495.

– déchets solides secs ou humides: filtres à cartouches; pièces métalliques, verrerie, chiffons, papiers, etc.

Les déchets de la première catégorie peuvent nécessiter un prétraitement avant leur enrobage.

2.1. Prétraitement

2.1.1. Concentrats d'évaporation

Le schéma de principe est représenté à la figure 1. Les concentrats sont collectés dans un réservoir où s'effectue notamment l'insolubilisation des radioéléments. Ces concentrats sont ensuite amenés à l'état sec, ce qui permet de réduire au maximum le volume à conditionner et à stocker. Le distillat obtenu, qui représente 5 à 10% de l'effluent primaire (avant évaporation), est, après contrôle d'activité, rejeté hors site ou réutilisé par le réacteur. La mise à l'état sec s'effectue dans un sécheur rotatif chauffé par un fluide caloporteur (vapeur, p. ex.).

La partie interne du sécheur est maintenue en légère dépression à travers une colonne de lavage destinée à retenir les entraînements éventuels de contamination par l'eau évaporée.

Le produit sec est déchargé, par un circuit étanche et en faible dépression (pour éviter toute dispersion d'activité dans l'atmosphère), dans les fûts maintenus eux-mêmes en dépression, soit directement, soit par l'intermédiaire d'une capacité tampon de volume calibré. Le produit sec obtenu a une humidité résiduelle de l'ordre de 1 à 2%. Les fûts sont ensuite amenés au poste de solidification.

2.1.2. Résines échangeuses d'ions

Le schéma de principe est donné à la figure 2.

Les résines sont collectées et stockées avec une partie de leur eau de transfert dans un réservoir en amont de la station de traitement de déchets.

La réduction de poids généralement recherchée par le procédé ne peut aller, dans ce cas, jusqu'à l'expulsion de l'eau incluse dans les résines échangeuses d'ions. L'expérience nous a montré en effet que leur dessèchement complet est une opération qui présente des risques notables d'inflammabilité dus à leur nature chimique, des nuisances dues aux produits de décomposition. De plus, c'est une opération longue (48 h environ pour 100 litres de résines gorgées d'eau) et qui n'a d'ailleurs pas l'intérêt de réduire substantiellement leur volume.

Nous avons donc développé le procédé de telle sorte que les résines échangeuses d'ions, avec leur eau incluse, puissent être enrobées grâce à un prétraitement chimique simple, effectué dans le réservoir de stockage lui-même pendant une heure environ.

Les résines échangeuses d'ions sont ensuite essorées – ce qui réduit le poids en ramenant leur taux d'humidité à 40 ou 60% environ, suivant qu'elles sont en grains ou en poudre – puis transférées au poste d'enrobage.

2.1.3. Boues de filtration

Ces boues proviennent de la filtration, sur filtres à précouches, des effluents liquides. Leur taux d'humidité est alors de l'ordre de 40 à 50% en poids.

2.1.4. Filtres à cartouches

Si les cartouches filtrantes sont délivrées en vrac à la station d'enrobage, il suffit de les rassembler dans un panier ou un support consommable; elles sont ensuite enrobées dans cet assemblage avec leur humidité résiduelle.

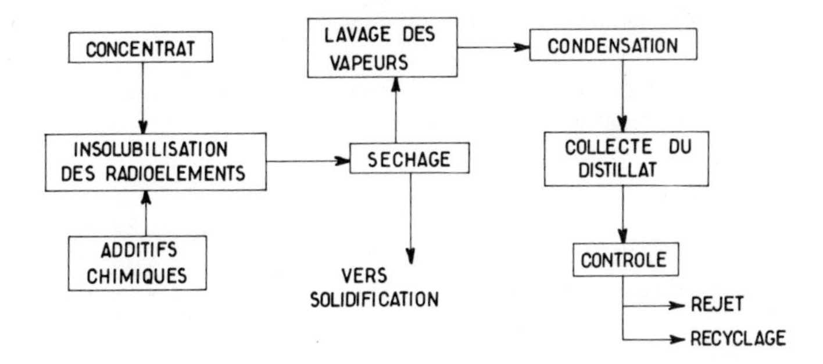

FIG.1. Prétraitement des concentrats d'évaporation: schéma de principe.

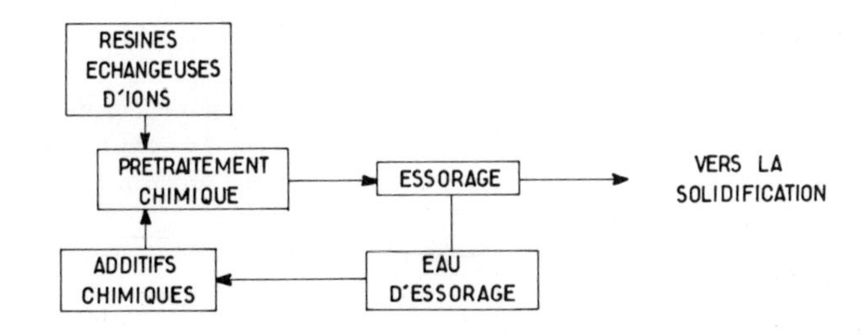

FIG.2. Prétraitement des résines échangeuses d'ions usées: schéma de principe.

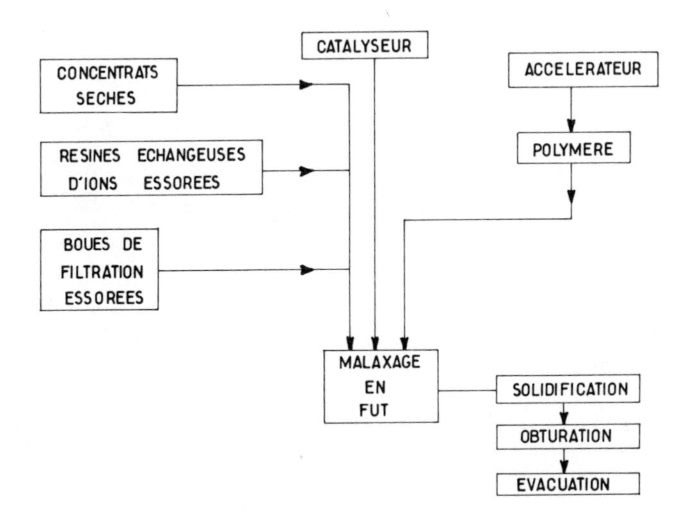

FIG.3. Enrobage des déchets: schéma de principe.

2.1.5. Déchets métalliques divers – verrerie

Amenés en vrac à la station de solidification, ils sont, comme les cartouches filtrantes, disposés dans un panier consommable avant leur enrobage.

2.2. Enrobage des déchets

L'enrobage des déchets se fait sans apport calorifique, directement dans les fûts de stockage final. La technique d'enrobage mise en œuvre diffère suivant la nature des produits à enrober.

2.2.1. Produits pulvérulents ou granuleux (concentrats desséchés, résines échangeuses d'ions, boues de filtration)

Le schéma de principe, représenté à la figure 3, correspond à l'enrobage des concentrats desséchés, des résines échangeuses d'ions et des boues de filtration.

Dans le fût contenant la quantité déterminée de déchets, on introduit le polymère préaccéléré. Ce mélange est homogénéisé par un malaxeur. La durée de cette opération, pour obtenir un mélange homogène, dépend de la densité des produits, de leur aptitude au mouillage, etc., et ne dépasse pas 20 minutes en général.

On introduit ensuite le catalyseur et le malaxage est maintenu jusqu'à un degré de gélification auquel correspond une viscosité donnée du mélange qui sert de point de consigne pour retirer le malaxeur. Le fût est ensuite enlevé du poste de malaxage. La prise totale du bloc (bloc dur) a lieu en 2 heures environ.

Le processus de polymérisation s'accompagne d'un dégagement de chaleur variable selon la charge. Il est possible d'ajuster l'évolution de la température, en jouant sur la formulation du mélange.

L'installation d'enrobage se compose uniquement d'un malaxeur en fût, pouvant travailler à plusieurs vitesses, doté d'un mouvement de monte et baisse. Cette installation peut être confinée dans une petite cellule maintenue en légère dépression pour diminuer le risque de contamination et évacuer les traces de vapeurs de styrène susceptibles de se dégager avant la gélification.

2.2.2. Déchets en panier

On malaxe dans le fût de stockage un mélange de polymère et d'une charge avec accélérateur et catalyseur. Le fût est enlevé du poste de malaxage et on introduit les paniers contenant les déchets. Les blocs ainsi obtenus sont très durs et ne peuvent être sciés qu'avec une meule diamant.

La formulation destinée à la polymérisation est ajustée pour tenir compte du temps de transfert du fût entre le poste de malaxage et le poste d'introduction du panier.

3. SURETE DU PROCEDE EN EXPLOITATION, POUR LE TRANSPORT ET LE STOCKAGE

La sûreté du procédé d'enrobage par résines thermodurcissables est à examiner pour l'exploitation de l'installation et pour le transport et le stockage (provisoire puis à long terme) des déchets radioactifs conditionnés par ce procédé.

Pour l'exploitation, il est généralement admis d'assurer la sûreté d'une installation en concevant, entre sources de risques d'une part et opérateur ou environnement d'autre part, des barrières successives.

En ce qui concerne le transport et le stockage, la qualité de la barrière constituée par le conditionnement est à évaluer, eu égard aux agressions dont il peut faire l'objet et aux conséquences qu'elles peuvent avoir pour le milieu environnant.

3.1. Sûreté en exploitation

Les risques sont ceux inhérents aux résidus radioactifs traités (notamment risques d'irradiation et de contamination) et ceux inhérents au procédé d'enrobage lui-même (notamment risques d'origine chimique liés à la nature même du matériau d'enrobage employé).

Il est intéressant d'examiner la contribution de l'équipement et du mode opératoire employés dans le procédé, à la «politique des barrières».

Nous examinerons, à titre d'illustration, le cas des concentrats d'évaporation et celui des résines échangeuses d'ions en indiquant comment sont réalisées les barrières contre la propagation de la contamination et l'irradiation.

3.1.1. Prétraitement des concentrats et résines échangeuses d'ions

3.1.1.1. Les concentrats d'évaporation sont desséchés jusqu'à obtention d'un produit sec pulvérulent.

Le confinement de la *contamination* est réalisé par l'ensemble du sécheur et de ses auxiliaires (pompe à vide à anneau liquide, condenseur, réservoir à condensats, etc.), et du réservoir ou fût de recueil des produits secs – ensemble maintenu en dépression: cette disposition permet d'assurer une barrière matérielle d'une très grande efficacité contre la contamination. Aucun élément contaminant ne se trouve en contact matériel avec le milieu extérieur: les fluides (de chauffage ou de refroidissement) circulent en circuit fermé. L'eau constitutive des concentrats est recueillie dans des réservoirs et est recyclée ou rejetée après contrôle de radioactivité.

Le contrôle de l'existence de la barrière peut être assuré simplement par la mesure de la pression (environ 40 à 70 mmHg) à l'intérieur de cet ensemble et le contrôle de son efficacité par la mesure de l'activité de l'air du local.

La source *d'irradiation* est constituée par le concentrat en cours de séchage et le produit sec collecté dans le fût de conditionnement. La barrière contre le risque d'irradiation est constituée par la protection biologique de la cellule contenant sécheur et fût.

Il existe cependant des types de sécheurs dans lesquels la quantité de concentrats en cours de traitement est assez faible (quelques litres) pour ne nécessiter qu'une protection de faible épaisseur; le fût de recueil des produits secs, contenant la totalité de l'activité, constitue alors la source d'irradiation principale à mettre sous protection.

3.1.1.2. Les résines échangeuses d'ions présentent des risques élevés de contamination et d'irradiation.

Pour éviter les risques de dispersion de la *contamination,* d'incendie ou de dégradation à la chaleur qu'entraînerait leur dessèchement, on maintient les résines échangeuses d'ions constamment *sous eau* pendant leur transfert hydraulique et le prétraitement chimique, ou *humides* (eau absorbée restant après l'essorage): il existe une barrière matérielle contine contre la propagation de la contamination, constituée par les réservoirs et tuyauteries.

Le risque *d'irradiation,* dont le niveau est important, nécessite un blindage localisé au niveau des réservoirs de stockage ou bien la mise de l'ensemble sous protection biologique.

3.1.2. Opération d'enrobage

Effectuée au poste de malaxage, l'opération d'enrobage dans le polymère comporte les risques radioactifs d'irradiation et de contamination propres aux déchets traités.

Elle comporte aussi les risques chimiques liés à la nature du produit d'enrobage et des réactifs (accélérateur, catalyseur) utilisés: risques d'incendie ou d'explosion, risques d'intoxication. Pour évaluer le niveau de ces risques, il convient de se référer aux caractéristiques de la résine d'enrobage (point éclair, tension de vapeur, limite inférieure et supérieure d'explosivité dans l'air, température d'auto-ignition, etc.) et des réactifs (catalyseur et accélérateur) utilisés (température de décomposition, point éclair, etc.). Ces caractéristiques sont données par les fournisseurs de ces produits et ont fait l'objet d'une étude de sécurité complète, confirmée d'ailleurs par l'expérience.

A titre d'exemple, le malaxage et la polymérisation ayant lieu à la température ambiante de 20°C, et la tension de vapeur du styrène étant de 4,53 mmHg, la teneur de vapeur dans l'air serait de l'ordre de 0,6% dans une atmosphère statique, donc plus faible que la limite inférieure d'explosivité (1,1% en volume à 20°C).

La simplicité de mise en œuvre du procédé, et notamment le fait que cette opération de malaxage et de polymérisation ne nécessite pas d'apport de chaleur, rend aisée la création de barrières contre les risques que nous venons d'énoncer.

La sûreté d'exploitation repose essentiellement sur les dispositions suivantes:

- le poste de malaxage où s'effectue le mélange polymère-résidus radioactifs est dans une enceinte ventilée (une dizaine de renouvellements horaires suffit); il est à noter que dès que la gélification commence, le styrène participe à la polymérisation et ne s'évapore plus;
- l'équipement électrique (moteurs, ventilateur, éclairage, etc.) est antidéflagrant;
- l'introduction du catalyseur et de l'accélérateur s'effectue en petites quantités et séparément pour éviter leur interaction.

Il est à noter que la barrière dynamique constituée par la ventilation et la mise en dépression de l'enceinte de travail (hotte, cellule ou autre) permet de contenir le risque de propagation de la contamination radioactive.

De même, il convient de remarquer que ce risque est notablement diminué dès que le polymère est introduit dans le fût au-dessus des déchets à enrober.

Les autres aspects importants relatifs à la sûreté ont trait aux conditions de stockage de la résine d'enrobage et des réactifs; ces conditions sont celles appliquées dans l'industrie des matériaux plastiques. La réglementation est précise à cet égard et spécifie les conditions de stockage en fonction des quantités dans un local ventilé, à distance fixée des locaux de travail, etc. (en France, loi de 1917 sur les établissements classés et tableaux annexes).

La barrière contre le risque d'irradiation est réalisée, par exemple, par les parois de la cellule munie de protections biologiques adéquates dans laquelle est installé le poste de malaxage. Suivant le niveau d'irradiation, c'est-à-dire l'activité β, γ des déchets radioactifs, le fût, après polymérisation, doit être mis dans une protection biologique conventionnelle, pour respecter les niveaux de débit de dose au contact pour le transport des blocs fabriqués.

3.2. Sûreté au transport et au stockage

Après enrobage, les produits obtenus sont des blocs de 100 ou 200 litres contenus dans des fûts standards. La sûreté du conditionnement, par enrobage sous résine thermodurcissable, s'évalue en considérant la résistance de cette barrière aux agents agressifs pour éviter la dispersion de la contamination radioactive dans le milieu en cas d'accident de transport ou lors d'un stockage provisoire ou à long terme. Les agents agressifs considérés sont la température, le feu, l'eau, l'irradiation, les variations climatiques, les éléments physico-chimiques et biologiques des sols (en cas d'enfouissement dans la terre).

3.2.1. Manutention et transport

Les caractéristiques suivantes du produit final enrobé sont à noter:

– *Résistance mécanique.* La tenue à la compression est supérieure à 2 $t \cdot cm^{-2}$; un fût de 200 litres doit pouvoir subir une chute de 10 mètres sur une dalle de béton, sans dommage pour le bloc contenu dans le fût.

– *Résistance au feu.* Pour simuler un incendie de stockage ou de transport, ont été exposés à des feux d'essence, au-dessus de la surface enflammée, dans la zone la plus chaude (700 à 900°C) des gaz de combustion, des blocs de 20 litres, instrumentés, et représentatifs par leur volume des blocs réels, et de composition variable: polyester plus charge de borates, de chlorures, de sulfates, de nitrates ou mélange de ces sels, ou enfin de résines échangeuses d'ions. Les expériences, réitérées, ont montré que:

- les blocs contenant sulfates, borates, chlorures ou résines échangeuses d'ions résistent parfaitement: au bout de 30 minutes de feu, seule la surface s'est trouvée détériorée sur une épaisseur de quelques millimètres; les blocs sont demeurés compacts et très durs;
- la présence de nitrates dans la charge n'entraîne pas de réaction explosive, même si la charge est constituée uniquement de ces sels (50% en poids du bloc); l'addition à la charge enrobée de composés minéraux tels que chlorure ou borate de sodium améliore la résistance au feu;
- avec une proportion de nitrate mélangé à du chlorure de sodium telle que l'on ait 50% de polyester, 10% de nitrate, 40% de chlorure, les blocs résistent au feu sans dommage majeur.

3.2.2. Stockage

La sûreté du stockage à long terme est jugée suivant les recommandations généralement adoptées, sur la base des propriétés physiques du produit final:

– *Résistance à l'irradiation.* Elle est d'au moins 10^{10} rad; à noter qu'une dose de 10^9 rad correspond à la dose infinie, absorbée par le polyester enrobant une activité de l'ordre de 200 Ci $\beta\gamma$ de produits de fission et d'activation provenant des effluents d'un réacteur à eau légère. A une dose intégrée de $5 \cdot 10^9$ rad, aucune dégradation mécanique n'est notée sur les échantillons enrobant sels minéraux ou résines échangeuses d'ions et soumis à l'irradiation; la résistance mécanique s'est maintenue.

– *Formation des gaz de radiolyse.* Les gaz de radiolyse (H_2, CH_4, CO_2, etc.) sont créés au cours de l'irradiation en quantité assez faible pour qu'ils n'entraînent pas de dégradation des éprouvettes soumises à irradiation et contenant des sels minéraux provenant de concentrats. Les expériences d'irradiation en cours d'échantillons de polyester enrobant des résidus de différentes natures ont pour but d'évaluer précisément, dans chaque cas, le dégagement des gaz inflammables, en vue d'estimer le risque éventuel créé au stockage en local clos d'un grand nombre de fûts d'enrobés: les résultats actuels permettent d'estimer à 20 litres le volume total de gaz dégagé en un temps infini d'un bloc de 200 kg enrobant pour moitié des sels minéraux; hydrogène et méthane ne représentent que un quart de ce volume.

– *Tenue en température.* Il n'a pas été noté de perte de poids notable pour des échantillons chargés de résidus minéraux et maintenus pendant plusieurs heures à 300°C. L'enrobage ne flue pas.

TABLEAU III. RESULTATS DES TESTS DE LIXIVIATION

En eau déminéralisée (conformément à la recommandation AIEA) [1,5]

Radioélément	Lixiviation (en cm par jour) au bout de		
	200 jours	600 jours	900 jours
Cs–137	$1,3 \cdot 10^{-7}$	$3 \cdot 10^{-9}$	$3 \cdot 10^{-9}$
Co–60	10^{-5}	$2 \cdot 10^{-7}$	$1,7 \cdot 10^{-7}$
Sr–90	$3 \cdot 10^{-6}$	$2 \cdot 10^{-6}$	$1,8 \cdot 10^{-6}$

En eau douce et en eau de mer

Radioélément	Lixiviation (en cm par jour) au bout de		
	200 jours	600 jours	750 jours
En eau douce			
Cs–137	$2,5 \cdot 10^{-8}$	$1,2 \cdot 10^{-8}$	$4 \cdot 10^{-9}$
Co–60	$2,2 \cdot 10^{-7}$	$1,8 \cdot 10^{-7}$	$1,1 \cdot 10^{-7}$
Sr–90	$7 \cdot 10^{-6}$	$2,6 \cdot 10^{-6}$	$2,8 \cdot 10^{-7}$
En eau de mer			
Cs–137	$2 \cdot 10^{-8}$	$5 \cdot 10^{-9}$	$3,2 \cdot 10^{-9}$
Co–60	$2 \cdot 10^{-7}$	$1,3 \cdot 10^{-7}$	$1,2 \cdot 10^{-7}$
Sr–90	$8 \cdot 10^{-6}$	$6 \cdot 10^{-6}$	$2,8 \cdot 10^{-6}$

Nota. Le pH des concentrats compris entre 7 et 11 n'a pas d'influence sur ces résultats.

– *Tenue aux agents climatiques.* Elle est vérifiée en vraie grandeur par exposition d'échantillor d'enrobés au gel et au dégel pendant une longue période en climat rigoureux et par des cyclages thermiques entre –20°C et +40°C en laboratoire en enceinte climatique.

– *Tenue à la lixiviation.* Les tests de lixiviation, effectués en application de la recommandatio de l'AIEA, ont porté sur des radioéléments sous forme de sels préalablement insolubilisés (avant dessèchement des concentrats), et de période assez longue pour être significativement intéressants du point de vue de la sûreté du stockage à long terme à l'air libre. Les résultats sont indiqués dans le tableau III.

– *Tenue à l'enfouissement direct dans la terre.* Elle est examinée pour évaluer les risques de dispersion de la contamination radioactive après disparition du fût métallique, au cas, bien improbab où le stockage s'effectuerait en tranchées creusées à même le sol, en terrain humide. Les conditions aérobies et anaérobies représentatives de l'enfouissement dans la terre ont été recréées.

Le polyester pur enterré pendant plusieurs mois n'a pas subi d'atteinte. Il en est de même des éprouvettes de polyester chargées de divers sels minéraux exposées aux bactéries en milieu de culture sec ou humide.

Cependant, pour les enrobés polyester + charge, il convient de tenir compte d'un autre phénomène, celui des échanges d'ions avec les sols, en particulier en terre humide; ce phénomène est lié à l'affleurement de la charge à la surface des blocs d'enrobés. Les expériences en cours semblent montrer l'opportunité de compléter les tests de lixiviation à l'eau par des essais tenant compte des échanges avec les milieux plus complexes tels que le sol.

Il est un fait que la nature même du sol intervient, ainsi que le taux d'humidité, et que pour chaque site des expériences de longue durée devraient précéder toute décision d'enfouissement direct.

3.3. Adaptation du procédé à des conditions particulières

Les caractéristiques du produit final citées ci-dessus ont été obtenues avec le polyester communément employé. Pour résoudre des problèmes très particuliers (de concentrats à très forte teneur en sels, et qu'on ne veut pas dessécher), ou pour répondre à des contraintes de sécurité plus sévères (tenue au feu supérieure à celle citée au paragraphe 3.2 et en particulier pour des enrobés à forte proportion de produits oxydants), le procédé peut utiliser des produits de base permettant de s'accommoder de quantités d'eau particulièrement élevées. Il peut aussi utiliser la technique du surenrobage, pour améliorer l'étanchéité à la migration ionique dans les sols. La mise en œuvre du malaxage ou de la technique du surenrobage se prête aisément à l'utilisation d'additifs tels que produits ignifuges, bactéricides ou fongicides qui peuvent apporter une défense supplémentaire contre les agents agressifs que des conditions de stockage particulières pourraient créer.

4. ASPECT ECONOMIQUE DU PROCEDE

Il peut être évalué, par rapport à d'autres, en considérant qu'il permet réduction de volume et de poids des déchets à stocker et simplicité des installations auxiliaires.

4.1. Réduction de volume et de poids

La réduction du volume des déchets (et donc du nombre de fûts à stocker) permet de réduire les dimensions des locaux des stockages provisoire et final et de diminuer le nombre de transports vers le stockage final. Elle intervient donc favorablement dans les coûts de construction, de transport et de stockage. Cet avantage est particulièrement marqué pour les concentrats d'évaporation.

Prenons comme exemple 100 m^3 de solution de régénération ayant une concentration de 10 $g \cdot l^{-1}$ en sulfate de sodium (cas des BWR). L'évaporation primaire aboutit à la production de 4 m^3 de concentrats à 250 $g \cdot l^{-1}$ de salinité, soit 1 tonne de sulfate de sodium. La solidification par le ciment de ces 4 m^3 correspond à 40 fûts à stocker, un fût de 200 litres n'absorbant que 100 litres de concentrats environ. L'enrobage en résine thermodurcissable des sels contenus dans ces 4 m^3 de concentrats conduit à la production de 7 fûts de 200 litres, l'enrobage se faisant avec 60% de sels minéraux.

Si l'activité enrobée ne nécessite pas de protection biologique pour le transport, il est intéressant de considérer la réduction de poids: 40 fûts de béton pèsent environ 22 tonnes, contre 2 tonnes pour les 7 fûts de résine thermodurcissable.

4.2. Simplicité des installations auxiliaires

Elle intervient sur l'aspect économique du procédé par les caractéristiques suivantes:

- la résine thermodurcissable peut être stockée soit en citerne, soit en fûts de 200 litres, à la température ambiante; une seule astreinte, qui concerne des volumes peu importants, est le maintien du catalyseur à une température de l'ordre de 4°C pour lui conserver sa réactivité dans le temps;

- les appareils utilisés pour le prétraitement et le malaxage sont des appareils industriels courants;
- la mise en œuvre du procédé se fait à la température ambiante;
- la ventilation de la cellule ne nécessite pas de dispositifs de traitement de l'air;
- grâce à la bonne tenue en température de l'enrobage, et notamment à l'absence de tout risque de fluage, il n'est pas nécessaire d'être exigeant en ce qui concerne les conditions d'ambiance des locaux de stockage, dont la conception se trouve ainsi simplifiée.

4.3. Coût de la matière de base

Il est un fait que le coût d'une matière organique de synthèse telle que celle utilisée dans le procédé est plus élevé que celui du ciment ou du bitume, mais on peut estimer que ce coût n'est pas prépondérant dans le prix de revient du traitement des déchets, si l'on tient compte des incidences financières favorables de la réduction des locaux de stockage et de la simplicité des installations auxiliaires.

Quoiqu'il en soit, ce type d'enrobage utilise un matériau de qualité, qui confère au conditionnement des caractéristiques de sûreté qu'il est bien justifié de rechercher pour la conservation à long terme des déchets radioactifs de faible et moyenne activité.

5. CONCLUSION

Les essais de qualification destinés à estimer les caractéristiques de l'enrobage et à vérifier la compatibilité avec les déchets de natures différentes, et l'expérience déjà acquise en exploitation permettent de définir assez clairement le domaine d'application de ce procédé d'enrobage: il paraît dès maintenant bien adapté aux déchets de toute nature, de faible ou moyenne activité $\beta\gamma$ des centres d'études nucléaires et des centrales de puissance à eau légère.

Ce domaine d'application est susceptible d'être élargi; des études en cours sont destinées à le vérifier.

REFERENCES

[1] CUAZ, D., THIERY, D., Procédé de conditionnement de déchets radioactifs: Brevets EN 71 090 057 (déposé le 16 mars 1971); EN 73 179 74 (déposé le 17 mai 1973); EN 73 40 005 (déposé le 9 novembre 1973); EN 74 18281 (déposé le 27 mai 1974).

[2] POTTIER, P., ANDRIOT, R., CUAZ, D., «Progrès dans les techniques de traitement et de conditionnement des effluents dans les centres de recherche», Peaceful Uses of Atomic Energy (Actes 4e Conf. Int. Genève, 1971) **11**, ONU, New York, et AIEA, Vienne (1972) 325.

[3] SOUSSELIER, Y., et al., «Conditioning of wastes from power reactors», ANS Winter Meeting, San Francisco, novembre 1973, **17**, TANSAO, 17.1.578 (1973).

[4] CUAZ, D., LIMONGI, A., THIERY, D., BAER, A., «Procédé d'enrobage de déchets radioactifs dans des résines thermodurcissables», VIIe Congrès Int. de la Société française de Radioprotection, Versailles, 1974, 769.

[5] HESPE, E.D. (Editor), «Leach testing of immobilized radioactive waste solids: a proposal for a standard method», At. Energy Rev. **9** 1 (1971) 195.

DISCUSSION

R. KROEBEL: I have several questions to ask you. How much waste has been treated by this method? And of what type is it? Is it correct that you need a hot cell facility or something similar for radiation shielding and ventilation?

The 'Gesellschaft für Kernforschung mbH', Karlsruhe, and Bayer A.G. in Germany have developed a similar process for the same kind of wastes and Steag-Essen has built a mobile pilot unit for fixing ion-exchange resins. In the last few months they have produced 600 drums, each containing 100 litres of radioactive material.

A. LIMONGI: We have prepared about 180 drums containing 60 or 100 litres each, incorporating radioactive evaporator concentrates after drainage. On the other hand, we have not made any radioactive ion-exchange resin blocks.

As regards the need for a hot cell, this is a matter of biological protection and should be assessed as a function of the radioactivity of the waste coated with the thermo-setting resin.

And as far as the radioactivity limits for the coated waste are concerned, I would say that, taking into account the results of behaviour under irradiation (to at least 10^{10} rads), it should be possible to coat with thermo-setting resin 200 Ci of beta-gamma activity in a 200 litre drum; at that radioactivity level one would have to provide for biological protection, which could be, for example, a hot cell with negative air pressure ventilation.

K. TALLBERG: Have you only been working with granular resin or have you also tried out microresin without encountering the problem of electrostatics when the water content is too low?

A. LIMONGI: We have carried out experiments both at the laboratory level and on a pilot scale (with non-radioactive material), in coating ion-exchange resins in the form of grains and powders, anionic and cationic resins and mixed beds, both fresh and used, with the thermo-setting resins. Since the ion-exchange resins are not dry, the problem of electrostatics does not arise.

K.A. GABLIN: How does your system solidify normal oil-contaminated wastes from floor drains in PWRs and BWRs and BWR sumps?

A. LIMONGI: We have tested the incorporation of oil and other organic fluids into other resin-coated waste at a rate of a few percent per mixture. We are studying effects on the behaviour of the resin-coated material for an organic fluids content as high as 10% of the waste coated.

N. VAN DE VOORDE: Mr. Gablin's question is very pertinent. Quite recently we received a boric acid effluent stemming from a reactor that contained oil amounting to about 10% of its volume.

MANAGEMENT OF ALPHA-BEARING WASTE (Session VIII)

Chairman:

N.J. KEEN, United Kingdom

IAEA-SM-207/46

OPERATIONAL TECHNIQUES TO MINIMIZE, SEGREGATE AND ACCOUNT FOR ALPHA-BEARING WASTES

L.F. JOHNSON
British Nuclear Fuels Limited,
Windscale Works,
Seascale, Cumbria,
United Kingdom

Abstract

OPERATIONAL TECHNIQUES TO MINIMIZE, SEGREGATE AND ACCOUNT FOR ALPHA-BEARING WASTES.

The paper presents an account of the measures being adopted at the Windscale Works of British Nuclear Fuels Limited to reduce arisings of plutonium-contaminated solid waste by a combination of good design, provision of equipment and suitable operating materials and education of operators, following detailed examination of the pattern of waste arisings. The segregation system for packaged waste is described and the reasons for it are explained. The various categories of waste are chosen according to the disposal, treatment and storage routes available. The routes presently available for drummed plutonium-contaminated waste include sea disposal via annual internationally-regulated dumps in the deep Atlantic, incineration and intermediate term storage pending the development of further treatment methods. The figures used for accounting are determined by instrumental non-destructive analysis in the central waste handling facility. The methods presently available are described with their limitations. Instruments which are expected to become available during 1976 are described briefly. These consist of a segmented gamma scanning instrument, an NNC 'random driver' neutron interrogation instrument and a means of determining the plutonium content of crated waste. A description is given of the data management system, which provides the information on plutonium losses to waste which is incorporated into the fissile material account for each area.

1. INTRODUCTION

Plutonium handling operations are carried out at the Windscale Works of British Nuclear Fuels Limited (BNFL). Activities include the separation of plutonium from spent reactor fuel, its preparation as fuel for thermal and fast reactors and associated development of alternative process routes.

Plutonium-contaminated materials arise in the classifications listed in Table I. The table shows the percentage of total waste each classification contains.

Waste is regarded as plutonium-contaminated if it is contaminated to a level exceeding 10^{-10} Ci/cm^2. This is assumed to include all waste from glove boxes, cells, plant access areas and certain designated laboratory fume hoods. In addition, it includes waste arising during tented maintenance operations, the cleaning up of spillages in operating areas, etc., where the surface contamination criterion applies.

Arisings in each of the last three years, represented by index values (1973 = 100), are shown in Table II.

TABLE I. Classification of waste types at Windscale

Classification	Contribution of classification to total waste
Fume hood waste (combustible with polythene)	0.5%
Fume hood waste (combustible - no polythene)	0.3%
Fume hood waste (non-combustible)	0.5%
Gloves	15.6%
PVC	27.7%
Tissues/swabs	8.1%
Process solids	1.7%
Floor covering and tenting	2.1%
Polybottles/polythene equipment	13.9%
Filter, gas or air	1.0%
Filter, liquor	3.4%
Rubber	New classification no data
Tools	3.8%
Small plant items	5.8%
Glass	4.7%
Tins/cans	10.2%
Fuel pins/cans	0.7%

TABLE II. Arisings of plutonium-contaminated waste at Windscale

Year	Total volume	Total weight	Weight of associated Pu
1973	100	100	100
1974	91	83	64
1975	86[a]	64	43

[a]excluding 20% of waste with no plutonium content detectable by gamma spectroscopy method used.

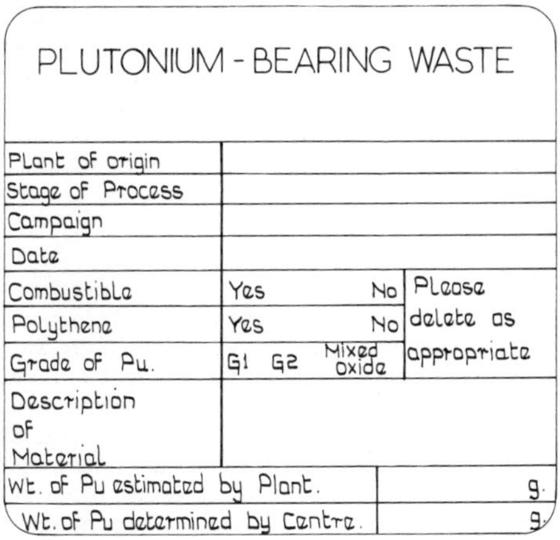

PLUTONIUM - BEARING WASTE

Plant of origin			
Stage of Process			
Campaign			
Date			
Combustible	Yes	No	Please delete as appropriate
Polythene	Yes	No	
Grade of Pu.	G1 G2	Mixed oxide	
Description of Material			
Wt. of Pu estimated by Plant.			g.
Wt. of Pu determined by Centre.			g.

FIG.1. Plutonium-contaminated waste package label.

2. MINIMISATION OF WASTE ARISINGS

2.1 Importance of minimising waste arisings

The effort to minimise waste arisings is regarded as important for the following reasons:

a. waste handling operations are reduced, minimising the associated hazard;

b. process costs are reduced;

c. waste handling treatment costs are reduced, eg eventual incineration;

d. losses of plutonium in waste are minimised, especially from fuel fabrication processes, so that the economics of the fast breeder principle may be fully demonstrated.

2.2 Identifying areas for action

All plutonium-contaminated waste arisings on the Windscale site have been handled centrally since early 1972. Records have been computerised since mid 1973. The data for each packet are recorded by the operator who generates it on a self-adhesive label, the form of which is shown as Figure 1. Numeric codes are used to indicate the type of waste and the point of origin. The waste type classifications used are given in Table I. The points of origin include individual process stages in all plants and laboratories where plutonium is handled, a total 150 separately identified sources.

Computer summaries have been produced which helped to identify clearly those areas giving rise to high waste volumes, or to waste with large amounts of associated plutonium. The summaries also show the types of waste arising in large amounts. This information has enabled a systematic approach to be made to the work of reducing waste arisings.

2.3 Waste reduction

The programme for waste reduction is at a comparatively early stage. Since late 1973, a working group has met to consider the steps required to be taken by plant management to reduce the arisings of plutonium-contaminated waste to a minimum. The group, containing managers from all plutonium plants, members of Windscale Technical Department and representatives of other plutonium-handling sites in the United Kingdom, also studies proposals concerning process or plant improvements that will assist in the longer term. A Code of Practice is being developed that will assist plant design, operating and maintenance teams to minimise waste arisings.

Some of the areas that have been identified already as having potential for further reduction in the contribution they make to waste arisings are:

use of liquor filters - where these are used to protect pumping systems, they can be eliminated by using ejectors or by removing the need to transfer liquid (eg minimising spills);

packaging materials - arisings can be reduced by the use of reusable containers;

ambidextrous gloves used on glove boxes - until glove-free plants are designed, usage of ambidextrous gloves can be minimised by selection of the most appropriate type for each glove box atmosphere, use of specially made PVC stubs for posting in operations and by replacing them by port seals at locations where frequent access is not required;

minimisation of posting operations - conventional bagging in and out techniques which give rise to PVC waste can, in some cases, be replaced by posting-in hatches, by engineered transfer systems or by glove box interconnection;

powder spillages - rather than transfer the powder to swabs which join the waste stream, operators are encouraged to recover powder for recovery by solution chemistry or, if possible, to return it to the process;

tools - plant design should enable safe tools to be used, and if these were made of corrosion resistant materials they could be retained in the glove box;

gas filters - arisings can be reduced by avoiding the use of high gas flows in dusty environments;

plant cleaning at change of enrichment of grade - the need for this should be minimised by careful campaign programming, but when it is necessary, as much material as possible should be recovered without being taken up on swabs (improved glove box vacuum cleaners are under development);

modifications to plants to reduce cell entries which give rise to waste;

washing of waste items to reduce the amount of associated plutonium.

That the programme for waste reduction is proving successful can be shown from the following data:

the weight of waste over the years 1973, 1974, 1975 has decreased progressively by 36% and the weight of associated plutonium by 57%.

2.4 Training

It is regarded as important that the need to minimise waste arisings is brought to the attention of all plant managers and operators.

Managers of plutonium-handling plants discuss the problems associated with waste minimisation both at the Plutonium Waste Reduction Working Group, described in Section 2.3, and at the Works' Technical Committee. They are presented with detailed computer-produced summaries each month, which enables performance in this matter to be assessed. A programme of work has been drawn up to reduce waste arisings in each plant, and the progress of this is periodically reviewed. All supervisors new to plutonium areas attend a course, one of the sections of which highlights the need to reduce waste arisings. Information is given about previous work and useful reference documents. Any waste packets containing particularly large amounts of plutonium are examined by the Waste Management Group, who report results of the examination to the plant manager concerned.

All operators and maintenance workers new to plutonium areas also attend a course, during which the need to achieve reduction in waste arisings is presented. Existing operators will have the opportunity to discuss the problem when they attend refresher courses.

3. SEGREGATION OF ALPHA-BEARING WASTES

3.1 Operations at point of origin

Reliance is placed on the operators who initially generate and package plutonium-contaminated waste to make a segregation between incinerable and non-incinerable items. A standard label (Figure 1) is applied at the point of origin. The information required includes an estimate of fissile material content for criticality control during interim storage. Plant operators and supervisors are trained in the necessary procedures.

TABLE III. Waste Categories used at Windscale

Category	Maximum Pu content	Contents	Destination
1	15 grams[a]	Non-combustible, no polythene[b]	Sea Disposal or Storage[c]
2	15 grams[a]	Combustible, no polythene[b]	Sea Disposal or incineration
3	15 grams[a]	Combustible, polythene[b]	Sea Disposal[d] or incineration
4	130 grams[a]	Non-combustible,	Storage[c,e]
5	200 grams[f]	Combustible, plutonium grade 1	Incineration
6	200 grams[f]	Combustible, plutonium grade 2	Incineration

a. limit chosen for reasons of criticality control during sea disposal operatio
b. polythene is segregated to assist sea disposal preparation operations
c. pending development of treatment methods for non-combustible materials
d. after shredding of polythene, in accordance with international regulations
e. remote possibility of sea disposal
f. limit chosen for reasons of criticality control during subsequent storage

3.2 Operations in Central Waste Handling Facility

Operators from the central waste handling facility collect waste from all the generating points. Each packet in turn is assessed for plutonium content (see Section 4.1). It is then routed into one of a number of disposal or treatment categories, the categories being chosen according to the disposal, treatment or storage options available.

The routes presently available for drummed plutonium-contaminated waste include sea disposal via annual internationally-regulated dumps in the deep Atlantic, incineration and intermediate term storage pending the development of further treatment methods.

The waste categories used are listed in Table III. The scheme, according to which each packet is routed into one of the categories, is illustrated as Figure 2. The system has been found to be flexible and able to respond quickly to any changes in waste treatment routes available. It takes account of the

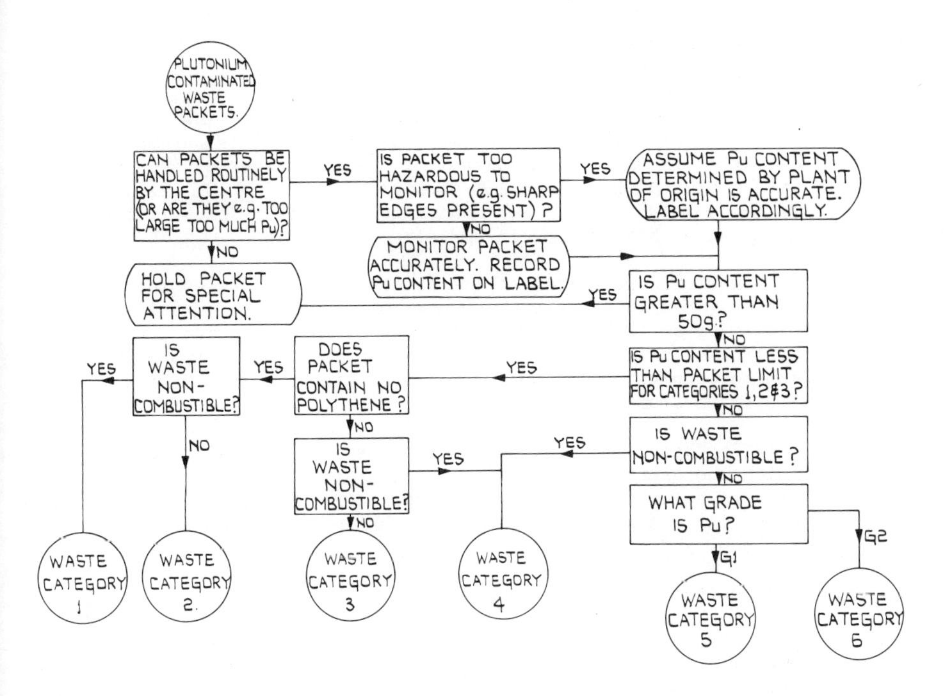

FIG.2. Windscale waste segregation system.

nature of the waste, grade and amount of plutonium in the packet. Different grades of plutonium are handled, because different irradiation histories give differing isotopic compositions.

Additionally, high plutonium-bearing, low volume waste packets are set aside in a priority category for distillation/calcination treatment within Windscale Technical Department to recover plutonium.

Consideration is now being given to enlarging the scope of the central waste handling facility to include simple waste treatment operations. For example, plastics might be shredded, waste compacted where possible, and, for appropriate categories, placed into concrete drums for sea disposal.

4. ACCOUNTING FOR PLUTONIUM-BEARING WASTE

4.1 Present instrumental methods

Although a preliminary estimate of plutonium content of each package is obtained in the plant of origin for nuclear safety purposes, the figures used for accountancy are determined by instrumental non-destructive analysis in the central waste handling facility. These figures are also used to ensure the nuclear safety of subsequent waste handling operations.

Packages are monitored for plutonium content by low resolution gamma spectroscopy using NaI (Tl) scintillation detectors. Each package is placed in turn into a fibreboard drum, 40 cm diameter by 60 cm long. The detector scans along the length of the horizontally rotating drum. The instrument is capable of measuring the plutonium content of small packages of soft waste, with a bias of less than 15% at a precision of ± 10% (2σ). Use of the 355-405 keV channel minimises the effects of variation in irradiation history and in age since separation.

A small proportion of the Windscale plutonium-contaminated waste arises in packets which are too large for the routine assay system or as 1000 cfm HEPA filters which are also too large.[1] A monitor is available, using the same principles as the package monitor, which can be used for larger items up to the size of 210 litre drum [1].

Packages with large amounts of associated plutonium, for which the gamma spectroscopic methods are not suitable, have been assayed using a calorimetric method originally developed for the determination of the plutonium content of reduction slags and residues [1].

4.2 Future instrumental methods

BNFL participates in a UK programme to develop suitable non-destructive measurement techniques for the plutonium content of waste. During the financial year 1976/1977, two improvements are expected to the instrumentation now available. These will help to overcome some of the sources of error in present measurements, namely:

a. expected future variation in plutonium isotopic composition, due to the reprocessing of fuel from a number of different reactor types,

b. presence of in-grown or concentrated 241-Am,

c. presence of fission products, the radiation from which interferes in the energy channel used at present,

d. absorption of gamma radiation by the waste matrix, significant with non-combustible wastes,

e. self-absorption of gamma radiation by the plutonium itself: the method of calibration presently used results in an over-estimate of plutonium content in most cases.

An instrument is being developed at UKAEA, Harwell, which will enable high resolution gamma spectroscopy to be employed using intrinsic Ge detectors. Improved geometrical arrangements will be achieved by scanning 4 cm segments of the rotating package in turn, an integrated result being calculated by linked computer. A correction will be made for matrix absorption

[1] cfm = $ft^3/min = 4.719 \times 10^{-4}\ m^3/s$.

effects by measuring the transmission of the 400 keV photons from a 75-Se source. System stabilisation and correction for dead time will be made with a 133-Ba source. The system will be based on work at Los Alamos Scientific Laboratory [2].

The computerised segmented gamma scan instrument will not be suitable in cases where the plutonium is present at high effective density, for example oxide pellets or centrifuge cake, because it will underestimate the plutonium content. The system is not suited either to determination of Pu in very hard waste, in which the transmission of the reference line is reduced to a few percent. It is hoped that Windscale will receive, under the auspices of the UK waste measurement programme, a National Nuclear Corporation 'random driver' system. This combines a passive neutron detection system with a neutron interrogation system for objects up to 20 litres in volume.

Plutonium-contaminated waste arises at Windscale in crates of varying dimensions. These contain, typically, glove boxes removed from process lines. Windscale Technical Department is now working on instrumental methods to enable plutonium contents to be determined. The technique is expected to include both high resolution gamma spectrosopy and neutron measurements, related to well-defined standards.

4.3 Use of data

When the plutonium content of a packet has been determined, it is written on to the packet label. Packet plutonium contents are entered on to a data sheet as the packets are placed into the 210 litre drums (one sheet for each drum filled).

The information on the data sheets is processed by computer. It is organised in a number of different ways; for example, as listings for each drum, and as listings of waste output for each area of plant. It is also listed to show plutonium outputs in waste for each accounting area. The Windscale Programmes and Technical Records Department then incorporates the information into the various area accounts.

REFERENCES

[1] McDONALD, B.J., FOX, G.H., BREMNER, W.B., Safeguarding Nuclear Materials (Proc. Symp. Vienna, 1975) **2**, IAEA, Vienna (in press).

[2] MARTIN, E.R., JONES, D.F., SPEIR, L.G., Rep. LA-5652-M (1974).

DISCUSSION

Y. NISHIWAKI: Table III of your paper shows that the maximum plutonium content of the waste for sea disposal is 15 g and that this limit was chosen for reasons of criticality control during sea disposal operations. May I ask what concentration this represents in curies per unit gross mass (in tonnes) of the packaged waste prepared for sea disposal in the United Kingdom?

L.F. JOHNSON: The maximum concentration of plutonium in packages prepared for sea disposal is 3 Ci (alpha) per tonne. Taking into consideration the composition of the United Kingdom sea disposal material, the specific alpha activity falls to 0.3 Ci (alpha) per tonne of waste. These figures may be compared with the maximum alpha concentration permitted by the London Convention, i.e. 10 Ci (alpha) per tonne.

J.B. LEWIS: The curie content of plutonium waste depends, of course, on its isotopic composition, i.e. on the burn-up. Most of the wastes disposed of hitherto have been of low burn-up and the plutonium has consisted primarily of ^{239}Pu. The plutonium wastes are handled at Harwell and in the final disposal operation a great deal of other waste will also be included.

A.F. PERGE: Mr. Johnson, you mentioned several times that the plant manager was involved in various types of meetings. Did you really mean the plant manager personally?

L.F. JOHNSON: Yes. The plutonium waste reduction working group has a representative from the management of each plant. After participating in discussions, the representatives approve the action to be taken in their own area. This action is reviewed at subsequent meetings.

A.F. PERGE: We have found that at one of our plants, namely the one generating 40% of plutonium-contaminated waste in the United States of America, the plant manager himself took an interest in how much waste was produced. In 15 months the waste volumes generated at that plant were reduced by 40%.

H. KRAUSE: Can you give us figures on the plutonium losses in the solid wastes from the total nuclear fuel cycle as compared with the plutonium losses in the hulls, in the high-level liquid waste, and in the medium-level waste?

L.F. JOHNSON: I am sorry, but I have no figures for plutonium losses at earlier stages in the reprocessing sequence.

E.P. UERPMANN: Do you know how many grams of plutonium have already been dumped in the North Atlantic?

L.F. JOHNSON: The present rate of about 5 kg/a has been reached from lower rates during previous years in which BNFL took advantage of this disposal route.

Since 1971, when BNFL first participated in the United Kingdom sea dumping operations, the total mass of plutonium disposed of from the United Kingdom has been about 28 kg. The disposal rate before 1971 was lower.

S.A. GÖKSEL: I should like to ask you a question about the waste categories used at Windscale. As shown by your Table III, there are two categories of wastes containing 200 g of Pu. Category 5 contains grade 1 plutonium and category 6 contains grade 2. In practice, how do you segregate these two categories of wastes? Do you have prior knowledge of the isotopic composition of the plutonium-containing wastes?

L.F. JOHNSON: The plutonium grade is checked at the nitrate solution stage by isotopic analysis. As waste arises after that, the grade is known from the point of origin, since the grade of plutonium at any particular stage of the process is always known. The grade is entered on the package label.

MANAGEMENT OF PLUTONIUM-CONTAMINATED WASTES AT THE PLUTONIUM FUEL FACILITIES IN PNC

K. OHTSUKA, H. MIYO, J. OHUCHI,
K. SHIGA, T. MUTO, H. AKUTSU
Plutonium Fuel Division,
Power Reactor and Nuclear Fuel Development Corporation,
Tokai Works,
Tokai-mura, Ibaraki-ken,
Japan

Abstract

MANAGEMENT OF PLUTONIUM-CONTAMINATED WASTES AT THE PLUTONIUM FUEL FACILITIES IN PNC.

A summary of experience in the management of plutonium contaminated wastes at the plutonium fuel facilities of the Power Reactor and Nuclear Fuel Development Corporation (PNC) is presented. Descriptions are given of waste generation, current handling techniques, plutonium measurement techniques and process techniques now under research and development such as an incineration system and closed-cycle liquid waste treatment by decomposition of the ammonium nitrate contained in the liquid.

INTRODUCTION

Over the last few years about 9.6 tonnes of mixed oxide fuels were fabricated for the deuterium critical assembly (DCA) and 1.6 tonnes for the JOYO reactor. The amount of plutonium-contaminated wastes has increased markedly along with these fabrication campaigns.

The paper describes the current situation concerning the management of plutonium-contaminated wastes and further developments expected in the future at the plutonium fuel facilities of the Power Reactor and Nuclear Fuel Development Corporation.

1. GENERATION OF PLUTONIUM-CONTAMINATED WASTES AND CURRENT HANDLING TECHNIQUES

Apart from the inside of glove boxes and hoods in which plutonium and uranium are handled, the operation areas have been maintained essentially at the zero-contamination level since the facilities began operation. Therefore, most of the solid wastes generated in the controlled areas have been disposed of as non-radioactive waste and only those generated in contaminated areas have been stored as radioactive wastes. In principle, these wastes are classified by origin into special containers, which are classified by colour separately as shown in Table I and treated as shown in Fig. 1.

Liquid wastes originating in the controlled areas have been treated as low level wastes, separated from non-process wastes and laundry wastes, which are sent to separate vessels and are monitored. These wastes, after being decontaminated, have been discharged below the rate of 200 m^3/month, as shown in Table II.

Exhaust gas from glove boxes, hoods, incinerator and controlled areas etc. is discharged to the atmosphere, after treatment with High Efficiency Particulate Air (HEPA) filters and monitoring.

TABLE I. CONTAINERS FOR SOLID WASTES

Standard containers		Wastes
20 litre cartons	(red)	Radioactive combustibles (paper, rags etc.)
	(green)	Radioactive combustibles (plastics, rubber)
	(white)	Radioactive non-combustibles (glass, metal etc.)
	(yellow)[a]	Non-radioactive combustibles (paper, rags etc.)
200 litre drums	(yellow)	Radioactive
	(blue)[a]	Non-radioactive (rubber, glass, metal etc.)
Paper bags	(light brown)[b]	Non-radioactive combustibles (paper, rags)

[a] Solid wastes generated in process rooms.
[b] Solid wastes generated in the corridors.
Steel containers and PVC-sheet are also used, but not yet standardized.

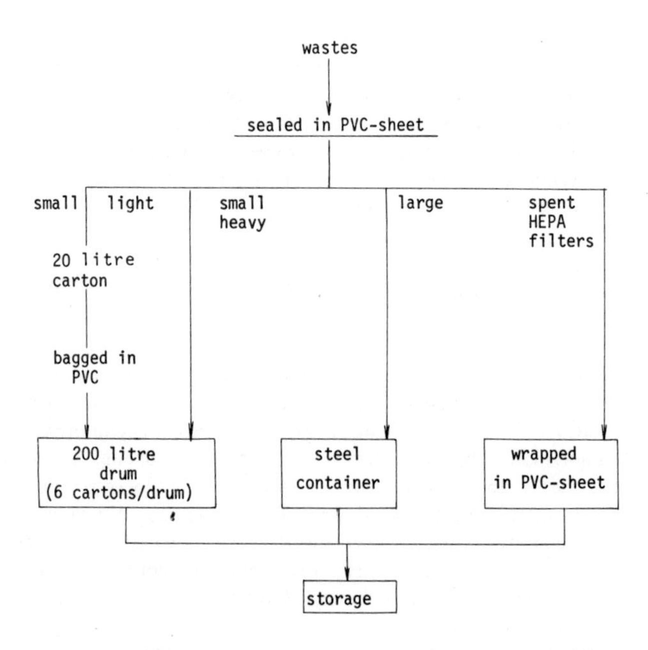

FIG.1. Treatment of plutonium-contaminated solid wastes at the plutonium fuel facilities of PNC.

TABLE II. PRESENT LIQUID WASTE TREATMENT PROCESS

Waste sources		Categories	Total alpha activity ($\mu Ci/cm^3$)
Chemical analyses ↓			
Wet recovery ⟶	decontamination		0.1 ~ 1.0
	↓		$< 1.5 \times 10^{-5}$
Hoods Decontamination rooms ⟶	⟶	LOW LEVEL	$< 1.5 \times 10^{-5}$ — treatment ⟶ discharge
Floor drain		NON-PROCESS	under detection limit ($< 3 \times 10^{-8}$) ⟶ treatment
Laundry room		LAUNDRY	under detection limit ($< 3 \times 10^{-8}$) ⟶ treatment

Note: Spent oils and solvents are stored.

1.1. Solid wastes

Over the last ten years the radioactive solid wastes at the facilities have accumulated to about 720 m^3 in net volume of the containers. The amount of waste generated each year is shown in Fig. 2. These wastes are stored in 8 storage depots with total net volume of 2000 m^3. The total amount of wastes contained in drums and spent HEPA filters occupy 65% and 26%, respectively.

Glove boxes are scrapped in the following steps:

(1) Bag out of the installed equipment

(2) Decontamination and painting of the glove box

(3) Loading of the box sealed in PVC sheet into a steel container.

The waste containers for the scrapped glove boxes had accumulated to about 37 m^3 by March 1975.

Table III shows the amount of radioactive solid wastes generated in the controlled rooms directly concerned with fabrication and inspection of the fuels for JOYO and DCA. The typical drum waste distribution in 1974 was as follows: ratio of paper to the total drum waste 29.8%; that of rubber and plastics 58.8% and that of metal and glass 9.1%.

Solid wastes generated in the corridors and process control rooms etc., where fissile materials are not handled, are disposed after checking for contamination. On the other hand, solid wastes generated in process rooms where fissile materials are handled are classified into two groups: combustibles mainly composed of paper, and others. The former, after being contained in 20-litre yellow cartons and temporary storage in the facilities, is hand-sorted and incinerated. Over the last ten years the amount of incinerated wastes has been 108 m^3. The other wastes, which are mainly rubber, plastics and glass etc., are put into 200-litre drums to be placed in the special outdoor storage. By March 1975 about 800 drums had been generated.

1.2. Liquid wastes

Liquid process wastes originating in chemical analyses and wet recovery processes at the Plutonium Fuel Development Facility (PFDF) and the Plutonium Fuel Fabrication Facility (PFFF)

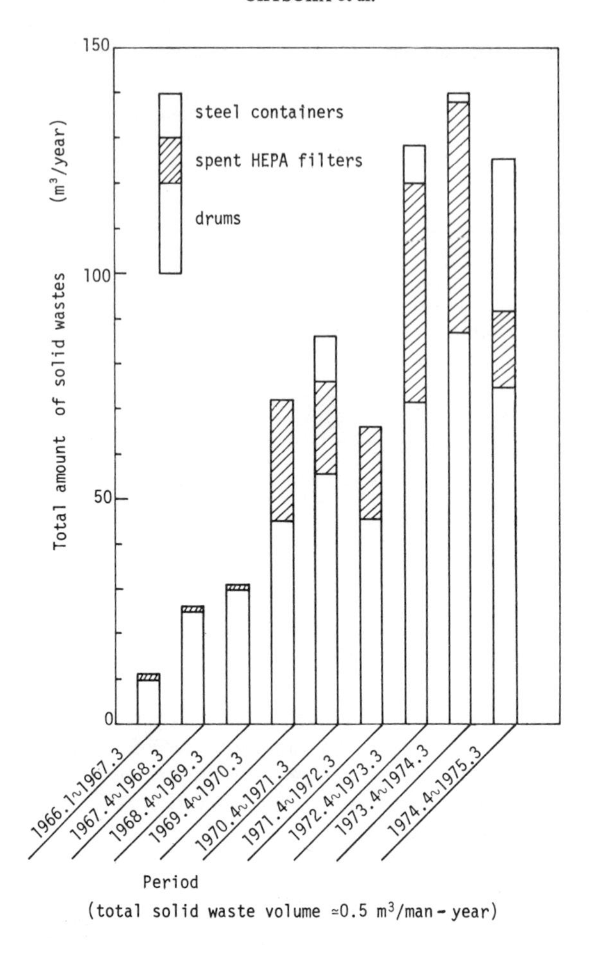

FIG.2. Total amount of plutonium-contaminated solid wastes generated annually at the PFDF and PFFF.

have been decontaminated by the following equipment. At the PFDF the liquid waste decontamination equipment consists of an evaporator and an active carbon bed with a capacity of 8 l/h. A decontamination factor (DF) of nearly 10^4 is obtained. At the PFFF the equipment consists of three flocculators and an active carbon bed etc. with a capacity of 30 l/h. A DF of 10^4 to 10^5 is obtained. By March 1975 27.8 m^3 of process wastes had been decontaminated at the PFDF and the PFFF. The waste treated in the above, together with the liquid waste originating from chemical hoods, has been treated as low level liquid waste.

The amount of liquid wastes generated at the PFDF and PFFF is summarized in Fig. 3.

1.3. Gaseous waste

The amount of exhaust gas was about 1.1×10^9 m^3/a until 1971 and then increased to 3.5×10^9 m^3/a with the start-up of the PFFF and other facilities. The amount of alpha-emitters contained in these gaseous wastes has always been below the detection limit (1×10^{-16} $\mu Ci/cm^3$).

TABLE III. AMOUNT OF PLUTONIUM-CONTAMINATED SOLID WASTES GENERATED ALONG WITH THE FABRICATION OF JOYO AND DCA FUELS

Period	Amount of mixed oxide fuels treated (1000 kg)	Amount of solid wastes generated (in 200-l drums) (m^3)
JOYO		
Nov. 1972 – Mar. 1973	0.04	1.2
Apr. 1973 – Mar. 1974	0.86	21.4
Apr. 1974 – Mar. 1975	0.70	10.0
Total	1.60	32.6
DCA		
Apr. 1972 – Mar. 1973	5.1	21.5
Apr. 1973 – Dec. 1973	4.5	15.2
Total	9.6	36.7

2. IMPROVEMENTS IN ALPHA-WASTE TREATMENT

2.1. Solid wastes

A major object in the near future is the stabilization of the plutonium-contaminated solid wastes after volume reduction in a retrievable condition. This may be attained, for example, by incineration of the combustible waste, melting and re-solidifying the plastic waste, compression of the bulk waste, and finally, melting and re-solidifying the glass and metal wastes.

2.1.1. Incineration of the plutonium-contaminated wastes

An incinerator with a capacity of ~ 20 kg/h was constructed in 1974 in the PFDF in order to reduce the volume of stored wastes by burning the plutonium-contaminated wastes.

The combustion chamber and post-combustion chamber consist of refractory bricks lining a stainless steel shell. Ignition is by a heavy-oil fuelled jet. The combustion chamber and post-combustion chamber are maintained by heavy oil at 200–800 and 800–1000°C, respectively. The off-gas treatment system consists of the post-combustion chamber with a silicon carbide brick bed, a high temperature filter (HTF) with 222 candle filters precoated with asbestos, a diluter for cooling the gas with air below 150°C, four high temperature HEPA filters, a heat exchanger for cooling the gas below the temperature of 60°C, and a wet scrubber for removing toxic elements such as hydrogen chloride and sulphur dioxide.

As shown in Fig. 4, the combustible wastes are hand-sorted to free them from non-combustibles, put into 20-litre cartons and hand-charged to the combustion chamber. To prevent contamination spreading in the operating room, this incinerator has four glove boxes for feed preparation, ash-removal, pulling out the HTF elements and exchanging the high temperature HEPA filters. Further care is paid to minimize the possibility of leakage of combustion products to the operating room through pressure fluctuations.

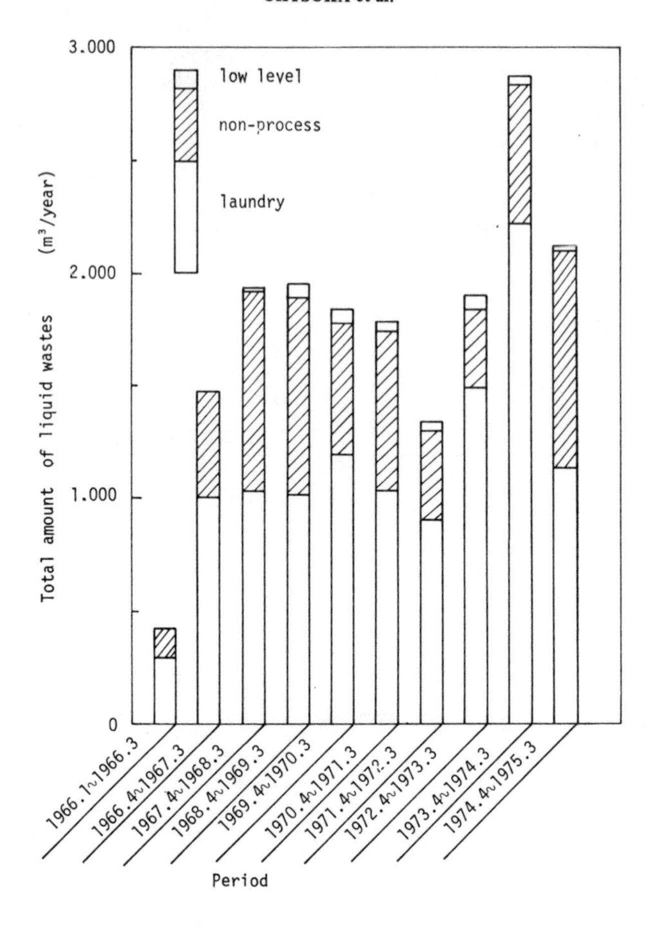

FIG.3. Total amount of liquid wastes generated annually at the PFDF and PFFF.

TABLE IV. SOLID WASTES FOR PRE-OPERATIONAL TEST OF THE INCINERATOR

Material	wt%
Polyethylene	48
Rags	23
Paper	23
Rubber	4
PVC	2
Natural uranium	(0.1)

TABLE V. RESULTS OF PRE-OPERATIONAL TEST OF THE INCINERATOR

Items	Results
Capacity	10 – 15 kg/h
U distribution	Ash $\simeq$ 100%, HTF elements = 0.1%, gas after HTF $<$ 0.001%
Volume reduction	1/85
Weight reduction	1/20 – 1/40

After the completion of this incinerator, pre-operational tests using artificially made uranium wastes shown in Table IV have been performed for testing and improving the equipment. Results of these tests are summarized in Table V.

2.1.2. Measurement of plutonium in the solid waste containers

(A) Drum scanning. The equipment itself is of the conventional type. It consists of a platform, on which a 200-litre drum is placed, and a single-channel gamma-scanning system. The platform is rotated 14 times uniformly during the counting period of 45 minutes. An NaI(Tl) detector is moved vertically during this period. A 2 mm thick lead shield covers the detector, eliminating some of the soft gamma rays of americium. The minimum quantity of plutonium detectable by the waste-drum scanner is 20 mg.

About 1700 drums have been measured since 1973. The percentage of the drums that contained less than 1 g of plutonium was 83% and that with more than 10 g was 0.5%.

(B) Carton scanning. It is considered better to classify the cartons at origin by their plutonium content before loading into drums. Equipment using the active neutron technique has been installed and is now being tested.

2.1.3. Treatment of spent HEPA filters

As a first step to develop a method of treating spent HEPA filters, a study of methods of melting and re-solidifying the glass fibre filter media has been started using a small electric crucible furnace.

2.2. Liquid waste

The development of a new closed-cycle liquid waste treatment system was started in 1975. In this system the solid waste would be minimized and the water treated for re-use. A feasibility study for such a closed-cycle system was performed before September 1975. Figure 5 shows a model flowsheet and the processes proposed for development.

The liquid process waste at the plutonium fuel facilities contains a large quantity of salt (~ 20 wt% of ammonium nitrate), which should be removed for full decontamination of the liquid waste. From the results of the feasibility study, a thermal decomposition process is considered to be the most promising process for development. In this process the liquid waste containing ammonium nitrate is decomposed to innoxious gas and the minimum quantity of solid residue.

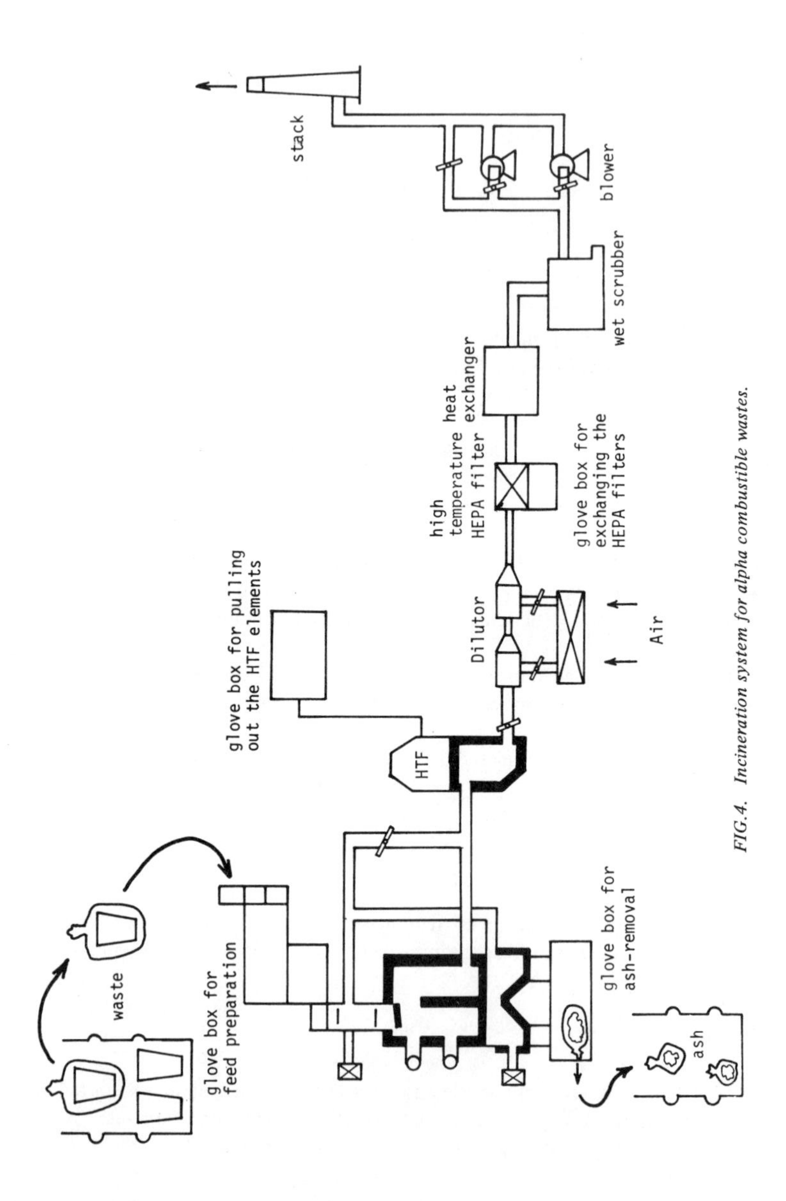

FIG.4. Incineration system for alpha combustible wastes.

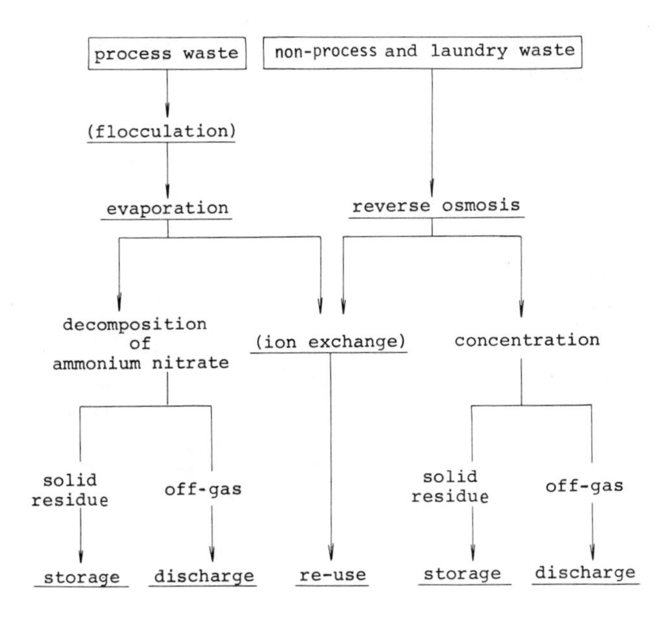

FIG.5. A flowsheet for a closed cycle system of liquid wastes.

3. SUMMARY

(1) Over the last ten years a total of 720 m^2 of solid waste was generated, sealed and stored within the facilities.

(2) About 2000 m^2 of liquid waste, mostly non-process waste, is discharged yearly after decontamination.

(3) An incinerator for combustible wastes including plastics and rubbers is now under pre-operational testing.

(4) Solid plutonium wastes in drums are measured by gamma-scanning, while another method using an active neutron technique is now under testing.

(5) A new liquid waste treatment programme for the closed cycle system was started in 1975.

DISCUSSION

H.W. LAHR: Have you had any trouble with the concentration and thermal decomposition of ammonium nitrate solutions? While doing this type of work, we have experienced two minor explosions in the evaporator during the concentration process. In any case, what was the concentration of your solutions?

K. OHTSUKA: No, we have not had any such problems as yet.

In the batch process, 150 litres of weak alkali liquid waste are concentrated to a volume of about 15 litres. The concentrated liquid is collected from several batches and evaporated again until the volume is reduced by half, after which it is cooled to room temperature to crystallize the salt; it is then stored, following filtration.

J.B. LEWIS: What levels of plutonium do you expect to use in the incinerator? We have found that combustible suspensions tend to build up on dry gas filter systems and that they sometimes catch fire. Have you had any such experience?

K. OHTSUKA: We will incinerate wastes with a surface dose rate lower than 50 mrad/h. Almost all combustible suspensions tend to burn out on the silicon carbide brick bed in the combustion chamber. We have not yet had any outbreaks of fire.

G. STOTT: Your paper, and the one presented today by Mr. Johnson (paper IAEA-SM-207/46, these Proceedings, Vol. 2) have an interesting point of similarity – both indicate the volume of rubber and plastic waste as being about 50 to 60%. Is any work being done with a view to reducing the volume of this type of waste after it has been collected, and before storage or disposal?

K. OHTSUKA: No, as yet we have not made any such study.

R. ALLARDICE: There is a joint UKAEA/BNFL working party, which co-ordinates the development work on plutonium-contaminated waste treatment methods within the United Kingdom. This working party has sponsored work aimed at reducing the amounts of rubber and plastic contaminated with plutonium at United Kingdom facilities, and seeking alternative materials that might be more amenable to treatment or destruction. The results of this development work will be applied in the design and operation of new and existing facilities.

IAEA-SM-207/86

CONSIDERATIONS ON NUCLEAR TRANSMUTATION FOR THE ELIMINATION OF ACTINIDES

W. BOCOLA, L. FRITTELLI, F. GERA,
G. GROSSI, A. MOCCIA, L. TONDINELLI
CNEN-CSN Casaccia,
Rome,
Italy

Abstract

CONSIDERATIONS ON NUCLEAR TRANSMUTATION FOR THE ELIMINATION OF ACTINIDES.

The sensitivities of actinide build-up in LWRs in relation to variations in capture cross-sections, capture resonance integrals and decay constants, have been calculated. The flowsheet of a chemical process, potentially capable of separating the actinides from high-level waste with the required efficiency, has been outlined. The actinides are assumed to be recycled in fast breeders in special fuel pins. The half-residence time of the actinides in fast breeders is about nine years. In a first approximation it appears that the risk associated with nuclear transmutation could be greater than the risk resulting from disposal in a very favourable geological formation.

1. INTRODUCTION

In the reprocessing of irradiated fuel associated with a large nuclear power industry based on fission reactors large amounts of high-level radioactive wastes are generated. Because of the special problems caused by the long-lived radionuclides involved, the long-term management of these wastes must be carefully planned. Theoretically a possible solution to the long-term disposal problem is to eliminate a significant fraction of long-lived radionuclides by neutron-induced transmutation. Nuclear transmutation is conceptually attractive since it would lead to a considerable reduction of the long-term hazard associated with radioactive wastes.

Usually the emphasis is on transmutation of the long-lived actinides, since burning of fission products (^{90}Sr, ^{137}Cs) in power reactors with present neutron fluxes does not seem to be practical. Furthermore, after a thousand years both of these isotopes will have decayed to near negligible levels, while the hazard of the total waste in the very long term from 1000 to 10^6 years or so is controlled by the actinide nuclides.

A waste management strategy based on nuclear transmutation of long-lived actinides consists mainly of the separation of all the actinides from the fission products and of their subsequent neutron-induced transmutation. Claiborne [1] has shown that long-term hazard reduction factors up to about 200 are possible with a 99.9% actinide extraction efficiency, with subsequent recycling of the neptunium, americium and curium in LWRs.

On the other hand, it is well known that the recycle of actinides in a Fast Breeder Reactor would increase the hazard reduction factors since the average fission-to-capture ratio of the actinides should be higher in a fast neutron flux. The main objectives of this study are to:

(a) Evaluate the sensitivities of final concentrations of Am and Cm isotopes with respect to nuclear data of their precursors in a typical LWR neutron flux
(b) Define a chemical process potentially capable of achieving the necessary separation efficiencies

(c) Assess the implications of burning the separated actinides in fast breeder reactors
(d) Discuss the overall risk implications of the nuclear transmutation strategy in relation to disposal of the unseparated wastes in geological formations.

2. SENSITIVITY OF ACTINIDE PRODUCTION ESTIMATES

One of the many questions hampering the evaluation of the long-term hazard potential associated with high-level radioactive waste is the reliability of current estimates of actinide inventories in spent fuel and solidified wastes. These estimates are a function of the irradiation history and the nuclear data used in the calculations.

A theoretical method has been developed to evaluate, with generalized perturbation techniques, the sensitivities of calculated quantities in respect to variations in the nuclear data [2,3].

The method permits a calculation of the nuclide inventories, hazard potentials and so on; however, in this paper only the sensitivities of actinide production in relation to uncertainties about the applicable nuclear data have been calculated.

The reference reactor assumed in the calculation of actinide production is a typical BWR with Pu recycle; its characteristics are shown in Table I.

The calculation assumes a 1-year irradiation with constant neutron flux and neutron energy spectrum.

TABLE I. DATA FOR THE REFERENCE BWR

Thermal power		2894 MW	
Specific power		26 MW/t	
Burn-up		28600 MW·d/t	
Fuel element		63 pins of UO_2 enriched with 2.75 wt% of PuO_2	
Pu isotopic composition (wt% of oxide)			
Pu-239	83.75	Pu-240	13.85
Pu-241	2.07	Pu-242	0.30

2.1. Theory

In matrix notations the burn-up equations, in the time interval $t_0 \leqslant t \leqslant t_F$, may be written as

$$\frac{dn}{dt} = An \; (n(t_0) = n_0) \tag{1}$$

Let us now consider an integral quantity Q, as given by the expression

$$Q = h^T n\,(t_F) \tag{2}$$

where h^T is a row vector representing assigned values. With respect to Eq. (1), the adjoint equation, by which the 'importance function' n^+ results defined, is:

$$-\frac{dn^+}{dt} = A^T n^+ \; (n^+(t_F) = h) \tag{3}$$

TABLE II. ACTINIDE PRODUCTION SENSITIVITIES RELATIVE TO NUCLEAR DATA

Nuclide (Q)	Precursosrs and nucl. itself (i)	Nuclear datum (q_i) $\sigma_{th}(n,\gamma)$	R.I.(n,γ)	$\lambda(\beta^-)$
Am-241	Pu-240	14.35	40.67	
	Pu-241	−0.98	−0.64	96.65
	Am-241	−6.84	−8.24	
Am-242-m	Am-241	37.42	45.08	
	Am-242-m	−10.88	−	
Am-243	Pu-241	39.12	47.08	0.38
	Pu-242	2.36	84.75	
	Pu-243			0.11
	Am-241	0.37	0.44	
	Am-242-m	0.59	−	
	Am-243	−1.00	−7.47	
Cm-242	Am-241	29.31	83.03	
	Am-242			0.55
	Cm-242	−0.17	−	−36.66[a]
Cm-243	Am-242			18.51
	Cm-242	99.86	−	
	Cm-243	−1.12	−1.46	−0.50[a]
Cm-244	Pu-243			0.18
	Am-242-m	0.37	−	
	Am-242			−0.15
	Am-243	10.21	75.63	
	Cm-244	−0.01	−	

[a] α-decay.

If after the time $t = t_0$ a perturbation δA into the matrix A is introduced, a change δQ of the quantity Q will generally follow. This change is given by the perturbation expression

$$\delta Q' = \int_{t_0}^{t_F} dt \; n^{+T} \cdot \delta A \cdot n \qquad (4)$$

The sensitivity of the quantity Q with respect to a variation of a nuclear datum q_i, to which the variation δA corresponds, is defined by

$$S_{Q,q,i} = 100 \, \frac{\delta Q/Q}{\delta q_i/q_i} \qquad (5)$$

2.2. Results

The method has been programmed for CNEN's computer IBM–370/168 in the PERSEO Code, which includes the decay chains of 26 isotopes of U, Np, Pu, Am and Cm, using the three-energy group formalism and the ORIGEN nuclear data library for LWRs [4,5].

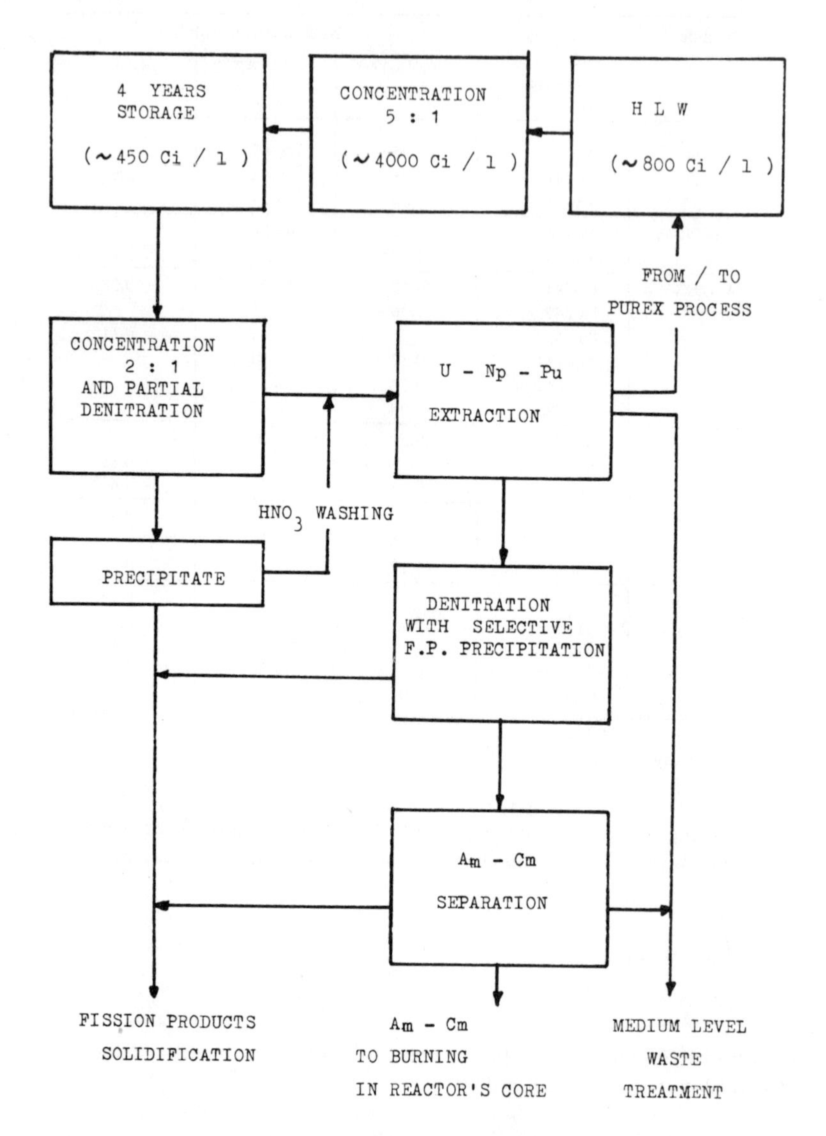

FIG.1. Stages of the separation process.

The sensitivities of the final concentrations of Am and Cm isotopes with respect to thermal capture cross-sections, capture resonance integrals and decay constants, relative to the nuclides themselves and to their precursors, have been calculated; a relative perturbation $\delta q_i/q_i = 0.2$ has been assumed.

Table II presents the results, giving, for each nuclear datum q_i of the precursor i, the relative sensitivity of the concentration of nuclide Q. In this table the sensitivities to (n, 2n) and (n, 3n) cross-sections are not reported. They have been calculated and fall in the range between 10^{-9} and 10^{-4}, thus, in this context, their consideration is not important. Some sensitivities have been omitted in Table II either because data are lacking in the ORIGEN library or because a rough estimate suggested that they would have a minor influence on the actinides production.

The general conclusions are:

(a) With few exceptions, the sensitivities relative to nuclear data are less important than uncertainties due to the neutron energy spectrum in the reactor;

(b) This calculation provides an indication of the nuclear data for which a better knowledge would give the highest return in order to reduce uncertainties about actinide inventories.

3. ACTINIDE SEPARATION

The feasibility of nuclear transmutation of the actinides depends on the efficiency of their separation from the other waste substances. Various chemical processes have been recently proposed to separate actinides from high-level waste solutions [6–8].

A complete separation flowsheet based on the integration of various proposals is shown schematically in Fig.1. On the basis of available information it can be stated, with a high degree of confidence, that this process could be technically feasible.

The aqueous reprocessing (PUREX) of one tonne of spent reference fuel (see Table I), 150 days after discharge from the reactor, produces about 5000 litres of high-level liquid wastes. To save storage volume these wastes are concentrated to a maximum specific activity of about 4000 Ci/l.

The evolution of the waste in function of decay time is shown in Table III. These data suggest that the storage of the wastes in liquid form for about four years would result in a significant benefit from the viewpoint of activity concentration and heat generation rate. A further concentration by a factor of 2 takes place at this stage, achieving a tenfold total volume reduction. This step can be used to adjust the solution to the chemical conditions required for the following extraction of U, Np and Pu.

TABLE III. AGEING OF HIGH-LEVEL WASTES FROM THE REFERENCE FUEL

Cooling time	Specific activity (Ci/t heavy metal)
150 days (after discharge)	3.94×10^6
2 years (after reprocessing at 150 days)	8.90×10^5
3 years	5.92×10^5
4 years	4.40×10^5
5 years	3.60×10^5
10 years	2.39×10^5

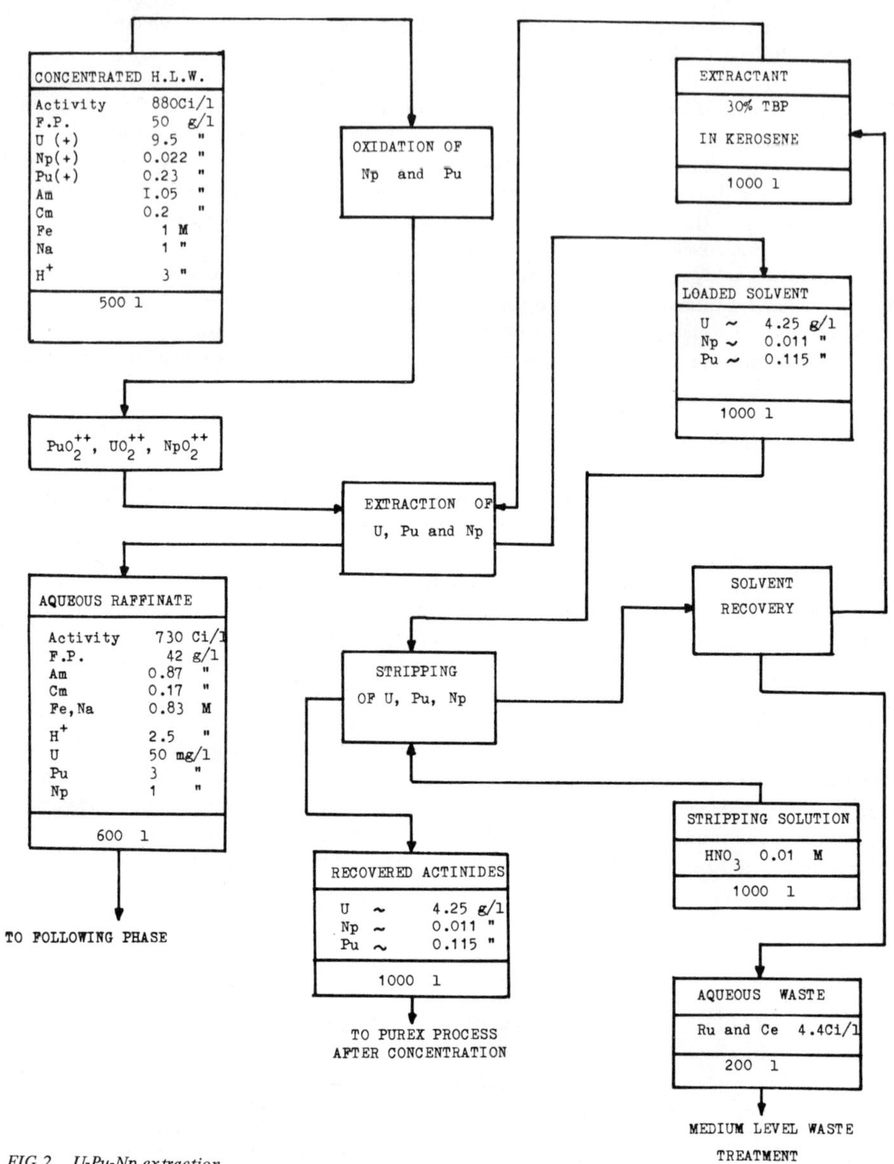

FIG.2. U-Pu-Np extraction.

(+) In the PUREX the overall losses of U, Pu and Np are considered to be recycled to the main waste stream. Assumed total losses are 0.5% for Pu and U and 10% for Np.

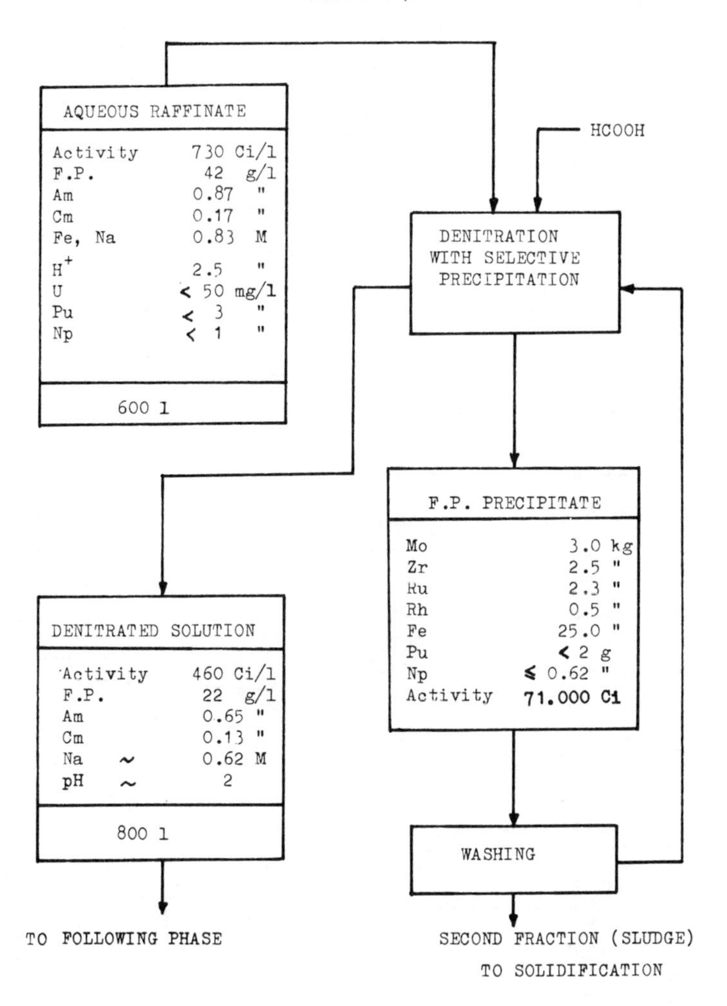

FIG.3. Denitration and selective fission product extractions.

The other main steps of the separation process are:

(a) Selective extraction and recycling to the PUREX process of U, Np and Pu. This extraction can be achieved with TBP, thanks to the high nitrate content obtained during the previous concentration. Furthermore, the removal of U, Np and Pu in this stage is very advantageous since the following steps will be performed at low acidity. (See Fig.2 for details.)

(b) Denitration and selective extraction of fission products. In this phase a fraction of the fission products and almost all the iron are precipitated to facilitate the subsequent separation of Am and Cm. (See Fig.3 for details.)

(c) Separation of Am and Cm. This step is an extraction with diethyl-exyl-phosphoric acid accomplished with a scheme similar to the Talspeak process [9]. This process separates the remaining fission products into two fractions. Details are shown in Fig.4.

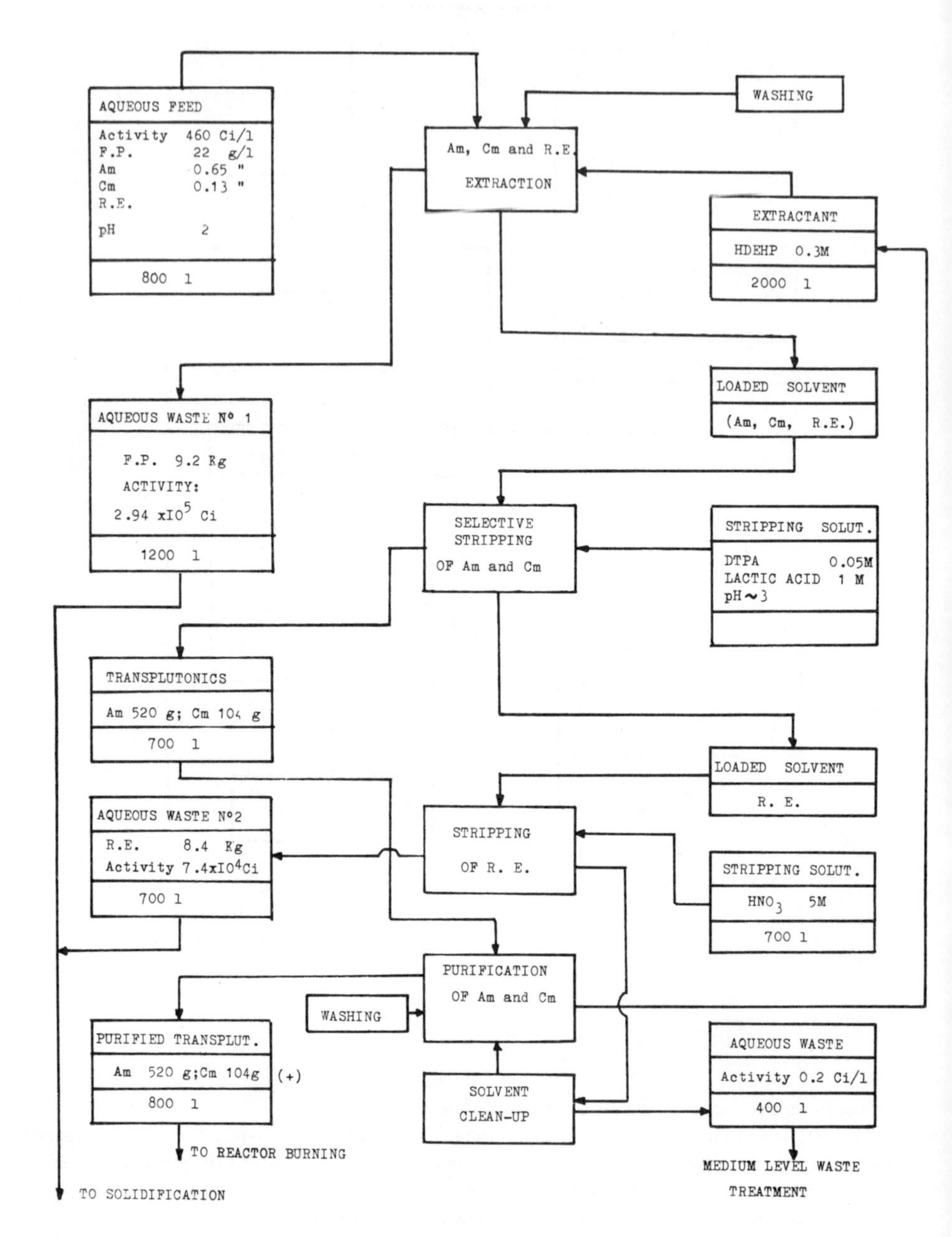

FIG.4. Am and Cm separation.

(+) Associated with 456 Ci of γ and 1.1×10^9 n/s.

TABLE IV. REQUIRED RECOVERY FACTORS TO LOWER WASTE HAZARD INDEX TO < 5% OF THAT FOR PITCHBLENDE

Actinide	Recovery factor from wastes
U	≥ 5
Np	≥ 20
Pu	≥ 50
Am	≥ 1000
Cm	≥ 1000

The whole chemical process produces a series of liquid and solid wastes containing the total $\beta-\gamma$ emitters which will be solidified by the appropriate techniques. The remaining actinide contamination in the high-level wastes should be lower than the values shown in Table IV. Claiborne [1] has indicated these values as the targets to bring the hazard potential in 1000-year-old waste to the same level as uranium ore.

4. ACTINIDE TRANSMUTATION

After the successful separation of the actinides it will be necessary to proceed to their transmutation. Elimination is theoretically possible in either one of the following types of facilities: thermal reactors, fast breeders, spallation reactors, thermonuclear reactors. This discussion is restricted to fast breeders because they seem to have some advantages in comparison with thermal reactors, thanks to the larger fission-to-capture ratio at higher neutron energies, and are much closer to availability than spallation and thermonuclear reactors.

Burning in fast breeders can be envisaged in special fuel elements, in special fuel pins, or by mixing the waste actinides with the normal fuel. However, the first possibility would introduce serious distortions in the neutron flux (ϕ), while the third alternative appears to create significant complications in the fuel fabrication industry; in addition it is basically less efficient since it requires the dispersion of the waste actinides in the fuel, with the subsequent separation of the remaining fraction from the total high-level waste stream. Therefore burning in special pins dispersed in the normal fuel elements has been assumed.

Since the term $\langle \sigma \varphi \rangle$ in burn-up equations is of the same order in thermal and fast reactors, the sensitivities of actinide production with respect to nuclear data are roughly of the same order too. This has been confirmed by calculations performed by the PERSEO Code with the ORIGEN fast cross-sections library. The results are in the course of publication elsewhere [10].

The transmutation efficiency of waste actinides in fast breeders is being evaluated by calculations based on the design of an actual power fast breeder. Some original *ad hoc* cross-sections have been developed and the appropriate self-shielding corrections for the specific fuel pins have been taken into account. The complete calculations and results will be published elsewhere [11].

The main assumptions are:

(a) Thermal power of reactor: 3000 MW
(b) Specific power: 80 MW/t
(c) Theoretical maximum burn-up of special pins: 120 000 MW·d/t
(d) ^{239}Pu in core: 5 t
(e) Mass of waste actinides in core: 30 kg

A preliminary calculation with the ORIGEN Code demonstrates that, after about 30 years of permanence in the reactor, 90% of the original amount of Am and Cm has been eliminated by transmutation. The resulting half-residence time is about nine years. This implies that, not considering problems connected with fuel management, about seven irradiation cycles would be needed to achieve a 90% reduction of Am and Cm. If it does not prove possible to reintroduce the special pins in the reactor without destruction, the burn-up would become equal to ~ 40 000 MW·d/t and the number of irradiation cycles would increase to ~ 20.

5. HAZARD CONSIDERATIONS

The critical question is whether a waste management strategy based on nuclear transmutation would actually be characterized by a lower overall risk than the reference management scheme, which is assumed to end with disposal of unpartitioned solidified wastes in deep geological formations.

For the time being any comparison of this kind will have a very low confidence level, in view of the still unknown factors about many steps of the transmutation strategy, but, even more, because of the many uncertainties about the geologic disposal option. Geological formations are extremely variable from the viewpoint of both intrinsic properties and general setting; consequently the reliability of waste containment in geological formations can vary within wide limits.

A recent study about the incentives for separating actinides from high-level waste has shown that, assuming that vitrified waste is disposed of in a geologic environment similar to the Hanford subsoil in the western USA, the benefit would be negligible [12]. Since the geological materials, currently considered as the best choice for waste disposal, namely: rock salt and plastic clays, are greatly superior to the formations underlying Hanford, the incentives for partitioning the actinides in the case of disposal in these formations should be even lower. These considerations are based on leaching and transport by groundwater as the mechanism responsible for entry of the actinides into the biosphere; in addition, the actinide-carrying groundwater is assumed to go through a significant thickness of porous materials characterized by fairly good ion exchange capacity.

With different assumptions, particularly with regard to the containment failure mechanism, the conclusions could be different. The impact of a giant meteorite is a typical example in this respect, since it could release a fraction of the activity by-passing all geologic barriers [13]. The probability of cratering to the depth of disposal is a function of the depth of the repository, but it is independent of location and geological conditions of the disposal formations. For a 600-m deep repository the probability of activity release as a consequence of a meteorite impact has been estimated as 1.6×10^{-13} per year.

Thus it is obvious that geologic disposal can reduce the risk associated with radioactive wastes to very low levels, but that a small risk will exist as long as the radioactive substances exist.

The actual hazard associated with radioactive substances is dependent on several factors, including the radiotoxicity of the nuclides and their biologic availability in the environment [14,15]. However, in the following discussion the environmental and biological aspects of the problem are essentially neglected. Thus the following considerations are an indication of a possible approach more than an actual hazard assessment.

If the actinides are separated and eliminated by transmutation, the long-term (> 1000 years) hazard potential of high-level waste is reduced by about two orders of magnitude [1]. In a first approximation it can be accepted that any accident happening to the actinide-free high-level waste (always after at least 1000 years decay) would have its consequences reduced by a factor of about 100.

It must be considered that a repository would contain, in addition to high-level waste, large volumes of alpha-bearing waste, with a total inventory of actinides roughly similar to that contained in the unpartitioned high-level waste. Thus, unless a 100-fold reduction of actinides in alpha-bearing

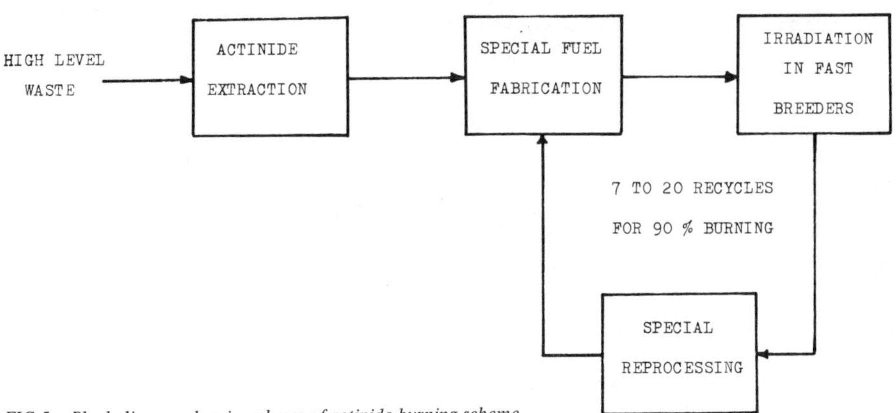

FIG.5. Block diagram showing phases of actinide burning scheme.

waste were also accomplished, any consideration of transmutation would be rather pointless. However, for the sake of discussion we will assume that decontamination of alpha-bearing waste by a factor of 100 would accompany the successful implementation of actinide transmutation.

The price for the reduction of the long-term environmental hazard would be the risk increment resulting from the series of operations required to separate the actinides and burn them. Figure 5 shows a scheme of the actinide transmutation concept. Each step presents a certain risk which, neglecting the environmental and biologic considerations, will be a function of the total quantity of actinides, the accident probability, and the fraction of activity that would be released. The accident probability for each operation can be assumed to be equal to the applicable value for the same operation involving normal fuel.

In conclusion, the hazard associated with nuclear transmutation can be expressed as:

$$H\ \text{n.t.} = Q_a\ pf\ \frac{T_{\frac{1}{2}}}{\ln 2}$$

where Q_a is the quantity of actinides, p is the probability of accident, f is the fraction of released activity, $T_{\frac{1}{2}}$ is the half-residence time of the actinides in the breeder.

For fuel fabrication, irradiation and reprocessing the product pf is in the same order of magnitude, i.e. 10^{-8} to 10^{-9} [16, 17]. Obviously in each cycle the actinides would spend some time off the reactor, therefore the effective mean time would become somewhat longer. Assuming $T_{\frac{1}{2}}/\ln 2 = \sim 15$ years, the total transmutation hazard can be expressed as:

$$H\ \text{n.t.} = Q_a \times 7.5 \times 10^{-8}$$

Handling and transportation will also present a certain risk. To calculate the transportation hazards the following values have been proposed [18]:

Probability of 'severe' accident: 10^{-8} mile^{-1}
Fraction of released activity: 10^{-3}

Since the length of transportation of the actinides could vary from very high values in the case of scattered facilities, to practically zero in the case of nuclear parks, no calculation will be attempted. As far as the handling hazard is concerned, it is clear that it would be directly proportional to the number of handling operations and therefore to the number of recycles.

The hazard of geologic disposal can be expressed as:

$$H_{g.d.} = Q_a \, p f \, \frac{T_{\frac{1}{2}}}{\ln 2}$$

In this case $T_{\frac{1}{2}}$ is the half-life of the mixture of actinides, and Q_a, p and f have the same meaning as above.

Since $T_{\frac{1}{2}}$ of the mixture of actinides is about 16 000 years, if we assume, in the time interval 1000 to 100 000 years, meteorite impact as the mechanism of containment failure and 10^{-2} as the fraction of released activity, the long-term hazard associated with geologic disposal is:

$$H_{g.d.} = Q_a \, 3.7 \times 10^{-11}$$

This conclusion seems to justify the suspicion that nuclear transmutation might increase the overall hazard associated with waste management, in comparison with geologic disposal in suitable formations. However, many geological formations might be less reliable containment media than it is assumed here; therefore it can be stated that meaningful comparison of the hazards associated with alternative waste management strategies would be possible only after the options had been defined in much greater detail.

6. CONCLUSIONS

Undoubtedly a better knowledge of nuclear data and, therefore, better estimates of actinide contents in spent fuel and in high-level waste are desirable. However, for most hazard assessments the calculated waste compositions are sufficiently reliable, since much greater uncertainties are attached to many steps of the various waste management strategies.

On the basis of available information, extraction of actinides from high-level waste with the required efficiency seems to be feasible. The envisaged process is based on current technology, but it is rather complex.

Burning in fast breeders is also considered feasible. It is proposed that the waste actinides should be kept separated from the normal fuel and that irradiation should take place in special fuel pins. The waste actinides in the assumed fast breeder would disappear with a half-time of about nine years. This value can be compared with the average half-life of the mixture of actinides in unpartitioned high-level waste, which in the time interval between 1000 and 100 000 years is in the order of 16 000 years.

Thus nuclear transmutation would result in a risk of much shorter duration. However, in the case of a very reliable geologic containment it is possible that the short-term risk increment due to the various phases of the nuclear transmutation strategy would exceed the long-term hazard associated with geologic disposal.

REFERENCES

[1] CLAIBORNE, H.C., Neutron Induced Transmutation of High-Level Radioactive Waste, ORNL-TM-3964 (1972).
[2] GANDINI, A., Perturbation methods in nuclear reactors from the importance conservation principle, Nucl. Sci. Eng. **35** (1969) 141.
[3] GANDINI, A., Time-dependent Generalized Perturbation Methods for Burn-up Analysis, CNEN Rep. RT/FI (75) 4 (1975).
[4] TONDINELLI, L., Il Programma PERSEO, to be published as CNEN internal Report.
[5] BELL, M.J., ORIGEN-The ORNL Isotope Generation and Depletion Code, ORNL-4628 (1973).
[6] BARTLETT, J.W., et al., Feasibility Evaluation and R & D Program Plan for Transuranic Partitioning of High-Level Fuel Reprocessing Waste, BNWL-1776 (1973).

[7] BOND, W.D., CLAIBORNE, H.C., LEUZE, R.E., Methods for removal of actinides from high-level wastes, Nucl. Technol. **24** (1974).
[8] MANNONE, F., CECIL, L., LANDOT, L., J.C.R. Ispra Establishment, to be published as EUR Report.
[9] KOCH, G., et al., "Recovery of transplutonium elements from fuel reprocessing high-level waste solution", Management of Radioactive Wastes from Fuel Reprocessing (Proc. IAEA/NEA Symp. Paris, 1972), OECD/NEA, Paris (1973) 1081.
[10] GANDINI, A., SALVATORES, M., TONDINELLI, L., New developments of generalized perturbation theory in the nuclide field (submitted to Nucl. Sci. Eng.).
[11] OLIVA, G., PALMIOTTI, G., SALVATORES, M., TONDINELLI, L., Trans-actinide elimination with burn-up in a fast power breeder reactor (submitted to Nucl. Techn.).
[12] BURKHOLDER, H.C., et al., Incentives for Partitioning High-Level Waste, BNWL-1927 (1975).
[13] CLAIBORNE, H.C., GERA, F., Potential Containment Failure Mechanisms and Their Consequences at a Radioactive Waste Repository in Bedded Salt in New Mexico, ORNL-TM-4639 (1974).
[14] GERA, F., JACOBS, D.G., Radioecology Applied to the Protection of Man and His Environment (Proc. Symp. Rome, 1971), **2**, CEC, Luxembourg (1972) 891.
[15] GERA, F., Geochemical Behavior of Long-Lived Radioactive Wastes, ORNL-TM-4481 (1975).
[16] Environmental Statement of LMFBR Program, WASH-1535 (1974).
[17] An Assessment of Accident Risks in U.S. Commercial Nuclear Power Plants, WASH-1400 (1975).
[18] Generic Environmental Statement Mixed Oxide Fuel (GESMO), WASH-1327 (1974).

DISCUSSION

Y. SOUSSELIER: As a comment, I would like to say that the evaluation of potentially harmful effects on the scale of a million years seems to me something of a philosophical problem, hardly to be analysed on the basis of scientific criteria. If we wish to avoid leaving any such effects for the world as it will be in a hundred thousand or a million years' time, we should look for solutions to the problem such as transmutation of the transuranics. But I do not think that the relevant decisions can be arrived at by a comparative risk analysis.

My question concerns neptunium. Your paper deals with the recycling of americium and curium, but does not refer to neptunium. What are you going to do with that element?

L. TONDINELLI: We feel that numerical comparison between the hazards associated with alternative waste management schemes is a necessary step, if rational decisions have to be taken.

As far as neptunium is concerned, we have assumed that uses will be found for this nuclide and that, therefore, it would not be burned along with the other actinides.

J. HAMSTRA: As a critical comment on your paper, I would say that the long-term hazard of nuclear wastes is governed by the effectiveness of the containment barrier around the disposal facility and its additional geochemical barrier.

Your conclusions are greatly influenced by the assumption that meteorite impact is a realistic containment failure mechanism for a disposal facility at a depth of 600 m. It is to be hoped that the Sandia group at present working on the New Mexico project will reassess the earlier assumptions made for this type of containment failure mechanism when they produce their final site safety analysis. Most other geological disposal concepts aim at burial at a depth greater than the 600 m mentioned in your paper. For example, the French study for ultimate disposal in granite envisages depths between 1000 and 1500 m. One may even go down further in a salt dome.

I think that if these disposal concepts were assessed correctly for comparison with nuclear transmutation, the conclusion you reach in your paper would have been a more explicit argument against the need for nuclear transmutation.

F. GERA: The paper represents an example of a risk comparison. The important point is the approach taken and not the numerical values arrived at. It is obvious that in the case of deeper disposal, the probability of containment failure due to meteorite impact would be even lower. On the other hand, in different geological conditions, other containment failure mechanisms might predominate; hence the conclusion is that a quantitative risk evaluation should be made for each proposed disposal site, and a comparison performed with all possible alternative waste management strategies.

J.O. LILJENZIN: At the Chalmers University of Technology, we have been working for some years along similar lines and have also tested some of our separation schemes with simulated waste solutions. I have the following questions in connection with what you have told us. First, have you considered the effects from closed cycles, i.e. neutron capture, beta decay, alpha decay, neutron capture cycles, in your calculations? Second, are you aware of the errors associated with the use of an ORIGEN-type code for actinide recycling to reactors? And, third, have you tested your proposed separation schemes in order to verify the assumed separation effects?

L. TONDINELLI: In answer to your first question, we have considered all possible closed cycles.

We are aware of the ORIGEN calculation limitations. As a matter of fact, at CNEN we are re-evaluating all nuclear data which sensitivity calculations have indicated to be critical.

In reply to your last question, the assumed scheme of separation is based entirely on the data in the literature.

K. SCHEFFLER: In as much as we are dealing with waste management in the uranium fuel cycle, we are looking at a new actinide transmutation cycle. How much beta/gamma waste arises from the whole transmutation process? Have you made any relevant calculations?

L. TONDINELLI: This information could be obtained from the calculations we carried out, but I am not able to recall the figures off hand. The quantity of actinides entering the reactor, however, is small compared with the normal fuel. Hence, the fission product increment due to actinide transmutation should be negligible.

P. CANDES: You base your arguments justifying transmutation of the actinides on considerations of risk reduction over the period of final storage, i.e. a period of the order of 100 000 years or more. I agree with Mr. Sousselier that risk reduction by a factor of about 100 is not very meaningful in relation to such a time span. Would not a decision to get rid of the actinides by dispatching them into outer space be a stronger argument in favour of reducing their quantity as much as possible by transmutation?

L. TONDINELLI: Both disposal into outer space and nuclear transmutation of actinides can reduce the long-term hazard associated with the remaining fraction of waste. The proposed disposal method for the separated actinides is not relevant; the point we are trying to make is that the total risk associated with geological disposal of the unpartitioned waste should be compared, quantitatively, with the total risk associated with every alternative disposal option.

ACTINIDE PARTITIONING
Arguments against

B. VERKERK
Reactor Centrum Nederland,
Petten (N.H.),
Netherlands

Abstract

ACTINIDE PARTITIONING: ARGUMENTS AGAINST.

The need for actinide partitioning is discussed on the basis of the radiotoxic hazard measure, the safety of disposal in a salt formation and the evaluation of the risk of 1000 year old high-level waste glass for the case of 100% availability for intake. It is shown that aged waste is a substance of only limited toxicity. This fact is the main argument to conclude that, except for the usual recovery of plutonium, we should not take the trouble to separate the actinides from fission products for the purpose of decreasing the long-term hazard of reprocessing waste.

1. NATURE OF THE PROBLEM

A number of years ago people began to realize that high-level radioactive waste from fuel reprocessing would not only be an extremely hazardous material for the first centuries of its radioactive decay, but that the persistent actinides would cause a hazard for up to a million years. This long-term hazard has forced us to think of their seclusion from the biosphere for periods that go far beyond man's imagination. As a way out it was suggested to separate the actinides from the fission products and to 'incinerate' the long-lived α-emitters in ordinary or special reactors. This would eventually make the fuel cycle more intricate and the repeated recycling necessary might also cause additional hazards. Therefore it would be of interest to develop criteria to decide whether actinide partitioning would be desirable or not.

While the discussion of the need of actinide separation and their burning in a reactor continued and laboratory work started, it became more and more clear that it would be very difficult to produce sound scientific and/or technical arguments either in favour of or against actinide separation. This is because the whole problem is in the first place a philosophical and ethical one and on that basis views of people may differ very much. Nevertheless, I will try to present arguments in order to indicate that it would be better not to separate the actinides from fission products if a reasonable disposal method is available.

2. THE RADIOTOXIC HAZARD MEASURE

In the USA, and especially at Oak Ridge, an extensive chemical programme is being carried out for the purpose of actinide separation [1, 2]. This work is motivated by saying that the long-term hazard of reprocessing waste as indicated by the radiotoxicity index or hazard index is too high. The long-term potential hazard is compared with the calculated hazard of pitchblende, the most radioactive mineral, and with the calculated hazard of average good quality uranium ore. At Oak Ridge it is considered that the hazard of pitchblende is higher than acceptable and that one should separate the actinides in order to arrive at a hazard level not far above that of uranium ore. For these two minerals hazard indices of 1×10^8 and 1×10^5 were used (vol. of water/vol. of waste or

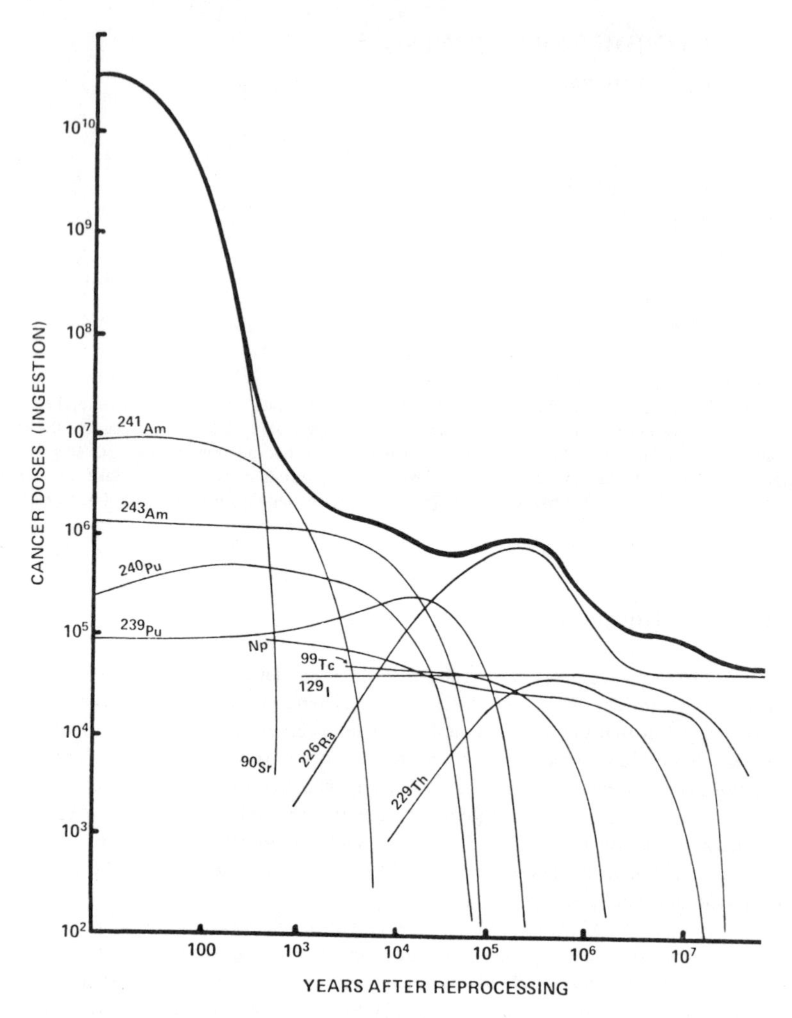

FIG.1. Time dependence of waste hazard expressed as number of cancer doses. Taken From Cohen [3].

ore). The goal of the work in Oak Ridge seems to be to arrive at a hazard index after actinide removal of the order of 10^6.

Figure 1 reproduces a hazard graph in which the hazard index is translated into a potential number of cancer doses that could be attributed to all the high-level waste from 400×10^6 nuclear kilowatt-years. This graph was taken from a report by Cohen [3] and it extends to 10^8 years after reprocessing. It can be seen that in the period between 10^5 and 10^6 years the effect of the grown-in ^{226}Ra dominates. The influence of ^{239}Pu is greatest at about 2×10^4 years and then quickly decreases.

The criterion of the hazard index is not sufficient in itself, as different pathways back to man from storage places may greatly influence the availability for uptake in the human body. Therefore the quality and long-term integrity of a repository began to play a dominating role in discussions on the actinide problem.

3. SAFETY OF SALT MINE DISPOSAL

Various authors have already indicated that salt-formation disposal of vitrified fission product waste containing the actinides and residual plutonium would be adequate for the long term. In the safety analysis prepared by Hamstra [4] an effort was made to show how far in the distant future the stability and the containment of the salt structure would last even under very bad accident conditions. It is difficult to prove what will happen to a salt dome in the distant future, but from the reasoning presented by Hamstra and also by other people, for instance Claiborne and Gera [5], containment failure of the geologic formation must always be viewed upon as happening in a geologic time span and in the geologic sense 10^5 years is only a very short period, so that we may accept that with a probability near to one the salt structure will persist for more than hundred thousand years.

This fact and the retardation in soil of actinides if a containment failure did in fact occur lead to the view that it is better to leave the actinides with the fission products. But this reasoning is not convincing enough. People do appreciate that after decay of the fission products the potential hazard has decreased by a factor of ten thousand but they fear that accidental return to the biosphere of part of the waste after periods of 1000 years or perhaps 100 000 years still constitutes a large and unacceptable hazard.

4. HAZARD ASSESSMENT OF OLD WASTE ON DOSE BASIS

To approach the problems from a completely different angle, we will try to estimate the risk of the high-level waste in the long term by considering the material as it is and the harm it would do, when eaten or inhaled.

As a starting point for our risk calculation, we assume that the high-level waste has been solidified as glass and that it originated from LWR's with a burn-up of 33 MW·d/kg. The wastes from reprocessing of 1 tonne of fuel are contained in 250 kg of glass. After 1000 years of decay the specific activities of the residual radionuclides in it, calculated with the ORIGEN code according to ORNL 4451 [6], are as shown in Table I. As usual, 0.5% of the plutonium and 100% of the other actinides produced are assumed to have gone to the waste.

To estimate the toxicity of old waste glass we assume that it has devitrified and pulverized during the 1000 years of ageing and that it is 100% available for intake by inhalation or ingestion, that means the powder lies on the table in front of one of our distant descendants. We will calculate the dose to the critical organ for all the radionuclides present when this material is ingested. Except for technetium the critical organ is bone.

The dose delivered till complete biological decay, or else over a period of 50 years, per single intake by ingestion of one gram is given by

$$D = 51.2 \times \frac{\Sigma\, EF(RBE)\, n}{m} \times \frac{T_b}{\ln 2} \times f_t \times C_R \qquad (1)$$

where D = dose (rems)

$\Sigma\, EF(RBE)n$ = effective deposited energy (MeV)

m = mass of critical organ

T_b = biological half-life (d)
(For $T_b \gg 50$ years the dose rate is taken to be constant)

f_t = fraction of intake going to critical organ
(for soluble compounds)

C_R = nuclide activity (μCi/g)

TABLE I. SPECIFIC ACTIVITY OF RESIDUAL RADIONUCLIDES IN HLW GLASS AFTER 1000 YEARS OF DECAY

Nuclide	Ci/kg	Nuclide	Ci/kg
Zr-93	7.5×10^{-3}	Pu-239	8.3×10^{-3}
Tc-99	0.057	Pu-240	0.033
Sm-151	1.6×10^{-3}	Am-241	0.166
Np-237	2.3×10^{-3}	Am-243	0.064
Pu-238	5.2×10^{-4}	Cm-242	1.2×10^{-4}

TABLE II. CALCULATED BONE DOSE IN rems OVER 50 YEARS FOR HLW NUCLIDES RESULTING FROM INGESTION OF 1 gram OF 1000-YEAR-OLD WASTE GLASS POWDER

Nuclide	50-year dose (rem)	Nuclide	50-year dose (rem)
Zr-93	2.7×10^{-5}	Pu-239	0.72
Tc-99[a]	0.013	Pu-240	2.9
Sm-151	1.2×10^{-5}	Am-241	15.5
Np-237	0.41	Am-243	5.8
Pu-238	0.047	Cm-242	–

[a] Kidney dose.

Results of dose calculations are shown in Table II. The numerical values used are all taken from ICRP publication 2 [7]. In calculating these doses it was assumed that during its passage through the GI tract 10% of the glass powder would become soluble in body fluids. This is of course an arbitrary assumption. If no solubility was assumed there would be no dose to bone at all and a suggestion of total solubility would also be rather unrealistic. The consequence of this uncertainty is that the calculated dose can only be considered as an order of magnitude value. The total dose, practically only caused by the actinides, amounts to 25.4 rems in 50 years or a dose rate of 0.5 rem/a. It is clear from the above that substantial amounts of old waste glass must be taken orally in order to show a harmful effect.

But for actinides the consequences of intake by inhalation are far greater. We, therefore, calculate doses to the lung and to the bone following inhalation of 1 milligram of 1000-year-old HLW glass powder (inhalation of gram amounts seems absurd). For simplicity we take all the actinides together, since their α-energies and effects are practically equal. The total dose to the lung per milligram inhaled is calculated by Eq. (1), where now the following numerical values apply:

Σ EF(RBE)n = 55 MeV; m = 1000 g; $f_t = 0.25 \times 0.6$;

$C_R = 0.275\ \mu$Ci/mg and T_b = 500 days

The total dose to the lung is 84 rem/mg inhaled.

TABLE III. CANCER-CAUSING AMOUNTS OF REACTOR Pu, ^{239}Pu, ^{241}Am AND ^{243}Am ACCORDING TO COHEN IN μg AND μCi

	Reactor Pu	^{239}Pu	^{241}Am	^{243}Am	μCi
Inhalation	260	1400	25	456	86
Injection into bloodstream	78	420	7.4	137	26
Ingestion with food	2.3×10^6	12×10^6	2.1×10^5	4×10^6	7.4×10^5

Of the material deposited in the lung 20% eventually reaches the blood and 45% of that amount is fixed in the bone [8]. Thus the calculated bone dose from the original intake by inhalation of 1 milligram of waste powder is 205 rems in 50 years or 4.1 rem/a. The values for lung dose and the following bone dose are more realistic than the calculated doses resulting from ingestion, because they are based on insoluble compounds such as PuO_2. Moreover, in the case of inhalation we may conclude that many milligrams must be breathed in before a harmful effect will result.

5. COMPARISON WITH DANGEROUS CHEMICALS

In the previous paragraph doses to individuals resulting from intakes of aged waste were estimated. Some indication about its toxicity may be gained from this information, but we would obtain a better insight if something could be derived about lethal effects, because we then could make a comparison with chemical toxicity.

In a recent report on the hazards involved in plutonium dispersal [9] Cohen calculates the quantities of plutonium (^{239}Pu as well as 'reactor Pu') that would lead to a 50% probability of cancer induction if this amount were inhaled or ingested (genetic risks of plutonium and other actinides are low compared to cancer risks and they will be neglected here).

We must be aware of the fact that the cancer-causing amounts are applicable only to a population, not to individuals and that the effect is not instant as for the case of lethal chemical poisoning. To translate the plutonium values to cancer-inducing amounts for americium, the dominating actinide after 1000 years, we have to change from micrograms Pu to microcuries and from there to micrograms Am. The great radiological similarity of both elements allows this to be done. Table III gives the values of cancer-causing amounts for plutonium and americium isotopes. The values from Table III and the specific activities shown in Table I allow us to calculate cancer-causing quantities for 1000-year-old HLW glass powder. Here again there is the uncertainty resulting from the unknown solubility of the glass powder. Values calculated for ingestion are based on the uptake of soluble compounds.

Cancer-causing amounts of old waste glass are:
for inhalation: 0.31 g
for ingestion: 2700 g

For comparison we may cite some values given by Cohen [9] for poisonous inhalants in terms of the amount that would be fatal to 50% of a number of people exposed for 4 hours (LC_{50}). Considering that the average breathing rate of an adult is 5 m^3 in 4 hours, the LC_{50} for old waste glass (for cancer induction) would be 60 mg/m^3, as compared with cadmium fumes 10 mg/m^3, phosgene 65 mg/m^3 and mercury vapour 30 mg/m^3. For ingestion of chemical poisons LD_{50} values are fractions of grams to some grams, so that the risk from eating 1000-year-old waste glass is negligible and could perhaps better be compared with the risk from eating sand.

Finally we will take a look into the very distant future and ask ourselves what the risk of 10^5-year-old waste glass will be, when according to Fig. 1 the ^{226}Ra grown-in will be the main radionuclide. The specific activity of the ^{226}Ra in the glass will then be 9.6×10^{-2} μCi/g and the dose to the bone in 50 years – again assuming 10% solubility – either following inhalation or ingestion of 1 gram, since the uptake fractions are nearly equal, is calculated to be 3.5 rems or a 70 mrem/a dose rate.

6. HAZARDS FROM EXTERNAL RADIATION

For completeness we also should briefly consider possible hazards of aged waste glass caused by gamma and X-rays and neutrons from (α, n) reactions with light nuclei and possibly from spontaneous fission. An estimate will be given of gamma and neutron dose rates per kilogram of old waste glass at 1 m distance. Since ^{241}Am and ^{243}Am are the dominating radionuclides after 1000 years, an estimate based on these two nuclides will be sufficient for our purpose.

The production of (α, n) neutrons is strongly dependent on the composition of the glass. Assuming a borosilicate glass we may take as the upper limit of neutron production rate the neutron yield of an optimized Am-B neutron source. ^{241}Am-B sources produce approximately 5×10^5 n/s per curie [10]. Therefore the production rate of 1 kg of old waste glass will be less than 1×10^5 n/s. The dose rate from these at a distance of 1 m will be of the order of 0.1 mR/h. The above Am isotopes do not exhibit spontaneous fission.

The gamma dose rate, not corrected for (significant) self-absorption, from the Am isotopes at 1 m/kg of old waste glass is estimated to be approximately 20 mrem/h.

7. CONCLUSIONS

The calculations above have shown that 1000-year-old high-level waste glass is a substance of limited toxicity. Our concern for safe disposal of HLW therefore should be much more concentrated on the short-term integrity of the repository than on the very long-term guarantee of seclusion. The actinides present in HLW from light water reactors do not constitute an unacceptable risk for man in the long term, even if under accident conditions some of the waste returned to the biosphere after decay periods of 1000 years or more.

REFERENCES

[1] BOND, W.D., LEUZE, R.E., Feasibility Studies of the Partitioning of Commercial High-Level Wastes Generated in Spent Nuclear Fuel Reprocessing: Annual Progress Report for FY-1974, ORNL-5012 (Jan. 1975).

[2] CLAIBORNE, H.C., Effect of Actinide Removal on the Long-Term Hazard of High-Level Waste, ORNL-TM-47 (Jan. 1975).

[3] COHEN, B.L., Environmental Hazards in High-Level Radioactive Waste Disposal, Institute for Energy Analysis, Oak Ridge Associated Universities (Mar. 1975).

[4] HAMSTRA, J., Veiligheidsanalyse voor ondergronds in een zoutkoepel opbergen van radioactief vast afval RCN-75-040 (Apr. 1975).

[5] CLAIBORNE, H.C., GERA, F., Potential Containment Failure Mechanisms and their Consequences at a Radioactive Waste Repository in Bedded Salt in New Mexico, ORNL-TM-4639 (Oct. 1974).

[6] Siting of Fuel Reprocessing Plants and Waste Management Facilities, ORNL-4451 (Jul. 1970).

[7] International Commission on Radiological Protection, ICRP Publication 2, Pergamon Press, Lond. (1959).

[8] ICRP Task Group on Lung Dynamics, Health Phys. 12 (1966) 173.

[9] COHEN, B.L., The Hazards in Plutonium Dispersal, University of Pittsburgh (Jul. 1975).

[10] 1974/1975 Catalogue of Radiation Sources for Laboratory and Industrial Use, The Radiochemical Centre, Amersham.

DISCUSSION

F. GIRARDI: I would agree with your statement that the occasional ingestion of one gram of aged vitrified waste lies within the present health physics standards and that aged waste, as a poison, has a limited toxicity as compared to many chemical poisons.

If geological containment failed, however, alpha emitters would be delivered to rather large environments and over periods of time that would most probably exceed the average human life. The risk should therefore be studied as the risk of chronic exposure of populations and not acute poisoning of individuals. The quantitative expressions are, of course, different and the numbers are much less reassuring.

I think that final statements on long-term hazards must be arrived at through careful study of the probability of actinides reaching the biosphere, of the critical pathways to man and the resulting doses to populations. Progress in such studies, as you know, is being made at various laboratories, including our own.

Regarding actinide partitioning, I think laboratory work on the subject should be continued as it could lead to a variety of alternative management strategies, which should then be carefully evaluated so as to choose, over the next few decades, optimized strategies taking into account the different geographical and sociological situations in various countries.

B. VERKERK: I would first draw your attention to the recent study made by Burkholder and co-workers at BNWL, already referred to at this session, in which a thorough assessment is made of the return to man of wastes disposed of underground and the resulting dose.

Further, I agree with your last statement that laboratory work on actinide partitioning should continue, but only on the basis of scientific interest. If it is considered a *sine qua non* for the further development of nuclear energy, I am strongly opposed to it.

J. PRADEL: I am grateful to you for having tried to define the importance of problems of waste storage, since we have sadly lacked this type of information so far at the meeting. Perhaps you will have helped us to avoid, under pressure from the technicians, doing things when there is no need and without orders of priority. However, I would like to make the following remarks. It would seem to me preferable not to use a given number of cancer cases to characterize the harmful effects of waste, although that might serve as a good unit if we need to sensitize public opinion. Accordingly, in the paper presented by Mr. Sousselier (paper IAEA-SM-207/28, these Proceedings, Vol. 1), of which I am co-author, we compare the harmfulness of wastes to that of pitchblende. Your arguments make it possible to place the problem in its correct perspective, but, in my opinion, they do not enable us to conclude that there is no point in separating the actinides. We arrive at the need for separation with a factor of 1000 if we are to make the harmful effects of 300 year old glasses equivalent to the effects of pitchblende; this, moreover, is only a basis for discussion. We should be aware of what a uranium deposit really amounts to, in the light of our present criteria for assessing waste storage sites: several 10 000 Ci of radium in a faulted terrain with water circulation and leakage in all directions (1000 Ci of radium dispersed in each cubic kilometre of granite around the deposit). I would suggest, finally, that the problem of actinide separation has been posed in somewhat the wrong way, because if we make the separation we then have no need for true geological storage systems – in 300 years' time we would have storage sites with a harmful effect comparable to a pitchblende deposit, where one could be working accidentally without excessive risk.

B. VERKERK: The comparison of the toxicity of waste with the hazard of pitchblende has already been used by other investigators. My purpose was to put forward a different method for expressing the possible risk of HLW, and I therefore used the concept that is described in detail in the paper.

W. HILD: I find your paper very interesting. I am always very impressed by calculations in favour of or against separation of the transuranics and their transmutation, but at the same time they increase my confusion.

As a waste manager faced with practical problems, I feel much more confident in the state of the art after hearing at this Symposium the papers on high-level waste fixation. There is enough evidence to show that glass is a safe containment for actinide elements; this has been demonstrated by extensive investigations at various research institutions. I really do not believe there is any need to separate these elements and create a set of new waste streams that pose additional problems. I would suggest keeping the alpha-emitters in the high-level waste and solidifying them together with the fission products.

Furthermore, the Oklo reactor phenomenon in Gabon has shown that nature offers an excellent containment system for transuranics, since the plutonium and actinides have remained in place after all this time, whereas the caesium and strontium have migrated.

Taking these two facts together, I am convinced we are doing the right thing in continuing our effort in the direction of including the transuranics in the solidification of high-level waste.

F. GERA: Mr. Verkerk, further to the statements made by several speakers in this discussion, I would like to point out that waste management strategies requiring a large number of operations on the waste and fairly long periods of storage in surface facilities are likely to result in significantly greater total risks than prompt disposal in deep geological formations with the requisite properties. This is because of the much greater reliability of the containment that deep geological formations have to offer due to their being able to withstand most credible accident conditions.

B. VERKERK: Thank you for the comment.

J.O. LILJENZIN: There seem to be two types of facts – actual facts and pseudo-facts – accepted as truths by the general public. The pseudo-facts are supposed to be truths, but are in fact false. One of the accepted pseudo-facts in Sweden is that actinides left in radioactive waste are hazardous. Scientists who try to disprove this are accused of seeking to hide the fact that there is no practical way of removing actinides from the waste. By demonstrating that removal is feasible we can overcome one of the major arguments against nuclear power.

On the other hand, I agree with the conclusions reached by Mr. Verkerk.

SOME ASPECTS AND LONG-TERM PROBLEMS OF HIGH-LEVEL AND ACTINIDE-CONTAMINATED SPENT FUEL REPROCESSING WASTES FROM THE U-Pu AND Th-U FUEL CYCLES

H.O. HAUG
Institut für Heisse Chemie,
Gesellschaft für Kernforschung mbH,
Karlsruhe,
Federal Republic of Germany

Abstract

SOME ASPECTS AND LONG-TERM PROBLEMS OF HIGH-LEVEL AND ACTINIDE-CONTAMINATED SPENT FUEL REPROCESSING WASTES FROM THE U-Pu AND Th-U FUEL CYCLES.

A comparison is made of the long-term radiotoxicity of different high-level and actinide-containing wastes from nuclear fuel reprocessing and of naturally occurring radioactive materials. The radiotoxicity of the radionuclide inventory in solid or solidified wastes fixed in a practicable insoluble matrix is compared with the radiotoxicity of the nuclide inventory of the same amount of uranium ores (0.2 wt% U), which consequently allowed the definition of a relative toxicity index. The comparison is extended to high-level waste disposed of in single bore holes in deep geological formations. The results indicate that after about 1000 years of decay the radiotoxicity of the controlling actinides in the total volume of the high-level waste disposal layer will not be higher than the radiotoxicity of the same amount of a low grade uranium ore deposit containing 0.2 wt% U. The concentration limitations preset by the single bore hole disposal geometry can also be met by similar disposal conditions for other actinide-containing wastes.

1. INTRODUCTION

Several radioactive waste streams generated at a variety of operations in the nuclear fuel cycle contain considerable amounts of α-emitting nuclides. Waste management systems are being developed that are capable of adequate protection during operational phases and provide the necessary long-term isolation for the long-lived isotopes to decay. For the assessement of the safety of proposed waste management operations or systems a methodology for risk analysis is being developed.

The assessment of the risk that identifies the most likely sequences of failure events (e.g. by fault-tree analysis) leading to possible release of radioactive material into man's environment and evaluates their probabilities is an important but difficult task for the years to come. Risk analysis starts with a definition of the characteristics of the wastes and of the disposal concept, i.e. the containment and confinement barriers. At the present stage we think that basic models and approaches may be useful in considering the magnitude and correlations of the problems.

2. COMPARISONS BETWEEN ACTINIDE AND RADIUM CONTAINING RADIOACTIVE MATERIALS

The long-lived α-emitting actinides, which are highly toxic, can be compared with the natural decay product ^{226}Ra, which is an α-emitter with 1620 a half-life having the lowest maximum

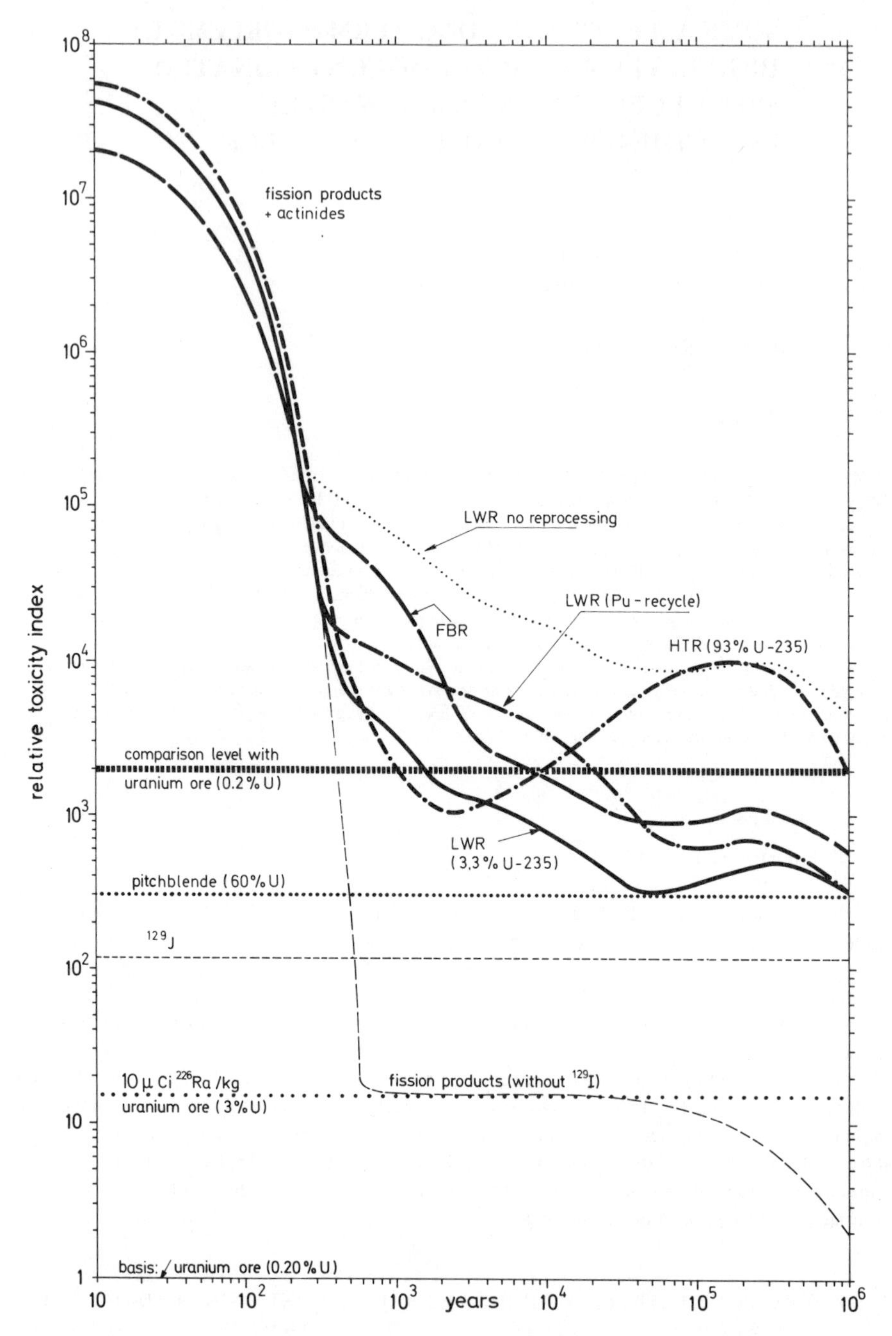

FIG.1. Relative toxicity index of high-level waste (HLW) from LWR, FBR, and HTGR fuel reprocessing and natural uranium ores.
Comparison level with uranium ore (0.2% U) designates comparison of total HLW disposal layer with the same amount of uranium ore deposit.

permissible concentration in water (MPC_w) of all radionuclides. The comparison can be made on three different levels:

(a) The maximum permissible concentrations of transuranium nuclides in water have been evaluated by comparison with the radiotoxicity of ^{226}Ra [1]. They are based on the maximum permissible dose for ^{226}Ra to the bones as the critical organ.

(b) This concept was extended by comparing the radiotoxicity of the radionuclide mixture (including the actinides) in solid or solidified wastes fixed in a matrix of very low solubility with the radiotoxicity of the nuclide inventory of the same amount of low-grade uranium ore. The radiotoxicity of naturally occurring uranium in equilibrium with its decay daughters is mainly determined by the toxicity of its ^{226}Ra daughter.

(c) Finally, the high-level waste disposal concept is considered, which comprises the final storage of the solidified high-level waste in single bore holes in a deep geological formation. The comparison is made between the radiotoxicity of the nuclide inventory of the total volume of the high-level waste disposal layer (i.e. the waste cylinders plus the surrounding rock) and the radiotoxicity of the nuclide inventory of the same volume of low-grade uranium ore deposit.

3. RELATIVE TOXICITY INDEX AND EFFECT OF RADIOACTIVE DECAY

Radiotoxicity of an isotope is caused by the characteristic radioactive decay properties of the isotope and the biochemical processes within the organism relevant for the particular element and its compounds. This is reflected by the maximum permissible concentrations in water (MPC_w) or in air (MPC_a) recommended by the International Commission on Radiological Protection [1] and issued in the national radiation safety regulations.

Several parameters (based on such MPC values) have been used in the literature in order to create a convenient scale for comparison of potential radiological hazards of radionuclide mixtures in wastes or naturally occurring materials. The potential hazard of radioactive waste is occasionally illustrated by the volume of water required to dilute the hypothetically dissolved waste nuclide inventory to drinking water tolerances. There is considerable resistance among scientists to accepting the resulting cubic-metre-of-water scale of the '(radiotoxic) hazard measure' [2, 3] or 'hazard index' [4] because of possible misunderstanding or intentional misleading interpretation. As a useful scale for comparison of potential radiotoxic hazards, we therefore have introduced a dimensionless Relative Toxicity Index (RTI) based on the radiotoxicity of low-grade uranium ore.

The RTI has been defined as the ratio of the water volume that could theoretically be contaminated to the MPC_w by the radionuclide inventory of a given amount of solid or solidified waste, and of the water volume that could be contaminated to the MPC_w by the radionuclide inventory of an equal amount of uranium ore containing 0.2 wt% uranium:

$$\mathrm{RTI} = \frac{\left(\sum\limits_i \dfrac{Q_i}{\mathrm{MPC}_{w,i}}\right)_{\text{waste}}}{\left(\sum\limits_j \dfrac{Q_j}{\mathrm{MPC}_{w,j}}\right)_{\text{uranium ore}}}$$

where Q_i = activity of nuclide i,
$MPC_{w,i}$ = maximum permissible concentration of the nuclide i in water.

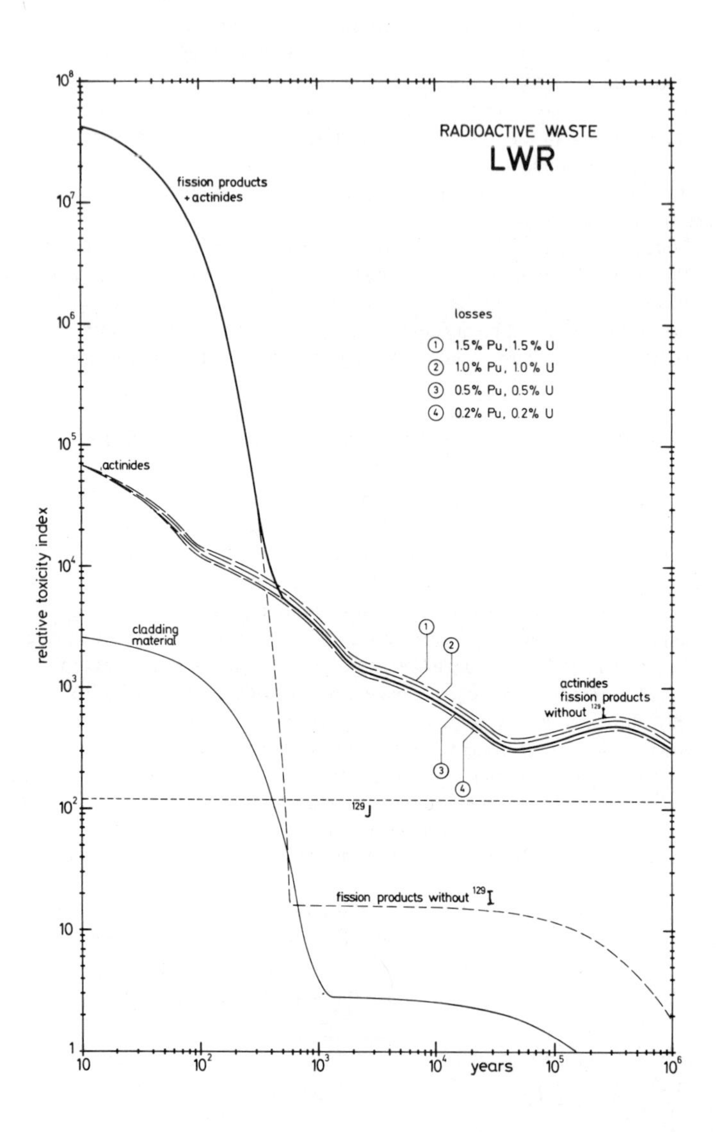

FIG.2. Relative toxicity index of HLW from LWR fuel reprocessing with different U and Pu losses (3.3% enriched U, burn-up 34 000 MW·d/t heavy metal, 150 days cooling time).

Radioactive material has been present in the earth's crust and surface, i.e. in man's environment, at all times in the form of uranium and thorium minerals and ores. The radiotoxicity of uranium ore deposits can therefore be considered a risk that is acceptable for man. The radiotoxicity of the naturally occurring uranium is controlled by its ^{226}Ra content, which is confined in the ore in a relatively insoluble form. Similar conditions will be found for actinide-bearing waste that is fixed in a practically insoluble matrix (glass, ceramics, etc.) and buried in deep geological formations.

For the calculation of the relative toxicity index of solidified high-level waste from fuel reprocessing a concentration of 25 wt% waste oxides (fission products, actinides, corrosion products) was assumed for the solid products (e.g. borosilicate glass). This corresponds to 180 kg for LWRs, 225 kg for FBRs, and 475 kg for HTGRs per tonne of heavy metal, or to 65, 80 and 170 l of borosilicate glass per tonne of heavy metal, respectively [5].

The results of the calculations of the RTI of high-level wastes of different fuel compositions in the U-Pu and Th-U fuel cycles are shown in Figs 1 and 2 [5]. The relative toxicity index is initially dominated by the fission products ^{90}Sr and ^{137}Cs. After about 400 years of decay the relative toxicity has decreased by more than three orders of magnitude and is controlled by long-lived actinides and their daughters. The RTI of LWR waste is predominantly determined by americium and curium and only to a minor extent by the initial Pu losses, as shown in Fig.2. The relative contribution of the actinide elements in the high-level wastes is illustrated by examples of the U-Pu (Figs 3 and 4) and the Th-U (Fig. 5) fuel cycles.

It should be noted that although much lower amounts of transuranium elements are produced in the Th-U fuel cycle, the long-term toxicity index is quite high because of the contribution of the 4n + 2 decay chain of ^{238}Pu and ^{234}U, resulting in relatively high concentrations of ^{226}Ra.

4. COMPARISON BETWEEN A HIGH-LEVEL WASTE DISPOSAL LAYER AND A URANIUM ORE DEPOSIT

It is common to the concepts for the final disposal of high-level waste in deep geological formations (e.g. a salt dome) that the waste will be

(a) Fixed in a practically insoluble matrix (like borosilicate glass or others),
(b) Contained in waste cans, and
(c) Placed in single bore holes at a certain distance to allow for the removal of the decay heat.

In a first approximation the waste, therefore, could be considered to be distributed in small units over a large layer of salt. A comparison is made between the radiotoxicity of the nuclide inventory of the total volume of the high-level disposal layer (i.e. the waste plus the embedding salt between the waste cylinders) and the radiotoxicity of the nuclide inventory of the same amount of low-grade uranium ore deposit containing 0.2 wt% uranium. The result is indicated in Fig. 1 by the broken horizontal line, which was obtained from a storage model assuming glass cylinders of 20 cm diameter and a distance of the bore holes of 10 m (corresponding to 400 t of salt per tonne of heavy metals from LWR fuels).

The results of the evaluation indicate that after about 1000 years of decay the radiotoxicity of the controlling actinides in the ultimate disposal layer will not be higher than the radiotoxicity of low-grade uranium ore of 0.2 wt% uranium, which even forms deposits on the earth surface. The differences between the different reactor fuels are small and would disappear with slight changes of the disposal geometry.

Considering therefore the total disposal layer after 1000 years there is no significant increase of the radiotoxicity level beyond comparable geological formations. Consequently, no necessity for actinide partitioning from high-level wastes can be seen.

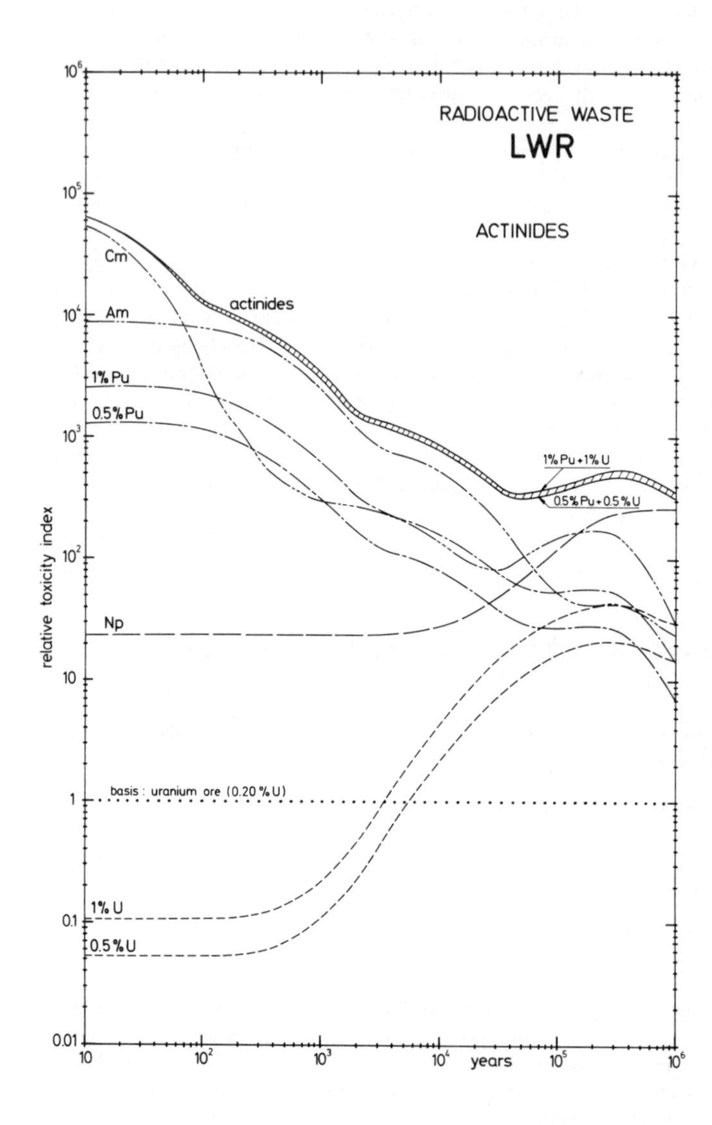

FIG.3. Contribution of single actinides to the relative toxicity index of HLW from LWR fuel reprocessing (3.3% enriched U, burn-up 34 000 MW·d/t heavy metal, 150 days cooling time).

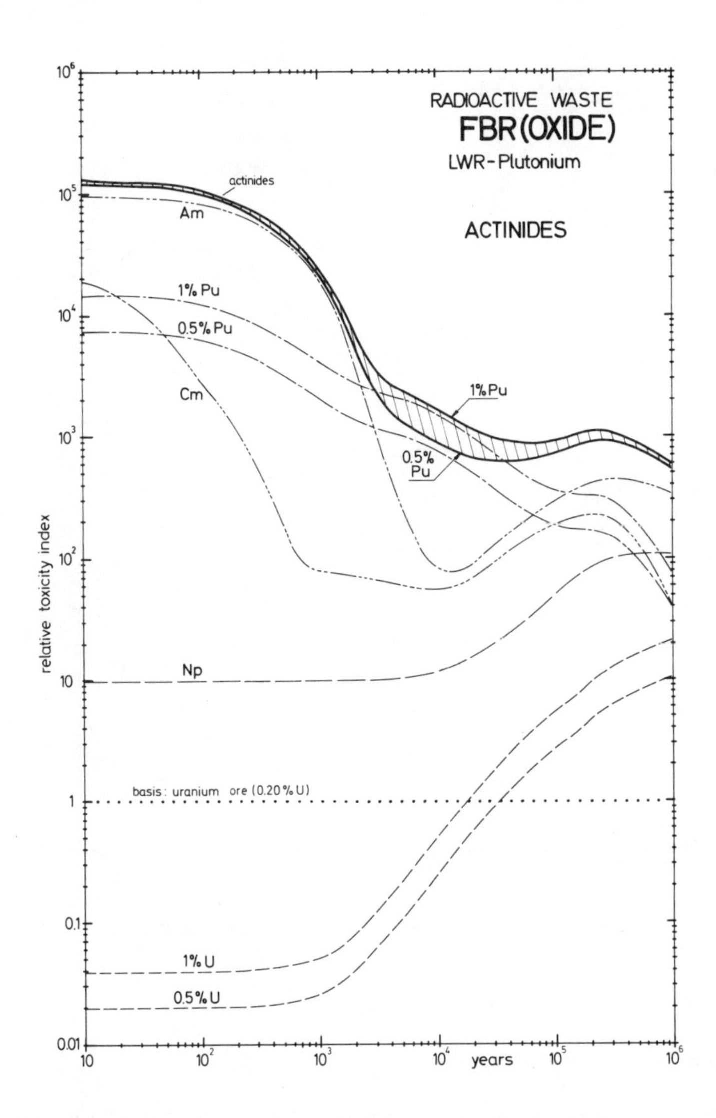

FIG.4. Contribution of single actinides to the relative toxicity index of HLW from FBR mixed core and blanket fuel reprocessing (average burn-up 34 000 MW·d/t heavy metal, 150 days cooling time).

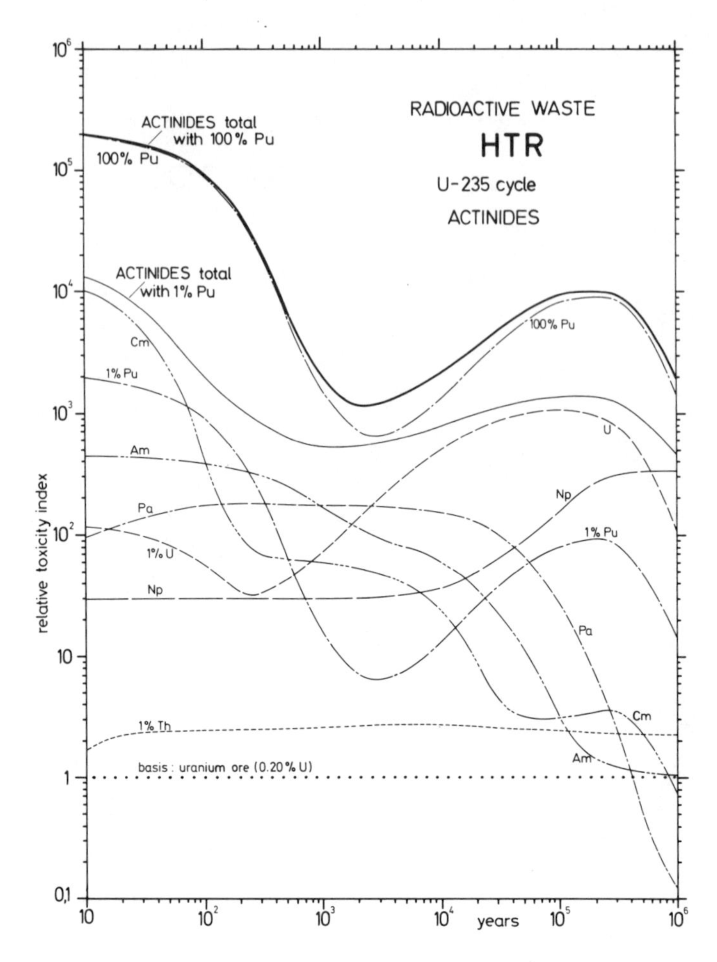

FIG.5. Contribution of single actinides to the relative toxicity index of HLW from HTGR fuel reprocessing (93% enriched U, burn-up 95 000 MW·d/t heavy metal, 270 days cooling time).

5. OTHER α-BEARING WASTES

Any single bore hole disposal geometry means a concentration limitation for the disposal layer. As has been shown for the case of high-level waste disposal, a concentration limitation automatically occurs through the requirements of decay heat dissipation.

Other α-bearing wastes arise at reprocessing and fuel refabrication waste treatment operations as miscellaneous liquid concentrates, or residues from incineration of burnable solid α-waste. Some concepts aim at the incorporation of the residues of these treatments into a glass or ceramic matrix [6]. Again it can be assumed that 20 to 25 wt% of the inorganic waste oxides are incorporated into the fixation product. Considering the plutonium contents of the incorporated material, the α-activity or the relative toxicity index, respectively, is in the same range as the α-activity or relative toxicity index of the high-level waste after 1000 years [6].

For final disposal the high α-bearing solidified wastes need no special heat removal like high-level waste. However, if the concentration limit preset by the high-level waste disposal geometry should not be exceeded by other wastes, a similar disposal technology using a single bore hole concept for this waste type would lead to a comparable radiotoxicity level.

REFERENCES

[1] Recommendations of the International Commission on Radiological Protection, ICRP-Publication Nr. 2 (1967).

[2] BELL, M.J., DILLON, R.S., The Long-Term Hazard of Radioactive Wastes Produced by the Enriched Uranium, Pu-^{238}U, and ^{233}U-Th Fuel Cycles, Rep. ORNL-TM-3548 (1971).

[3] HAMSTRA, J., Nucl. Safety **16** (1975) 180.

[4] BLOMEKE, J.O., NICHOLS, J.P., McCLAIN, W.C., Physics Today **26** 8 (1973) 36.

[5] HAUG, H.O., Anfall, Beseitigung und relative Toxizität langlebiger Spaltprodukte and Actiniden in den radioaktiven Abfällen der Kernbrennstoffzyklen, Rep. KFK-2022 (1975).

[6] SCHEFFLER, K., RIEGE, U., HILD, W., JAKUBICK, A.T., Zur Problematik der sicheren Beseitigung α-haltiger Abfälle aus Wiederaufarbeitung und Brennelementfertigung am Beispiel des Langzeitverhaltens hochaktiver Gläser, Rep. KFK-2170 (1975).

DISCUSSION

F. GERA: I would like to make two comments. First, the uranium concentration figure of 0.2% that you quote, which is typical of many uranium ores, cannot be considered as a safe concentration in all cases. Many uranium-bearing rocks, in fact, contain groundwaters with radium concentrations that are much higher than the limit. Hence the approach should be to assess possible release mechanisms, pathways and doses to the critical group concerned.

Second, I believe that the assumption of the dilution of waste by salt is not acceptable. The solubilities of salt and glass are so different that in the case of dissolution by groundwater, salt would be removed more readily and phase separation would certainly take place. Even in that case, a more realistic transport mechanism should be assumed.

R. KROEBEL: I also have a comment to make. I think we can conclude from the papers presented today by yourself, Mr. Verkerk and Mr. Merritt, that actinide separation is more of a philosophical problem than a necessity. Our disposal schemes show that after a lapse of about 1000 years the residual risk is tolerable, even if the actinides are only slowly dissolved.

The experience gained in Canada, moreover, shows that actual leach rates are much lower (by several orders of magnitude) than the figures reported from hot, laboratory experiments.

J.B. MORRIS: I very much agree with your approach in comparing a waste disposal site with a uranium mineral deposit. This is surely an analysis that the public will accept, since it knows that man has lived with the natural earth for .... I don't know how many, but let us say 2000 to 4000 million years. If we now suddenly decide that certain areas of the earth are hazardous because of uranium deposits, we will have no great philosophical difficulty in adjusting our life style to suit the situation. If a man-made product that is no more hazardous than a natural material is placed in the environment, why should we suddenly find difficulty in accepting such a practice? Are we now supposed to say that God made a mistake 4000 million years ago and that the Garden of Eden should not have been made toxic by so much natural uranium?

R.W. BARNES: I believe that members of the public are wondering what solutions may be available today for the safe management of radioactive wastes from spent fuel. I think we can answer with confidence that the disposal of such waste in geologically stable strata is within our technological reach today. I agree, however, with Mr. Sousselier's comment earlier today that separation of the actinides would open the way for other methods of management. But I do not think we could recommend actinide separation as an approach available at the present time, though possibly we could for the future. I also subscribe to the view that risk analysis of the schemes will be required to demonstrate their acceptability.

My second comment, Mr. Haug, refers directly to your paper. During a recent study we made at Ontario Hydro we reviewed the methods available for describing the toxic nature of spent fuel. We felt that the relative toxicity approach, though helpful, did not fully cover the subject and was not easily understood by the general public. The hazard to man represented by a substance is dependent on the mode of uptake and its availability to man – on its quantity, chemical form and pathway. At the present time we are working on an approach which involves comparing the hazard of spent fuel with other known chemical and biological toxins.

H.O. HAUG: Several of the arguments presented here deal with what is known as risk analysis. Risk analysis is under development in the Federal Republic of Germany for different parts of the waste management process, including final geological disposal. It includes fault-tree analysis for the identification of the most likely sequences of failure events leading to possible release of radioactive material, the pathway by which released material reaches man, and the dose to man, as well as evaluation of the probability of the release sequences. Until the results of such risk analyses are available, some basic models can serve to illustrate the magnitude of the problem, to clarify possible licensing problems, and to eliminate some of the objections raised in public debate.

K. KÜHN: Your paper is very valuable in that it makes a first approach. Naturally, further work needs to be done on risk analysis. But we have a number of people in Germany who are willing to undertake risk analysis in connection with a disposal site in a geological formation, so perhaps in a year from now we can report the results.

C.J. JOSEPH: Did you consider the non-reprocessing scheme?

H.O. HAUG: Yes, I did. The results are shown in Fig. 1 of the paper for high-level waste from LWR fuel. The relative toxicity index of non-processed fuel is higher by a factor of about 20.

D.W. CLELLAND: In relation to the last two papers and the discussion on them, it is interesting to note that not one expert here today has expressed the view that it will be necessary in the future to separate the actinides from the high-level waste.

GEOLOGIC DISPOSAL
(Session IX)

Chairman:

A.F. PERGE, United States of America

INVESTIGATIONS SUR LA POSSIBILITE DE REJET DEFINITIF DE DECHETS RADIOACTIFS DANS UNE FORMATION ARGILEUSE SOUTERRAINE

R. HEREMANS, P. DEJONGHE, J. WILLOCX
Centre d'étude de l'énergie nucléaire,
Mol

M. GULINCK, P. LAGA
Service géologique de Belgique,
Bruxelles,
Belgique

Abstract–Résumé

INVESTIGATIONS INTO THE POSSIBILITY OF FINAL DISPOSAL OF RADIOACTIVE WASTES IN AN UNDERGROUND CLAY FORMATION.

The final disposal of long-lived and very long-lived radioactive waste is a problem that will have to be solved within the next few decades. One possible solution is to bury the solidified waste in a carefully chosen geological formation. With this purpose in mind, the Nuclear Research Centre has begun a research and development programme aimed at evaluating the suitability of a clay bed situated in the Campine region of Belgium. The paper briefly describes the general geological and hydrogeological aspects of a site selected in the region of Mol, where the Centre is located. In addition to the initial results of field experiments, it also reports on laboratory analyses and tests of the chemical and mineralogical compositions, and likewise on the mechanical properties and ion-exchange capacity of clay collected during geological drilling. The work at present under way should provide a set of scientific data that will assist in assessing the validity of this solution to the problem of the final disposal of solidified radioactive wastes.

INVESTIGATIONS SUR LA POSSIBILITE DE REJET DEFINITIF DE DECHETS RADIOACTIFS DANS UNE FORMATION ARGILEUSE SOUTERRAINE.

L'élimination définitive des déchets radioactifs de longue et très longue période est un problème qui demande à être résolu dans les prochaines décennies. Une solution possible est l'enfouissement de ces déchets solidifiés dans une formation géologique judicieusement choisie. Dans ce but, le Centre d'étude de l'énergie nucléaire a entrepris un programme de recherche et développement destiné à évaluer la convenance d'une couche argileuse située en Campine belge. Le mémoire décrit brièvement les aspects géologiques et hydrogéologiques généraux d'un site sélectionné dans la région de Mol où le Centre est implanté. En plus des premiers résultats expérimentaux obtenus sur le terrain il est également fait part des analyses et expériences de laboratoire concernant les compositions chimiques et minéralogiques ainsi que les propriétés mécaniques et le pouvoir d'échange ionique de l'argile recueillie au cours d'un forage géologique. Les travaux en cours doivent fournir une série d'informations scientifiques qui contribueront à l'évaluation de la validité d'une telle solution au problème du rejet définitif de déchets radioactifs solidifiés.

INTRODUCTION

La mise en place progressive d'une industrie nucléaire en Belgique a incité le Centre d'étude de l'énergie nucléaire (CEN) de Mol à examiner les possibilités de rejet définitif de déchets émetteurs α et de moyenne ou haute radioactivité $\beta\gamma$ produits sur le territoire national.

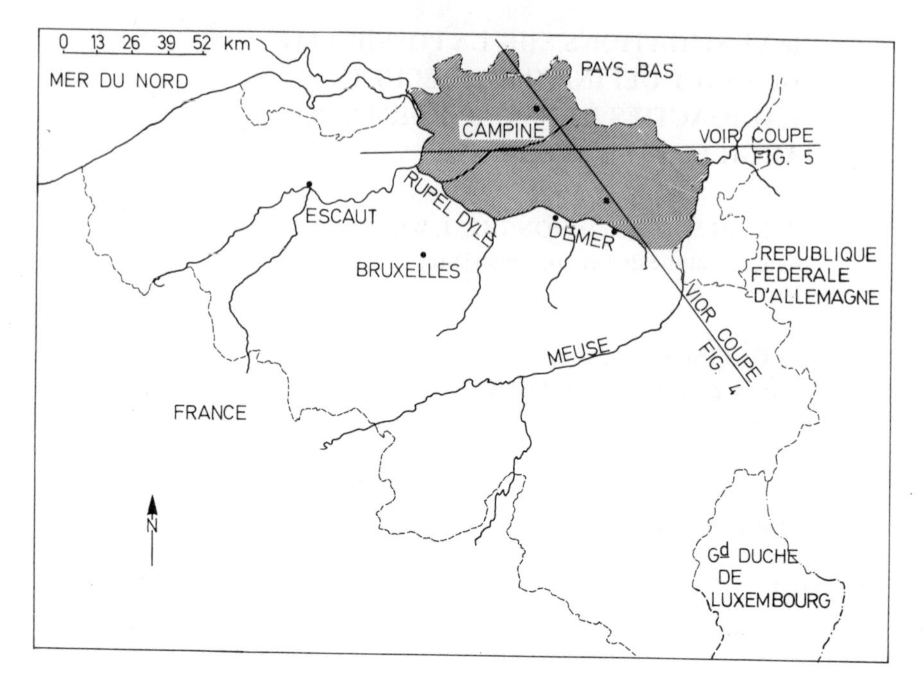

FIG.1. Situation géographique de la Campine en Belgique.

Il est couramment admis dans les milieux scientifiques et techniques concernés que des formations géologiques judicieusement sélectionnées peuvent jouer le rôle de barrière contre la dispersion dans la biosphère des radionucléides solides qui y seraient entreposés. Etanchéité à l'eau, plasticité et stabilité tectonique sont, parmi d'autres, trois critères très importants à prendre en considération pour le choix d'une formation géologique destinée au rejet de déchets. Il est reconnu que les dômes salins et les couches d'argile sont très intéressants de ce point de vue. Les premiers sont inexistants en Belgique; par contre diverses couches argileuses relativement étendues et d'une épaisseur suffisante sont connues de longue date et situées à des profondeurs techniquement accessible dans certaines régions, entre autres dans la Campine belge. C'est donc sur cette base qu'un premier programme d'études et de mises au point couvrant une période de cinq années a été établi. Le but du présent mémoire est de décrire l'approche du problème telle qu'elle a été envisagée et entamée par le CEN/SCK.

Les résultats des travaux présentés dans les pages qui suivent doivent aider à faire une première évaluation des risques nucléaires liés au rejet de déchets radioactifs de diverses natures dans la formation argileuse prospectée et des possibilités de créer des galeries ou cavités souterraines dans cette formation.

1. ASPECTS GENERAUX DE LA GEOGRAPHIE PHYSIQUE ET DE LA GEOLOGIE

Le CEN/SCK est implanté à Mol, en Campine, dénomination donnée à une région d'environ 4500 km^2 située dans le nord-est de la Belgique et délimitée à l'ouest par l'Escaut, à l'est par la Meuse au sud par le Rupel, le Demer et la Dyle et au nord par la frontière avec les Pays-Bas (fig.1).
Le relief est plat, l'altitude au-dessus du niveau de la mer variant le plus souvent entre + 20 et

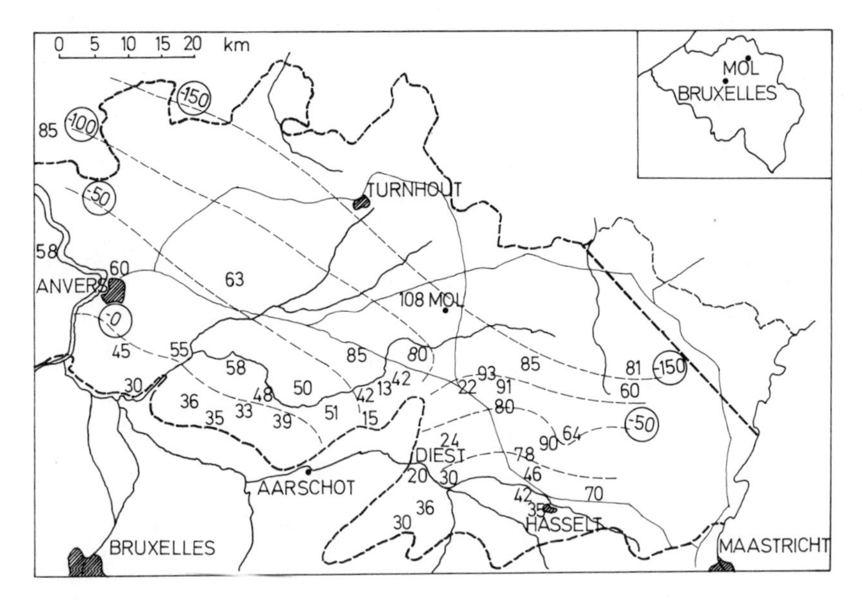

FIG.2. Etendue de la formation argileuse de Boom, profondeur (chiffres entourés d'un cercle) et épaisseur de la couche (en mètres).

+ 80 m. Le sol, généralement sablonneux, est pauvre du point de vue agricole, particulièrement dans la partie nord où 30% environ de la superficie totale est couverte de bruyères et de pinèdes. La densité de la population est de l'ordre de 200 habitants par km^2.

Du point de vue géologique, cette région est bien connue grâce à des forages réalisés jadis lors de la prospection du bassin houiller situé à l'est. La surface du socle paléozoïque est plane, avec un léger pendage en direction du nord-nord-est. Celui-ci est recouvert par des formations d'âge secondaire et tertiaire, essentiellement d'origine marine. Les dépôts d'âge permo-triasique et jurassique y sont inexistants, sauf dans la région orientale faisant partie du graben de Roermond. Les dépôts tertiaires sont formés d'une alternance de sables et d'argiles. La première couche d'argile que l'on rencontre en profondeur est l'argile de Boom (septariën ton) faisant partie de l'étage rupélien et située dans l'Oligocène. Cette argile est recouverte essentiellement par des sables d'âges miocène, pliocène et pléistocène. En se basant sur les résultats d'anciens forages faits en Campine, on pouvait prévoir que dans la région de Mol, l'argile de Boom présenterait un caractère massif, sur une assez forte épaisseur (~90 m) (fig.2). A plus grande profondeur existe une autre formation argileuse relativement épaisse, notamment l'argile d'Ypres ou argile des Flandres (Eocène moyen). Mais ici, on pouvait s'attendre à rencontrer une argile silteuse avec intercalations de sable très fin.

Au sud, dans la région Aarschot-Diest, l'argile de Boom est partiellement et même complètement érodée par un large chenal rempli de dépôts miocènes (fig.2), de direction NNE. Dans la partie NE de la Campine, en bordure du graben, une série de failles mineures, dont le rejet n'est pas exactement connu, affecte le recouvrement mio-pliocène et fatalement aussi le toit de l'argile de Boom. Ces diverses failles ont travaillé jusqu'au Pléistocène ancien mais sont maintenant inactives.

Depuis le début du siècle, époque à laquelle les premières observations scientifiques furent effectuées, aucun séisme sérieux n'a été enregistré dans cette partie du pays. En ce qui concerne le passé plus lointain, il est probable que toute secousse importante qui aurait eu lieu aurait laissé des traces dans les chroniques et écrits anciens, ce qui n'est pas le cas à notre connaissance.

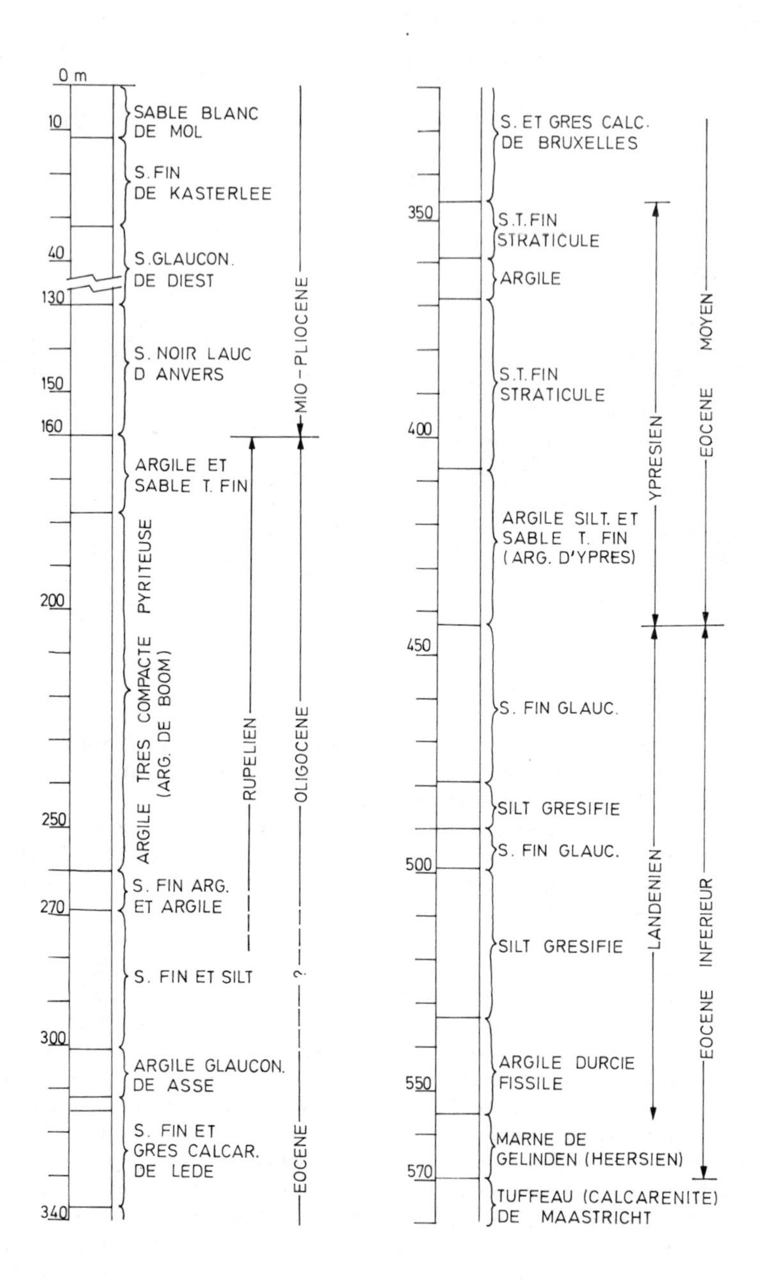

FIG.3. Coupe stratigraphique du forage de Mol.

2. FORAGES GEOLOGIQUE ET GEOTECHNIQUE ET MESURES GEOPHYSIQUES SUR LE SITE PRECONISE

La nature des diverses couches géologiques et leur épaisseur respective sont les premières informations indispensables pour l'appréciation du site choisi. Un forage de reconnaissance a donc été effectué jusqu'à une profondeur de 590 m, de manière à recouper le toit du Crétacé. La technique de forage utilisée a permis d'extraire mètre après mètre des carottes continues d'un diamètre de 10 cm. Comme il sera mentionné dans les paragraphes 4 à 7, un très grand nombre d'analyses diverses a été fait sur ces échantillons de sol. Une coupe stratigraphique résumée du forage est représentée sur la figure 3. Il ressort d'un premier examen que l'argile de Boom y est homogène et consistante sur une épaisseur de l'ordre de 100 m. La formation argileuse d'Ypres par contre est beaucoup moins homogène. Pour cette raison tous les efforts sont actuellement concentrés sur la première formation.

En insérant les informations recueillies dans le réseau des données existantes en provenance de forages plus anciens il a été possible d'établir diverses coupes passant par le forage du site choisi. Ces coupes sont données dans les figures 4 et 5. Dans une zone circulaire de 30 km de diamètre avec comme point central le forage du site de Mol, la formation argileuse de Boom a ainsi été bien délimitée et on peut considérer que sa configuration dans l'espace est suffisamment connue. Toutefois, dans un stade ultérieur, des forages de confirmation seront sans doute nécessaires à proximité immédiate du site choisi dans le but de déceler de petites anomalies locales toujours possibles.

Afin de limiter l'étendue des contrôles à faire dans ces forages complémentaires il a été procédé à un sondage géophysique dans le puits de forage géologique. Les variations de résistivité, de densité, de rayonnement γ, de diamètre et de température ont été enregistrées en fonction de la profondeur. Après corrélation de ces données avec les résultats d'analyses d'échantillons de sol effectuées en laboratoire, on espère pouvoir juger, par assimilation, de la nature et des propriétés générales des couches géologiques rencontrées à ces endroits.

Complémentairement au forage géologique un second forage géotechnique est en cours. Son but est de recueillir des échantillons non perturbés d'argile de Boom. En effet si les analyses courantes peuvent se faire sur des carottes de terrain recueillies au cours du premier forage effectué suivant une technique simple, il est indispensable de déterminer les caractéristiques mécaniques sur des échantillons de sol qui n'ont subi aucune déformation au cours de leur prélèvement. La connaissance des propriétés mécaniques de l'argile est indispensable, comme on le verra plus loin, pour une évaluation exacte des techniques à mettre en œuvre pour le creusement éventuel de galeries souterraines.

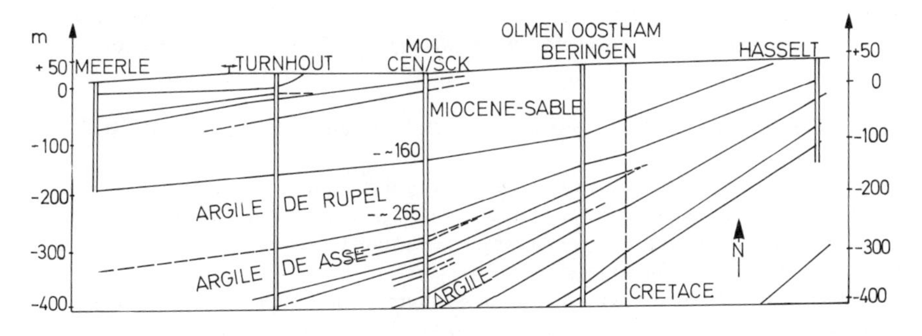

FIG.4. Coupe géologique NNO-SSE passant par le forage de Mol.

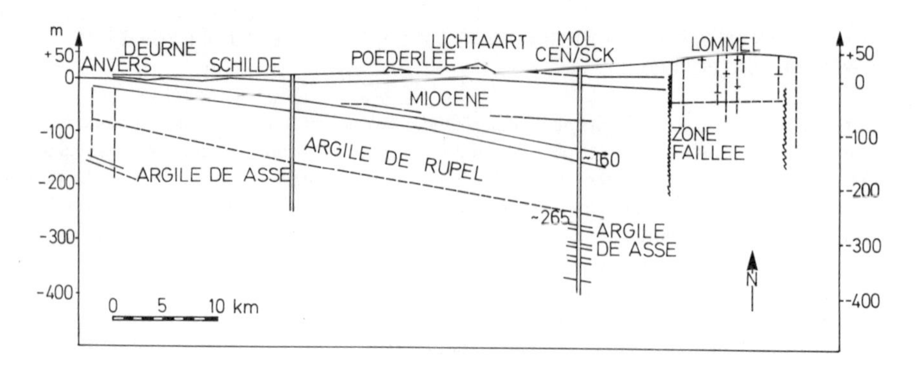

FIG.5. Coupe géologique O-E passant par le forage de Mol.

3. RECONNAISSANCE HYDROGEOLOGIQUE

La protection des nappes phréatiques contre la pollution par des radionucléides est un problème vital. Du point de vue de la captation d'eau, la situation en Campine se résume comme suit:

- les sables néogènes recouvrant l'argile de Boom renferment une nappe phréatique d'intérêt primordial pour toute la région; le niveau de cette nappe se trouve à faible profondeur (1 à 2 m) sous la surface du sol;
- sous l'argile de Boom existent plusieurs nappes captives non exploitées dans la région de Mol; plus au sud cependant, ces nappes se trouvent à des profondeurs accessibles et sont donc utilisées pour des usages privés et industriels.

Les zones d'alimentation de toutes ces nappes se trouvent en bordure sud de la Campine. Pour évaluer le danger potentiel de contamination de ces eaux souterraines, il convient de procéder à des études théoriques et expérimentales très poussées. Diverses mesures ont donc été prises:

- dans une zone circulaire de 30 km de diamètre, toutes les données hydrologiques en provenance des forages géologiques et des puits de captation d'eau sont rassemblées et collationnées;
- dans les puits de forage du site préconisé, des piézomètres ont été installés dans les principales nappes aquifères en vue de mesurer les variations de pression hydrostatique en fonction du temps;
- comme il est mentionné dans le paragraphe 6 un modèle mathématique pour l'étude du mouvement des eaux souterraines et de la migration des ions en solution est en voie d'élaboration.

Au stade actuel des études, il apparaît par l'examen des carottes que l'argile de Boom ne referme aucune intercalation perméable aquifère; tout au plus quelques zones d'argiles silteuses ou de silts très argileux ont été détectées. Une circulation macroscopique d'eau au sein de cette formation est donc inexistante. Les mesures géophysiques effectuées confirment cette affirmation.

4. ANALYSES CHIMIQUES ET MINERALOGIQUES DES ECHANTILLONS DE SOL

Si l'examen visuel des carottes de forage par des géologues spécialisés en la matière permet d'apprécier la nature et l'homogénéité des couches géologiques, il faut néanmoins confirmer et compléter ces informations par des méthodes d'investigation plus scientifiques et notamment par des analyses chimiques et minéralogiques.

Le terme «argile» est très général; il s'adresse à un type de matériau terreux naturel, de fine granulométrie, possédant une certaine plasticité lorsqu'il est mélangé avec des quantités limitées d'eau. Ainsi définie, l'argile n'a pas de composition chimique précise; c'est un mélange en proportions variables de différents composants. La fraction granulométrique inférieure à 2 μm de ce mélange est généralement dénommée «fraction argileuse», dont le composant principal est un silicate d'aluminium hydraté présentant une grande variété de structures. La composition granulométrique, le pourcentage des différents composants présents, la structure du minéral argileux sont tous des éléments qui caractérisent les propriétés de l'argile. Comme, dans le cas présent, il convenait de connaître ces propriétés sur l'épaisseur totale de la formation considérée, il a donc fallu effectuer un grand nombre d'analyses diverses sur des échantillons prélevés tous les mètres, soit environ 110 échantillons au total. Les déterminations suivantes sont en cours sur la majorité des échantillons:

- perte de poids après attaque au perhydrol (détermination des matières oxydables, entre autres les matières organiques);
- perte de poids après attaque à l'acide chlorhydrique (destruction des carbonates et des sels d'acides faibles);
- analyse granulométrique;
- analyse chimique quantitative des composants principaux: SiO_2, Al_2O_3, et des composants secondaires (impuretés): Ca, Fe, Mg, K, etc., et dosage de l'eau;
- détermination des minéraux non argileux tels que feldspath, mica, quartz, etc.;
- détermination de la présence relative des minéraux argileux: illite, montmorillonite, chlorite, etc.

On ne se propose pas de donner dans ce mémoire des résultats détaillés. Toutefois, pour illustrer ce paragraphe quelques valeurs expérimentales ont été mises sous forme de graphique (fig.6) et de tableaux (tableaux I et II). Une interprétation correcte des données ne sera possible que lorsque tous les résultats de toutes les analyses seront disponibles.

5. PROPRIETES PHYSIQUES ET MECANIQUES DES ARGILES

Pour l'évaluation des moyens techniques à mettre en œuvre pour la réalisation de galeries ou de cavités dans une couche d'argile souterraine, il est indispensable de connaître aussi certaines des propriétés physiques et mécaniques de cette argile. Une première série de déterminations peut se faire sur des échantillons prélevés au cours du forage géologique. Ce sont notamment les mesures de répartition granulométrique et des limites de consistances (limites d'Atterberg). Ces données permettent de définir, en fonction de la teneur en eau, les propriétés plastiques des argiles et leur résistance à la compression. Une seconde série d'essais doit se faire sur des échantillons prélevés selon une technique spéciale (mentionnée dans le paragraphe 2 sous la dénomination: forage géotechnique). Elle se rapporte à des mesures de densité, de perméabilité et de résistance au cisaillement.
Toutes ces informations doivent, pour être représentatives de toute l'épaisseur de la couche d'argile, être obtenues à partir des échantillons prélevés à différentes profondeurs.

Les figures 7 et 8 donnent les premiers résultats de laboratoire obtenus concernant la granulométrie et les limites d'Atterberg; il en ressort que de ces points de vue la couche d'argile est bien homogène entre environ – 175 m et – 265 m. Il est intéressant de mentionner que les pics qui apparaissent vers – 235 et – 245 m dans les courbes de la figure 7 correspondent à de minces lentilles d'argile silteuse qui avaient été décelées lors de l'examen visuel des carottes et qui ont spécialement été prélevées pour analyse.

6. PROPRIETES D'ECHANGE IONIQUE DES ECHANTILLONS DE SOL

Si les coefficients de perméabilité des argiles consistantes, et donc les valeurs de la vitesse de déplacement de l'eau dans ces milieux, sont généralement très faibles, elles ne sont cependant pas absolument nulles. La perméabilité des argiles compactes est cependant difficile à déterminer.
Il semble en particulier qu'il existe un seuil dans la différence de pression hydrostatique en deçà duquel aucune circulation n'a lieu. Vu les temps géologiques dont il faut tenir compte pour

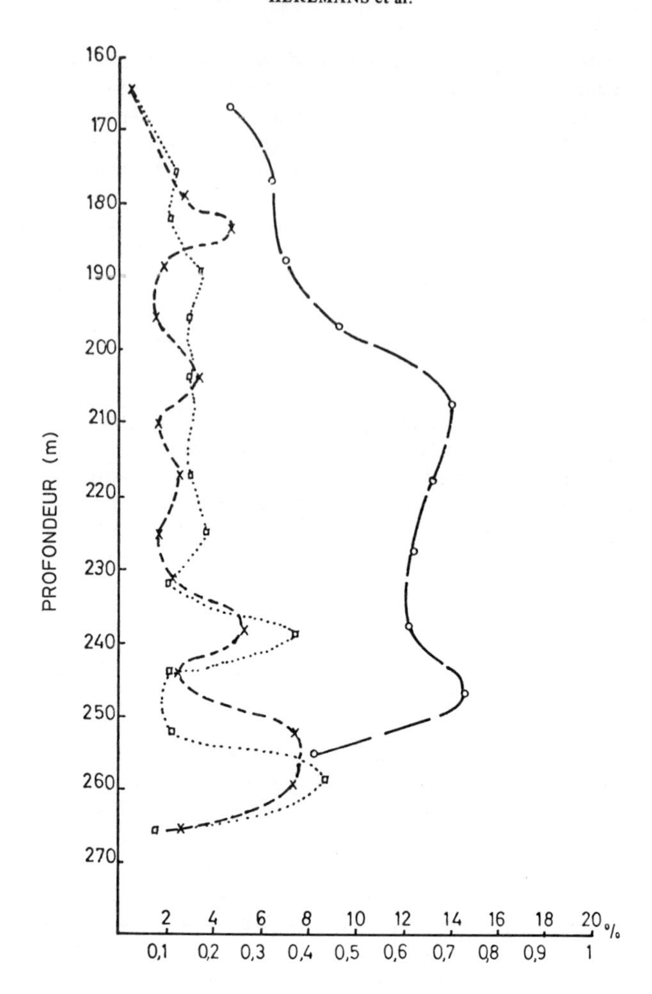

FIG.6. Variation du pourcentage de perte de poids après attaque de l'argile par HCl et H_2O_2 et variation de la teneur en Fe^{2+} en fonction de la profondeur de prélèvement.

Echelle 0–20: { % perte de poids après attaque par HCl ---X---
% perte de poids après attaque par H_2O_2 --------□--------

Echelle 0–1: % de Fe^{2+} ——○——

l'évaluation des risques dans le problème du rejet des déchets radioactifs à très longue période, il est prudent de tenir compte d'un déplacement non négligeable de l'eau dans une telle formation. Supposer d'autre part que certains radionucléides en provenance des déchets solides pourraient au cours des siècles se dissoudre progressivement et être entraînés par l'eau en mouvement vers la biosphère est une hypothèse à ne pas exclure a priori. Dans le cas des argiles la bonne capacité d'échange ionique constitue une importante barrière complémentaire contre la migration des éléments dissous.

TABLEAU I. RESULTATS DE QUELQUES ANALYSES CHIMIQUES EFFECTUEES SUR L'ARGILE DE BOOM

Composants (% de poids sec) \ Profondeur (m)	201	206	210	214	218
SiO_2	51,0	46,0	48,0	47,5	46,0
Al_2O_3	15,0	13,5	18,0	17,0	15,0
Fe_2O_3	3,5	3,5	3,8	3,8	3,9
K_2O	2,6	2,5	3,0	2,7	2,5
CaO	0,9	0,5	0,5	0,6	3,4
TiO_2	0,6	0,6	0,7	0,7	0,6
Na_2O	0,3	0,3	0,4	0,4	0,3
Non définis (matières organiques Mg, P, Mn, etc.)	26,1	33,1	25,6	27,3	28,3
Fe^{2+} (% de poids sec)	0,39	0,43	0,43	0,47	0,55
Fe^{2+}/Fe^{3+}	11	14	18	21	22

TABLEAU II. POURCENTAGE DES DIVERS MINERAUX ARGILEUX PRESENTS DANS LA FRACTION < 2 μm DE L'ARGILE DE BOOM

Minéraux argileux	Quantité (%)
Illite	~ 25
Smectite (beidellite)	~ 20
Vermiculite	~ 30
Interstratifié (illite, montmorillonite)	~ 15
Chlorite Interstratifié (chlorite, vermiculite)	} ~ 10

Sur la base des considérations précédentes un large programme d'expériences en laboratoire et in situ a été mis sur pied. Dans une première phase de travail les coefficients de distribution (quantité pondérale du radionucléide fixé par rapport à sa concentration initiale dans la solution), ont été déterminés. Pour cette étude il a été considéré que:

- les éléments Sr, Cs, La, Ba, Eu, Pu et Am étaient représentatifs des produits de fission et des actinides;
- ces éléments seraient présents dans les déchets rejetés sous forme d'oxyde;
- les eaux de contact pourraient être l'eau de pluie, l'eau de mer ou les eaux des couches aquifères les plus proches de la formation argileuse;
- la capacité d'échange ionique des échantillons de sol pourrait varier en fonction de leur profondeur de prélèvement.

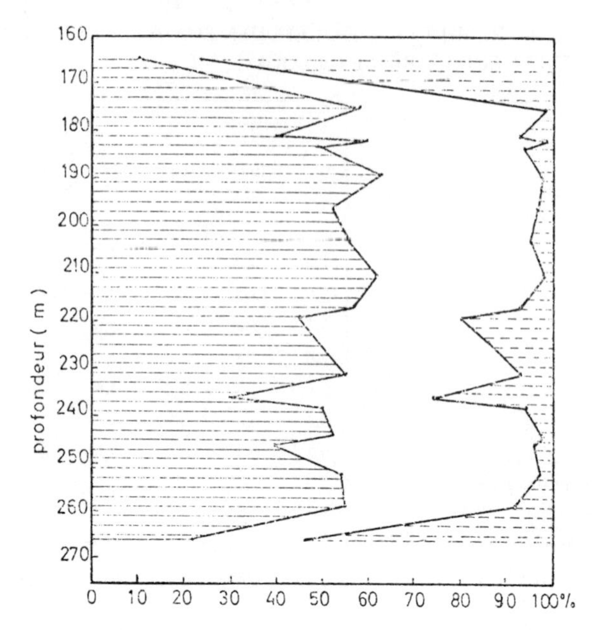

FIG. 7. Pourcentage de la fraction granulométrique de l'argile en fonction de la profondeur de prélèvement. < 2 µm; > 2 µm, < 60 µm; > 60 µm.

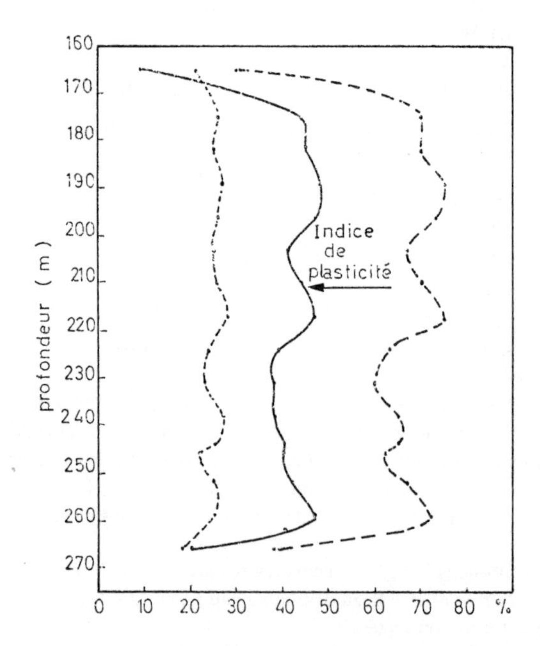

FIG. 8. Limites d'Atterberg de l'argile en fonction de la profondeur de prélèvement.
------ limite d'écoulement; --------------- limite de plasticité.

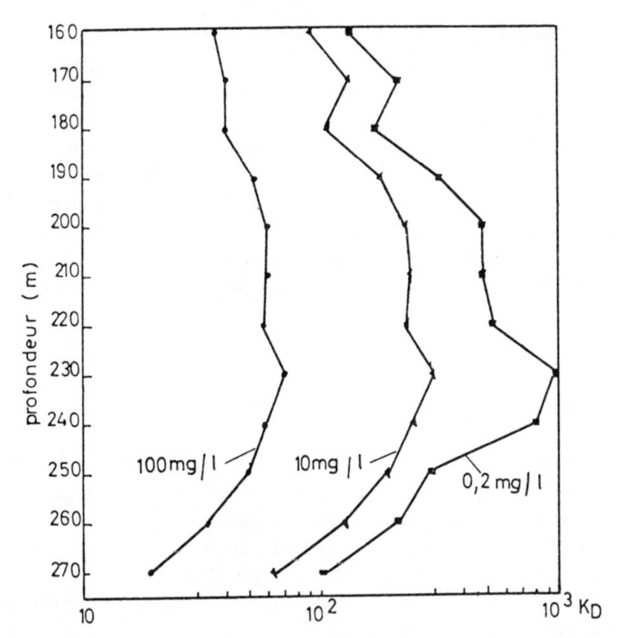

FIG.9. Coefficient de distribution (K_D) en fonction de la profondeur de prélèvement de l'argile pour des concentrations différentes de Sr dans l'eau de la nappe phréatique.

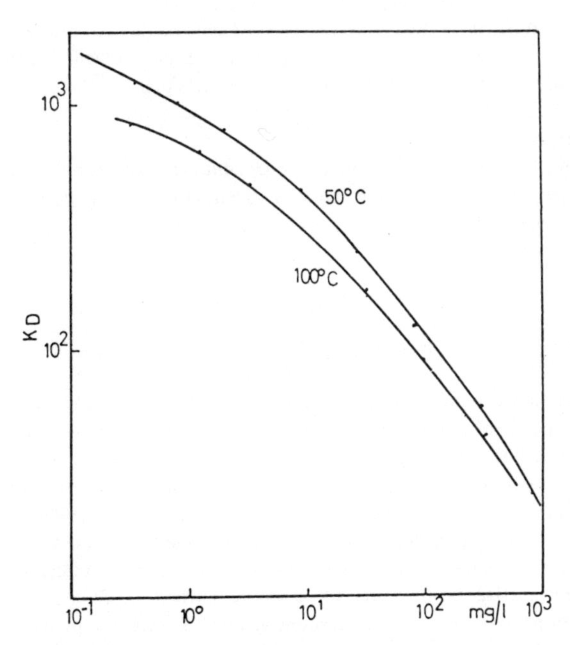

FIG.10. Coefficient de distribution (K_D) en fonction de la concentration de Cs dans l'eau de la nappe phréatique pour un échantillon d'argile préchauffé à 50° et 100°C.

Dans une seconde phase l'influence d'un traitement thermique, de l'irradiation γ et de l'effet combiné de ces deux facteurs sur le pouvoir d'absorption des argiles sera étudiée. Dans ce but des irradiations à la bombe au ^{60}Co d'échantillons portés à une température au-dessus de l'ambiante sont en cours. Comme troisième phase de travail il est prévu d'établir un modèle mathématique dans lequel pourront être introduits les résultats des analyses de laboratoire et les données hydrologiques recueillies sur le terrain. Ultérieurement des expériences de confirmation in situ seront entreprises.

Les premiers résultats de laboratoire font apparaître clairement l'importance de cette étude. De la figure 9 par exemple on peut conclure que, pour le Sr tout au moins, l'absorption varie en fonction de la profondeur et passe par un maximum aux environs de – 230 m. La figure 10 montre qu'un chauffage préalable de l'argile diminue son pouvoir d'échange ionique vis-à-vis du Cs.

7. ANALYSES, RECHERCHES ET ETUDES DIVERSES

Dans les paragraphes 4, 5 et 6 il a surtout été question des analyses et mesures systématiques qui sont faites dans le but de déterminer la composition chimique et minéralogique et les propriétés physiques, physico-chimiques et mécaniques de l'argile en fonction de la profondeur de prélèvement dans la couche considérée. Parallèlement, d'autres expériences de laboratoire et d'autres études ont été entreprises dont certaines ont trait à l'interaction possible entre l'argile et les matériaux qui y seront introduits, et d'autres sont d'un intérêt plus général. Quelques exemples cités ci-après illustrent les problèmes qui peuvent se poser:

- sous l'effet combiné de la température et de l'oxygène de l'air certaines matières organiques ou impuretés minérales peuvent donner lieu à la formation de gaz corrosifs; dissous dans l'eau contenue dans l'argile les acides ainsi formés pourraient progressivement corroder les emballages des déchets et promouvoir le passage des radionucléides dans le milieu;
- le champ de rayonnement γ dû à la présence d'éléments hautement radioactifs peut donner lieu à des produits de radiolyse dont la nature et le comportement doivent être connus;
- un inventaire quantitatif des éléments radioactifs naturels présents dans les diverses formations géologiques et dans les nappes aquifères est une information précieuse pour le contrôle ultérieur d[u] site et de son environnement.

D'autres sujets de travaux expérimentaux pourraient être signalés mais leur énumération serait forcément incomplète car il est inévitable que de nouvelles questions se posent au fur et à mesure qu[e] le problème sera mieux cerné.

8. CONCLUSIONS ET PROGRAMME FUTUR

Une évaluation correcte des possibilités techniques de rejet de déchets radioactifs solides en formations géologiques profondes et des dangers qui pourraient en découler au cours des siècles pou[r] notre biosphère ne peut se faire sans disposer de résultats expérimentaux recueillis sur le terrain et en laboratoire. C'est dans cette optique que le CEN/SCK a entrepris les travaux qui ont été brièvem[ent] décrits. Ces travaux seront poursuivis encore durant des mois afin que soient obtenues des informati[ons] suffisantes pour permettre, après dépouillement et mise en corrélation, d'avoir une vue claire des différents aspects du problème. Si les conclusions du rapport qui sera rédigé à la fin de cette phase du programme sont positives et moyennant l'accord des autorités compétentes, il est prévu de procéder à la création dans l'argile de Boom d'une galerie ou cavité qui pourrait se situer entre 200 et 250 m de profondeur. Ce site souterrain serait utilisé comme laboratoire expérimental et installation pilote durant plusieurs années.

REMERCIEMENTS

La réalisation d'un programme de cette ampleur nécessite la collaboration d'un grand nombre de spécialistes dans les disciplines les plus diverses. Les auteurs tiennent à remercier tous ceux qui, en mettant leur compétence et leur savoir à disposition, participent ainsi activement à ces travaux.

DISCUSSION

Y. SOUSSELIER: In view of the problems of heat release from the very high-level waste (fission products) and the advantage of the adsorption capacity of clay, which you emphasize, do you envisage the storage of waste containing mostly transuranics or of very high-level waste?

And do you not think that mining operations required to establish a storage site in clay formations might present difficult technological problems?

R. HEREMANS: In Belgium we are looking into both the disposal of waste containing transuranics and the disposal of fission product waste from the first extraction cycle of a reprocessing plant.

From the point of view of the heat given off by fission products, clay is certainly not an ideal medium. The limitations that would have to be imposed in such a case are being studied in detail through laboratory experiments and theoretical calculations and also in tests which are to be carried out in the sub-soil.

As far as the general strategy for waste management is concerned, in Belgium the prevailing ideas tally with those which you have yourself expressed during this symposium, namely the storage of fission products in solution for 5 to 10 years; solidification and bituminization of these fission products, followed by surface storage in engineered structures for a few decades; and, finally, burial underground, with possible recovery, over a long period, or else terminal disposal.

In answer to your second question, evaluation of the technological problems of digging drifts or tunnels in clay cannot be pursued further until the results of mechanical and physical tests carried out on undisturbed samples taken from the sub-soil are available towards the end of 1976. Nevertheless, it can be said that certain Belgian firms have acquired experience in this field through the excavation of numerous tunnels in the industrial region of Antwerp.

H. KRAUSE: Despite the fact that we prefer salt formations in Germany for the final disposal of radioactive wastes, we are always glad when other techniques are investigated. I share Mr. Sousselier's concern, however, about the mechanical stability of clay and would stress that this problem needs thorough attention.

J. HAMSTRA: Mr. Heremans, with reference to the possible pathways by which radionuclides may reach the environment in the case of a containment failure, I would like to ask whether the inclination of the Boom clay layer affects groundwater movement above the layer, and, if so, should we consider the pathways back to the biosphere to be geographically directed towards your northern borderline?

R. HEREMANS: There is no relationship between the inclination of the Boom clay and the groundwater circulation. In the region of Mol, the natural drainage occurs in a westerly direction, towards the Schelde. There is no risk of contamination of the groundwater in the south of the Netherlands. The confined water layers below the Boom clay north of Mol are of no commercial use.

P. CANDES: Regarding the transport and diffusion of the most significant radioisotopes towards the surface, more especially the transuranics such as plutonium and americium, you consider that these elements are likely to be found in the oxide form. Do you think that this approach is conservative enough from the standpoint of safety? After all, a certain percentage of transuranics could also be found in more complex chemical forms that might be less favourable from the point of view of ion exchange and which might therefore migrate more rapidly.

R. HEREMANS: To determine the distribution coefficients we have assumed in our present experiments that all elements are initially in oxide form. Depending on the final decisions to be taken with regard to the conditioning of various types of waste, additional tests will be carried out with allowance for the chemical form in which the elements are present.

DISPOSAL OF LONG-LIVED RADIOACTIVE WASTES IN ITALY

F. GERA, G. LENZI, L. SENSI
CNEN-CSN Casaccia,
Rome

G. CASSANO
CNEN-CRN Trisaia,
Matera,
Italy

Abstract

DISPOSAL OF LONG-LIVED RADIOACTIVE WASTES IN ITALY.

The CRN Trisaia, located in Basilicata in Southern Italy, is the reference site for CNEN's research and development programme on disposal of long-lived radioactive wastes in clay formations. The Trisaia Centre is in an area characterized by relative tectonic stability, despite its proximity to the Apennine orogen. The site is located a few kilometres from the Ionian Sea on a Quaternary terrace, a few metres above the elevation of the coastal plain. The terrace materials, which are mainly sand, gravel and conglomerate, are underlain by 800 to 1000 m of marly clays. Some sand lenses are contained in the clay formation nearby the site, particularly between 350 and 500 m in depth; however, thick sections of homogeneous clays have been crossed by all exploratory boreholes drilled in the area. The marly clays are Plio-Pleistocene and contain about 70% clay minerals and about 20% calcium carbonate. Their colour is blue-gray at depth, while at the surface they are yellow as a result of weathering. The permeability is of the order of 10^{-8} cm/s. A heating experiment was performed with an electrical heater placed at a depth of 8 m. However, the results are somewhat questionable due to the defective plugging of the holes where the heater and the thermocouples were positioned. A second heating experiment is now planned; it should employ three heaters at a somewhat greater depth; both pressures and temperatures will be measured. Many additional geological, engineering and safety studies are planned on the Trisaia clay formation.

1. INTRODUCTION

All countries with sizeable nuclear programmes are confronted with the problem of the safe disposal of long-lived radioactive wastes. High-level wastes, because of the high heat generation rate, present the most difficult technical problems.

It is assumed that long-lived radioactive wastes require containment for a time period of the order of one hundred thousand years. Deep geological formations with appropriate characteristics are currently considered as the best option for the disposal of long-lived radioactive wastes [1]. The critical property of a formation suitable for radioactive waste disposal is the absence of circulating groundwater. Many geological formations are practically free from groundwater circulation; however, only plastic materials possess the capability to accommodate some deformation without losing their overall imperviousness to water. Consequently rock salt and plastic argillaceous sediments are considered the most promising geologic materials for disposal of long-lived radioactive wastes.

Salt formations have been extensively studied in the United States of America and in the Federal Republic of Germany and the basic data necessary to design a demonstration facility for disposal in salt are now available. On the other hand, very little work has been performed on disposal into argillaceous formations. This consideration, added to the plentifulness of clay formations in Italy, has lead to a research programme on disposal in such formations.

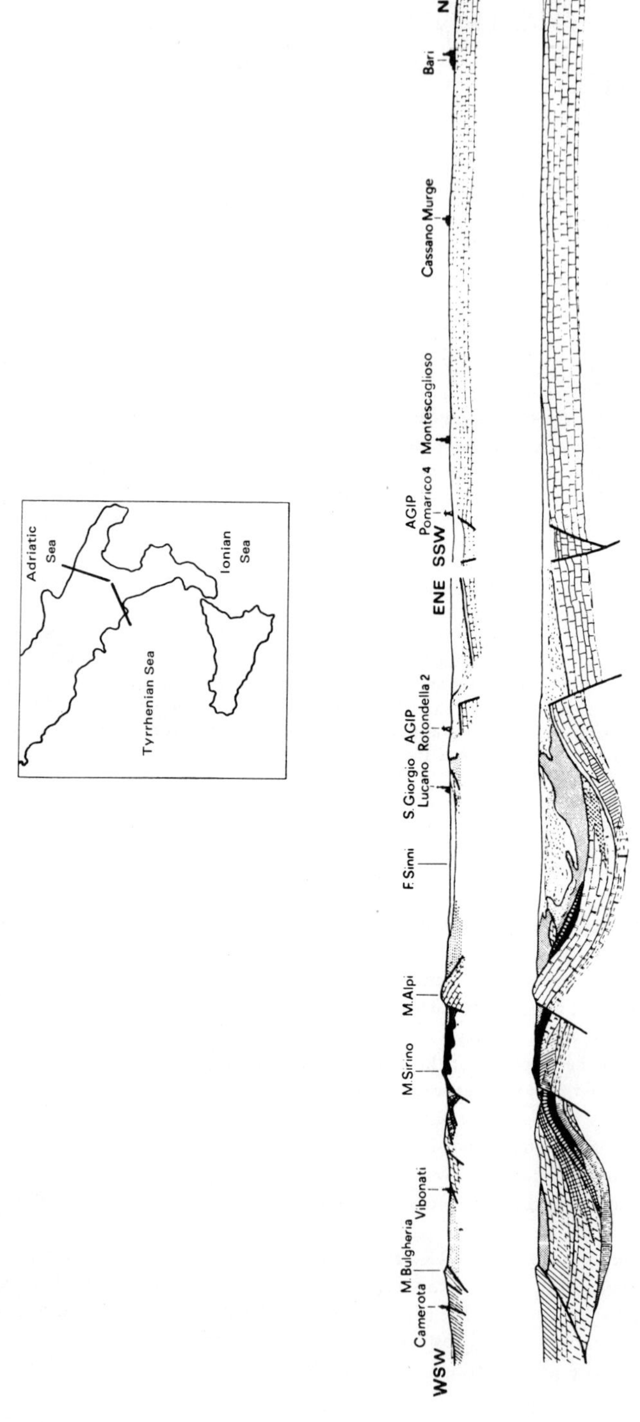

FIG.1. General geologic section through the Southern Apennines (from Ref. [2]).

Since the CRN Trisaia of CNEN, located in Basilicata, is underlain by a thick formation of plastic clays it was decided to perform the *in situ* experiments of this programme there. Consequently the Trisaia Centre will be the reference site for the research programme on disposal in clays; however, this does not imply any commitment of this site for a possible future waste disposal facility.

2. REGIONAL GEOLOGY

The geographical configuration of Basilicata is determined by the presence of the Apennines on the Tyrrhenian side and by their gradual lowering towards the plains bordering the Ionian Sea. In Southern Italy three structural elements of regional significance can be recognized: (1) the Apennine orogen, (2) the Bradanic basin, and (3) the Apulian foreland [2].

The Apennine orogen is characterized by a series of 'units', which include a great variety of lithologic materials. These units constitute several 'nappes' which have been emplaced mainly during the Oligocene and the Miocene. The total thickness of these units is of the order of 15 000 m.

The Bradanic basin has a basement formed by the Mesozoic limestones which outcrop in the Apulian foreland. The basin is formed by a series of faults that cause a stepwise lowering of the basement approaching the Apennines. The basin has been filled in Pliocene times by nappes and terrigenous sediments derived from the orogen. The total thickness of sediments in the Bradanic basin is of the order of 3000 m.

The Apulian foreland is formed by a sequence of neritic limestones with a thickness of several thousand metres.

The area of the Trisaia Centre is contained in the Bradanic basin. Figure 1 shows a general geologic cross-section from the Tyrrhenian Sea to the Adriatic Sea through the Southern Apennines, crossing the Sinni River valley. Significant tectonic activity in the Quaternary is indicated for areas of the Apennines and of the Bradanic basin. For example, marine sediments of the early Pleistocene have been found in the nearby Calabrian Apennines about 1000 m above mean sea level, and the oldest terraces along the northern part of the Gulf of Taranto are found at about 400 m above mean sea level. This value provides an indication of the uplift of the area since the Sicilian period.

Erosion rates in Basilicata can be rather high. For example, it has been calculated that in some parts of the watershed of the Sinni River the rate of erosion is in the order of 20 mm/a [3]. This value applies to areas characterized by higher elevation, steep slopes, and the outcropping of materials belonging to the chaotic complex.

There is evidence that in the Quaternary period the surface hydrology of Ionian Basilicata has been affected by block tilting and differential uplifting [4]. On the whole, the region cannot be considered as tectonically inactive; however, the geologic evidence shows that in the last 125 000 years the sea-level changes in the Gulf of Taranto can be completely accounted for by glacio-eustatic oscillations [5].

The data on historical earthquakes confirm the relative stability of the coastal area since in the period from 1893 till the present it has not been affected by any earthquake with intensity greater than 6 degrees on the Mercalli scale [6]. Figure 2 shows the seismic map of the region under consideration.

3. LOCAL GEOLOGY

The CRN Trisaia is located about five kilometres from the coast of the Gulf of Taranto, in close proximity to the Sinni River. The stratigraphy at the site is well known thanks to some exploratory wells for oil and gas that were drilled in the area by AGIP. The most useful is well Nova Siri Scalo 1, which is about 1 km south of the fence of the Centre and reaches the depth of 1991 m [7]; its location is shown in Fig. 3.

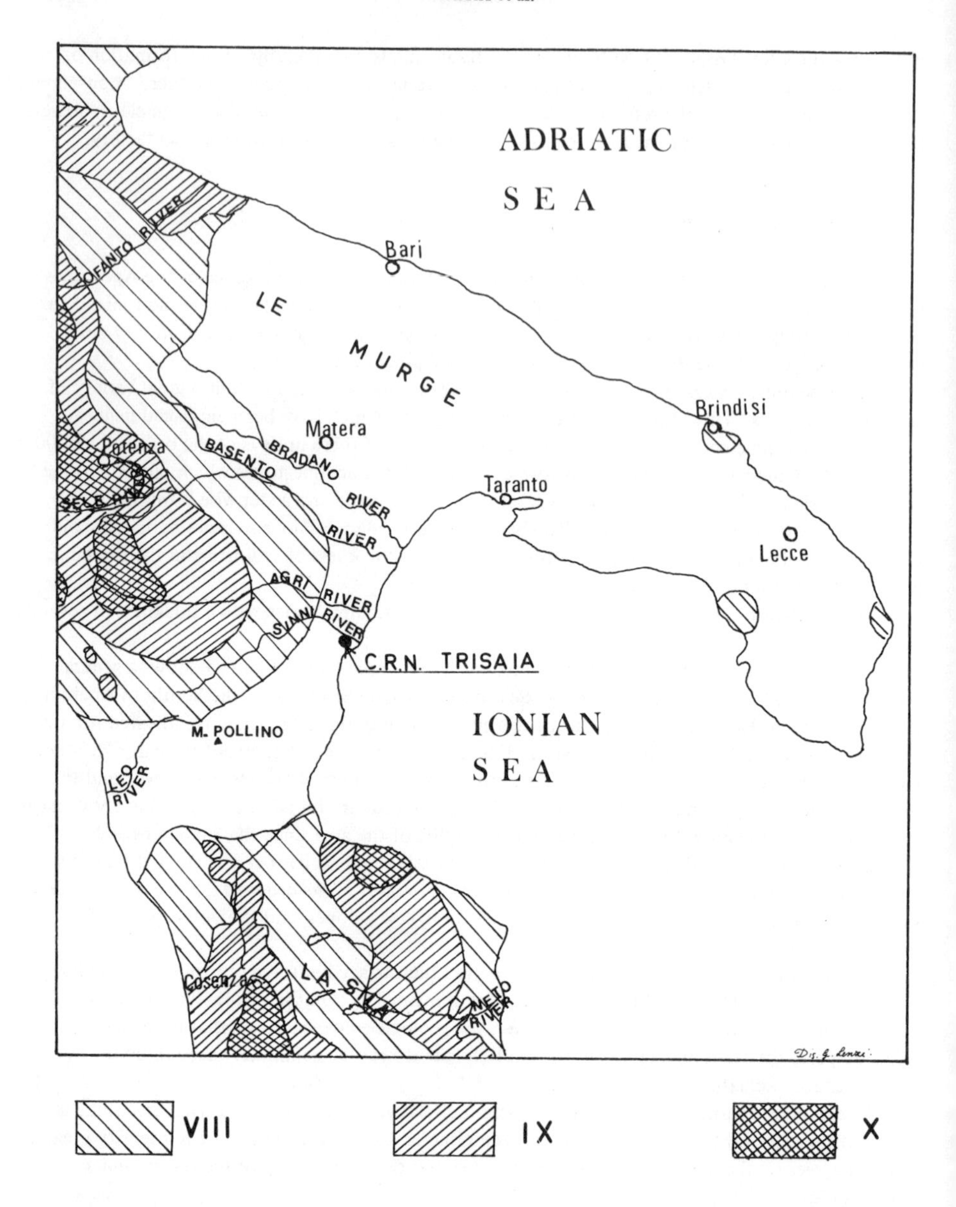

FIG.2. Seismic map of southeastern Italy (Mercalli Scale) and location of CRN Trisaia (Redrawn from CNEN – Laborat. Geominerario).

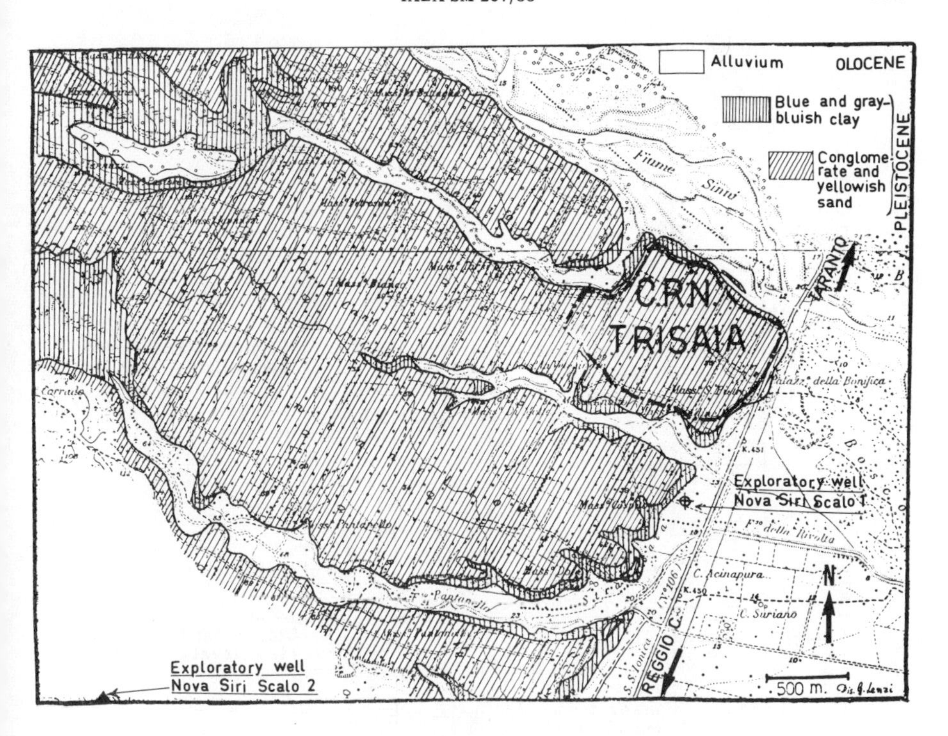

FIG.3. Geologic map of the Trisaia region.

The Centre is on a terrace at an elevation of a few tens of metres above mean sea level. The thickness of the terrace materials is extremely variable; in the area of the Centre it ranges from 0 to about 10 m [8]. The terrace materials are yellow sands, gravels and conglomerates, occasionally containing thin lenses of silt and clay.

The terrace at the Centre is underlain by a thick formation of marly clays. The clays are massive and gray-bluish in colour; the age is Plio-Pleistocene. The total thickness of the clay formation at the site is in the order of one thousand metres. Sand lenses usually containing saline water, but occasionally some methane, are present in the clay formation, particularly at depths between 350 and 500 m [7]. The stratigraphy of the marly clays is shown in Fig. 4.

The marly clays are underlain by the chaotic complex (Argille Scagliose, marls, sandstones, with many fragments of other formations). The chaotic complex is an allochthonous formation, resulting from gravitational sliding in the early Pliocene. The chaotic complex outcrops extensively several kilometres west and southwest from the Centre [7]. Figure 5 shows a schematic cross-section illustrating the relationships between the Plio-Pleistocene formations and the chaotic complex.

The AGIP well Nova Siri Scalo 1 was terminated within the chaotic complex. From other oil wells existing in the area it is known that the chaotic complex overlies more marly clays of the early Pliocene. The autochthonous clays are underlain by Upper Cretaceous limestones of great thickness. The Cretaceous formation is known to be intersected by many faults with large displacements.

The rates of erosion in the vicinity of the Centre are very low, due to the proximity of the base level, the lack of relief and the resistance of terrace materials. Some channel cutting takes place during the violent storms typical of the area, particularly if the cuts have reached the marly clays. It is obvious that future significant deepening of stream channels will be possible only if the sea level is greatly lowered or the area is greatly uplifted.

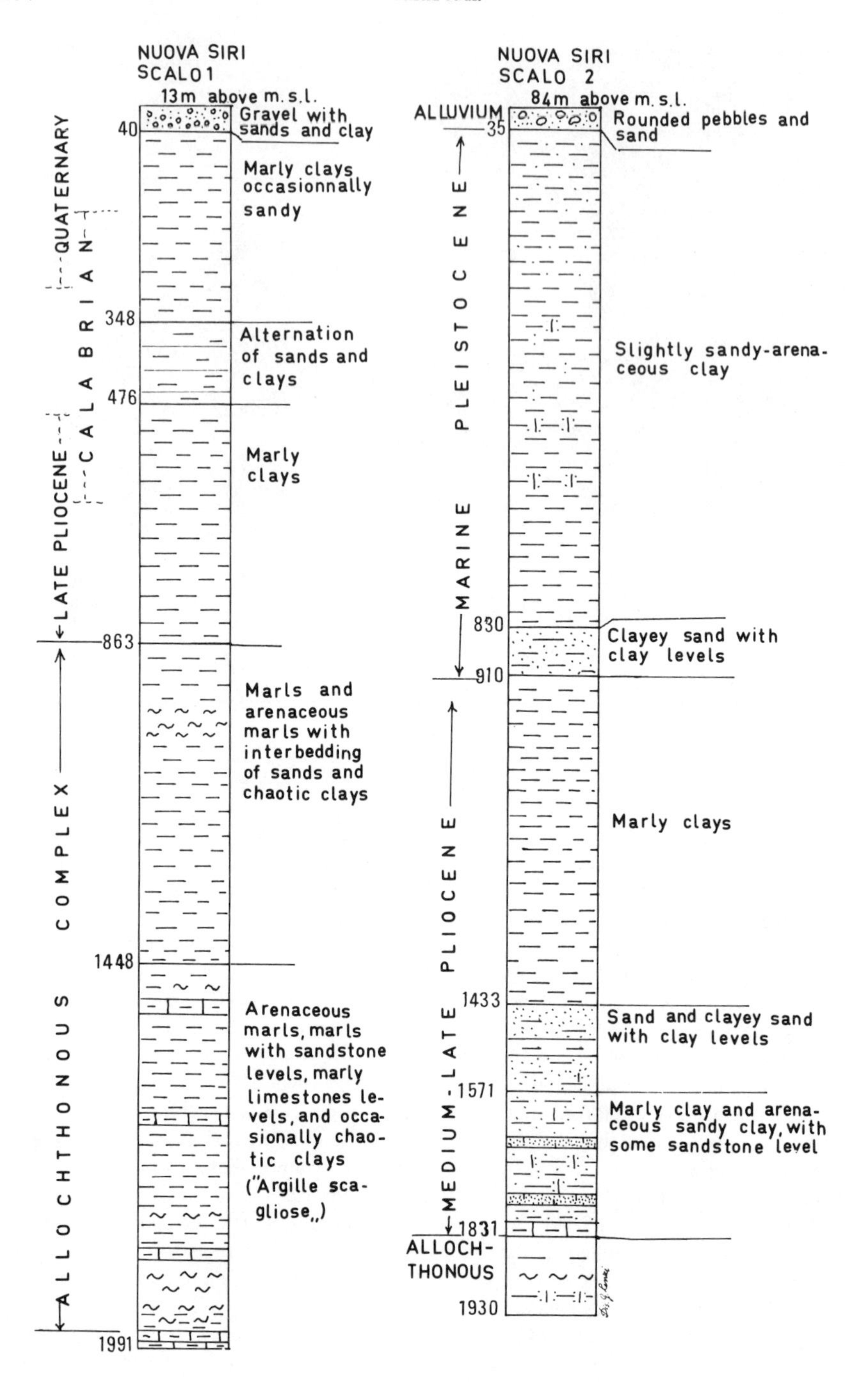

FIG.4. Stratigraphy in proximity of CRN Trisaia.

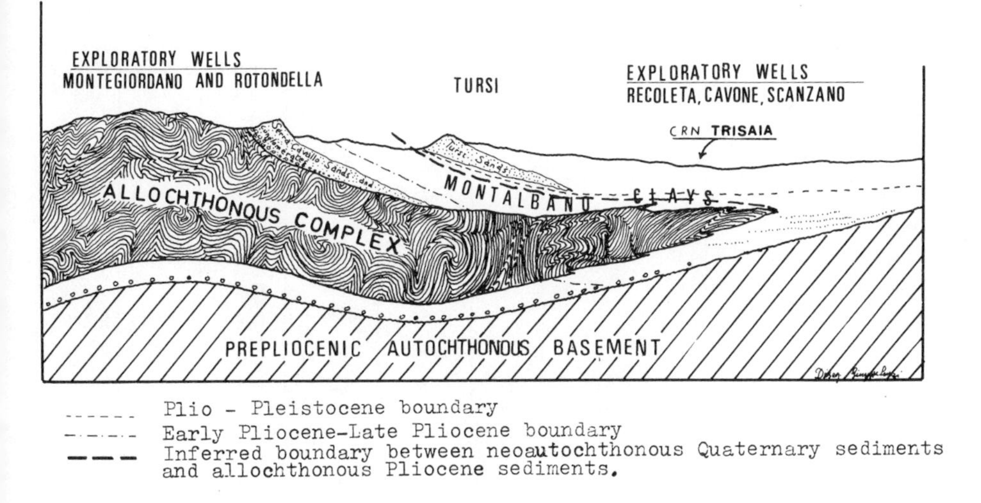

FIG.5. Relationship between the Plio-Pleistocene formations and the chaotic complex (Redrawn from Ref. [7]*).*

4. MARLY CLAYS

The Plio-Pleistocene clays underlying the Trisaia Centre have been investigated on a number of occasions. As a result a good deal of information is already available. However, all the data refer to observations of the upper section of the formation. The deepest available samples have been taken about 50 m below the top of the clay formation [9].

The following mineralogical composition (by weight) is the average of 15 samples taken in the area surrounding the Centre: quartz 12%; plagioclase < 2%; carbonates 20%; clay minerals 66%.

The composition of the clay fraction is (average of 11 samples): illite 60%; montmorillonite 13%; kaolinite + chlorite 27%.

The chemical composition of samples which had been dried at 110°C was (average of 4 samples):

	% (by weight)
SiO_2	49.85
TiO_2	0.74
Al_2O_3	13.33
Fe_2O_3	3.61
FeO	1.60
MnO	0.05
MgO	2.78
CaO	10.51
Na_2O	1.36
K_2O	2.53
P_2O_5	0.18
H_2O	4.34
CO_2	8.57
Total	99.45

TABLE I. GEOTECHNICAL PROPERTIES OF MARLY CLAYS

Sample depth (m)	Granulometric analysis (%)		Grain density (g/cm^3)	Bulk density (g/cm^3)	Water content (%)	Liquid limit (%)	Plastic limit (%)	Plasticity index (%)
	$\emptyset < 0.02$ mm	$\emptyset > 0.02$ mm						
12.75	82	18	2.71	1.61	25.50	50.98	20.93	30.05
13.50	88	12	2.72	1.63	24.50	44.70	21.40	23.30
14.85	87	13	2.70	1.59	25.60	48.60	23.76	24.84
20.85	77	23	2.70	1.78	18.50	39.40	21.62	17.78
33.85	84	16	2.72	1.59	24.50	49.30	26.69	22.91
35.50	88	12	2.70	1.73	20.30	44.40	20.00	24.40
35.85	81	19	2.71	1.71	21.00	39.45	20.00	19.45
41.75	85	15	2.71	1.61	23.50	44.15	22.22	21.93

The water loss during drying at 110°C was about 16% (by weight). The content of uranium and thorium was determined as about 3 and 10 ppm, respectively. The ion-exchange capacity was determined in the laboratory, with results ranging between 10 and 50 meq/100 g. The very low permeability is an essential property for a geological formation that has to provide long-term containment of radioactive wastes. An *in situ* permeability test which used ^{82}Br as tracer gave a value in the order of 10^{-8} cm/s [9].

Table I shows the results of some geotechnical analyses. No correlation with depth can be recognized for any one of the observed properties. This is probably due to the insufficient depth range over which the samples have been taken.

The ultimate strength of unconfined samples of marly clays falls in the range 2 to 7 kg/cm^2 [10].

5. THERMAL STUDIES

The high heat generation rate of high-level waste is a serious complicating factor for any geologic disposal scheme. CNEN has already conducted some preliminary studies on the heating of the Trisaia clays both in the laboratory and *in situ.* The laboratory experiments essentially consisted in inserting an electrical heater in a large block of clay and in measuring the temperature distributions at various times and under various heating conditions [11].

At the Trisaia site an electrical heater simulating waste with an initial power density of 40 W/l was introduced in the clays at a depth of 8 m. The total initial power of the heater was about 1 kW. The heater was surrounded by thermocouples placed both on the heater itself and in observation wells. The peak temperature in the clays exceeded 500°C. After about one year the experiment was terminated by mechanical failure of the heater [11].

The results of this experiment are somewhat questionable since plugging of the holes was not effective and heat removal by convection is likely to have taken place. After breakdown of the heater samples of heated clays and the heater itself were recovered. The samples from the immediate proximity of the heater are obviously baked. The mineralogical analyses, conducted by X-ray diffraction, have shown the following changes as a result of heating:

(a) A certain reduction of calcium carbonate content, particularly in the fraction < 2 μm.

(b) The appearance of an amorphous residue and gypsum, the latter probably from oxidation of FeS_2 impurities and reaction with $CaCO_3$.

(c) The clay fraction became composed almost completely of illite with seriously damaged crystalline structure. No montmorillonite and only traces of kaolinite and chlorite were found in the altered clays.

The thermal conductivity of the water-saturated clays at ambient temperature is assumed to be about 3.8×10^{-3} cal/cm·s·°C. For the dewatered clays the thermal conductivity is significantly lower, a value of $\sim 2.0 \times 10^{-3}$ cal/cm·s·°C can be tentatively taken. The thermal diffusivity is taken to be about 5.0×10^{-3} cm^2/s. The specific heat of the water saturated clays is taken as 0.36 cal/g·°C.

Using the thermal properties listed above a theoretical calculation was performed for the case of emplacement of a waste cylinder identical to the ones assumed in the Oak Ridge calculations for disposal in salt [12]. With these assumptions, the theoretical maximum temperature in the clays is almost 800°C and is reached about 15 months after emplacement. It is clear that, in comparison with salt, clays will require either smaller waste containers or older waste. In addition, the calculation indicated that over a period of about two years no temperature increase takes place in the clays at a distance from the source of greater than 15 to 20 m.

Therefore, if the heaters are surrounded by at least 20 m of clay the medium can be considered as infinite, provided the experiment does not last more than two years. This has been considered in the design of the second heating experiment, which will use multiple heaters placed more than 20 m below the top of the clays. The objective of the second heating experiment is to measure both temperatures and pore fluid pressures.

The evolution of pore fluid pressures should provide an insight in the behaviour of pore fluids when confined clays are heated. To obtain representative pressure measurements it will be essential to achieve perfect plugging of all boreholes. The current plan is to produce layered plugs composed by clay, expanding cement and epoxy resin.

6. FUTURE STUDIES

To demonstrate the feasibility of high-level waste disposal in a particular geological formation it is necessary to perform many studies and, eventually, to proceed to the construction of an actual demonstration facility.

For the time being CNEN's programme plan covers only the investigations required to qualify the marly clays, underlying the Trisaia Centre, for siting a radioactive waste repository. In the event of a positive outcome of the many studies envisaged, the logical follow-up would be to build a pilot-plant repository; however, no decision on this matter is yet available.

The main tasks in the programme plan are:

(1) **Geological studies:**

(a) General geological and geophysical studies of the area
(b) Assessment of maximum extent of possible geologic changes over the next 100 000 years
(c) Drilling and coring of the clay formation, including geophysical logs and hydraulic testing
(d) Mineralogical, chemical and geotechnical analyses of the core
(e) Laboratory and *in situ* experiments on the effects of heating confined clays
(f) Experimental studies of radiation effects on clays, including energy storage (Wigner effect)
(g) Study of faulting in clay formations
(h) Study of known instances of naturally heated clays.

(2) **Engineering and development studies:**

(a) Assessment of alternative waste emplacement concepts
(b) Thermal and mechanical analysis of alternative waste geometries in the formation

(c) Development of plugging techniques capable of preserving long-term waste containment
(d) Investigation of waste-clay interactions
(e) Study of systems to receive, move and position the waste containers
(f) Design of a pilot-plant repository.

(3) **Hazard analysis**:

(a) Short-term hazard assessment
(b) Long-term hazard assessment.

The time necessary to perform the site qualification studies has been estimated to be between five and ten years, depending on the amount of funding available.

7. CONCLUSIONS

On the basis of general considerations the formation of massive marly clays that underlies the Trisaia Centre of CNEN in Basilicata in Southern Italy appears a very promising medium for the emplacement of long-lived radioactive wastes.

The relative proximity to the Apennines and the evidence of Quaternary tectonic activity in the area are not believed to constitute insoluble difficulties since the thickness and plasticity of the formation should ensure preservation of waste containment even in the event of faulting. The combination of very low permeability and high ion-exchange capacity ensure that, even in the event of waste leaching, no significant migration of hazardous radionuclides would take place.

An important question is the behaviour of pore fluids in the period of greatest heating of the clays. An *in situ* experiment planned for the coming year should contribute to the solution of this problem.

Assuming a favourable outcome of the various studies, sufficient funding of the programme and prompt political decision making, a pilot-plant repository could become available by the middle eighties.

ACKNOWLEDGEMENT

Gratitude is expressed to Professor P. Paolo Mattias, Institute of Applied Geology, Department of Engineering (S. Pietro in Vincoli), University of Rome, for performing the X-ray diffraction analyses of the heated clay samples.

REFERENCES

[1] GERA, F., JACOBS, D.G., Considerations in the Long-Term Management of High-Level Radioactive Wastes, ORNL-4762 (1972).
[2] D'ARGENIO, B., PESCATORE, T., SCANDONE, P., Moderne vedute sulla geologia dell'Appennino (Proc. Conf. Rome, 1972), Atti della Accademia Nazionale dei Lincei, Rome (1973) 49.
[3] COTECCHIA, V., VALENTINI, G., Geologia, erosione, condizioni di stabilità e possibilità di sbarramento della media valle del fiume Sinni, Geol. Appl. Idrogeol. **1** (1966) 179.
[4] DEMANGEOT, J., Moderne vedute sulla geologia dell'Appennino (Proc. Conf. Rome 1972), Atti della Accadem Nazionale dei Lincei, Rome (1973) 215.
[5] COTECCHIA, V., MAGRI, G., Gli spostamenti delle linee di costa quaternarie del Mare Ionio fra Capo Spulico Taranto, Geol. Appl. Idrogeol. **2** (1967) 3.
[6] IACCARINO, E., Attività sismica in Italia dal 1893 al 1965, RT/GEO (68) 14 (1968).

[7] MOSTARDINI, F., PIERI, M., Note illustrative della carta geologica d'Italia, Foglio 212, Montalbano Ionico, Servizio Geologico d'Italia, Rome (1967).

[8] CASSANO, G., LENZI, G., MARZOCCHI, A., Indagine preliminare su alcuni tipi litologici del C.R.N. della Trisaia in relazione alle possibilità di contaminazione delle falde acquifere e delle acque di superficie, RT/PROT (69) 22 (1969).

[9] CNEN, Laboratorio Geominerario, Contratto EURATOM 004-65-3 WASI, Vol. 1, Rome (1973).

[10] DEL PRETE, M., VALENTINI, G., Le caratteristiche geotecniche delle argille azzurre dell'Italia sud-orientale in relazione alle differenti situazioni stratigrafiche e tettoniche, Geol. Appl. Idrogeol. **6** (1971) 197.

[11] FERRO, C., et al., Management of Radioactive Wastes from Fuel Reprocessing (Proc. Symp. IAEA/NEA Paris, 1972), OECD/NEA, Paris (1973) 887.

[12] CLAIBORNE, H.C., Personal communication, Oak Ridge National Laboratory (1975).

DISCUSSION

K. KÜHN: Figure 4 of your paper shows that you expect to find clay at your site, even at depths of about 2000 m. Is it not probable that some degree of diagenesis may have occurred at these depths and that you will encounter shale or even slate rather than clay?

And a second question is, have you made any tests yet on samples from around the heaters — for example, diffusibility or permeability tests?

F. GERA: The consolidation of clays increases with depth. At depths below 1000 to 1500 m, however, we do not expect any significant changes in the properties of the clay. A definitive answer will be provided by the core hole which we plan to drill fairly soon.

In reply to your second question, we have only had time so far to perform X-ray diffraction analyses.

T. WESTERMARK: I have a question regarding the studies of radiation effects that you have planned. I am interested in the role of radiation not only in the repository you are considering in Italy, but also in the Belgian one, and the Swedish one which, according to our geologists, will be built in primary rock (gneiss or granite). Since gamma doses may stem predominantly from ^{137}Cs and may attain 10^{11} rads in the space between the canisters and the rock or clay, as well as in the neighbouring clay mineral, I would like to ask whether your programme will deal with radiolytic gases, such as hydrogen, and whether it will include the possible change in redox potentials that may influence the migration of actinides, should such occur. Will it also include the effect of radiation on the corrosion of cladding materials?

F. GERA: We plan to study all the effects of radiation on clays, both in laboratory and *in situ*. I should point out, however, that although radiation effects may be drastic in the immediate vicinity of the waste canisters, they should not affect the formation as a whole to an extent where the reliability of the containment might be reduced.

W. HILD: You stated that the clay you investigated lost up to 16 wt% H_2O on heating. Can you give us an idea where this water goes when in contact with a heat-producing container? Do you expect changes in the adjacent layer due to this migration? Furthermore, in your heat experiment you noticed the decomposition of carbonates and formation of $CaSO_4$. Do you have any idea as to the corrosion that these reactions might cause on your waste package containers?

F. GERA: The planned heating experiment is supposed to provide answers to exactly these questions. I should be able to give you a reply within the next two or three years.

S.A. MAYMAN: What will you fall back on should you find, following your investigations, that clay is not a suitable formation for disposal? Have you considered any other possibilities?

F. GERA: There are some salt formations in Italy, which in some cases are well known, and in other cases have only recently been encountered during the drilling of oil and gas wells. Generally speaking, the tectonic situation appears to be less favourable than in the United States of America or the Federal Republic of Germany, though these formations might provide a viable alternative. We intend to assess the Italian salt formations regardless of the outcome of the clay programme.

J. HAMSTRA: For purposes of a hazards evaluation, one needs to predict future seismic and tectonic stability of the area around a high-level waste repository. Do you consider the Mediterranean area generally acceptable from the point of view of long-term tectonic behaviour in this sense?

F. GERA: The Mediterranean area is undoubtedly marked by appreciable tectonic activity. However, plastic formations should be able to contain the waste even in the event of tectonic movement, providing they have the requisite extension and thickness. The right approach is to assess the maximum geological change that could take place over the next 100 000 years and to design the disposal system in such a way that containment would be preserved even in extreme cases. We intend to try some geological forecasting in this respect.

IAEA-SM-207/44

DISPOSAL OF RADIOACTIVE WASTES PRODUCED IN NUCLEAR INSTALLATIONS IN THE GERMAN DEMOCRATIC REPUBLIC

D. RICHTER, W. KÖRNER
Nuclear Safety and Radiation Protection Board,
Berlin,
German Democratic Republic

Abstract

DISPOSAL OF RADIOACTIVE WASTES PRODUCED IN NUCLEAR INSTALLATIONS IN THE GERMAN DEMOCRATIC REPUBLIC.

Studies on methods for the disposal of radioactive wastes applicable to the territory of the German Democratic Republic were made in connection with the prospective use of nuclear power. The investigations included the following methods: storage above ground or immediately below ground level, storage in salt formations and mine galleries, and injection into porous subsurface rock. From a comparison of the economic and nuclear-safety aspects of various variants it follows that the secondary use of an abandoned salt mine as central repository for radioactive wastes represents a favourable and prospective solution for the GDR over the next decades. At present the central repository is under construction. It was subjected to a thorough safety assessment according to the regulations on the licensing of nuclear installations in the GDR.

1. INTRODUCTION

As in other countries with small resources of conventional energy carriers, the energy demand, especially that for electric energy, will in future be met to an increasing extent by nuclear power plants in the German Democratic Republic. Proceeding from the requirement of observing the permissible radiation exposure limits both for individuals and the population as well as of preserving the purity of water, air and other natural resources, on the one hand, and from the requirement of optimizing power economy on the other, the necessity results, from a prognostic aspect and on a national scale, to assess beforehand the problems arising from the production of radioactive wastes in the operation of nuclear installations.

In the GDR the Nuclear Safety and Radiation Protection Board is the responsible regulatory body for all questions of safety in the use of nuclear energy and of protection from ionizing radiation, among others for issuing regulations, granting licences and the performance of inspections. This responsibility also covers the following fields [1]:

(a) Setting permissible limits for the discharge of radioactive substances with off-gases and waste waters into the environment;

(b) Deciding on methods for interim storage and ultimate disposal of solid, liquid and gaseous radioactive wastes of all categories of activity;

(c) Stipulation of regulations for central collection, processing and disposal of radioactive wastes arising from defined areas;

(d) Assessing all questions of formation, handling, processing and interim storage of radioactive wastes within the licensing procedure for nuclear installations.

In accordance with the 'Decree of Radiation Protection Licensing for Nuclear Installations' of the GDR [2] at the stage of siting approval of a nuclear installation the Nuclear Safety and Radiation Protection Board has to decide on the method to be applied in the disposal of radioactive wastes for

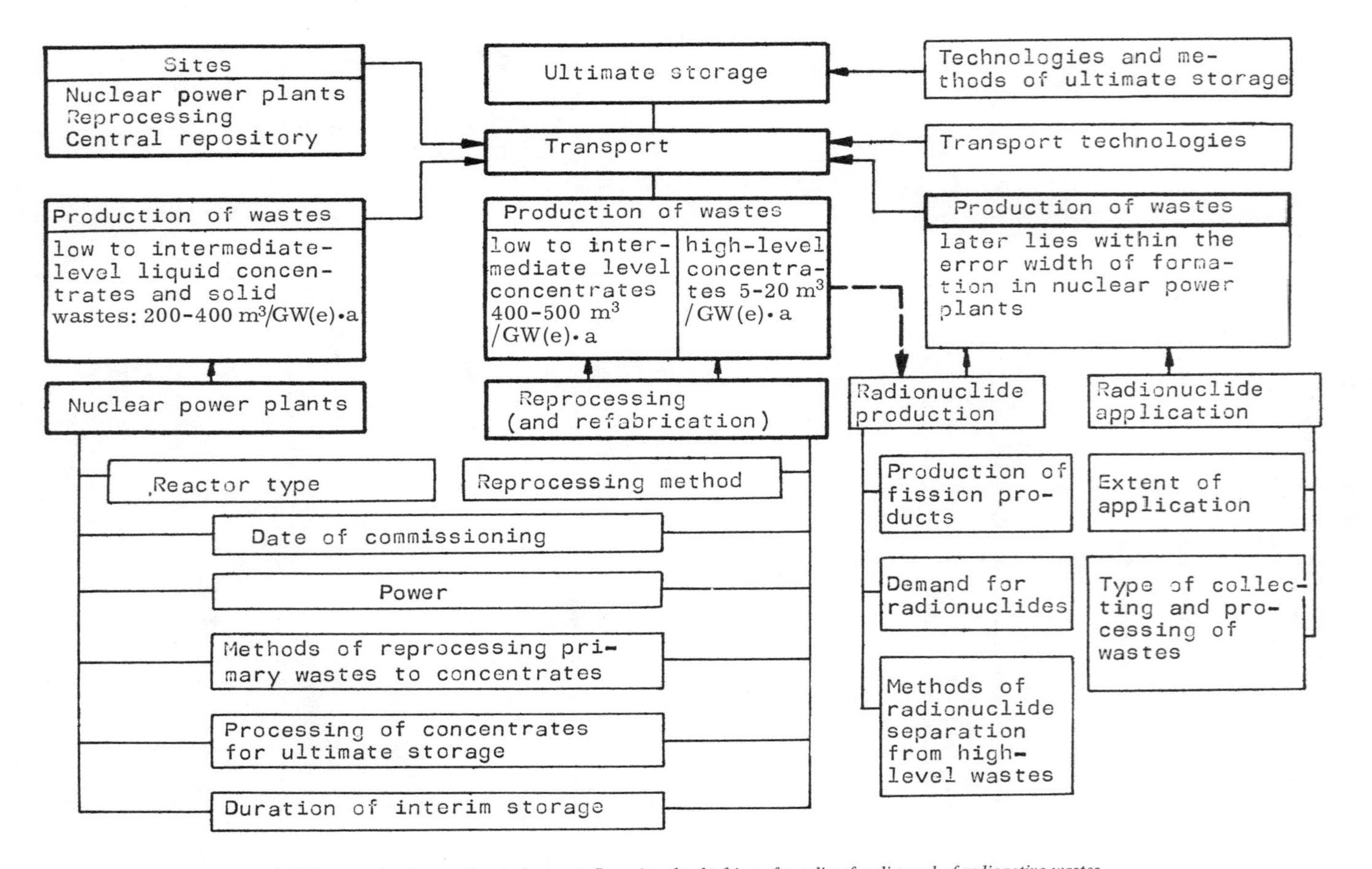

FIG.1. Rough scheme of main factors influencing the drafting of a policy for disposal of radioactive wastes.

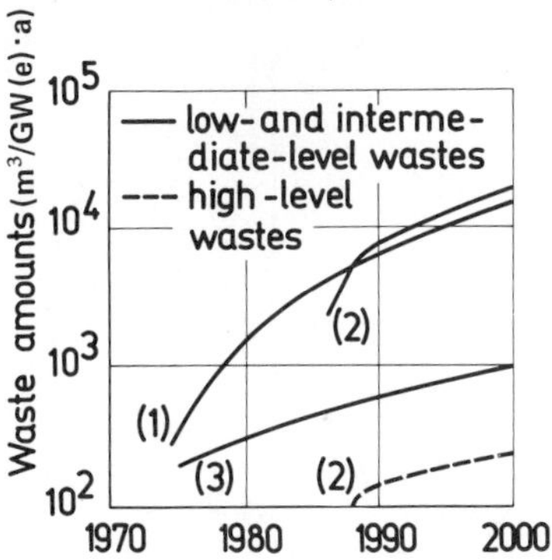

FIG.2. Production of radioactive wastes for a variant of the use of nuclear energy: (1) From nuclear power plants; (2) From spent fuel reprocessing; (3) From the production and the use of radioisotopes.

this installation. For this reason, some years ago a systematic study of the possibilities of disposal of radioactive wastes on the territory of the GDR was made to analyse the possible consequences of an expanding use of nuclear energy to GDR territory in the next few decades, to indicate disposal methods applicable from the nuclear safety and economic aspects and, if necessary, to make decisions on favourable regulations on an all-state scale [3–5].

2. STUDIES OF POSSIBILITIES OF DISPOSAL OF RADIOACTIVE WASTES FROM NUCLEAR INSTALLATIONS IN THE GDR

2.1. Initial parameters

The production of radioactive wastes in the industrial use of nuclear energy is determined by the installed electric power, type and mode of operation of nuclear power plants, by the methods for processing radioactive media and, especially, by the time and type of spent fuel reprocessing. The factors essentially influencing the establishment of a system of radioactive waste disposal are listed in Fig. 1.

Proceeding from the installed nuclear power plant capacity of ca. 2 to 5 kW per head of population to be expected for industrial states by the year 2000, the production of radioactive wastes and the concomitant inventory of radioactive substances and its hazard potential has been assessed for the GDR for various variants of the use of nuclear energy within the next few decades. Pressurized-water reactors of Soviet production with a power at 440 and 1000 MW will be used. The operation of fast breeder reactors and reprocessing plants will be taken into consideration at a later period, roughly from 1990 onwards. For the assessment of the generation of radioactive wastes shown for a variant in Fig. 2, the amounts and types of radioactive wastes given in Table I, which arise per GW(e) of installed electric power and year in the operation of nuclear power plants and reprocessing plants (Purex process) and which have to undergo interim or ultimate storage, have been taken as a basis. The relatively large amounts of liquid radioactive wastes of low to intermediate-level activity mentioned for nuclear power plants arise because only small amounts of excess waters with a low content of radioactive substances, which, as a rule, fall considerably below the permissible environmental capacity, are discharged into the environment. Normally, solid low- to intermediate-level wastes are taken unprocessed to ultimate storage, liquid low- to intermediate-level waste concentrates are stored after solidification by concrete or by bitumen.

TABLE I. ESTIMATED QUANTITIES OF SOLID AND LIQUID WASTES FOR ULTIMATE DISPOSAL FROM NUCLEAR POWER PLANTS AND SPENT FUEL REPROCESSING RELATED TO AN INSTALLED ELECTRIC CAPACITY OF 1000 MW AND ONE YEAR OF OPERATION (m^3/GW(e)·a)

Nature of the wastes		Pressurized water reactor		Fast breeder reactor (sodium cooled)		Fuel reprocessing (Purex process)	
Composition	Activity category[a]	(m^3/GW(e)·a)	Radioisotopes	(m^3/GW(e)·a)	Radioisotopes	(m^3/GW(e)·a)	Radioisotopes
Solid wastes (ion-exchange resins, disused equipment, paper, plastics, junk)	Low level (low activity)	80	^{60}Co, $^{55,59}Fe$, ^{51}Cr, ^{54}Mn	100	^{22}Na, ^{60}Co, $^{55,59}Fe$,	80	Up to 1% of the fission product inventory of the fuel
	Intermediate level (medium activity)	20	Fission products, mainly ^{137}Cs, $^{95}Zr/Nb$	50	^{51}Cr, fission products, transuranium elements	120	
Solidified/ vitrified wastes contained in steel cylinders	High-level (high activity)	–	–		–	1–5 (20)	99% of the fission product 0.5–1% of the Pu and 100% of the Am and Cm contained in the fuel
Liquid wastes (concentrates from waste water treatment, decontamination etc.)	Low-level intermediate-level	160 40	As above	70 30	As above	150 100	As above

[a] Low-level: 10^{-8}–10^{-5} Ci/l or Ci/kg, intermediate-level: 10^{-4}–10 Ci/l or Ci/kg, high-level: 10 Ci/l or Ci/kg.

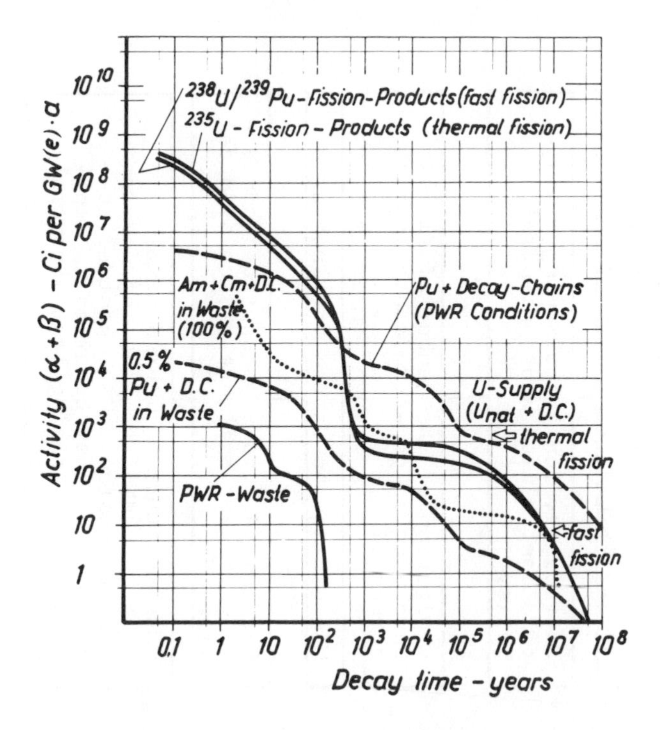

FIG.3. Radioactivity of fission products and transuranium elements in spent fuel elements, reprocessing wastes and wastes from nuclear power plants related to an installed electrical capacity of 1000 MW and an operation time of one year.
Conditions: typical operating conditions of a pressurized water reactor (fission of ^{235}U; thermal efficiency 33%, neutron flux $5 \times 10^{13} n \cdot cm^{-2} \cdot s^{-1}$; irradiation time 3 years) and of a sodium cooled fast breeder reactor (fission of ^{239}Pu and ^{238}U; thermal efficiency 40%, neutron flux $5 \times 10^{15} n \cdot cm^{-2} \cdot s^{-1}$, irradiation time 3 years); 300 days full load per year [5].

A survey of the radioactivity of fission products, transuranium elements and their decay products produced per 1000 MW installed electric capacity and year and contained in spent nuclear fuels, wastes of reprocessing and wastes from nuclear power plants is shown in Fig. 3 as a factor of decay time up to some million years. For comparison, the radioactivity of the daughter products of the nuclear fuel uranium 'burnt' to generate this energy is given.

For the production and use of radionuclides an increasing development is taken as a basis, which, however, from the quantitative aspect lies within the variation width of the amounts of radioactive wastes formed in nuclear installations, especially in years to come.

2.2. Possible methods of ultimate storage of radioactive wastes

Proceeding from the basic assumption that the radioactive wastes arising in the use of nuclear energy on the territory of the GDR will also have to be disposed of on GDR territory, the possibilities of the disposal of liquid and solid low- to intermediate-level and high-level wastes have been studied for the practice of industrial waste disposal. Methods that might have a high radiological impact on the environment (e.g. discharge of liquid wastes to surface waters including coastal waters, seepage of

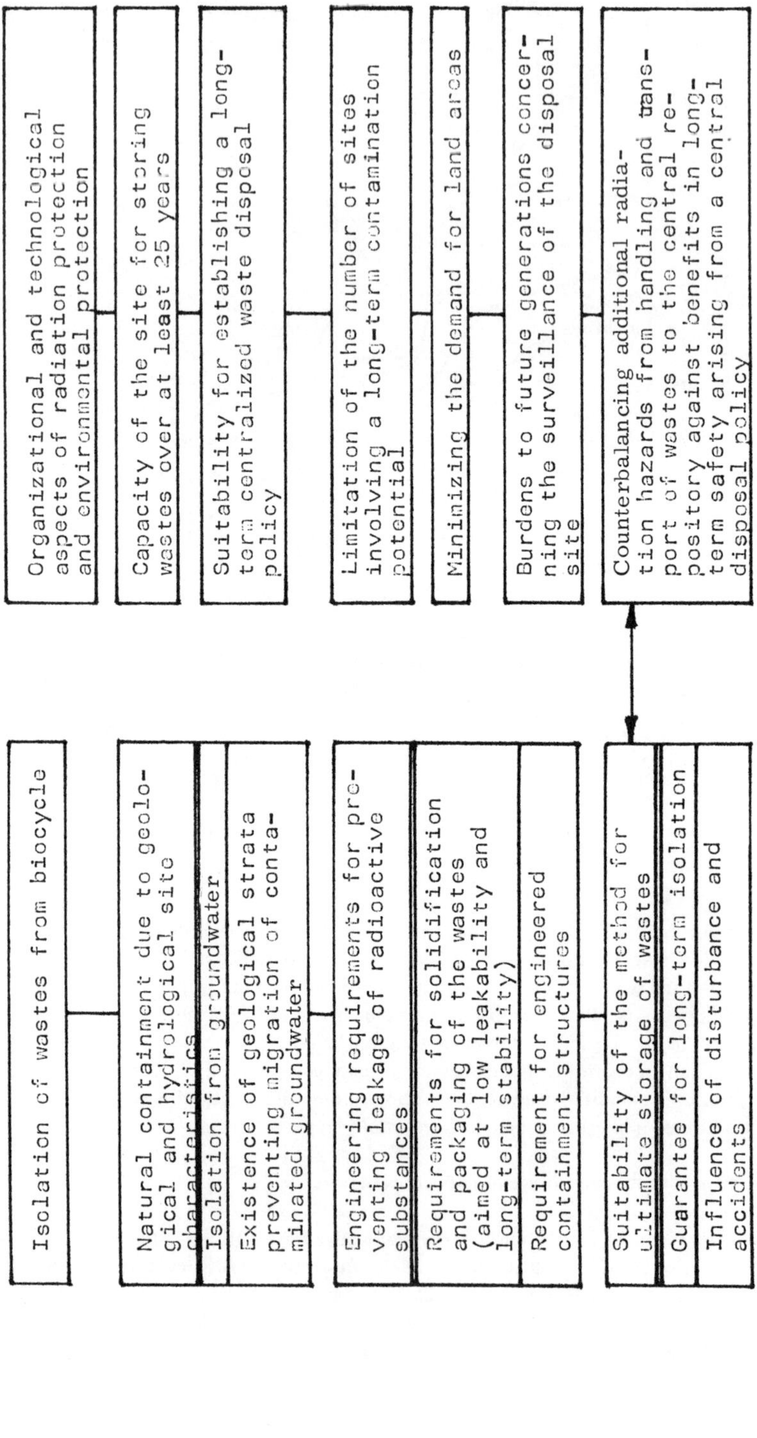

Isolation of wastes from biocycle

Natural containment due to geological and hydrological site characteristics

Isolation from groundwater

Existence of geological strata preventing migration of contaminated groundwater

Engineering requirements for preventing leakage of radioactive substances

Requirements for solidification and packaging of the wastes (aimed at low leakability and long-term stability)

Requirement for engineered containment structures

Suitability of the method for ultimate storage of wastes

Guarantee for long-term isolation

Influence of disturbance and accidents

Organizational and technological aspects of radiation protection and environmental protection

Capacity of the site for storing wastes over at least 25 years

Suitability for establishing a long-term centralized waste disposal policy

Limitation of the number of sites involving a long-term contamination potential

Minimizing the demand for land areas

Burdens to future generations concerning the surveillance of the disposal site

Counterbalancing additional radiation hazards from handling and transport of wastes to the central repository against benefits in long-term safety arising from a central disposal policy

FIG.4. Criteria for comparative assessment of methods for disposal of radioactive wastes.

liquid wastes into the ground near to the surface), that require international agreement (e.g. dumping into the ocean) or are far from feasible or safe in application (e.g. storage in polar ice, transport into outer space or transmutation to stable or short-lived nuclides in the case of highly toxic wastes) have been excluded from consideration.

The following methods were assessed:

(a) Disposal on the ground or directly below ground level:
   (i) disposal of solid or solidified low- to intermediate-level wastes in hydro-isolated concrete trenches or in tanks on the site of a nuclear installation
   (ii) storage of highly active wastes in tank facilities or concrete vaults
(b) Storage of solid and solidified low- to intermediate-level wastes in mine galleries (caverns in sedimentary rock) above groundwater level
(c) Storage in salt formations:
   (i) storage of solid and solidified radioactive wastes of all categories in abandoned salt mines or in abandoned parts of operating mines
   (ii) development of a new salt mine for storage of highly active wastes
   (iii) storage of low- to intermediate-level wastes (solid and liquid) in specially developed salt caverns
(d) Injection of liquid wastes into porous rock of deeper strata
(e) Injection of gaseous wastes into porous rock of deeper strata.

The technological components necessary to implement methods (a) to (c) (transport to repository, solidification methods, repository equipments, demand for land etc.) are included in the assessment.

2.3. Assessment criteria

The disposal methods were assessed qualitatively according to the criteria listed in Fig. 4 for their suitability from the radiation and environmental protection aspects [5, 6]. Based on total expenses either in terms of prime costs or social expenditures, a rough economic comparison was made by means of a computer program between several variants of combinable parts of technologies of processing, of transport and of ultimate storage of radioactive wastes [7].

2.4. Results of variant analyses

Figure 5 gives a survey of the analysed main variants of ultimate storage of radioactive wastes in the GDR. Organizationally, central storage of radioactive wastes of all categories, on-site disposal of all wastes and mixed variants of centralized and decentralized systems for the storage of different types of wastes have to be distinguished.

The results of a comparative economic assessment of variants for a period of 25 years and the conditions in the GDR have been listed in Table II.

There is essential evidence that both main variants – central storage of all wastes in an abandoned salt mine and on-site disposal of wastes in concrete trenches or special tanks – become equivalent at a definite level of waste production and over longer periods and that, at higher rates of waste production and over periods of use longer than 25 years, central ultimate storage shows distinct economic advantages. As expected, mixed variants of centralized and decentralized storage systems show higher expenses than the above-mentioned main variants.

As the schematic survey of the distribution of Zechstein salt in Fig. 6 shows, there are generally favourable geological conditions in the GDR for storing radioactive wastes in salt deposits [8, 9]. There are economic advantages for the secondary use of mines that are soon to be abandoned, since the cavities for storing low- to intermediate-level wastes already exist and favourable preconditions for guaranteeing mine safety can be expected. Such conditions often no longer hold for mines abandoned longer ago. The development of a new salt mine and, if necessary, the development of

Variant I	Central ultimate storage of wastes of all categories in salt mines(abandoned mines, abandoned parts of operating mines, sinking of new shaft)
Variant II	Storage of wastes of all categories on site: ultimate storage of low-to intermediate-level wastes in hydro-isolated concrete trenches, interim storage of high-level wastes in tanks
Variant III	Central ultimate storage of high-level wastes in salt mines
	Ultimate storage of low to intermediate-level wastes in hydroisolated concrete trenches on site
	Central ultimate storage of high-level wastes in salt mines
	Central ultimate storage of low-to intermediate-level wastes in mine galleries
Variant IV	Central ultimate storage of low-level wastes in salt mines
	Interim storage of high-level wastes in tanks
	Central ultimate storage of low-to intermediate-level wastes in mine galleries
	Interim storage of high-level wastes in tanks
Variant V	Ultimate storage of high-level wastes in salt mines
	Ultimate storage of low-to intermediate-level solid wastes on site in concrete trenches
	Ultimate storage of low-to intermediate-level liquid waste concentrates in salt caverns

FIG.5. Assessed main variants of ultimate storage of radioactive wastes in the GDR.

a cavern made by solution mining require very high investments, which seem justifiable for waste disposal only if the mined salt or salt solutions can reasonably be used in other branches of industry The discharge of salt solutions into rivers must be excluded for reasons of environmental protection. Central ultimate storage in an abandoned salt mine calls for relatively high initial investments and expenses for transportation (30 to 50% of the overall costs).

For the disposal of radioactive wastes of all activity categories, total expenses arise for the main variants, which lie around 1 to 2% of prime costs of power generation and thus seem economically justifiable. If only low- to intermediate-level wastes from nuclear power plants are considered,

TABLE II. RESULTS OF THE COMPARATIVE ECONOMIC ASSESSMENT OF SOME VARIANTS FOR THE ULTIMATE DISPOSAL OF RADIOACTIVE WASTES

Reference points: generation of radioactive wastes according to Table I and Fig. 2, time period considered: 25 years, averaged distance between nuclear installations and the central repositories 250–300 km.

Variant	Disposal method: For low- to intermediate-level wastes	Disposal method: For high-level wastes	Cost index[a]
Central repository	Storage in an abandoned rock salt mine		1
On-site disposal	Storage in hydro-isolated concrete trenches on the site	Storage in tanks on the site	0.9
Mixed variant	Storage in hydro-isolated concrete trenches on the site	Storage in a newly developed salt mine	1.5
Two central repositories	Storage in mine galleries in limestone	Storage in a newly developed salt mine	1.9
Mixed variant	Storage in an abandoned salt mine	Storage in tanks on the site	1.3
Mixed variant	Storage in mine galleries in limestone	Storage in tanks on the site	1.3
Mixed variant	Disposal into newly developed salt caverns	Storage in tanks on the site	1.4
Two central repositories	Disposal into newly developed salt caverns	Storage in a newly developed salt mine	1.9

[a] Prime costs of the respective variant related to the prime costs of a central repository using an abandoned salt mine.

the establishment of a central repository entails higher expenses than on-site disposal, assuming favourable hydrogeological conditions for on-site disposal. These, however, will be compensated by a prolongation of the utilization life of the salt mine considerably beyond the period of 25 years on which the above-mentioned economic assessments are based.

From the aspect of nuclear safety and radiation protection, all variants discussed above can be implemented, if the site-specific conditions, which also influence expenses, are allowed for. Storage in salt formations poses the smallest risk and also makes safe storage of highly active wastes and α-bearing wastes possible. As a result of salt mining in the GDR, there are favourable conditions for the application of this method.

Because of the existence of the salt deposits over geological periods, long-term hydro-isolation of the wastes stored can be expected. Rock salt has high strength, good heat conductivity and very small porosity. However, each site of such a central repository must be thoroughly analysed from the aspect of mining and nuclear safety (see section 4.3).

It should be pointed out that when implementing a central waste disposal policy the siting of other nuclear installations is facilitated, since the suitability of the site no longer depends on the requirement for ultimate disposal of wastes on the site. From the aspect of landscape preservation

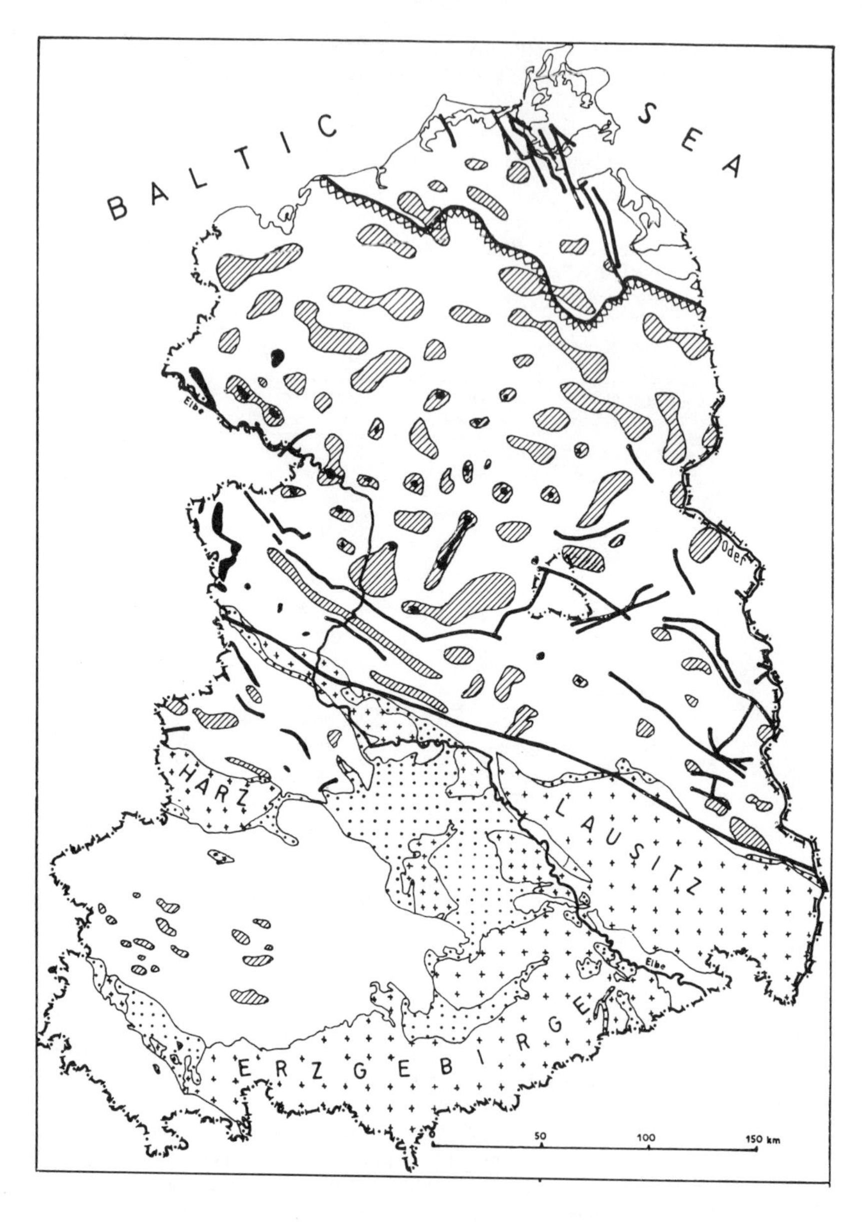

FIG.6. Schematic survey of Zechstein distribution in the GDR (according to authors' collective 1968, Meinhold 196.

- *Variscous basic rock and plutonites* } *mainly near to surface*
- *Variscous molasses* } *mainly near to surface*
- *Zechstein, mainly below large rock burden of rupelian clay*
- *Northern distribution limit border*
- *Zone of disturbances*
- *Salt diapirs*
- *Salt domes*

and environmental protection, central ultimate storage is of advantage because, apart from the mine premises, no land is occupied or needed for protective zones. An important task from the nuclear-safety and economic aspect is the organization of the transport system from the nuclear installation to the central repository.

The expenditure for the disposal of low- to intermediate-level wastes in hydro-isolated concrete trenches or tanks largely depends on the site conditions. Waste repositories in river valleys, i.e. areas that are often used for drinking water production, require special safety precautions to prevent the spread of radionuclides, e.g. in case of flooding and tank leakage, so that the prime costs mentioned in the economic section may rise by a factor of 2 to 3. If this method is applied, higher demands have to be made on the leaching behaviour of wastes.

The GDR possesses some mine galleries in limestone or sandstone accessible at surface level, which would make the storage of solid low- to intermediate-level wastes possible above the groundwater level. As these cavities are often not sufficiently isolated from groundwater, increased demands on the properties and packaging of wastes arise. Because of the smaller size of the cavities, the utilization life of galleries is also often shorter than that of a salt mine.

The overall results of the assessment of the various variants studied give unequivocal evidence of the benefits arising in the long run from the implementation of a centralized waste disposal policy by using an abandoned rock salt mine.

Injection of liquid wastes into deeper strata, which has not been included in the above considerations, has been practised for many years in the GDR for the disposal of waste waters from the potash and petrol industries. For the disposal of radioactive waste waters increased safety precautions are required, including extensive exploration of the store before use, calculation and control of the migration of radionuclides and pretreatment of the waste water to prevent damaging interaction with rocks and stratal waters. Porous and fissured rocks of the deeper strata that are isolated by huge water-repellent rock burdens from groundwater in use are suitable as storage levels. Studies showed that in some areas of the GDR, though mostly not identical with appropriate power plant sites, there are some possibilities of applying this method [8]. The injection of liquid radioactive wastes may be of interest later as a variant for the disposal of large amounts of liquid wastes, e.g. wastes containing tritium.

Isolation of gaseous radioactive wastes containing long-lived fission gases from the biocycle seems not to be necessary for the next few decades. Preliminary studies on the possibilities of storing gaseous wastes in the ground on the territory of the GDR [10] show that, as far as any necessity will arise, certain possibilities for using such methods exist with advantages from the aspect of safety and economy compared with methods of storing such wastes at surface level. However, extensive investigations to explore a site and to solve the engineering problems are necessary to guarantee safe use [11].

3. THE PRESENT SYSTEM OF RADIOACTIVE WASTE DISPOSAL IN THE GDR

The radioactive wastes from the 70 MW(e) Rheinsberg Nuclear Power Plant, which has been in operation for ten years, are disposed on site. Solid wastes are disposed to concrete trenches. After interim storage in stainless steel tanks, liquid wastes (evaporator concentrates) are mixed with cement, and the waste-cement mixture allowed to solidify as a monolith in a trench surrounded by concrete walls.

The waste management of the Greifswald Nuclear Power Plant and the nuclear power plants planned has already been orientated towards central ultimate storage in a salt mine. Until transport to the central repository, liquid and solid wastes are stored in tanks for the time being.

At present wastes from radionuclide production and application are centrally collected by a service group working at the Nuclear Safety and Radiation Protection Board [12]. These wastes are treated using flocculation and ion exchange for radioactive waste waters, evaporation for liquid

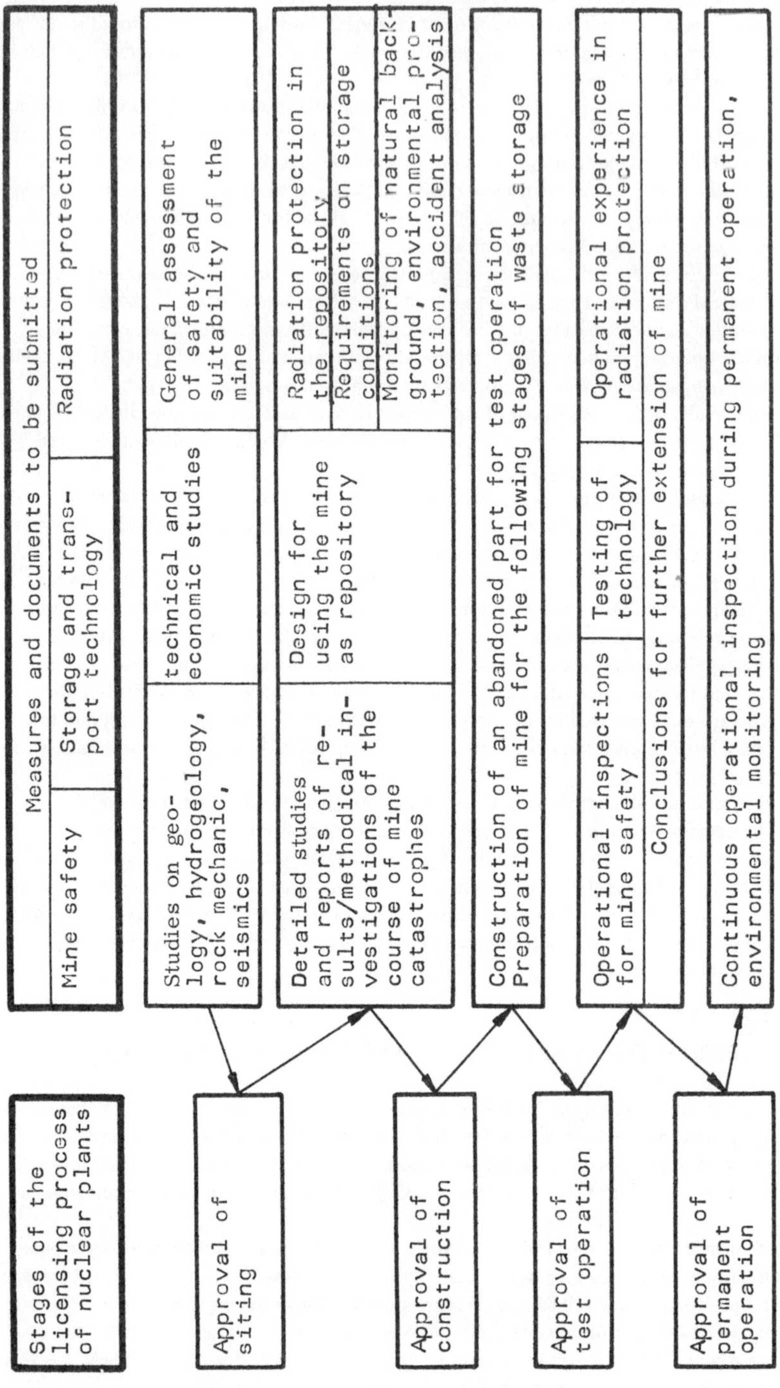

Stages of the licensing process of nuclear plants

Approval of siting

Approval of construction

Approval of test operation

Approval of permanent operation

Measures and documents to be submitted

Mine safety | Storage and transport technology | Radiation protection

Studies on geology, hydrogeology, rock mechanic, seismics | technical and economic studies | General assessment of safety and suitability of the mine

Detailed studies and reports of results/methodical investigations of the course of mine catastrophes | Design for using the mine as repository | Radiation protection in the repository; Requirements on storage conditions; Monitoring of natural background, environmental protection, accident analysis

Construction of an abandoned part for test operation
Preparation of mine for the following stages of waste storage

Operational inspections for mine safety | Testing of technology | Operational experience in radiation protection

Conclusions for further extension of mine

Continuous operational inspection during permanent operation, environmental monitoring

FIG. 7. Licensing process of radiation protection for a central repository of radioactive wastes in a salt mine.

wastes and ion-exchange regenerates, cementation or bituminization for evaporator concentrates and sludges and baling for solid wastes, and the waste concentrates are stored in 200-l steel drums in a sandstone gallery or in the open air. After commissioning the central repository, the inclusion of wastes arising from radionuclide application and production in this central system of waste disposal is envisaged.

4. THE CENTRAL REPOSITORY

4.1. Selection of a salt mine

Within the comparative studies mentioned in section 2.4, the secondary utilization of ten salt mines already out of operation or planned to be decommissioned by 1980 was assessed using the following main criteria:

(a) Total cavity volume, form of cavities and possibility of applying planned disposal technologies
(b) Technical state and mine-technological conditions
(c) Geographical position, transport communications
(d) Expense
(e) Present conditions of utilization.

After mine safety had been investigated in greater detail the Nuclear Safety and Radiation Protection Board had the most suitable mine taken over for secondary utilization and for extension as a central repository for radioactive wastes by the responsible branch of industry.

4.2. Short description of the central repository [13]

The former rock salt mine Bartensleben is easily accessible to traffic in the Magdeburg district. It has been developed in a cleft diapir of the upper Aller Valley. Below a rock burden of sands, clay and loam (ca. 200 m), are salt rocks (mainly rock salt of the Stassfurt type) with a depth of more than 500 m. The mine had operated for more than 50 years until decommissioning. Its distance from neighbouring mines is sufficiently far to satisfy safety considerations. The depth of the pit is 520 m. Four main horizons were developed at depths from 380 to 500 m. The excavations made by chamber working have the following dimensions: length 100 m, width 30 m and height 30 m. The cavity volume totals 5×10^6 m^3, 20% of which is utilizable for storing low- to intermediate-level wastes without larger expenditure. If required, new excavations can be made for highly active wastes.

As a result of technical and economic investigations for optimizing the system of waste transport from nuclear power plants to central repository, a container system has been developed. Depending on dose rate, the wastes are transported in various shielded containers, which are again contained in large containers for carriage by public transport.

In the central repository the shielded containers are transported underground. It is envisaged to dump solid wastes into the storage cavity from a transport gallery above. Underground, the liquid waste concentrates are mixed with hydraulic binders, pumped into the storage cavity and allowed to solidify there. Furthermore, the stacking of solid wastes, e.g. of drums, is envisaged. 2700 drums with low-level wastes were stored in 1972 using conventional mine technology.

After suitable structural alterations and the necessary radiation-protection adaptation, the storage of wastes from nuclear power plants will be started by way of an industrial test operation, 3 cavities below the 4th gallery with a volume of 40 000 m^3 being utilized.

4.3. Investigations into the safety of the central repository

The central repository has the nature of a nuclear installation and is, therefore, subject to licensing for radiation protection in the following phases according to the Decree for Licensing Nuclear Installations in the GDR (see Fig. 7):

Siting approval
Construction approval
Commissioning approval
Approval of permanent operation.

To obtain approval of the site the geological and hydrogeological conditions of the mine were given thorough consideration to demonstrate the safety of the mine over a long period. The leakage of radioactive substances from the mine is considered impossible if the static stability and dry state of the mine are guaranteed. Contact of wastes with groundwater, which would form a hazard to environmental media, would be possible if water penetrated into the mine, becoming contaminated by the wastes, and come into contact with the rock burden. Static stability can be diminished by fresh water inflow, whereby new access channels may develop. Therefore, hydrogeological hazards along with changes in the geomechanical conditions are of great importance for the suitability assessment of the central repository. In the case of high-level waste disposal the influence of heat and radiation on the salt rock has to be allowed for as well. For suitability assessment the following complexes were taken into account:

(a) Mine technological conditions:
Extension of the mining area, existing bores, state of shaft and excavations
(b) Geological investigations:
Structure of the rock burden, structure and rock content of the deposit, distance of excavations from water-bearing strata
(c) Hydrological investigations:
Groundwater dynamics in the rock burden, inflows into the mine and composition of waters, hazard from neighbouring mines
(d) Geomechanical investigations:
Determination of the stability of excavations by laboratory investigations and measurements in the mine
(e) Seismic investigations:
Analysis of the seismic activity of the area and its influence on static stability
(f) Investigation of long-term safety:
Behaviour of the mine after decommission, assessment of the probability and impact of disturbances on the environment.

After the submission of studies and the carrying out of special investigations (measuring programmes), it was assessed that the high safety factors of the Bartensleben mine suit it as a central repository for radioactive wastes.

The experience gained in the selection and suitability assessment of the central repository of the GDR was summarized in 'CMEA Methods of Investigations on Sanitary-Hydrogeological and Radiation-Protection Reasons for the Safety of Disposal of Radioactive Wastes in Salt Formations' [14].

5. CONCLUSIONS

As a result of studies from the nuclear-safety and economic aspect, the central ultimate storage of radioactive wastes of all activity categories in salt formations was laid down as the future disposal method for the GDR. After extensive investigations of the site, especially with respect to geomechanical and hydrogeological safety, a former salt mine was selected to be adapted as central repository and subjected to the licensing procedure for nuclear installations. The central repository has initially been planned for the storage of low- to intermediate-level wastes.

With the construction of a central repository, a solution was found to the disposal of radioactive wastes that meets the prospective demands of the development of nuclear energy in the GDR.

REFERENCES

[1] Verordnung über den Schutz von der schädigenden Einwirkung ionisierender Strahlung – Strahlenschutzverordnung – vom 26.11.1969, Gesetzblatt der DDR Teil II Nr. 99, p. 627; Erste Durchführungsbestimmung zur Strahlenschutzverordnung vom 26.11.1969, Gesetzblatt der DDR Teil II Nr. 99, p. 635.

[2] Anordnung über die Erteilung der Strahlenschutzgenehmigung für Kernanlagen – Kernanlagen-Genehmigungsanordnung – vom 4.12.1970, Gesetzblatt der DDR Teil II Nr. 102, p. 697.

[3] FISCHER, W., KÖRNER, W., RICHTER, D., Variantenuntersuchungen zur Auswahl des Verfahrens der Endlagerung radioaktiver Abfälle aus Kernanlagen in der DDR, SZS-138.

[4] RICHTER, D., "Regulatory aspects of environmental protection with regard to radioactive wastes and effluents produced by nuclear installations in the GDR", 6e Congres International sur les Tendances Nouvelles en Radioprotection (Proc. Congr. Bordeaux, 1972) Paris, Societé Francais de Radioprotection (1972) 1001.

[5] RICHTER, D., "Criteria for the assessment of possible methods for the long-term storage of radioactive wastes in the GDR", 7e Congres International de Societé Francais de Radioprotection, Versailles 1974, 679.

[6] RICHTER, D., KÖRNER, W., Strahlenschutz und hydrogeologische Anforderungen an die Lagerung radioaktiver Abfälle in der DDR, SZS-148, p. 134.

[7] FISCHER, W., BUHL, H., Rechenmodell zur Auswahl von Varianten der Beseitigung radioaktiver Abprodukte aus Kernanlagen, Kernenergie **15** (1972).

[8] ADAM, C., KÖRNER, W., RICHTER, D., Möglichkeiten des Versenkens flüssiger radioaktiver Abfälle in Gesteinsschichten des tieferen Untergrundes der DDR, Geol. Wissenschaften **2** (1947) 1017.

[9] RICHTER, D., Zur Nutzung von Hohlräumen in Salzformationen für die Endlagerung radioaktiver Abfälle in der DDR, Kernenergie **11** (1968) 241.

[10] SCHULZ, H., Lagerung gasförmiger radioaktiver Abfälle im Untergrund, SAAS-139, p. 1.

[11] DDR-Delegation in der Ständigen Kommission des RGW für die friedliche Anwendung der Atomenergie, "Methodik der Untersuchungen zur Einschätzung der geologischen Sicherheit bei der unterirdischen Lagerung nichtkonzentrierter radioaktiver Gase" (unpublished draft in Russian).

[12] Richtlinie für die zentrale Erfassung radioaktiver Abfälle vom 4.6.1974, Mitt. des SAAS 1975, Nr. 5.

[13] KÖRNER, W., EBEL, K., RICHTER, D., "Aufgaben des Strahlenschutzes bei der Zwischen- und Endlagerung radioaktiver Abfälle in der DDR", Vortrag zur RGW-Konferenz zu Strahlenschutzproblemen in Kernkraftwerken, Usti nad Labem, CSSR, 8.-12.9.1975 (in Russian).

[14] Ständige Kommission des RGW für die friedliche Anwendung der Atomenergie, "Methodik der Untersuchungen über die sanitär-hydrogeologische und strahlenschutzmäßige Begründung der Sicherheit bei der Lagerung radioaktiver Abfälle in Salzformationen", Berlin 1974 (in Russian).

DISCUSSION

K. KÜHN: You indicated that in the Bartensleben salt mine there is a heavily folded anhydrite seam. Is there, or was there, any brine or water in the 'main anhydrite', and have you made any tests?

D. RICHTER: During our investigations of the mineral composition of the deposit special emphasis was layed on the anydrite layers embedded in the rock salt. The 'main anhydrite' and the rock salt in the neighbourhood of it have been found to be dry.

J. HAMSTRA: It should be pointed out that the German Democratic Republic is the first country to specify its investigation methods for a disposal site in a salt formation. I think that we could all gain considerably from international co-operation in this respect. Obviously, it would allow a greater uniformity of approach.

The question I wish to ask is, up to what weight and size can you ensure vertical transport at the mine?

D. RICHTER: As far as I remember, the hoisting equipment in the shaft is designed to raise 20 tons.

H. KRAUSE: I see from your paper that in the future you are going to store high-level waste in the salt mine; you also mentioned there were a large number of sizeable chambers already in existence. Are you at all worried that the decay heat liberated from the high-level waste might affect the mechanical stability of these rooms and internal structures?

D. RICHTER: The geochemical stability of the part of the mine you refer to has been evaluated for the storage of low- and intermediate-level waste. For high-level waste storage – at least on an industrial scale – we shall have to develop new excavations in other parts of the mine.

RECENT RESULTS AND DEVELOPMENTS ON THE DISPOSAL OF RADIOACTIVE WASTES IN THE ASSE SALT MINE

K. KÜHN, E. ALBRECHT, H. KOLDITZ,
K. THIELEMANN
Institut für Tieflagerung,
Gesellschaft für Strahlen- und
Umweltforschung mbH,
München

W. DIEFENBACHER, W.J. ENGELMANN,
H. KRAUSE, M.C. SCHUCHARDT,
E. SMAILOS
Abteilung Behandlung radioaktiver Abfälle,
Gesellschaft für Kernforschung mbH,
Karlsruhe,
Federal Republic of Germany

Abstract

RECENT RESULTS AND DEVELOPMENTS ON THE DISPOSAL OF RADIOACTIVE WASTES IN THE ASSE SALT MINE.

Disposal of radioactive wastes into rock salt formations has been pursued in the Federal Republic of Germany since 1963. The Asse Salt Mine serves as the national R&D facility. 62 000 containers with low-level wastes were disposed of from April 1967 till February 1976. In addition, 746 drums with intermediate-level wastes found their final place in the mine. Both disposal techniques are under continuous development. One further R&D project is at present under construction, while a second one is in the planning phase. The Asse prototype cavity will test the possibility of direct disposal of intermediate-level wastes without shielding through a small diameter shaft into a cavity. This project will start operation in early 1979. Within the same time all necessary investigations and developments will be performed in order to start a test disposal operation with solidified high-level wastes in the early eighties.

1. INTRODUCTION

In 1957 the National Academy of Sciences – National Research Council of the United States of America published a report in which the disposal of high-level radioactive wastes into geologic formations was recommended [1]. One of the important sentences from that report reads as follows: "*Disposal in cavities mined in salt beds and salt domes is suggested as the possibility promising the most practical immediate solution of the problem.*" This idea was taken up in Germany in the early sixties when the problem of radioactive waste disposal arose for the first time. Because of the favourable geological situation in Germany, the Bundesanstalt für Bodenforschung (German Geological Survey) recommended in 1963 to the then responsible Bundesministerium für Atomkernenergie (Federal Ministry for Atomic Energy) in an expert's report on the different possibilities for the safe disposal of radioactive wastes that attention be focussed from the beginning on disposal in rock salt formations deep underground [2].

Production in the Asse Salt Mine was shut down in 1964 for economic reasons. After short negotiations the government of the Federal Republic of Germany purchased the Asse Salt Mine in 1965 and transferred it to the Gesellschaft für Strahlen- und Umweltforschung mbH München (GSF) to collaborate with the Gesellschaft für Kernforschung mbH, Karlsruhe (GfK), in developing and testing methods for the disposal of radioactive wastes in salt formations.

At about the same time another R & D programme was started, namely to look into the possibility of using a solution cavity in a salt dome for radioactive waste disposal. This programme was initiated in co-operation with Euratom.

Besides the main attention on radioactive waste disposal in salt formations, all further international developments to use other geologic formations for that purpose were closely and attentively followed in Germany. This paper reports on recent results and developments on the disposal of radioactive wastes in the Asse Salt Mine.

2. USE AS R&D FACILITY

Location, geologic setting and mining conditions of the Asse Salt Mine have been published in a number of papers, especially at earlier IAEA/NEA-Symposia [3-5], and so these data will not be reiterated here. But before giving the recent results of disposal operations some general remarks have to be made. As briefly indicated in the introduction, the Asse Salt Mine was a potash and rock salt producing mine from 1906 until 1964. Therefore, no site selection factors for a radioactive waste repository were applied to this plant. The cut and the shape of the underground workings were clearly specified by the former mining operations. In consequence, a great number of chambers exist today which were created by the exploitation of salt. On the other hand, the available underground space to provide new chambers or even more new panels for disposal operations is not unlimited.

There are some additional disadvantages that restrict the development of the Asse Salt Mine as a final repository for large quantities of especially high-level and alpha-bearing wastes:

(1) It is not completely possible with the currently available methods of measuring and calculating to prove and to predict for some thousand years the rock-mechanical stability of a complicated underground system of chambers, pillars and safety roofs situated above and beside each other.

(2) There is a thick seam of carnallitic potash salt in the Asse anticline, composed of roughly 52% carnallite, 30% rock salt, 14% kieserite, and some minor constituents. Carnallite is the hydrated chloride of potassium and magnesium with the chemical formula $KCl \cdot MgCl_2 \cdot 6H_2O$. By heating this mineral to temperatures above 110°C dehydration water is set free, at temperatures above 165°C sometimes even gaseous hydrochloric acid. In addition, carnallite creeps extraordinarily under stress thus having a very low strength. Its solubility in water and in a NaCl-saturated solution is fairly high.

(3) From our knowledge of the history of potash and salt mines in Germany we cannot completely exclude that the Asse Salt Mine will be flooded at some time. This happened in 1906 to the mine Asse I, which is situated 1.5 km to the west. However, the reasons for this flooding were inadequate knowledge of geology and mining and thus human failure.

(4) There is a small occurrence of $MgCl_2$-saturated brine in the mine.

All these disadvantages in no way harm a normal salt mine but are not favourable for a repository for high-level and alpha-bearing wastes. On the other hand, there are also quite a number of advantages that suggest the Asse Salt Mine as an excellent R&D facility and a successful prototype repository:

(1) Exploration did not have to start from the surface.

(2) The mine with shaft and underground workings was already in existence. Thus, operations could be started after only a short period of reconstruction and modification.

(3) It is possible to dispose of, and not only to store, low- and intermediate-level wastes without an undue risk to the environment.

(4) 'Public acceptance' was no sensitive keyword in 1965. In spite of some minor demonstrations at the time, the public today has completely accepted the Asse Salt Mine as an important research facility within the nuclear fuel cycle. In particular, those people in the neighbouring communities are living quite normally with 'their plant'.

These general remarks on the Asse Salt Mine can be summarized by saying that it is an old mine that will certainly not be the repository for all categories and all volumes of radioactive wastes generated in the Federal Republic of Germany. But it is and still will be for quite a number of years the national R&D facility where disposal technologies can be developed and proved, and where valuable experience of all different safety aspects can be gained. In addition, low- and intermediate-level wastes can be disposed of. Thus, the Asse Salt Mine is an indispensable facility within the nuclear fuel cycle.

3. DISPOSAL OF LOW-LEVEL WASTES

Disposal of low-level radioactive wastes was started in April 1967. The results of the first test disposal phases were reported at the IAEA/NEA Symposium at Aix-en-Provence in 1970 [6]. Thus, they are given here in a very condensed form.

The first chamber was filled with 6400 200-l drums, stacked four high. In the next three chambers about 25 900 drums were disposed of by laying horizontally from wall to wall in rows of ten drums on top of each other. The entrance tunnels of these chambers were then sealed.

Since Fall 1974 a new disposal technology has been used. The drums are transported from the pit bottom to the selected disposal chamber by a scoop-tram. The entrance tunnel to the disposal chamber is no longer situated on the floor of the chamber but in its upper quarter. A ramp of loose salt is banked up into the chamber. The drums are dumped onto this salt ramp by the scoop-tram. They chute and roll down the ramp without getting broken. From time to time the drums are covered with loose salt.

Though this method may seem a little careless at first glance, it has a great number of advantages:

(1) The dose rate to the disposal personnel is much lower than when stacking the drums.

(2) The necessary dressing of the old chambers is much smaller. Only the ceiling above the immediate working area has to be dressed.

(3) More chambers in the mine can be used for disposal, especially on the higher levels because of the easier dressing.

(4) The danger of the displacement of possible contamination is much smaller than before as the direct disposal area is entered neither by vehicles nor by persons.

(5) The chambers can be completely filled with drums and salt, thus enlarging the stability of pillars and walls.

(6) In the event of flooding the loose salt offers a large suface for the process of concentration.

(7) A substantial increase in efficiency can be achieved. In addition, fewer personnel is necessary and standard mining equipment can be used.

Two chambers on the 750 m level have already been filled with this technique and a third one is at present being filled as the first chamber on the 725 m level. Experience so far has been excellent, without the smallest mishap.

In total, 62 000 containers with low-level radioactive waste were disposed of from 4 April 1967 to 29 February 1976. This filled seven chambers on the 750 m level, the eighth one being in operation. Of this total number 58 000 were drums and 4000 'lost concrete shielding containers', which will be explained later. The operational and safety records can be called very good without any restriction. It is hoped that the same can be said in a few years from now, clearly knowing that the waste volume to be disposed of will steadily increase.

As already outlined in Ref. [6], conditions of acceptance were established for the low-level radioactive wastes that can be disposed of in the Asse Salt Mine. These conditions of acceptance were revised in 1975 in co-operation with the licensing authorities and the waste producers, the revision being based on the experience gained through eight years of operation. The most important changes or completions will be given here.

In addition to the 200-l drum with its three varieties (iron-hooped drum, drum with reinforced seams, and simple sheet metal drum) a 400-l drum may now be used with the same varieties. Furthermore, the waste can be packed in a 200-l drum, which in turn can be inserted into a 400-l drum with the annulus and the top and bottom parts cast with concrete [7, 8]. An additional new type of packaging, called the 'lost concrete shielding container', is a prefabricated reinforced concrete container with a wall thickness of 20 cm. After insertion of a 200-l drum, the annulus and top part are again cast with concrete. This 'lost concrete shielding container' can be used in two varieties: one consisting of normal concrete with a total weight of 2.5 tons, and a second one consisting of barium-sulphate concrete with a total weight of 5.0 tons.

The basic concept of the conditions of acceptance has not been changed in the revised form. This concept is that the better the conditioning of the waste and the more stable the containers, the higher is the admitted radioactivity. This admitted radioactivity was set up for different waste categories, different waste conditioning methods and different waste containers. This time not only the admitted β-γ-activity was set up, but simultaneously an attempt was made to fix an upper limit for the admitted α-activity [9]. It is beyond the scope of this paper to give all the details. It should only be mentioned that the admitted activities range between 0.1 and 25 Ci per container for β-γ-emitters and between 0.001 and 10 Ci per container for α-emitters.

Another concept was also kept, namely that the containers with low-level radioactive waste can be transported and handled without additional shielding. Thus, the normal dose rate of the containers must not be higher than 200 mrem/h on their surface. Exceptionally, 10% of one shipment may also have dose rates up to 1000 mrem/h on the surface of the containers, this exception not being applicable to the 'lost concrete shielding containers'.

These revised conditions of acceptance became effective on 1 January 1976 [10]. Another new arrangement became effective at the same time. Whereas all waste producers could deliver their wastes for disposal free of charge up to the end of 1975, they now have to pay a disposal fee. An attempt was made to fix differentiated fees by following the same guidelines as for the conditions of acceptance, taking into account in addition the necessary handling time. The fees were calculated in relation to operational economy, neglecting the costs for the already existing and thus available disposal space. They are differentiated in the final table for different weights and dose rates. The cheapest disposal fee is DM 150.- for a 200-l drum weighing up to 700 kg and with a surface dose rate not higher than 100 mrem/h. The most expensive disposal fee comes to DM 3700.- for a 400-l drum weighing up to 1250 kg and with a surface dose rate up to 1000 mrem/h.

In conclusion, it can be stated for the disposal of low-level radioactive wastes that a certain degree of routine operation has been reached. Nevertheless, further R&D work is also necessary in this field in order to achieve higher efficiencies and to improve the long-term safety of the disposal facility.

4. DISPOSAL OF INTERMEDIATE-LEVEL WASTES

The technical details of the first test disposal installation for intermediate radioactive wastes in the Asse Salt Mine were all given in Refs [5, 6] and so only some comments on the experience gained hitherto will be made here.

A total number of 746 drums were disposed of from August 1972 to February 1976. The slope of the filling cone in the disposal chamber is somewhat gentler than expected, so it might well be possible to dispose of more than the 10 000 anticipated drums in the chamber with a volume of 8000 m^3. In spite of using the multiple transport container of type 7 V [7, 8] for transportation between the Karlsruhe Nuclear Research Centre and the Asse Salt Mine, the disposal capacity of this system is limited. The round trip of a single shielding container being used in the Asse Salt Mine takes 26 minutes, so that this system certainly does not qualify for large throughputs. On the other hand, the operational and safety records are excellent. There has been no technical difficulty, the dose rates to the personnel were below the susceptibility of the dose meters, and even the filters of the ventilation system showed no contamination. Further R&D-work should therefore be invested to modify this system to a higher capacity.

5. DISPOSAL OF HIGH-LEVEL WASTES

As already stated, the Asse Salt Mine will not be used as a repository for high-level and alpha-bearing wastes but will only serve as a test facility for that purpose. No great effort has been made up to now in the direction of the disposal of high-level wastes. This has mainly been limited to the development of computer programs for heat dissipation problems, which was predominantly done at the Technical University of Aachen and at the Bundesanstalt für Bodenforschung in Hannover, and to some tests in the mine with simulating the decay heat production by electrical heaters. Another deficiency in this regard is that no real solidified high-level radioactive waste was and is available, not even for test disposal operations.

In the last few months, however, an exactly defined R&D programme was formulated, which will be performed within the 'Programme on Radioactive Waste Management and Storage' of the Commission of the European Communities from 1976 to 1979. The goal of this programme is to be ready for a test disposal operation of solidified high-level wastes in the Asse Salt Mine in the early eighties. The main objective of the programme are:

(a) To perform all relevant laboratory and *in situ* tests for, e.g., heat dissipation, rock-mechanical questions, corrosion tests, etc.

(b) To improve the above-mentioned computer programs and to develop new ones for rock-mechanical questions

(c) To design and specify the test disposal configuration

(d) To construct the disposal site in the mine with the necessary chambers, pillars and boreholes

(e) To design, construct and install all necessary technical equipment like transport container, underground transport vehicle, borehole equipment, suitable devices to retrieve the waste canisters, measuring and monitoring equipment

(f) To elaborate an exact schedule for the performance of the test disposal operation.

Parallel to the laboratory and technical efforts safety studies and risk analyses have to be elaborated in order to get the necessary licences from the authorities.

At present, negotiations with the commission of the EC are under way. It is hoped that the commission as well as the other member countries of the EC will agree to the German programme so that it can still be started this year (1976).

6. ASSE CAVITY PROJECT

As mentioned briefly in the introduction, a second R & D programme was started simultaneously with the Asse Salt Mine project, namely the investigation of using a solution cavity for radioactive waste disposal. This cavity would be washed out in a virginal salt dome near the North Sea coast. Because of a number of difficulties (site selection, absence of political decision, public acceptance, excessively raised prices for the land) this project could not be realized in its original form. It finally was decided to transfer this project to the site of the Asse Salt Mine.

This decision brought a number of disadvantages, but also of advantages. The former were:

(1) Because there is no big river in the neighbourhood of the Asse Salt Mine the cavity cannot be washed out but has to be mined conventionally.

(2) The cavity is situated within the mine. Therefore the mine authorities demanded that access to the cavity must be inspected by man. This had the consequence that the originally planned small-diameter borehole had to be enlarged to a drilled shaft.

(3) The disposal technology had to be changed completely.

(4) In consequence of all these changes, the originally planned costs for this project rose many-fold.

Compared with this, the following advantages are at hand:

(1) The drilled shaft for the cavity can be used simultaneously as an emergency exit for the miners should the only main shaft of the Asse Salt Mine be blocked for some reason.

(2) There was no difficulty in buying the necessary land or in getting public acceptance.

(3) Because the cavity will be mined conventionally and so can be entered by man – until disposal operations begin – there is the unique chance to perform an extensive rock-mechanical *in situ* investigation programme. This will be done.

(4) Because of the time delay the disposal technology could be modified and adapted to the more urgent problem of the disposal of intermediate-level wastes compared to the original plan, which only was assigned for the disposal of low-level wastes.

(5) The whole infrastructure of the Asse Salt Mine can be used simultaneously for the cavity project.

(6) The salt mined by the construction of the cavity can be used in the mine for the disposal of low-level wastes.

The Asse cavity project is a system completely closed in itself in spite of being situated within the Asse Salt Mine. The drilled shaft with a diameter of 1.5 m in its upper part is situated 50 m west of the main shaft. After intersecting the existing mine workings at the 490 m and 750 m levels this shaft will have a final depth of 959 m. These intersections will be locked completely during normal operation and can be opened only in a case of emergency. The cavity itself will have a volume of 10 000 m^3 and the form of an elongated spheroid. The vertical height will be 36 m and the diameter 24 m.

This cavity project has the pure character of a R & D facility. Therefore, the volume of the cavity is limited. The technical concept of this R & D facility provides for the unloading of the intermediate-level waste drums from the multiple transport container in a new annex to the shaft shed, their shielded transport into a shielded charging cell on top of the drilled shaft, and their single transport in an unshielded hoisting cage through that drilled shaft to an unloading facility at a depth of 926 m. At this point the drum will be automatically discharged from the hoisting cage and will fall free 69 m to the bottom of the cavity. To avoid contamination of the environment the cavity project will be equipped with two separate ventilation systems, each of which is completely closed in itself.

The construction phase of this cavity project was started in August 1974 with drilling the shaft. This shaft has now reached the 750 m level. The next construction phases will be the sinking of the lower part of the shaft to its final depth of 995 m, the mining of the cavity itself, and the construction of the shaft shed on the surface. The time schedule for this R & D facility foresees start of operation in early 1979. The total costs of the Asse cavity are at present estimated to amount to about DM 40 million.

Besides the design and technical development work, investigations were performed in order to determine the necessary product quality standards. Detailed results have been reported in the series of annual progress reports [9]. As an example, the investigation of conditions of disposal for bituminized waste products in the prototype cavity showed that there are specific activity limits to be met in order to avoid the formation of an ignitable gas-air mixture by radiolysis.

The calculations considered different disposal times, space factors and ages of the fission products. The production rate of hydrogen increases with increasing values of the specific activities, with increasing age of the fission products and with increasing space factors. For example, the limits for the specific activity are for a space factor of 25% 0.9, 0.5 and 0.3 Ci/l for an age of 0.5, 1 and 2 years, respectively. For a space factor of 50% the corresponding values are 0.3, 0.16 and 0.1 Ci/l.

7. OUTLOOK

The Asse Salt Mine serves as the national R & D facility for the disposal of radioactive wastes in the Federal Republic of Germany. To close the nuclear fuel cycle for a 50 000 MW(e) power plant capacity until 1985 it is intended to construct a 'nuclear fuel cycle park', which will contain a commercial fuel reprocessing plant, a plutonium fuel element fabrication plant, all waste treatment plants, and the repositories for different categories of radioactive wastes. Because of the latter condition the site of this nuclear fuel cycle park will be located on top of a salt dome. Detailed site investigation studies have already been started. To reach the mentioned goal the results of all R & D work in the Asse Salt Mine are of outstanding importance.

REFERENCES

[1] National Academy of Sciences – National Research Council, The Disposal of Radioactive Waste on Land, NAS-NRS Publication 519, Washington (1957).

[2] MARTINI, H.J., Bericht zur Frage der Möglichkeiten der Endlagerung radioaktiver Abfälle im Untergrund der Bundesrepublik Deutschland, unpublished expert's report by the Bundesanstalt für Bodenforschung, Hannover (1963).

[3] KÜHN, K., "Geo-scientific investigations in the Asse II Salt Mine", Disposal of Radioactive Wastes into the Ground (Proc. IAEA/NEA Symp. Vienna, 1967), IAEA, Vienna (1967) 509.

[4] KÜHN, K., et al., "Disposal of solidified high-level radioactive wastes in the Asse Salt Mine", Management of Radioactive Wastes from Fuel Reprocessing (Proc. IAEA/NEA Symp. Paris, 1972), OECD, Paris (1972) 917.

[5] KÜHN, K., "Asse Salt Mine, Federal Republic of Germany – Operating facility for underground disposal of radioactive wastes", Underground Waste Management and Artificial Recharge, **2**, AAPG, Tulsa (1973) 741.

[6] ALBRECHT, E., et al., "Disposal of radioactive wastes by storage in a salt mine in the Federal Republic of Germany", Management of Low- and Intermediate-Level Radioactive Wastes (Proc. IAEA/NEA Symp. Aix-en-Provence, 1970), IAEA, Vienna (1970) 753.

[7] GfK – GSF, Endlagerung radioaktiver Abfälle – Jahresbericht 1972, KFK 1862 – GSF-T 45, Karlsruhe (1974).

[8] GSF – GfK, Endlagerung radioaktiver Abfälle – Jahresbericht 1973, GSF-T 52 – KFK 2105, München (1974).

[9] KRAUSE, H., RUDOLPH, G., Jahresbericht 1974 der Abteilung Behandlung Radioaktiver Abfälle, KFK 2212, Kralsruhe (1975).

[10] GSF, Bedingungen für die Lagerung von schwachradioaktiven Abfällen im Salzbergwerk Asse, Gesellschaft für Strahlen- und Umweltforschung, München (1975).

DISCUSSION

F.E. RIEKE: Thank you for your excellent presentation on developments in the Asse salt mine. It only makes a fleeting reference to tests in the mine with simulated levels of decay heat. Many countries, including the United States of America, are apparently looking into the use of stable salt beds for long-term storage of high-level wastes.

I should like to ask whether, over the past few years, your tests have given rise to any serious doubts about the safety of indefinite-term storage in salt beds for solidified, thermally hot, nuclear wastes containing substantial levels of radioactivity? Briefly, is the salt bed repository your site of choice for such waste storage?

Second, is the 'roll-in' disposal method unduly wasteful of space? Substantial voids can obviously occur between the randomly arranged drums.

K. KÜHN: None of our experiments or tests have shown any reason for restricting the use of salt deposits, either salt domes or salt beds, for the final disposal of radioactive waste.

Your statement about the 'roll-in' method is essentially correct. But since there are a large number of rooms available, and since we fill up the voids with crushed salt wherever they occur, we do not consider this a wasteful process.

Furthermore, we see an advantage in this practice in terms of long-term safety, since it provides large amounts of salt for saturation of the water should the mine be flooded.

J.B. LEWIS: What is the radiation dose received by the operators during disposal of low-level wastes, bearing in mind that you allow up to 10% of the containers with 1000 mrem/h on the surface? You also refer to the dose rates to personnel engaged in the disposal of medium-level waste as being below the response of the dose meters. What does that mean ?

K. KÜHN: All low-level wastes are disposed of in a common chamber. This means that a maximum of one drum in ten has a surface dose rate of up to 1000 mrem/h. I cannot give you the figures for the dose rates you mention. The dose received by the operating personnel is lower than 40 mrem/month · man.

A.F. PERGE (*Chairman*): In Section 6 of your paper you say that you have not been able to wash out and discharge the salt into rivers in the neighbourhood of Asse. But would you in any case consider the disposal of salt from such an operation into a river? Would that be allowed?

K. KÜHN: Quite a large number of cavities have been washed out (solution mined) for crude oil storage near the mouths of the larger German rivers – the Ems, Weser and Elbe – which flow into the North Sea. The brine originating from such operations could be fed into these rivers, though obviously only with strict observance of the relevant regulations.

P.A. BONHOTE: I am interested to hear that you are levying charges on producers of waste. Has there been any objection to such charges, and have the producers taken steps to reduce the amount of their waste as a result?

K. KÜHN: No, there has not been any objection and in any case the waste producers knew about the charges before they were introduced.

Most of the producers tried to deliver as much waste as possible to the salt mine before the end of 1975 when the charges were imposed. I do not know whether they have taken steps to reduce their waste production, but it does not seem to be the case so far.

THE WATERPROOF GEOMETRY OF SALT DOMES, AN IMPORTANT SAFETY ASPECT FOR THEIR EXPLOITATION FOR ULTIMATE DISPOSAL OF HIGH-LEVEL WASTE

J. HAMSTRA
Reactor Centrum Nederland,
Petten N.H.,
Netherlands

Abstract

THE WATERPROOF GEOMETRY OF SALT DOMES, AN IMPORTANT SAFETY ASPECT FOR THEIR EXPLOITATION FOR ULTIMATE DISPOSAL OF HIGH-LEVEL WASTE.

Salt domes used as possible containmnet structures for radioactive waste repositories can only fail in the long term through the continuous action of groundwater. The potential failure mechanism will not be the groundwater penetrating into the repository area, as long as the salt dome still remains in contact with its salt matrix. The buoyant force of the sediments overlying and surrounding the salt dome will, under certain stress conditions, enable salt to migrate towards the dissolving surface at such a rate that the salt dome will maintain its piercement shape, notwithstanding the continuing dissolving action. If this salt migration should prove to be a containment failure mechanism, it may only affect the waste repository to a limited extent and after a very long period of time. Model studies might reveal whether it is still realistic to adapt the layout of the waste repositories inside salt domes to the salt migration pathways.

1. INTRODUCTION

The existence of salt domes below ground in the north and east of the Netherlands made it worthwhile to evaluate the radiotoxic hazards arising from the disposal of the radioactive waste produced by the Dutch nuclear energy programme into one of these salt structures.

A safety analysis was made for a hypothetical salt dome, which originated from a matrix bed of salt situated 3000 to 3500 m below surface and whose top is now situated 250 m below surface [1]. Of this salt structure at least 200 m of rock salt was assumed to be maintained as a containment shield around both a dumping cavity for low- and intermediate-level wastes, and a burial horizon mined in this salt structure 800 m below surface for the disposal of high-level waste (see Fig. 1).

This safety analysis led to the conclusion that only a continuous groundwater attack into either the top of the rock salt structure or its flank could lead to a containment failure in the long term, such that the buried radionuclides would come into contact with groundwater and thus be transported back to the biosphere.

Dissolving calculations were made, based on the assumption that any water attack would result in the development of a solution cavity in the rock salt structure. Several counteracting effects were not taken into account in these calculations, two of which are worth mentioning. First, rock salt deforms readily under relative low stresses, resulting in a tendency to close any solution cavity that might originate. Secondly, the rock salt was considered to be pure halite. The anhydrite and other insoluble impurities present in the rock salt will stay behind at the dissolving surface, establish a protective layer, and thus slow down the solution rate. Notwithstanding that, the assumed penetration of groundwater through the rock salt containment shield around the

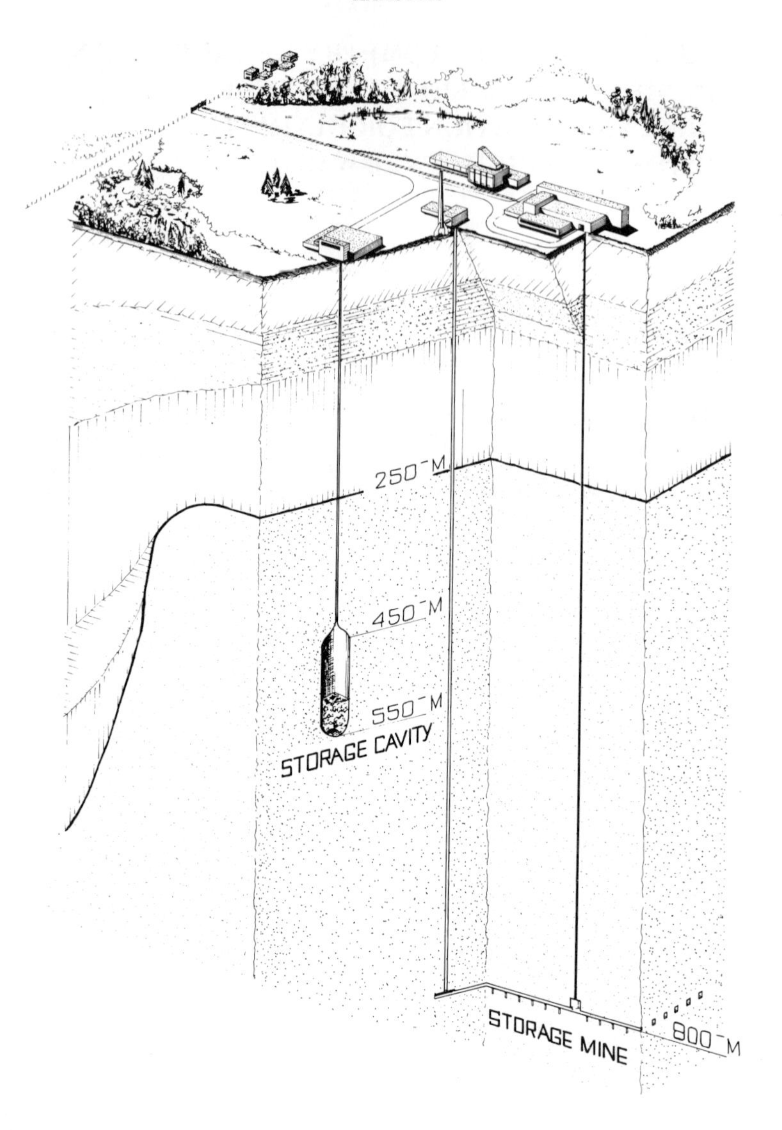

FIG.1. Layout of waste disposal facilities over-ground and in a salt dome.

buried radioactive wastes was calculated to take at least 20 000 years for a sidewall attack and at least 170 000 years for an attack at the top of the rock salt mass.

This report is a first attempt to demonstrate that virtually no water attack can create a local solution cavity in the flank or in the top of a salt dome as long as the salt dome still is in contact with its salt matrix bed. Any water attack will disturb the prevailing stress conditions in the rock salt near the area under attack. This change in stress conditions may initiate a migration of salt towards that area at such a rate that the geometry of the salt structure will not be affected by the water attack. It is this self-healing quality of the rock salt that establishes the waterproof geometry of salt domes.

2. SALT DOME GEOMETRY

In the process of dome formation the rock salt is subjected to very high stresses, undergoing severe plastic deformation when intruding into the overlying sediments. The pressure exerted by the rising salt on the adjacent strata will create a confinement reaction from the surrounding and overlying sediments.

The geometry of a well-developed salt dome that still is in contact with its matrix bed may be considered as stable so far as its bottom part and flanks are concerned. If domal growth has not yet terminated or possibly might be rejuvenated, the upward migration of salt from the salt matrix bed into the domal structure mainly will affect the geometry at the top of the salt dome.

The Netherlands' salt domes under consideration for disposal of radioactive wastes all have the form of a truncated cone with an oval cross-section. Some of them have a slight lateral spreading at the top. At least one of them has an anhydrite caprock layer of some 80 m thickness overlying the rock salt and indicating an exposure to dissolving water at an earlier period. So far as is known, none of them can be shown to have been exposed to dissolving water through a caprock shield on one of its flanks.

3. INTERNAL STRUCTURE AND STRESS CONDITION OF THE SALT DOMES

In the Netherlands no salt mining is done that can display the vertical or almost vertical folding to be expected in diapiric salt structures. Considering the principal mechanism of deformation of rock salt to be a translation gliding along the crystal planes, an overall vertical orientation of the salt grains may nevertheless be assumed, at least along the flanks of the salt dome.

Although, broadly speaking, a salt dome may be considered to be one mass of halite, it should be recognized that the impurities contained in the halite and the presence of more ductile evaporites such as potassium and magnesium chlorides may affect the structural behaviour locally. The rock salt mass may be assumed to have a density of 2.16 g/cm^3, whereas the density of the overlying and surrounding sediments may, according to Nettleton [2], increase with depth, as shown in Fig. 2. From this density contrast the overpressure can be derived at which the halite crystals within the domal structure are kept in a hydrostatic stress state. This hydrostatic overpressure, which increases with the depth, is also plotted in Fig. 2.

As the creep rate at a given stress state may be assumed to increase with the temperature, it is interesting to note that in one of the Netherlands' salt domes rock salt temperatures were measured of 26.7°C at 312 m depth and of 33.9°C at 685 m depth [3]. Figure 2 gives a rock salt temperature line, derived from these two temperature measurements and extrapolated down to 3000 m depth. This temperature curve not only is anomalous in comparison to the surrounding rocks, it also marks the absence of internal friction as a consequence of salt movement. At the prevailing temperatures, up to about 40°C at 1000 m depth, any deformation by slip in the upper part of the domal structure will have a work-hardening effect on the halite crystals, which need recrystallization before they may slip again.

Because of the geologic time-scale on which diapirism and thus all other salt migration within the domal structure may be assumed to take place at these relatively low temperatures, the rock salt may be assumed to be in an unstrengthened stress state due to relaxation or recrystallization. According to Balk [4], a threshold shear stress of 30 to 50 kgf/cm^2 may be assumed, beyond which the salt will deform plastically, even at temperatures of 30 to 40°C, that is at a depth of 500 to 1000 m. A pressure differential of 60 to 100 kgf/cm^2 created in the domal salt structure at such depths would thus result in a certain creep rate. Figure 2 shows that if a dissolving action at the flank of the salt dome takes place beyond some 500 m in depth, a drop in confinement pressure can be expected that will cause the pressure differential required to start salt migration.

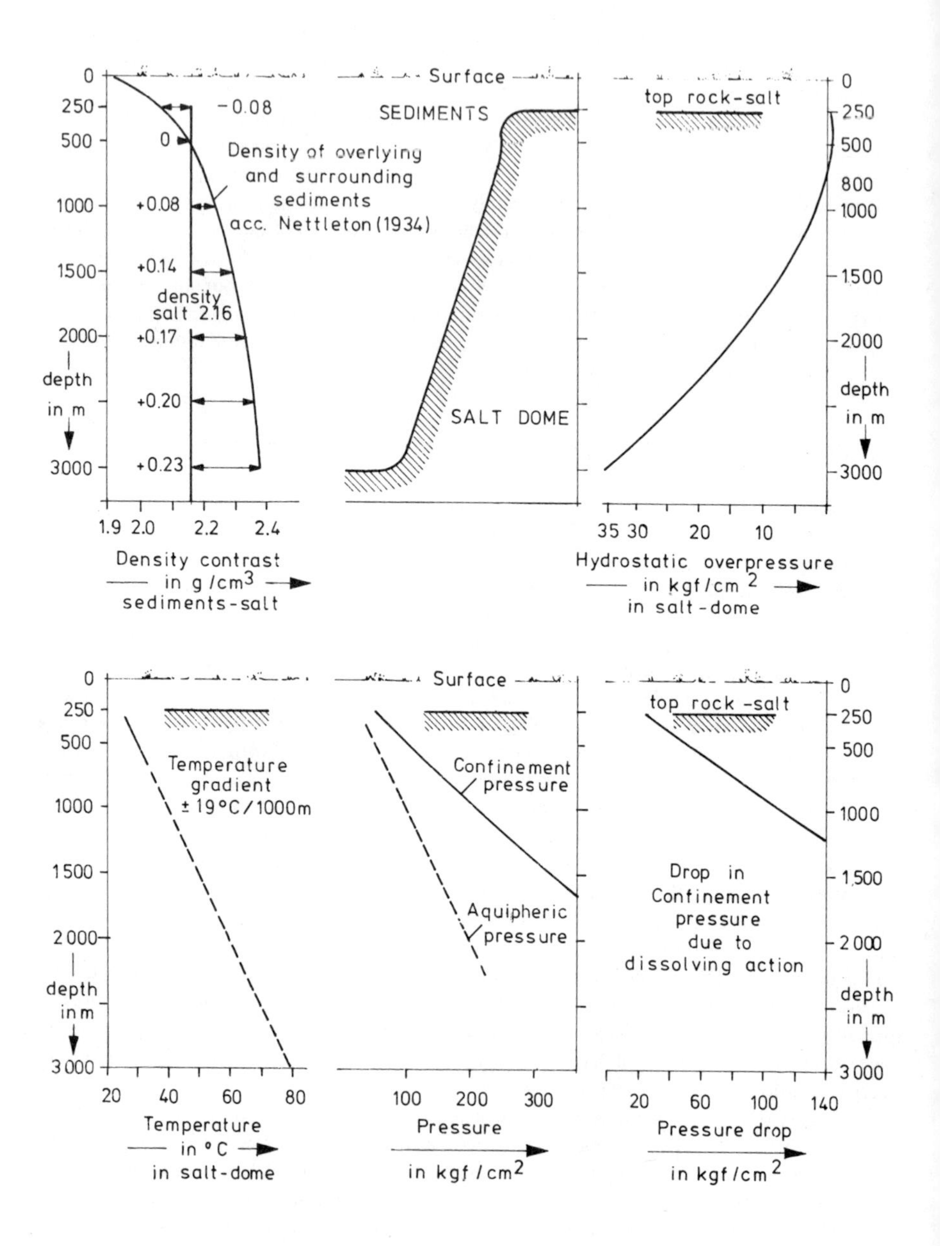

FIG.2. Variation of salt dome confinement characteristics with depth.

A creep theory, based on the intersection of moving dislocations and on the dynamic recovery by cross-slip of screw dislocations as a creep-controlling mechanism, was derived for rock salt by Le Comte [5].

4. SALT DISSOLUTION BY GROUNDWATER CONTACT

Gera [6] placed much emphasis on an adequate hydrolic integrity of salt domes destined to contain a radioactive waste repository. The presence of thick layers of caprock on top of a great number of salt domes and thinner caprock shields at the flanks of some salt domes indeed offer conclusive proof of salt dissolution due to groundwater contact during the slow process of diapir growth.

Whether or not dissolution is occurring at a specific salt dome is difficult to establish, because in general little is known about groundwater movement at greather depths in the vicinity of the salt dome. The hydrologic integrity of a salt dome can only be established by studying in considerable detail the characteristics of the groundwater flow around the salt dome and possibly by checking actual dissolution. Up till now no attention has been paid in this respect to the self-healing qualities of a rock salt structure, whereby under certain conditions it will react to each water attack by a migration of salt towards the dissolving surface.

No saline water was observed with any certainty above the salt domes under consideration in the Netherlands [7]. If groundwater comes into contact with a salt structure in the near future, the rate of salt dissolution will be a function of the amount of fresh water per unit of time that may act as transport mechanism for the dissolved salt. In the region under consideration groundwater moves at a rate of about 20 m/a at 200 m depth, which rate gradually decreases to about 2 m/a at 1000 m depth.

5. STRESS CONDITIONS AT THE DISSOLVING SURFACE

Prior to any mass transport of salt due to the dissolving action of groundwater the stress conditions at the outer surface of the salt structure may be assumed to be in equilibrium with and determined by the stratigraphic pressure at that depth in the surrounding sediments. The stratigraphic load is passed through the grain-structure of the surrounding sediments into the rock salt outer surface. This confinement load determines the hydrostatic stress state of the halite crystals prior to an exposure to groundwater finding its way towards the rock salt outer surface through the pores of the stratum.

As a consequence of a dissolving action, mass transport of salt will take place from the salt surface in contact with the groundwater into the aquifer. Instantaneously, however, this action will change the stress condition of the halite crystals at the dissolving surface. The original confinement pressure, determined by the stratigraphic pressure at that depth, will become replaced by the hydrostatic pressure present in the aquifer at that same depth. The stratigraphic pressure is determined by the density of the overlying sediments, which may vary from 1.9 to 2.4 g/cm^3. The hydrostatic pressure in the aquifer, however, will be determined by the density of the water, which may vary from 1 g/cm^3 for fresh water to 1.3 g/cm^3 if the groundwater becomes a saturated brine. Figure 2 shows the order of magnitude of the drop in confinement pressure that may be expected to arise from a dissolving action. It is evident that both the surrounding sediments and the rock salt structure may react to such a pressure drop at their interface. Beyond a certain depth the ability of salt to flow in response to load differentials may result in a migration of salt towards the area of lower pressure at an earlier stage than that where the confining consolidated rocks will be able to subside. In a salt dome that still is in contact with its salt matrix bed there will always be the buoyant force of the heavier sediments surrounding the salt structure and overlying

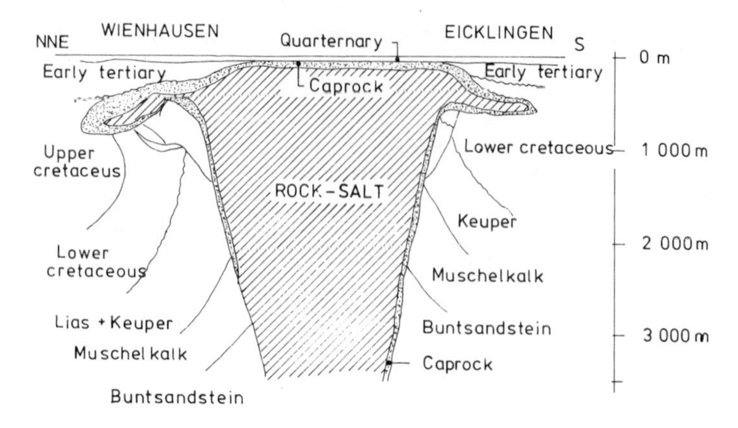

FIG.3. Wienhausen salt dome in North-West-Germany. (According to Schott (1956).)

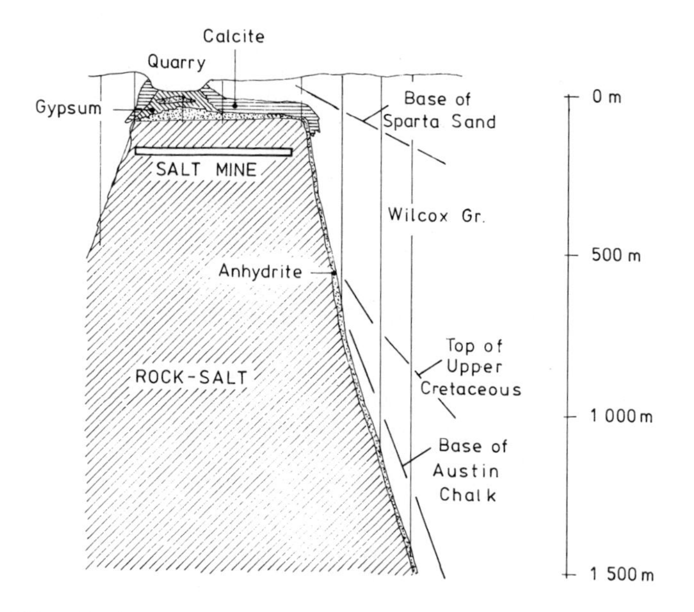

FIG.4. Winnfield salt dome in North Louisiana (USA). (According to Belchic (1960).)

the salt matrix. Beyond a certain depth this buoyant force, responsible for the hydrostatic overpressure within the salt dome, will be the mechanism responsible for the self-healing quality of salt domes. Although detailed information on dome geometry and caprock formation is mainly restricted to the upper part of the salt structure, the two following examples of dissolving action on the flank of salt domes show that an anhydrite sheath forms within a geometry, which may be assumed to be the original piercement shape. If this assumption is correct, both examples indeed demonstrate the waterproof geometry of a salt dome. The first example, shown in Fig. 3, is the lower part of the Wienhausen dome. The mushroom type of this dome should be left out of consideration because it was formed after this diapir reached the surface in the Upper Cretaceous. The second example, shown in Fig. 4, is the Winnfield dome.

6. SALT MIGRATION TOWARDS THE DISSOLVING SURFACE

In evaluating the possible pathway for the rock salt to migrate towards a dissolving surface it should be recognized that both the mass transport of salt into an aquifer due to a dissolving action and the migration of salt towards the dissolving surface as a consequence of the drop in confinement pressure are very slow processes.

Large-scale salt migration or salt flow will be affected by the amount and nature of the impurities contained in the salt and the presence of other more ductile evaporites. Bands or even layers of these other evaporites will be present in a salt dome, folded during the diapiric stage predominantly in the vertical direction. Following this preferential orientation, present in the salt structure from earlier intrusion movements, the buoyant force may supply the salt to compensate the stress differences predominantly from underneath.

The waterproof geometry of a salt dome can, however, only be considered as an important safety aspect for a waste repository situated in that dome if the salt migration that effects this impermeability itself does not turn out to be a potential containment failure mechanism.

7. SALT MIGRATION AS A POTENTIAL CONTAINMENT FAILURE MECHANISM

From the preceding considerations it will have become evident that it is no longer realistic to assume that water may penetrate salt domes and reach the buried radionuclides at their disposal horizon. On the contrary, it should be realized that at sufficient depth serious dissolving actions may trigger salt movement towards the dissolving surface and that the buried waste canisters could be carried with the salt flow towards the dissolving surface.

This may prove to be less hazardous than the water penetration assumed in the safety analysis [1]. The migration time for the first waste canisters to reach the outer surface of the salt structure may easily be a factor of 10 more than the time calculated for the water from a side-wall attack to penetrate into the waste repository area.

It is evident that during and after that migration period far more salt will have to be dissolved than the amount of rock salt originally available around the waste canisters in the repository area.

It is questionable whether sufficient groundwater will be available to effect the mass transport of such amounts of salt. If so, the dilution rates will again be far more favourable than calculated in the safety analysis.

In this respect model studies will have to reveal whether it is realistic to adapt the layout of the high-level waste repositories inside salt domes to the salt migration pathways.

REFERENCES

[1] HAMSTRA, J. Veiligheidsanalyse voor ondergronds in een zoutkoepel opbergen van radioactief vast afval, RCN-75-040 (1975).
[2] NETTLETON, L.L., Fluid mechanics of salt domes, Bull. Am. Assoc. Pet. Geol. **18** 9 (1934) 1175.
[3] HARSVELDT, H.M., personal communication (1973).
[4] BALK, R., Structure of Grand Saline salt dome, Van Zandt County, Texas, Bull. Am. Assoc. Pet. Geol. **33** 11 (1949) 1791.
[5] LE COMTE, P., Creep in rock salt, J. Geol. **73** (1965) 469.
[6] GERA, F., Disposal of radioactive wastes in salt domes, Health Phys. **29** (1975) 1.
[7] Vooruitzichten tot ondergrondse verwijdering van afvalstoffen in Nederland, Rapport Centrale Organisatie T.N.O., Delft (1973).

DISCUSSION

D. RICHTER: Do you think that the temperature rise due to the storage of high-level wastes may adversely affect the self-healing mechanisms of the salt dome?

J. HAMSTRA: No, on the contrary, a rise in temperature of the rock salt as a consequence of storing high-level waste will favourably affect the creep phenomenon and, consequently, the ability of the rock salt to flow.

Y. NISHIWAKI: The exact standardization of radioactive waste categories may be a difficult and complex problem. In discussing the effects of heat from radioactive wastes, however, it is important for the people working in waste management to have a common description of the different waste levels. In Table I of his paper (IAEA-SM-207/44, these Proceedings, Vol.2), Dr. Richter gives a very clear-cut definition of low-, intermediate- and high-level wastes in the German Democratic Republic, according to which the high-level waste is defined as waste with a specific activity higher than 10 Ci/l (or Ci/kg). I would like to ask whether the definition of high-level waste in the Netherlands and in the Federal Republic of Germany is the same as in the GDR?

J. HAMSTRA: No, under the Netherlands waste disposal project high-level waste means solidified reprocessing waste, since it is this that will be shipped back to our country by the fuel reprocessing plant. The same applies in the Federal Republic of Germany.

H. KRAUSE: As a general comment, I would say there is no doubt that radioactive waste can be stored safely for decades in engineered storage facilities. It might even be cheaper to store the waste in such structures rather than in deep geological formations, provided the amounts involved are small. However, I should state that it is very important to know as soon as possible how the waste is to be finally disposed of and what specifications will apply to its acceptance for the selected type of disposal. It will not otherwise be possible to establish an optimum waste treatment system. I am convinced that some of the waste treatment systems would take on a somewhat different aspect if the requirements for final disposal were changed.

F.C.J. TILDSLEY: Mr. Hamstra, you mention two mechanisms by which the deposited waste may come into contact with groundwater; in one the water finds its way to the waste and in the other the waste moves to the water. The latter process will take at least 10 times as long as the former. What sort of time-scale do you have in mind?

J. HAMSTRA: Both mechanisms should be considered in process on a long-term time-scale of 100 000 or more years, or in other words, on a geological time-scale.

RADIOACTIVE WASTE DISPOSAL PILOT PLANT CONCEPT FOR A NEW MEXICO SITE

W.D. WEART
Sandia Laboratories,
Albuquerque, New Mexico,
United States of America

Abstract

RADIOACTIVE WASTE DISPOSAL PILOT PLANT CONCEPT FOR A NEW MEXICO SITE.

Twenty years of investigation have shown that disposal of nuclear wastes in deep salt formations is the safest means of isolating these wastes from the biosphere for the extremely long period of time required. A large-scale demonstration of this capability will soon be provided by a Radioactive Waste Disposal Pilot Plant (RWDPP) to be developed in southeastern New Mexico. Initially, the pilot plant will accept only ERDA generated waste; high-level waste from the commercial power reactor fuel cycle will eventually be accommodated in the pilot plant and the initial RWDPP design will be compatible with this waste form. Selection of a specific site and salt horizon will be completed in June 1976. Conceptual design of the RWDPP and assessment of its environmental impact will be completed by June 1977. Construction is expected to start in 1978 with first waste accepted in 1982. The present concept develops disposal areas for all nuclear waste types in a single salt horizon about 800 metres deep. This single level can accommodate all low-level and high-level waste generated in the United States of America through the year 2010. A major constraint on the RWDPP design is the ERDA requirement that all waste be 'readily' retrievable during the duration of pilot plant operation.

INTRODUCTION

For nearly twenty years the United States Energy Research and Development Administration (ERDA) and its predecessor, the Atomic Energy Commission (AEC), has pursued a research effort investigating the physical phenomena accompanying the emplacement of radioactive wastes in salt (halite). The program has explored such diverse effects as solid diffusion, thermal and radiation effects, two-phase fluid migration, and mechanical behavior. This research has not revealed any phenomena which would preclude the use of geologic salt formations for the disposal of radioactive waste. Consequently, ERDA is now initiating a large-scale demonstration of radioactive waste disposal in a bedded salt formation in the State of New Mexico. Initial site selection efforts for the pilot plant were conducted for ERDA by the Oak Ridge National Laboratory (ORNL), and an area in the Delaware Basin was selected for further study. In April 1975, ERDA assigned to Sandia Laboratories the responsibility for all aspects of conceptual design, development and implementation of a Radioactive Waste Disposal Pilot Plant (RWDPP) to be located in southeast New Mexico about 50 km east of the city of Carlsbad. This task includes continuation of site characterization studies essential to preparation of an Environmental Impact Statement.

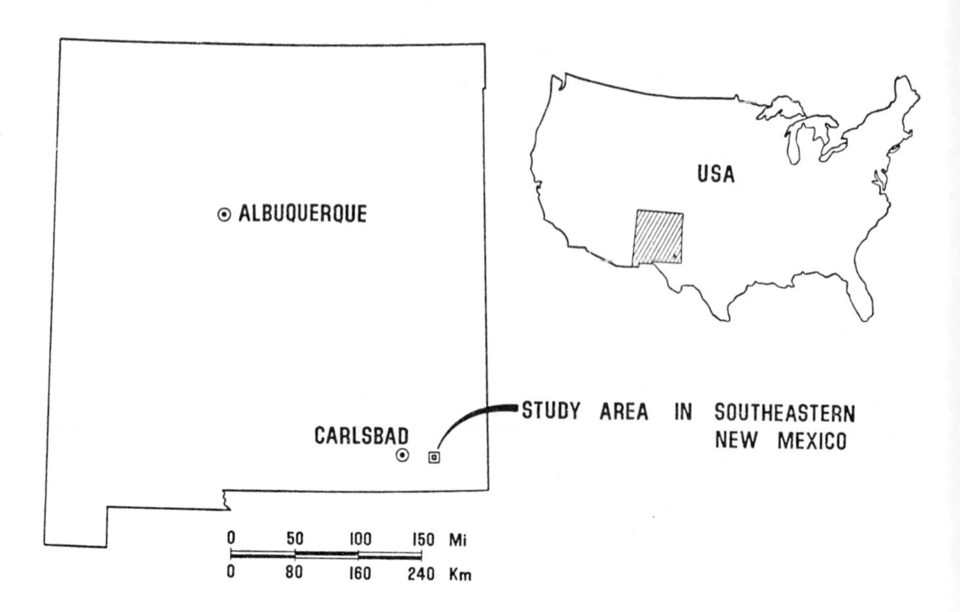

FIG.1. Study area.

PRESENT STATUS AND CURRENT ACTIVITIES

Site Selection and Confirmation

A major effort is now in progress to select a specific site location for the RWDPP. The required area (20 km^2) consists of 7.7 km^2 for the repository itself and a buffer zone approximately 1.6 km in width surrounding the repository. While this size repository would suffice to accommodate all commercially generated radioactive waste, through at least the year 2010, ERDA intends to establish eventually additional radioactive waste disposal sites at other geographic locations in the United States.

Sandia Laboratories is continuing exploration in the Delaware Basin with additional core drilling and seismic reflection analyses. Figure 1 indicates the location of this exploratory work. In late 1975 it was determined that the site tentatively chosen by ORNL failed to meet the criteria for relatively flat bedding and predictable structure. Rather than the gentle dips of one to two degrees common in the evaporites of the Delaware Basin, the basal salt member of the Salado Formation possessed considerable flow structure and, along with the underlying anhydrite, exhibited dips exceeding 70°. The anhydrite was severely fractured and formed a significant reservoir for pressurized brine containing dissolved carbon dioxide (CO_2) and hydrogen sulfide (H_2S) gases. Seismic data indicate that this is a local phenomenon to be expected in a narrow band paralleling the front of the buried Capitan Reef.

Current activities to confirm acceptable geology are located about 10 km southwest of the earlier drilling. Existing drill hole data forecast suitable geologic and hydrologic conditions; seismic profiling and exploratory drilling are now underway to confirm these expectations. A cross-section through the area now under examination is shown in Figure 2. The lowest salt bed in the Salado Formation, at depths between 750 and 900 m, is the primary candidate for the repository.

Other criteria were considered in locating the present area of investigation - this area is at least 1.6 km from the western dissolution front of the Salado Formation; if past dissolution rates continue, this assures that the repository will not be breached by the dissolution front for more than a half-million years. Acceptable depth of the basal salt in the Salado -- the unit most desirable for high-level waste because of its thickness and purity -- was restricted to less than 900 m. Site selection criteria arbitrarily forbid drill holes within the repository which penetrate the evaporite sequence and extend into underlying aquifers; such hole penetrations are tolerable but undesirable within the buffer zone. Although current development efforts provide a high level of assurance that such holes can be plugged to preclude water migration for as long as the salt beds exist, such plugging is expensive and has not yet been tested under field conditions. Dissolution analyses indicate that, even if some borehole circulation should develop and persist, a buffer of 1.6 km is adequate to isolate the waste for the time required. The area presently under investigation can accommodate a repository site and associated buffer zone which are unperturbed by deep drill holes.

Potential conflict with the recovery of natural resources is an important consideration in locating the repository site. Extensive potash and oil and gas production occur in this general area of southeast New Mexico. On the basis of known structure and exploratory drilling, the present area minimizes the potential for this conflict. Since the repository horizon is nearly 300 m below the productive potash zone and more than 2000 m above the greatest gas potential, it is very likely that both resources could be exploited without compromising the long-term integrity of the repository. The deeper resources could be produced by drilling through the evaporite sequence outside the buffer and deviating the holes below the salt beds to reach under the repository.

The present schedule for exploratory drilling and hydrologic testing will permit tentative selection of a repository site in June 1976, provided no unacceptable conditions are encountered. Site-specific studies necessary for the conceptual facility design and environmental impact analyses will begin at that time.

CONCEPTUAL DESIGN

The general aspects of conceptual facility design which do not require site specific properties are well advanced. Many facets of the design are influenced by ERDA's policy regarding radioactive waste disposal. Other criteria regarding waste characterization and packaging are only now being formulated.

In its initial operation as a pilot plant, the RWDPP will accommodate only ERDA-generated wastes -- both low-level and intermediate-level waste. The latter category includes the liquid and solvent extraction wastes now stored in tanks at the Savannah River and Hanford facilities. (While often referred to as high-level wastes, these wastes are much lower in thermal power and radiation effluent than the solvent extraction wastes from the

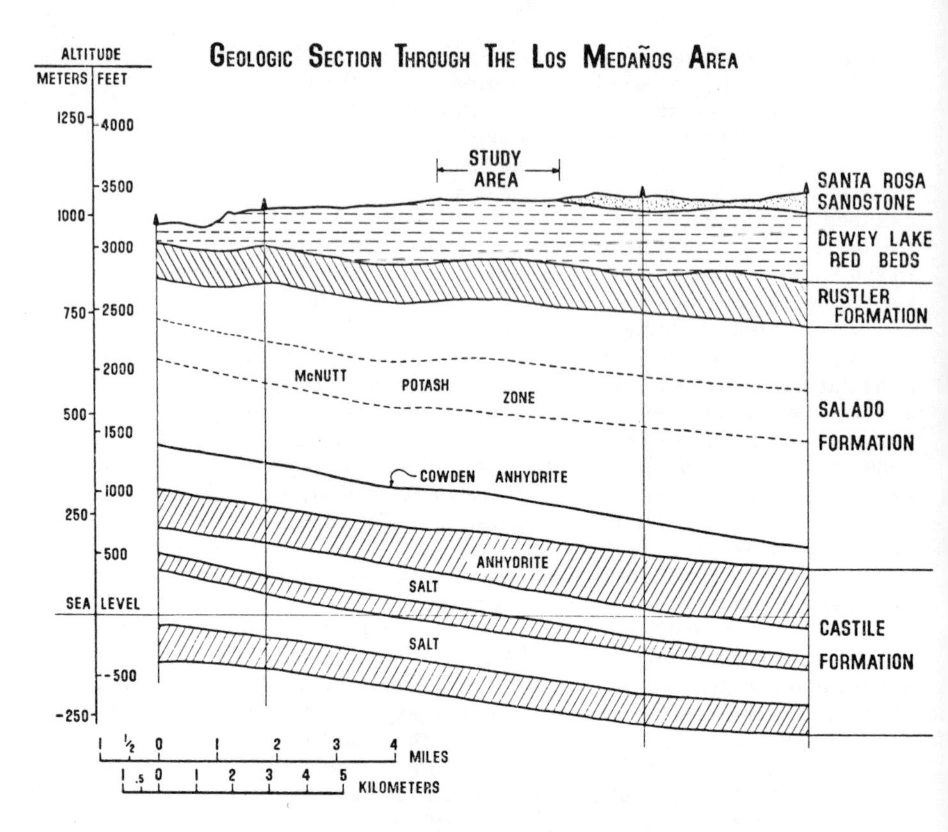

FIG.2. Geologic cross-section.

power reactor fuel cycle.) As in the case of commercial wastes, all ERDA wastes will be reduced to an essentially non-leachable solid form before shipment to the repository.

The facility design will be compatible with the storage of commercia high-level wastes, and in situ experiments intended to define the long-term behavior of the wastes in the salt storage environment will begin as soon as progress on underground construction permits. By the time commercial reprocessing plants begin operation, the RWDPP will be able to accept high-level wastes on a "demonstration" scale.

During the pilot plant phase, waste quantities will be limited to those necessary to demonstrate operational safety and to address technical questions which can only be resolved in large-scale experiments. Duration of the pilot plant phase required to provide these assurances for low-level wastes is expected to be only a few years. Adequate in situ examination of phenomena associated with high-level waste disposal may require a much longer time - perhaps ten years or more - due to the time required for thermally driven behavior to manifest itself.

An ERDA policy which severely impacts the pilot plant concept is the commitment to provide "easy" or "engineered" retrievability of waste throughout the duration of the pilot plant stage. The capability to retrieve waste will be demonstrated at intervals during pilot plant operation. High-level waste presents the most severe retrieval problem because of the thermal and radiation effects on the salt and on the waste package. Most of the high-level waste canisters will be designed and/or emplaced in a manner which prolongs their integrity, thus facilitating their recovery. Since this mode of emplacement does not provide a complete simulation of the storage environment characteristic of a fully operational repository, a limited but statistically significant number of regular steel canisters will be placed directly into salt. On a still smaller experimental scale, deliberately degraded waste forms will be placed in direct contact with the salt in order to accelerate the observation of any migration of radionuclides that may occur. These latter emplacement modes imply a retrieval method based on overcoring the waste -- in effect, recovering a "salt canister" containing the waste.

After successful completion of the pilot plant demonstration, the RWDPP will be converted to full-scale repository operation and will be licensed by the Nuclear Regulatory Commission (NRC). To facilitate this conversion, applicable NRC standards will be met in the initial pilot plant design. In the fully operational repository phase, waste recovery will still be possible - but difficult and expensive.

Acceptance criteria related to waste form and packaging are currently under discussion with prospective producers of the various waste types. Present RWDPP concepts require that all wastes be solid, "non-combustible", and stabilized in some fashion to retard dispersal and leaching should the container be breached. Low-level waste will undoubtedly be incinerated, and it will probably be stabilized in concrete or bitumen. Two ERDA containers, 2.4 m^3 fiberglass-coated plywood boxes and 200-liter steel drums, will serve as prototypes for commercial low-level waste packages.

It is expected that most of the intermediate-level waste will be packaged in cylindrical steel canisters. The zircalloy cladding hulls from light water reactor fuel elements constitute a special category of intermediate-level waste. Uncompacted, they require an inordinately large storage volume in the repository, but volume reduction is difficult and expensive because of the pyrophoric nature of zircalloy. Both mechanical compaction and smelting are under consideration; volume reduction factors for the two processes are six and nine, respectively.

The liquid high-level waste will be calcined to a dry powder and stabilized in a low leachability solid such as zinc borosilicate glass. Solidified high-level waste will be packaged in stainless steel canisters about 30 to 40 cm in diameter and 3 to 4.5 m in length.

The age and thermal power density of high-level waste is very important to the design of the underground storage facilities. Age, measured in terms of the time since the spent fuel was discharged from a reactor, determines the effective half-life of the waste. This, in turn, determines the total energy deposited in the salt over a given interval of time and thereby influences the macroscopic behavior of the salt bed. On a mesoscopic scale, decrepitation phenomena establish the maximum permissible temperature (~250°C) in the salt. That temperature, for a given canister configuration, depends on the thermal power density of the waste. During the first five years or so of repository operation, all the high-level waste will be at least ten years old. For ten-year-old waste in the form of zinc

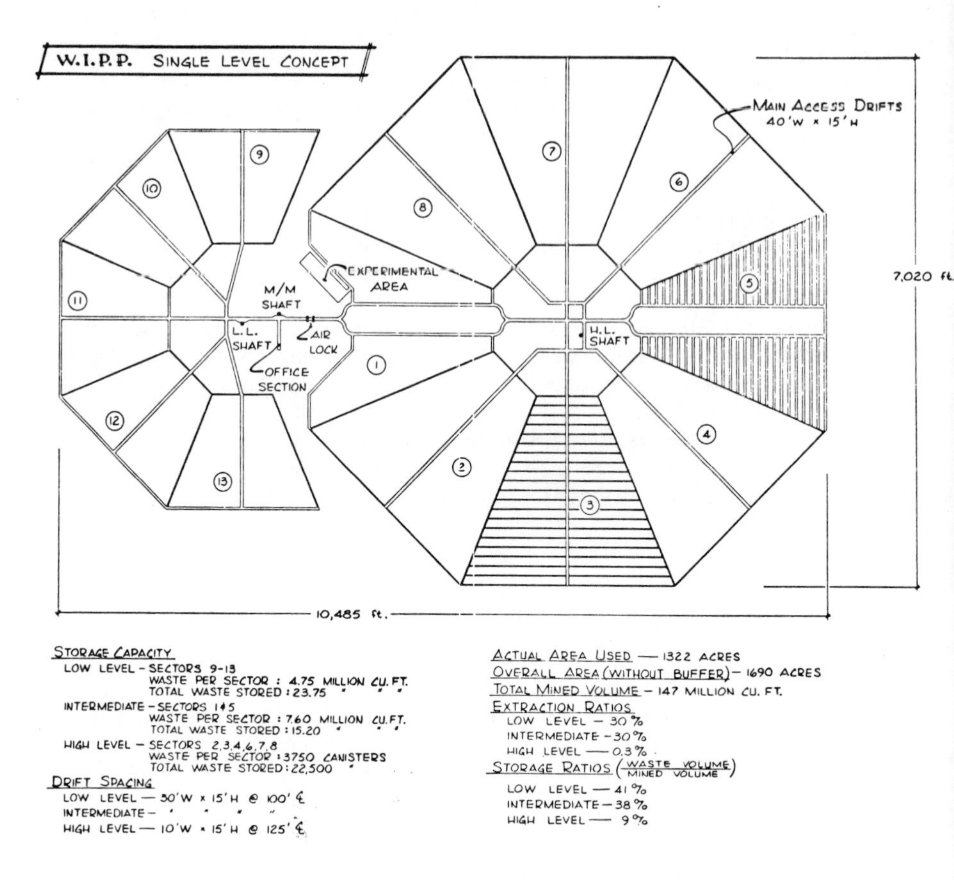

FIG.3. Tentative concept for underground facilities layout.

borosilicate glass, the thermal power density will be of the order of 25 W/l and the acceptable areal power density will be about 370 kW/ha.

Of the several waste categories for which the repository is being designed, only the low-level waste is amenable to handling by "contact" methods; all other waste types require shielding and/or remote handling. It is logical, therefore, to divide the repository into two facilities: a "cold" facility for the low-level waste and a "hot" facility for all other wastes. The present design concept develops both the low-level waste and intermediate- and high-level waste repositories in the same salt horizon at a depth of about 800 m. As shown in Figure 3, the two storage areas will be served by a common man-and-materials shaft but will be separated by air locks. The two areas will have independent ventilation systems. Separate waste transport shafts, communicating with completely isolated surface facilities, will be required.

Because of its heat generation, the high-level waste will ultimately require more storage area than other waste types. If all the waste from ERDA facilities and from the domestic power reactor fuel cycle is stored at the RWDPP, the storage areas required for the low, intermediate and

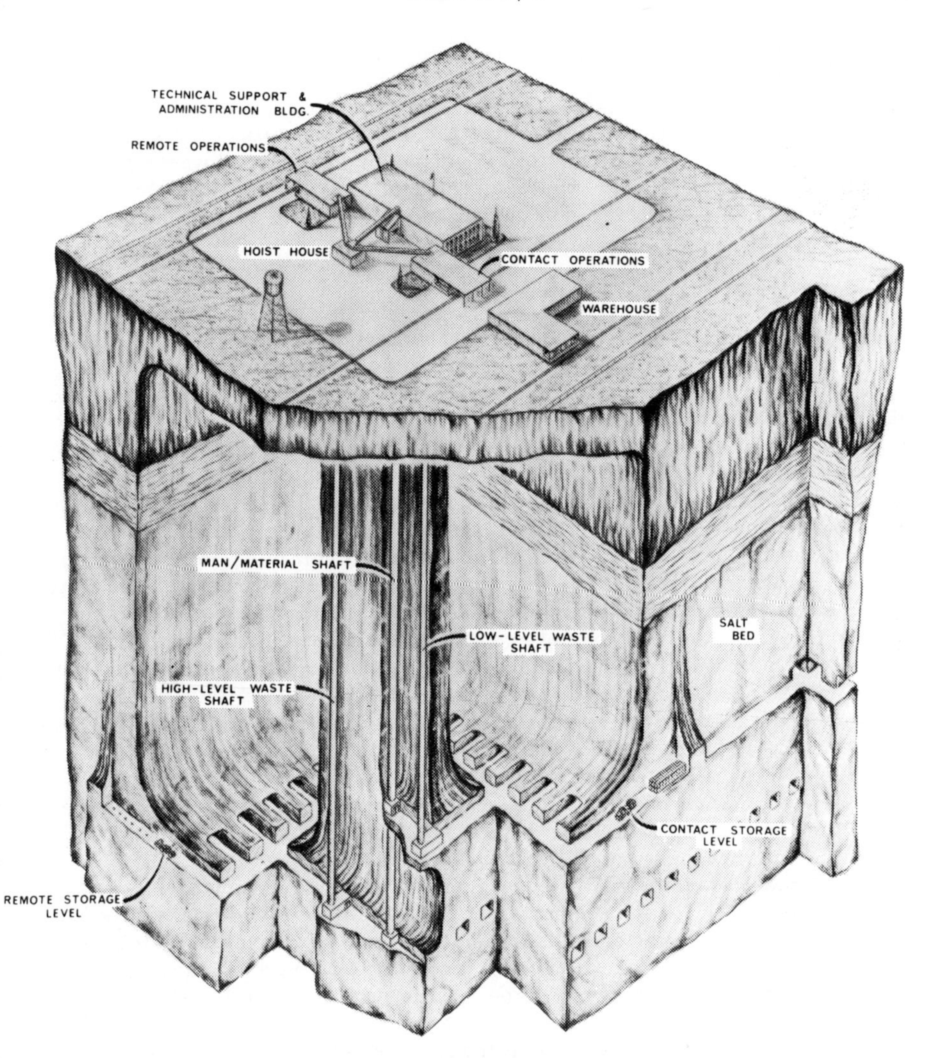

FIG.4. RWDPP bi-level terminal storage.

high-level wastes generated through the year 2000 are estimated to be 100, 90 and 270 ha, respectively. During the pilot plant operation it will be necessary to develop only a portion of one of the sections illustrated.

Should additional storage be required for any reason, the less pure salt at higher horizons can be developed to accept low-level waste without any problem. While these units are less suited to a high-level waste pilot plant, operational experience and experiments could develop the knowledge needed to allow future utilization of these less pure salts for high-level waste disposal. A conceptual view of such a two-level facility is illustrated in Figure 4.

The low-level and "hot" surface facilities will accommodate waste arriving by both rail and motor truck. Temporary surface storage will be provided for limited quantities of waste to allow for surges in shipment and for operational downtime. Provision for decontaminating shipping casks and waste containers and facilities for "overpacking" ruptured waste packages will be necessary. Except for the treatment of site-generated decontamination wastes and air filters, no other waste processing will be carried out at the RWDPP.

ENVIRONMENTAL IMPACT CONSIDERATIONS

Sandia is preparing a Draft Environmental Statement for ERDA which will set forth the anticipated -- as well as the unexpected, but possible -- impacts of the RWDPP on the natural environment. A detailed discussion of the phenomena that accompany the disposal of radioactive waste in these salt beds is an essential part of the document. Due to the persistence of the wastes it is necessary to consider events which could lead to dispersal of radioisotopes over the next few hundred thousand years. Since it is not possible to guarantee human surveillance of the site for hundreds much less thousands of years, geologic disposal in bedded salt does not rely on maintaining human control. It has been established, and will be detailed in the Draft Environmental Impact Statement, that natural geologic processes will not destroy the integrity of this repository for the period of concern. The hydrologic conditions at the tentative site are being established in considerable detail to permit quantification of the consequences of an inadvertant human penetration of the repository. The leachability, in brine, of various waste forms is being studied and, coupled with ion-exchange data now being developed, will allow examination of the transport of significant radionuclides in the specific environment of concern. Catastrophic events, such as meteor impact or breaching of the salt beds by major faulting, are among the improbable occurrences that will be considered.

Credible accidents that could occur during the operational phase of the repository will also be quantified. These failure modes are dependent on the details of the facility design and have not yet been completely identified. Two extreme conditions which will have to be considered are fire (surface and underground) and mine flooding due to the failure of a shaft liner.

The Draft Environmental Impact Statement will also address the social and economic impacts of the RWDPP. Alternatives to the proposed facility will be discussed and the cost/benefit aspects of the repository considered.

The completed Draft Environmental Impact Statement will be distributed to concerned government agencies, environmental organizations and individuals for comment. Public hearings on the facility and its impact will be held in several localities to assure that the general public has an opportunity to participate in the deliberations. An acceptable Environmental Impact Statement is an essential document in the process of obtaining approval for construction of the RWDPP. Equally important is a continuing dialogue, on a local and national level; with environmentally-concerned organizations. These groups, and the general public as well, will be informed early and often about the RWDPP so that their concerns can be considered in a timely manner. Full disclosure and cooperation with local and state governments

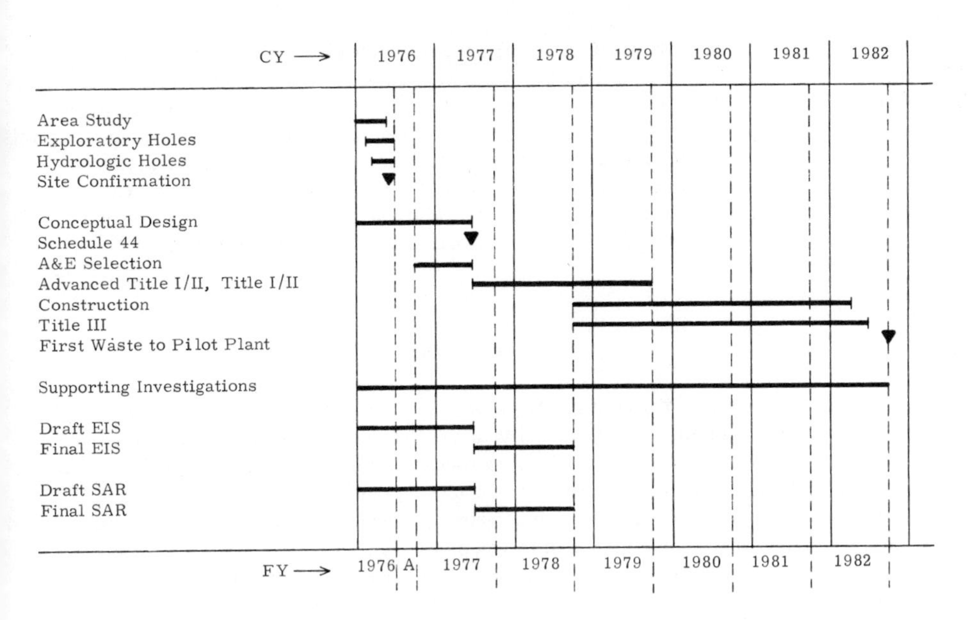

FIG.5. RWDPP milestone chart.

has resulted in a positive attitude toward this project in the State of New Mexico. Peer review by a committee of New Mexico scientists advises the State administration on the technical aspects of the RWDPP. This has been an excellent arrangement since the committee provides a focus for communication with the technical community within the State.

FUTURE ACTIVITIES

Site selection studies have been accelerated to allow a tentative confirmation by June 1976 that the geology, hydrology and other site criteria are acceptable. Site-specific characterization will be obtained from at least three core holes to 1000 m depth and from several hydrology test holes drilled to the top of salt at about 300 m. Rock mechanics testing on selected core samples will establish its strength and creep properties and relate these properties to information existing for other salt bodies.

Biological communities and ecological relationships will be established for the site chosen. Generalized ecology studies which have been underway for nearly a year will guide a more intensive effort. Meteorological data accumulation will be continued and correlated with regional data from surrounding permanent stations.

Some of the major milestones and their dates are shown in Figure 5. Selection of a specific site location and horizon will permit completion of the conceptual design for the RWDPP. This design, which will form the main basis for cost estimates, will provide necessary input to many of the Draft Environmental Impact Statement considerations. The Conceptual Design Report and the Draft Environmental Impact Statement, both to be completed in June 1977, are prerequisites to receiving authority to begin construction in 1978. Access to the repository salt horizon in 1981 and

1982 will allow initiation of several in situ experiments which do not involve radioactive waste. First contact with low-level radioactive waste is anticipated in late 1982. High-level waste pilot plant operations could commence as early as 1985. The ultimate scope and operational lifetime of this repository will depend upon the availability of other geologic disposal sites now being investigated and upon new disposal technologies now under study. If necessary, this one site could accommodate all commercial radioactive waste generated in the United States well into the 21st Century.

DISCUSSION

Y.P. MARTYNOV: You gave a figure of 25 W/l as the thermal power density for the high-level waste. Don't you think this figure a little high? Do you envisage additional ventilation requiring a heat removal shaft or is the heat removed by the thermal conductivity of the soil alone?

Secondly, have you any information on the method described at the Paris meeting in 1972 – I believe it comes under your Plowshare programme – whereby high-level waste is 'self-incorporated into rock? And what is your opinion of this method?

W.D. WEART: The level of 25 W/l thermal power density is the maximum considered desirable for disposal in bedded salt. It depends on the canister size and the canister spacing. From the economic standpoint as large a canister and as heavy a waste loading as possible are desirable.

Regarding your second question, there are two methods proposed that may fall into the category to which you refer. The first, often called 'deep rock melt disposal', is being studied on a modest scale by Sandia Laboratory to establish its feasibility and to permit the design of practical experiments. The second method involves the use of cavities formed by nuclear detonations as reservoirs for high-level waste. The liquid wastes would be placed in the cavity, would self-boil and ultimately fuse the surrounding rubble. This concept is still considered a potential alternative for radioactive waste disposal. However, it is not at present a main focus of attention in the United States waste management programme.

B. LOPEZ-PEREZ: Let me first say that I am asking my question as a layman in geology. Mr. Perge told us on the first day of this symposium that in the United States of America a broad research project on all types of rocks was about to be launched. In spite of this, in the papers presented at this session dealing with the disposal of waste in geological formations, only salt and clay deposits have been considered. I think we all agree that salt beds are, by a sizable margin, the first choice. And do we consider that clays are the second choice? Are we therefore to consider that all other geological formations are classed together as a third option, as suggested by the experts

W.D. WEART: More is known about the phenomena accompanying waste disposal in salt than in other geological media. We are convinced that salt can safely be used as a repository medium and that the operational needs are such that an adequate permanent disposal method should soon be demonstrable. This does not mean, however, that salt is the only medium into which waste can be safely placed. The economics of waste shipment dictate a need in the United States for research into geographically distributed waste repositories. To meet this need other salt formations and other geological formations will be evaluated and tested. It is not necessary that the first pilot plant site should be the best (although it may possibly be), but it is essential to demonstrate that at least one adequate and acceptable method of waste disposal does in fact exist.

F. GERA: Replying to Mr. López-Pérez's question, I would say that at this time it is impossibl to give any order of preference of the various types of geological formation, since too little is known about the geological materials which have been proposed as alternatives to rock salt. Even when dat on other geological media have been obtained, it will make more sense to discuss specific sites rather th rock types. It can be stated in general, however, that plastic media such as rock salt and clay should be able to provide more reliable containment than hard rock since they appear to withstand faulting without acquiring secondary permeability.

My question to you, Mr. Weart, relates to the target disposal horizon in the New Mexico salt formation. It was originally intended to leave a significant thickness of salt below the disposal zone. I see that the disposal horizon has now been moved further down, in close proximity to the underlying anhydrite. Since it would appear advantageous to have the waste surrounded by considerable thicknesses of salt in all directions, what is the reason for this change?

W.D. WEART: The salt layer below the Cowden anhydrite is much purer, has fewer clay interbeds and therefore presents fewer problems for the thermal and structural integrity of the repository. There are still several hundred feet of salt below the repository in the Castile formation, and this provides an adequate barrier. But thermal and structural calculations are being performed on higher levels and we believe that higher horizons of less pure salt will prove to be completely adequate for a repository.

Nina V. KRYLOVA: Do you intend to build a solidification plant near the repository, or will the waste be brought in after solidification elsewhere?

W.D. WEART: The ultimate answer to your question has yet to be provided. For the near future, however, all waste will be solidified elsewhere and then shipped to the repository.

SEA DISPOSAL
(Session X)

Chairman:

B. VERKERK, Netherlands

ENVIRONMENTAL SURVEYS OF TWO DEEPSEA RADIOACTIVE WASTE DISPOSAL SITES USING SUBMERSIBLES

R.S. DYER
Office of Radiation Programs,
USEPA,
Washington, D.C.,
United States of America

Abstract

ENVIRONMENTAL SURVEYS OF TWO DEEPSEA RADIOACTIVE WASTE DISPOSAL SITES USING SUBMERSIBLES.

This paper discusses the use of the manned submersible ALVIN and the unmanned submersible CURV III to survey disused U.S. radioactive waste dumpsites at depths of 2800 m in the Atlantic ocean and 900 and 1700 m in the Pacific Ocean. Data are presented showing the presence of plutonium contamination in sediments collected within a cluster of both intact and breached packages at the 900 m dumpsite area in the Pacific. The level of $^{239,240}Pu$ contamination in surface sediments is shown to be from 2 to 25 times higher than the maximum expected concentration that could have resulted from weapons testing fallout. Data are also presented confirming the presence of ^{137}Cs contamination in the 2800 m Atlantic dumpsite with concentrations ranging from 3 to 70 times higher than the maximum expected fallout concentration. Packaging records, photographs of the coring operations, and information written on the waste packages themselves establish a direct relationship between the radioactive waste packages and the contamination in the adjacent sediments. Mechanisms of vertical redistribution of caesium and plutonium are discussed. Bioturbation is suggested as the most probable factor in the redistribution of plutonium in the 900 and 1700 m Pacific-Farallons subsites, while leaching from packages and periodic burial of the leached wastes by lateral sediment transport may account for the deep vertical distribution of ^{137}Cs in sediments at the 2800 m Atlantic site. The potential significance of this radionuclide contamination of the dumpsites is discussed in terms of the evidence for possible physical transport of the radioactive material and the presence of edible marine species.

INTRODUCTION

There is a growing interest nationally and internationally in sea disposal as a waste management alternative to land burial of low-level nuclear wastes. With increased competing demands for a decreasing amount of available land, several nations are looking towards the oceans to solve their low-level radioactive waste disposal problem.

In 1946, the United States started ocean dumping of low-level radioactive wastes under licensing authority of the Atomic Energy Commission (AEC). Most of the wastes were dumped between 1946 and 1962. In 1962, AEC contractors turned to land disposal. From 1962 to 1970 there was a phasing out of ocean dumping of radioactive wastes. Three sites received the majority of all wastes dumped.(1) These are detailed in Table I. The radioactive wastes were generally packaged in 55-gallon drums filled with concrete or occasionally other experimental matrices and dumped at depths ranging from 900 m to 3800 m. Two of the sites are located in the Atlantic Ocean off the Maryland-Delaware coast while the other site is located in the Pacific Ocean off San Francisco, California, near the Farallon Islands. The Farallon Islands site contains two subsites at 900 and 1700 m depths. These three major dumpsites received more than 90 percent of the

TABLE I. PRIMARY U.S. RADIOACTIVE WASTE DUMPSITES (2), (3), (4)

Site	Coordinates	Depth (m)	Distance from Land (km)	Years Dumpsite Used	Estimated No. of 55-Gallon Drums Dumped	Estimated Activity in Drums at Time of Packaging (Ci)
Atlantic	38°30'N 72°06'W	2800	190	1951-56 1959-62	14 300	41 400[a]
Atlantic[b]	37°50'N 70°35'W	3800	320	1957-1959	14 500	2 100
Pacific						
Farallon Island (Subsite A)	37°38'N 123°08'W	900	60	1951-1953	3 500	1 100
Farallon Island (Subsite B)	37°37'N	1700	77	1946-1950 1954-1965	44 000	13 400

[a]This does not include the pressure vessel of the N/S Seawolf reactor with an estimated induced activity of 33 000 Ci.

[b]This site has not yet been investigated by the Environmental Protection Agency.

55-gallon packages and 95 percent of the estimated activity dumped. The Pacific-Farallons site was surveyed twice, once in 1957[5] and again in 1960[6]; the 2800 m Atlantic site was the subject of a 1961 survey.[7] Over 11,000 underwater photographs were taken in these previous Atlantic and Pacific dumpsite surveys but no packaged radioactive wastes were identified. From 1961 to 1974 no further site investigations were conducted. In 1974 the U.S. Environmental Protection Agency (EPA) initiated the surveys which are the subject of this paper

Environmental concern for the ocean dumping of packaged low-level radioactive wastes has again been brought to focus in the United States with the development of two regulatory documents: (1) the 1972 International Convention on the Prevention of Marine Pollution by Dumping of Wastes and Other Matter, and (2) the U.S. Marine Protection, Research, and Sanctuaries Act of 1972. Both the Convention (now a treaty) and the U.S. legislation prohibit ocean dumping of high-level radioactive wastes. The U.S. Environmental Protection Agency has been designated to administrate the domestic legislation through development of a permit program for controlling ocean dumping of all wastes including nuclear wastes. As a first step in developing effective controls on any ocean dumping of low-level radioactive wastes, and in order to assess the effectiveness of past packaging techniques, it was necessary to determine the fate of radioactive waste packages dumped in deepsea disposal sites in past years and to make preliminary determinations concerning the distribution of any released wastes. This would require a new survey approach if the actual waste packages were to be found and a meaningful radiological survey conducted. This paper describes a survey method using submersibles by which radioactive waste packages have been located and positively identified at the Pacific-Farallons 900 m and 1700 m depths and the Atlantic 2800 m depth. Radioanalytical data from the 900 m and 2800 m surveys is presented confirming the release of low levels of plutonium-238, plutonium-239, 240, and cesium-137 from some of the waste packages. This information is then related to the condition of the radioactive waste packages observed within the 900 m and 2800 m sites and to the general dumpsite characteristics.

SURVEY METHODS

Submersibles

Pacific

An unmanned, tethered, surface-controlled vehicle called the CURV III (Cable-Controlled Underwater Recovery Vehicle) was selected both for the 1974 survey at the Farallon Islands 900 m subsite and the 1975 survey at the Farallon Islands 1700 m subsite. CURV III is operated by the U.S. Naval Undersea Center and is essentially a rectangular metal framework with syntactic foam blocks for balance and buoyancy. CURV III has a depth capability of 2300 meters and is equipped with two movable television cameras with water-corrected lenses and a 35 mm color camera with synchronized strobe. Auxiliary lighting is provided by mercury vapor spotlights. It has a sonar system capable of scanning an area of 120° and detecting 55-gallon radioactive waste drums at 400 m. For these surveys it was fitted with an electronically triggered Shipek sediment grab sampler for bulk samples. The manipulating arm is non-articulated and can rotate 360° through a single plane. The manipulator has interchangeable tools and was used alternately with a claw and a cruciform coring device.

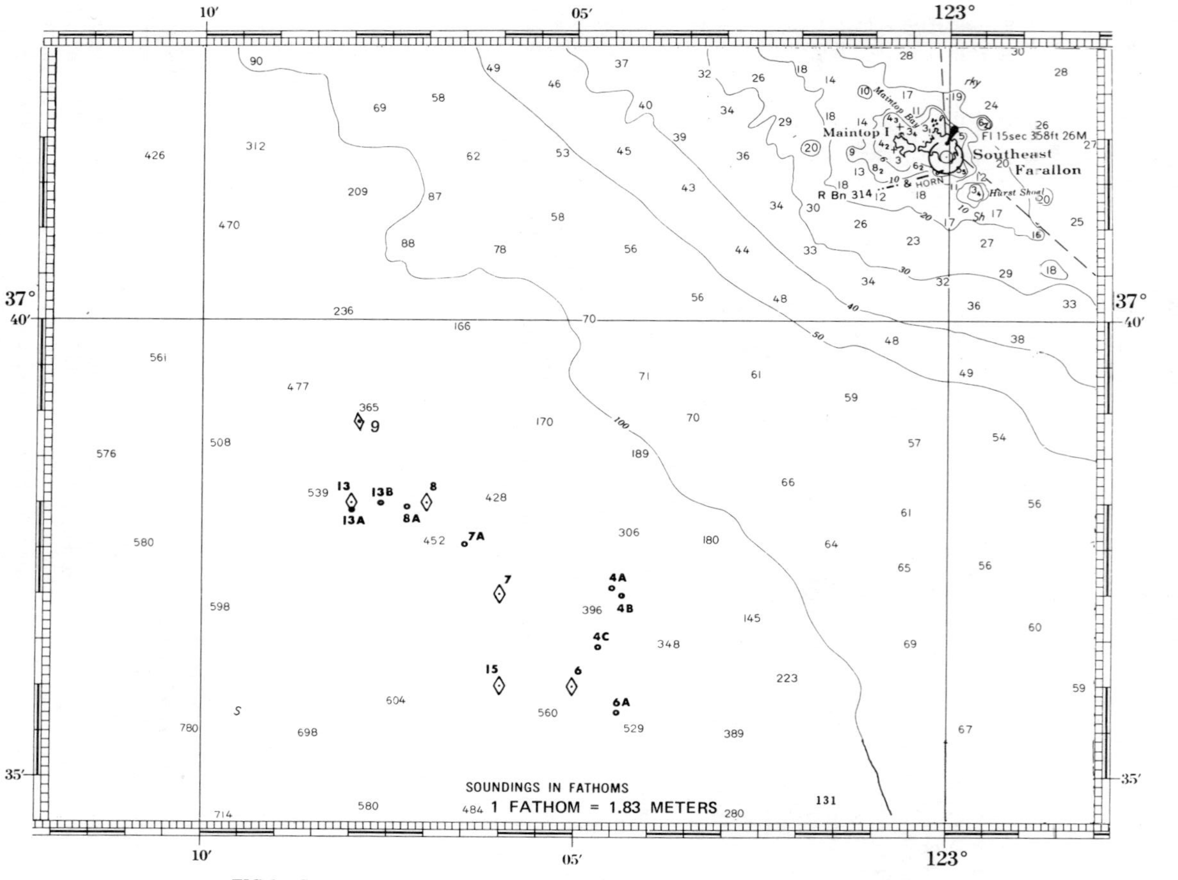

FIG.1. Stations investigated at Farallon Islands 900 m radioactive waste dumpsite.

Atlantic

The 2800 m Atlantic dumpsite was surveyed using the manned, untethered submersible ALVIN. ALVIN is operated by the Woods Hole Oceanographic Institution for the U.S. Navy. It carries a crew of three and has a depth capability of 3800 m. It comes equipped with: (1) a CTFM sonar system that can detect 55-gallon radioactive waste containers at a distance of 350 m, (2) two 35 mm color cameras with synchronized strobes, (3) a fixed television-videotape system, (4) thalium iodide and incandescent external lighting systems, (5) an articulated manipulating arm with six degrees of movement, pivotal wrist motion through 360°, and a lift capacity exceeding 25 kilograms, and (6) variable sampling tools for use on the manipulating arm.

Sediment Coring

The Pacific-Farallons sites were sampled using a standard Shipek grab and a cruciform coring device consisting of four polycarbonate core tubes mounted at right angles to one another in a metal frame bolted to the manipulating arm. Each core tube was equipped with a one-way finger closure to prevent sediment loss. Cores were 3.8 cm in diameter x 48 cm deep.

The Atlantic 2800 m site was sampled using polycarbonate core tubes 6 cm in diameter x 35 cm deep. Box core samples were taken using a special Hessler-modified Ekman box corer equipped with a T-handle for use with the submersible manipulator.

Radioanalysis

All sediment samples collected in the Farallon Islands 900 m subsite in 1974 were routinely gamma scanned for ^{137}Cs without chemical separation and analyzed for ^{90}Sr, ^{230}Th, ^{232}Th, ^{234}U, ^{235}U, ^{238}U, ^{226}Ra, ^{238}Pu, and $^{239,240}Pu$. Grab aliquot samples were analyzed on board ship by personnel of the EPA Las Vegas Radiation Facility to make preliminary determinations of the presence of any significant amounts of radionuclide contaminants. All core samples and remainders of grab samples were frozen and brought back to the lab for analysis. The wet, dry, and ash weights of the sediment samples were taken and sample preparation and analyses were conducted by the EPA Environmental Monitoring and Support Laboratory according to standard methods as outlined in the EPA Handbook of Radiochemical Analytical Methods.(8) Plutonium analysis was performed based on procedures developed by Talvitie.(9),(10)

Sediment samples collected in the 2800 m Atlantic dumpsite during the summer of 1975 were also analyzed for ^{90}Sr, ^{137}Cs, ^{230}Th, ^{232}Th, ^{234}U, ^{235}U, ^{238}U, ^{226}Ra, ^{238}Pu, and $^{239,240}Pu$. The cores were sectioned into 5 cm lengths immediately upon being brought to the surface and were analyzed on board ship using a multi-channel analyzer and internal proportional counter to determine if any specific area or group of containers investigated contained significant activity which would warrant further investigation during a subsequent ALVIN dive. Samples were reanalyzed in the laboratory. All shipboard and laboratory radioanalyses were performed by EPA's Eastern Environmental Radiation Facility according to standard analytical techniques.(11) Analysis for ^{137}Cs was performed first on a NaI multichannel gamma analyzer and the more active samples were re-analyzed using a GeLi detector system. No preliminary chemical separation was performed.

Site Selection - First Survey

The Farallon Islands 900 m subsite was selected for the first survey in August, 1974 based upon four inclusive factors: (1) it was the only site used

TABLE II. ^{238}Pu AND $^{239,240}Pu$ IN SEDIMENT AT FARALLON ISLANDS 900 m RADIOACTIVE WASTE DUMPSITE

(pCi/Kg Dry Weight)

Core No.	Sample Identification	Core Section (cm)	^{238}Pu			$^{239,240}Pu$			$^{238}Pu/^{239,240}Pu$		
			A[a]	B[b]	C[c]	A	B	C	A	B	C
134273	Station 13A[d] 37°38'N 123°08'W 28 August 1974 Depth = 920 m	0-5	<4.3	0±1[e]	3.8±0.9	26±3	23±3	46±4	---	---	0.08±0.02
		5-10			Lost			---			---
		10-15	<2.5	0±3	---	<4.2	0±3	0.9±0.2	---	---	---
		15-20			---			0.5±0.1			---
134274	Station 13A[d] 37°38'N 123°08'W 28 August 1974 Depth = 920 m	0-5	140±7	160±6	146±14	360±14	447±13	482±48	0.39±0.02	0.36±0.01	0.30±0.04
		5-10			225±23			532±77			0.42±0.06
		10-15	3.8±1.8	2.2±2.2	2±0.4	8.8±2.5	6.5±2.6	4.0±0.5	0.43±0.24	0.34±0.36	0.50±0.12
		15-20			---			1.4±0.4			---
134277 Shipek Grab	Station 13A[d] 37°38'N 123°08'W 28 August 1974 Depth = 920 m	0-15	13±2	16±1	123±12	330±11	357±6	397±40	0.04±0.006	0.05±0.004	0.31±0.04
134271	Station 13A[d] 37°38' 123°08'W 28 August 1974 Depth = 920 m	0-10	<28*	4.1±1.2	NA[f]	66±17*	55±4	NA	---	0.07±0.02	---
		10-20	<39*	0.0±1.4	NA	32±14*	7.4±2.3	NA	---	---	---
134272	Station 13A[d] 37°38'N 123°08'W 28 August 1974 Depth = 920 m	0-10	<39*	2.7±2.7	NA	91±20*	50±5	NA	---	---	---
		10-20	<21*	0.0±1.4	NA	<31*	1.4±1.4	NA	---	---	---

134270	Station 15 37°36'N 123°06'W 27 August 1974 Depth = 929 m	0-5	<26*	NA	5.9±2.0	67±14*	NA	80±9	---	---	0.07±0.03
		5-10			2.2±0.5			62±4			0.04±0.01
		10-15	<29*	NA	---	63±22*	NA	15±4	---	---	---
		15-20			---			0.6±0.3			---
134269	Station 15 37°36'N 123°06'W 27 August 1974 Depth = 929 m	0-10	<29*	0.0±3.6	NA	29±12*	0.0±3.6	NA	---	---	---
		10-20	<21*	NA	NA	32±14*	NA	NA	---	---	---
134276 Shipek Grab	Station 15 37°36'N 123°06'W 27 August 1974 Depth = 929 m	0-15	<28*	NA	NA	53±14*	NA	NA	---	---	---
134275 Shipek Grab	Station 6 37°36'N 123°05'W 26 August 1974 Depth = 945 m	0-15	<19*	NA	1.4±0.1	55±11*	NA	38.9±1.4	---	---	0.04±0.003

[a] Environmental Monitoring and Support Laboratory, EPA

[b] LFE Environmental Analysis Laboratories

[c] Woods Hole Oceanographic Institution

[d] The five sediment samples at Station 13A were taken at different locations among a cluster of about 150 radioactive waste packages occupying an area approximately 150m x 300m.

[e] counting error for all samples ± 1 σ

[f] NA - Not analyzed

* Analysis of 1g sediment sample; all other Pu analyses on 10g sample (B) or 40-50g sample (C)

TABLE III. ^{137}Cs IN SEDIMENT AT THE FARALLON ISLANDS 900 m RADIOACTIVE WASTE DUMPSITE

Core No.	Sample Identification	Core Section (cm)	^{137}Cs (pCi/kg Dry Weight)		
			A	B	C
134273	Station 13A 37°38'N 123°08'W 28 August 1974 Depth = 920 m	0-5	ND[a]	NA	49±8
		5-10			6±15
		10-15	ND	NA	7±4
		15-20			1±4
134274	(Same as above)	0-5	ND	0.0±180	18±7
		5-10			5±5
		10-15	ND	0.0±135	1±3
		15-20			-2±3
134277 Shipek Grab	(Same as above)	0-15	ND	0.0±90	29±2
134271	(Same as above)	0-10	ND	0.0±135	NA
		10-20	ND	0.0±90	NA
134272	(Same as above)	0-10	ND	0.0±90	NA
		10-20	ND	0.0±45	NA
134270	Station 15 37°36'N 123°06'W 27 August 1974 Depth = 929 m	0-5	ND	NA	110±10
		5-10			77±8
		10-15	ND	NA	6±8
		15-20			-1.4±9.1
134269	(Same as above)	0-10	ND	0.0±90	NA
		10-20	ND	NA	NA
134276 Shipek Grab	(Same as above)	0-15	120±50	NA	NA
134275	Station 6 37°36'N 123°05'W 26 August 1974 Depth = 945 m	0-15	ND	NA	41±3

[a]ND - Not detected.

Note: footnotes (a) - (f) of Table II also apply to Table III.

exclusively for dumping radioactive wastes, (2) it was only used from 1951-1953 thereby allowing estimations of the rate of observed effects such as biological fouling, (3) the approximate number of containers and the estimated activity at the time of packaging were documented (Table I), (4) and the depth was only about one-half the maximum operating depth of the CURV III submersible. This would permit a better evaluation of the CURV III's operating capabilities before subjecting it to the greater rigors of the Farallon Islands 1700 m subsite. It should be recognized that while the precise depths and locations of actual radioactive waste packages are casually introduced in this paper, this information was not known prior to these EPA surveys. In fact the largest obstacle to the success of these recent surveys has been the uncertainty regarding just where the radioactive waste packages had actually been dumped.

RESULTS AND DISCUSSION

Three dumpsite areas have been investigated: (1) the 900 m Pacific-Farallon subsite in 1974, (2) the 1700 m Pacific-Farallon subsite in 1975, and (3) the 2800 m Atlantic site in 1975. Radioanalyses of sediments from the 900 m subsite and preliminary radioanalyses of sediments at the 2800 m location have been completed.

Farallon Islands 900 m Subsite

An EPA Technical Note[12] has been issued presenting both historical and technical information on this survey operation. This operations report includes the coring procedures, at-sea radiation monitoring and analysis program, sample-handling procedures, detailed bathymetric map, operations log, and photographic documentation of the 55-gallon radioactive waste packages.

Six core samples and three Shipek grabs were collected during the 1974 survey. Figure 1 shows the location of all stations investigated. Core and grab samples were collected at Stations 6 and 15 to be used as 'controls'. At Station 13A a large cluster of radioactive waste packages was located consisting of approximately one hundred and fifty 55-gallon drums located in an area 150 m by 300 m. Tables II and III present the results of radioanalyses for ^{238}Pu, $^{239,240}Pu$, and ^{137}Cs in the sediment samples at Stations 13A, 15, and 6. No ^{90}Sr contamination was found at the 900 m dumpsite.

Radioactive Contamination

The results of three independent analyses are presented in Columns A through C of Tables II and III. Column A presents the initial EPA results, Column B gives the results of the crosscheck analyses on the same vertical 20 cm core splits or replicate Shipek grab aliquots, while Column C presents more detailed analyses of the other one-half of the vertical split using chemical separation techniques and larger sample sizes. The radiochemical methods for plutonium analyses employed in Table II are comparable: Column A analyses are based on a modified Talvitie[10] method, Column B analyses employ a modified method of Chu[13], and Column C analyses follow the method of Livingston, Mann, and Bowen.[14] The method of solubilizing plutonium in sediment varies between Columns A-B, and Column C. Column C values are based on initially leaching the plutonium from the sediments with strong acid while Columns A and B employ a total sample dissolution method. However, since plutonium escaping from radioactive waste packages in the ocean would be expected to be finely divided, then both acid leaching and total dissolution of a sediment sample should yield similar results. That this is the case is substantiated by the close comparison of ^{238}Pu and $^{239,240}Pu$ values in all three columns (A-C) for Core No. 134274 of Table II.

It is clearly evident in Table II that both ^{238}Pu and $^{239,240}Pu$ have been released from the radioactive waste containers found in Station 13A. Average fallout values for $^{239,240}Pu$ in sediments in the 35°-40°N latitude at depths of 700 m to 1300 m vary between 4.5 and 18 pCi/kg dry weight in the top 0-5 cm of sediment, and 0.5-7 pCi/kg dry weight in the 5-10 cm core section.(15) In Table II, the $^{239,240}Pu$ values in one Shipek grab sample (No. 134277) and in the top 10 cm of one core (No. 134274) clearly exceed the upper limit of the expected range of fallout values by greater than an order of magnitude. The $^{238}Pu/^{239,240}Pu$ is also significantly elevated above the expected range of 0.05-0.08. These findings are not totally unexpected when we consider Joseph's statement in a 1957 report that an estimated 30 curies of long-lived alpha activity was dumped off San Francisco, California, between 1946 and 1953.(2)

It is interesting to note that effectively all the ^{238}Pu and $^{239,240}Pu$ contamination in the cores at Station 13A was restricted to the top 10 cm of sediment. This is similar to observations made on the vertical fallout distribution of $^{239,240}Pu$ in both shallow and deepwater sediments.(16),(17) But the core with the highest plutonium contamination (No. 134274) exhibited higher ^{238}Pu and $^{239,240}Pu$ concentrations in the 5-10 cm section. This sugges the possibility that the plutonium release may have occurred many years before perhaps immediately following the initial implosion or pressure-deformation of the package. Redistribution of the plutonium is probably taking place both downward and back towards the sediment surface, but again it is strikingly limited to the top 0-10 cm of sediment. It should also be assumed that a sharp vertical concentration gradient exists down the 5-10 cm section of Core No. 134274 to account for the 1-2 orders of magnitude decrease between the 5-10 cm and 10-15 cm sections. It is doubtful that the estimated sedimentation rate of 1.6 cm per 100 years(18) in this subsite area could significantly alter the vertical distribution that has been observed in Core No. 134274. However, there is good initial evidence suggesting that bioturbation could act as a plutonium distribution mechanism in these sediments. Preliminary examination of the sediment grabs indicates high biological activity especially in terms of the polychaete infaunal population and fecal pellet density in the top section of sediment. In addition, photographic documentation indicates an extensive epifaunal assemblage. Quantitative data from both the 900 m and 1700 m subsites is being compiled.

One other plutonium redistribution factor should be considered -- the physical burial of the material by sediment flow or slumping down steep grades Station 13A was located towards the upper side of a 100 m rise and the entire 900 m site was characterized by small hills of 50 m to 100 m relief. Direct observation using the TV system on the CURV III submersible indicated that sediment was moving downslope and building up slightly around some containers on steeper areas of the slope. Lastly there is some evidence of plutonium sediment penetration beyond 10 cm in Core 134274 but it is reasonable to surmise that this may also be the result of slight physical contamination duri extrusion of the core or slight disruption of the core and porewater content across the 10 cm interface during freezing and thawing.

The core and grab samples at Station 13A were all collected near visibly imploded containers within a fairly restricted survey area. However, only two samples showed significantly elevated ^{238}Pu and $^{239,240}Pu$ concentrations; while three other samples had moderately elevated $^{239,240}Pu$ values. On a very preliminary basis this could suggest that the transport of the plutonium in this site is fairly well localized. As a contradiction, however, Core No. 134270 at Station 15 and the Shipek grab at Station 6 both exhibited an elevat $^{239,240}Pu$ concentration; Station 15 is 3.2 km SE of Station 13A and Station 6 is approximately 4.2 km SE of Station 13A. No radioactive waste packages

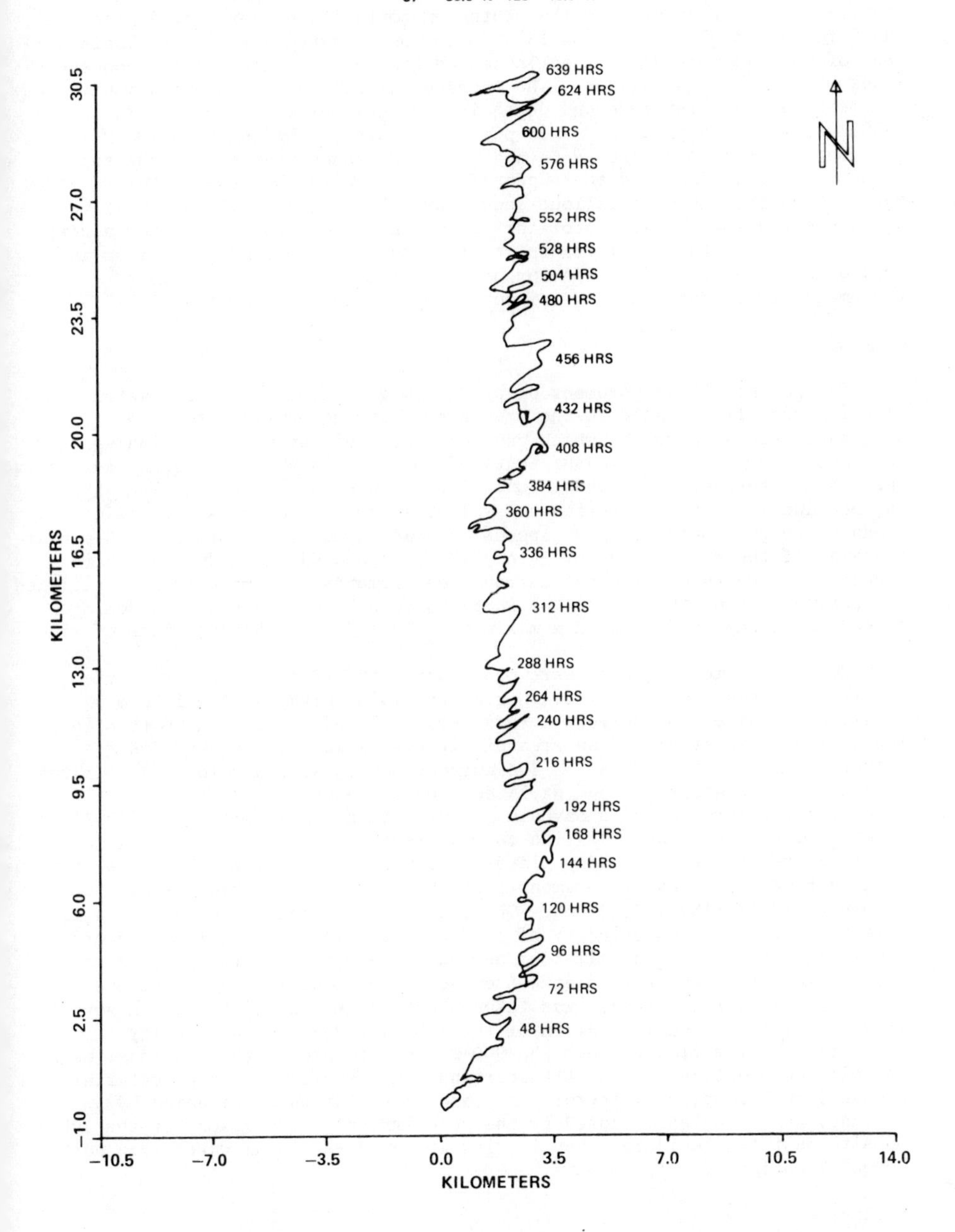

FIG.2. 27-day current metre record at the Farallon Islands 1700 m radioactive waste dumpsite.

were sighted either visually or on sonar at Station 15 or Station 6 yet the $^{239,240}Pu$ contamination is present. It appears that this is a separate source of plutonium contamination since the $^{238}Pu/^{239,240}Pu$ at these stations is significantly different from the plutonium-contaminated cores at Station 13A; and Core No. 134270 at Station 15 is also contaminated with ^{137}Cs (Table III). None of the sediment samples collected at Station 13A showed ^{137}Cs concentrations which exceeded fallout. The average fallout ^{137}Cs concentration in the sediments at this latitude and depth is 9-77 pCi per kg dry weight in the 0-5 cm sediment section, and 2-23 pCi per kg dry weight in the 5-10 cm sediment section.(15) The ^{137}Cs and $^{239,240}Pu$ concentrations in the top 0-10 c of Core No. 134270 exceed the expected fallout range, but the $^{238}Pu/^{239,240}Pu$ ratio is in the expected fallout range at 0.07. Two possible alternatives are suggested for the source of this contamination -- this is either a very low-level manifestation of transport of radioactive contamination from an area other than at Station 13A, or perhaps it is a localized fallout phenomenon from, for example, rainout from a Nevada weapons test.(19)

Currents

The potential for movement of plutonium or cesium from a deepwater dumpsite is generally related to the presence of strong physical transport processes such as currents which can move the sediment material binding these radionuclides. Very few measurements of currents in any deep-ocean radioactive waste dumpsites have been made. One of the few was conducted by Kautsky(20) in 1966 and 1968 in the Iberian abyssal plain of the northeast Atlantic Ocean where horizontal current speeds were measured at a depth of 5300 m near the site of the Nuclear Energy Agency (NEA)-sponsored ocean dumping operations. The length of time the current measurements were made is not specified but during most of the period the currents were below 1-2 cm/s, seldom exceeded 4-5 cm/s and reached a maximum of 10 cm/s for short periods of time.

No current measurements were made during the 1974 EPA survey at the Farallon Islands 900 m subsite. During the 1975 survey at the 1700 m Farallons subsite four current meters were emplaced 2 m off the bottom in a square 1.6 km apart around an area of sighted radioactive waste packages centered at 37°38'N, 123°18'W. The current meters were developed at Scripps Institute of Oceanography and utilized a potentiometric compass for directional reference and a Savonius rotor with a sensitivity of 0.5 cm/s for velocity determinations.(21) The current meters were emplaced on August 21, 19 and recovered twenty-seven days later. Figure 2 presents the vector plot minus tidal effect of the water movement past the meter for the twenty-seven day record at 37°38.5'N, 123°18.0'W. The transport is clearly north with a mean direction of 004° at a velocity of 1.33 cm/sec. The highest speed recorded was 16.5 cm/sec during a half-hour period. The water generally moved east-west during the tidal period at 4-8 cm/sec with excursions of about one km. Although this current record was taken about 14 km west of the 900 m subsite, it still gives a good indication of the likelihood of a low-velocity directional water flow through the 900 m dumpsite area with occasional higher velocity movement capable of transporting finer surface sediment materials. However, such complexing factors as seasonal variation accompanied by upwelling, and turbulence caused by the more irregular topography at the 900 m subsite must be considered when looking at net directional water transport beyond a period of a few months.

Packaging

All of the packages examined in the 900 m Farallon Islands subsite were 55-gallon mild-steel drums which had lain there for 21-23 years. All of the steel drums sighted exhibited some surface corrosion but probing with the

CURV III manipulator indicated that the metal was still sound. There were no signs of scouring of the metal such as might be expected from turbidity currents. Some of the packages were rolled over by the CURV III to expose the metal surface which had been in contact with the sediment. Although the underside of the metal containers was coated with a black sulfide layer, indicative of anoxic conditions, the metal tested firm. All of the packages examined, except one, were either filled with concrete or had 15-30 cm concrete plugs. None of the packages examined showed evidence of a pressure equalization device. Consequently, the most common failure mode was hydrostatic implosion in the center of the package where the radioactive wastes had been compressed. Compression in the center of a 55-gallon drum would tend to splay the metal outward at the end of the package creating a gap between the concrete cap and the chime of the drum. This would then provide a direct pathway between the package contents and the surrounding water. The range of conditions of the barrels examined at the 900 m subsite is shown in twenty color photographs contained in the EPA survey operations report.(12)

One 55-gallon drum was visibly breached and did not appear to be capped with concrete at the top or bottom. The metal container was brittle and broke off easily and the corrosion appeared to have occurred both externally and from within. Close examination of the interior of the package(12) revealed what appeared to be a tar liner with most of the waste contents dissolved away. One possibility as to the nature of this package comes from an examination of early packaging techniques. Constant experimentation with radioactive waste solidification materials was conducted during the early 1950's. One of the packaging methods consisted of solidifying a liquid radioactive waste of pH 5-12 with a gelling agent such as fibrous $CaSO_4$-gel or a corn-starch gel packaged within a tar-lined 55-gallon drum. The tar liner served not only to protect the metal container from the potentially corrosive action of the wastes but to keep the gel from sweating when in contact with air. These gels were soluble when immersed in water for extended periods. The breached radioactive waste package examined in this recent survey could approximate the possible fate of a package prepared in the above manner, especially if the contents dissolved away exposing both sides of the metal to corrosive attack by seawater. This corrosion could also be accelerated by early failure of the tar liner thus exposing the metal to potentially corrosive action by the waste itself.

Biota

Biological variety in the 900 m subsite was quite high although no comprehensive characterization of the species present was attempted. The presence of the sable fish Anoplopoma fimbria is noted as this is a commercially important demersal food fish caught off the California coast at depths up to 800 m. Another fish commonly seen around the radioactive waste packages was the deepsea sole, Embassichthys bathybius. This is not commercially important but is occasionally caught by trawling. Since it is usually caught when the flesh is in the jellied condition (high liquid, low protein induced by spawning), it is not generally eaten.(22)

One other interesting biological observation was noted. Some of the containers at the 900 m site were seen to have many large white vasiform sponges attached, growing to a height of 1 m. The sponges are in the class Hexactinellida and are a new undescribed genus.(23) It is interesting to speculate on the relative contribution the secretions from sessile fouling biota, such as a sponge with a large holdfast attachment area, might have on the ultimate breach of a metal container.

Atlantic 2800 m Dumpsite Survey

This radioactive waste dumpsite is centered at 38°30'N, 72°06'W and occup an area of 256 km^2. It is approximately 15 km SE of an actively-used industri waste dumpsite (Deepwater Dumpsite 106) which has been the subject of extensiv baseline studies during the past two years.[24] Because of the proximity of the two sites, much of the oceanographic data which has been collected for the industrial waste dumpsite is applicable to both sites.

During the period July 25-27, 1975, three dives in the deep submersible ALVIN were conducted in the 2800 m radioactive waste dumpsite. The purpose was to locate radioactive waste packages, document the condition of the containers and the type of packaging used, and take sediment samples around th packages to determine whether they had released any radioactivity and, if so, what isotopes were released. The survey was confined to the NW quadrant of the dumpsite since there was not enough time to survey the entire dumpsite are and since this was the area of the shortest direct route from the ship loading dock used for the past disposals. The NW quadrant is characterized by large boulders, a rocky ridge, and munitions. This made sonar scanning for the radioactive waste packages difficult since they were masked by the terrain and gave the same sonar signal return as a munitions package. Eventually ten packages were located. All but one were 80-gallon drums rather than the 55-gallon drums reportedly dumped at this site. The packages were formed by welding approximately one-half of a 55-gallon drum to the end of another 55-gallon drum to increase its length. The packages were clearly labeled, according to the requirements at that time, to indicate the most hazardous isotope, the cubic volume of wastes, the dose rate at the package surface and at one meter at the time of packaging, and the package number. None of the packages seen in this site had recorded dose rates higher than 35 mR per hour at the surface of the package.

A series of 7 tube cores and 3 box cores were collected near the sighted packages and an additional 10 cores were collected in the immediate vicinity and up to a distance of 20 kilometers from the area. Not all of the analyses of the tube core and box core sections have been completed but Table IV presents the radioanalytical results of those cores which exhibited the highest shipboard activity or those cores which might have been expected to show contamination because of their very close proximity to the waste packages

Radioactive Contamination

Table IV clearly demonstrates the presence of ^{137}Cs contamination in thre sediment cores at this site. The contamination varies between 3-70 times the maximum expected ^{137}Cs concentration that could result from weapons testir fallout at this latitude and depth of water. The ^{137}Cs fallout concentration ranges between 30-70 pCi per kilogram dry weight in the top 0-5 cm of sediment and 4-25 pCi per kilogram dry weight in the 5-10 cm section.[15] Two of the three contaminated sediment samples were collected adjacent to breached containers (see data in Table IV for Core No. 585/T-2 and 585/B-2). Although analyses of all sediment samples have not yet been completed, this data strongly suggests that the ^{137}Cs sediment contamination is attributable to a failure point in the packaging, or leaching from the concrete matrix. No evidence was seen of hydrostatic implosion in any of the packages. There is good evidence, deduced from identifying information on the concrete caps of the packages, that leaching may be the source of contamination in at least two of the packages where there was either no sign of direct package failure (Core No. 584/T-3), or the failure point was visible only in the concrete cap (Core No. 585/T-2). The general absence of macrofaunal fouling on the concret caps allowed us to decipher the information on the caps which indicated that

TABLE IV. ^{137}Cs IN SEDIMENT AT THE ATLANTIC 2800 m RADIOACTIVE WASTE DUMPSITE

(pCi/kg Dry Weight)

Core No.	Sample Identification	Location of Core Relative to Radioactive Waste Package	Core Section (cm)	^{137}Cs [a,b]
584/T-3	ALVIN Dive 584	5 cm from concrete cap where	0.0- 0.5	<100
	38°-30.7'N	metal-concrete junction of	0.5- 7.5	500±50
	72°-09.4'W	package interfaces with the	7.5-15	200±50
	26 July 1975	sediment (Fig. 3).		
	Depth = 2829 m			
584/T-4	ALVIN Dive 584	60 m NW of the package where	0- 7.5	<100
	38°-30.7'N	core 584/T-3 taken.	7.5-12.5	<100
	72°-09.4'W		12.5-17.5	<100
	26 July 1975		17.5-22.5	<100
	Depth = 2828 m			
585/T-1	ALVIN Dive 585	5 cm from concrete cap where	0-7	<100
	38°-30.2'N	metal-concrete junction of	7-12	<100
	72°-09.4'W	package interfaces with the	12-15	<100
	27 July 1975	sediment. Package appears intact.		
	Depth = 2819 m			
585/T-2	ALVIN Dive 585	Core taken 5 cm from package	0- 0.5	2500±100
	38°-30.2'N	in sediment directly below	0.5-5	4800±100
	72°-09.4'W	visible crack in concrete cap.	5-8	1600±100
	27 July 1975		8-15	1100±100
	Depth = 2822 m		15-20	<100
585/T-6	ALVIN Dive 585	10 cm from concrete cap.	0- 7.5	<100
	38°-30.25'N		7.5-15	<100
	72°-09.45'W		15-20	<100
	27 July 1975		20-25	<100
	Depth = 2827 m			
585/B-2 Core liner #1	ALVIN Dive 585	Opposite end of above package;	0-5	210±50
	38°-30.25'N	this end has metal cap which	5-8	120±50
	72°-09.45'W	appears to have corroded through.		
	27 July 1975	Core taken 10 cm from metal cap.		
	Depth = 2827 m			

[a] Analyses performed by Eastern Environmental Radiation Facility, EPA

[b] counting error = ± 1 σ

FIG.3. Radioactive waste package investigated at the Atlantic 2800 m dumpsite.
(Note the capability for precise positioning of the tube core and box core relative to the radioactive waste package.)

in all cases ^{60}Co was the most hazardous isotope in the package. However, only ^{137}Cs contamination was detected in any of the cores analyzed. Subsequently it was determined that the barrels contained demineralizer resins from research and prototype reactors. These resins were encapsulated in a stainless steel tube and the resulting container was too long to be packaged in a 55-gallon drum hence an 80-gallon drum was fabricated. It was also a practice to occasionally slurry the concrete packaging material with low-level radioactive waste liquids, many of which contained ^{137}Cs. Thus, the ^{137}Cs contamination detected in the sediment samples would not have necessarily arisen from breaching of the encapsulated wastes but could have been released by continuous leaching from the concrete in contact with the seawater. The possibility of long-term leaching from the waste packages coupled with our direct observations suggesting horizontal sediment transport would then explain the relatively deep penetration of ^{137}Cs into the sediments -- that is, the contaminated surface sediment is periodically buried by translocated sediment which in turn is contaminated by the continued leaching.

Figure 3 illustrates the precise nature of the coring operations. In this case ^{137}Cs contamination was found in the tube core sample to a sediment depth of more than 7.5 cm although no ^{137}Cs was found in the box core. The lack of ^{137}Cs contamination in the box core may be related to the fact that the box core pressure release flaps performed improperly during this dive. This resulted in core washout during the submersible ALVIN recovery in choppy seas, a problem rectified before the third and last ALVIN dive. With the exception of ^{137}Cs, no other radioactive contaminants have been detected to date in cores from the 2800 m dumpsite.

Currents

Most of the packages examined were deeply buried in the sediment and the sediment was scoured out along the sides of the package and piled up at the ends. Since the sedimentation rate is low in this region at approximately 0.5-3 cm per 100 years[25] and since the radioactive waste packages have only lain on the bottom for approximately fifteen years, then the sediment buildup cannot be attributed to direct deposition. However, this sediment buildup pattern would occur if there was a strong current capable of horizontal sediment transport and scour.

An experimental dye-string current meter array for use with the ALVIN manipulator was tested in the dumpsite area. The dye exhibited strong directional flow in a SSW direction from a height of ten cm to 3 m above the sediment surface, with slightly faster dye movement at the top of the array. Deep current measurements north and east of this 2800 m dumpsite have exhibited strong contour currents with measured flow rates of 10 to 22 cm/sec in a west to southwest direction.[26],[27] These currents are strong enough to erode and transport sediment. This 2800 m radioactive waste dumpsite is at the top of the continental rise and would be expected to show the presence of the south and west moving current flow. The potential for shoreward transport of surface contaminated sediments from the radioactive waste dumpsite warrants further attention.

Benthic Fish

Trawls were conducted around the four perimeters of the radioactive waste dumpsite in May, 1974. The results are reported in detail in a baseline studies report,[24] but the most significant observation was that the numerical abundance and biomass of fishes sharply decreases below 2200 m in the general areas of the radioactive waste and industrial waste dumpsites. The predominant fish caught in the 1974 trawls and seen in the July, 1975 ALVIN dives in the radioactive waste dumpsite was Nematonurus (Coryphaenoides) armatus which is closely relate to the shallower-water commercial species Coryphaenoides rupestris. Nematonuru armatus is an edible fish but has not yet been commercially exploited since it lives in water deeper than the present trawling capabilities of the major world fishery fleets.

An interesting observation was made during the ALVIN dives regarding the feeding habits of Nematonurus armatus. These fish were often seen rooting in the sediments adjacent to the radioactive waste packages. This would not be unexpected since the packages attract deposit and filter feeders which serve as food for this benthic fish. However, the continual rooting and feeding action in contaminated sediments around the radioactive waste packages could significantly mix and redistribute the radionuclides over many years, a possible method of bioturbation on a larger scale.

CONCLUSIONS

Surveys of deepsea radioactive waste disposal sites are feasible using both manned and unmanned submersibles. The unique capability of submersibles to obtain precisely-positioned and photographically documented sediment samples near breached and intact containers has been demonstrated. This has enabled us to relate the ^{238}Pu and ^{239,240}Pu contamination in the Pacific-Farallons 900 dumpsite and the ^{137}Cs contamination in the 2800 m Atlantic dumpsite directly to the presence of radioactive waste packages. Although the presence of these radioactive waste dumpsite contaminants has not yet been translated into any health risks to man or to the marine environment, it underscores the need to substantiate hypothesized release and transport events in the deep-ocean by actual in-situ studies before ocean dumping of radioactive wastes becomes more widespread.

Few current measurements have been taken around deepsea radioactive waste dumpsites, but our preliminary evidence indicates the presence of weak but directional currents in the 1700 m Pacific subsite and stronger currents in the deeper 2800 m Atlantic site. The long-term directionality and speed of currents in a dumpsite area should be known before any extensive use of a site is envisioned.

The biological abundance and diversity in the 900 m site was high. The presence of the commercially-exploited sablefish in the 900 m Pacific site, in an area contiguous to a fishing zone, should exclude this site from any further consideration as a radioactive waste dumpsite.

The extensive hydrostatic implosion of containers observed at the 900 m and 1700 m Pacific-Farallons subsites indicates a definite need for pressure equalization devices on radioactive waste packages destined for deepsea disposa The waste packages should be filled as homogeneously as possible with a rigid matrix material such as concrete, and any remaining air voids must be pressure-equalized during descent to the sea floor. The mild-steel containers examined all sites had immersion times ranging from 13-23 years with considerable evidence of surface corrosion and blistering, especially at the 2800 m site. So far, however, none of these containers show signs of having been breached solely from external corrosive forces. For those countries advocating a policy of continued containment rather than dispersion and dilution of the dumped radioactive materials after they reach the sea floor, information is needed on corrosion rates of various metals in deepsea high-pressure conditions Since the metal sheath significantly reduces the surface area of the matrix exposed to the leaching effects of seawater, the advantages of using more corrosion resistant alloys should be considered, especially for packaged longer-lived radioisotopes.

In all deepsea monitoring and survey operations conducted to date, the analytical results have been compared to the background or fallout concentrations of particular nuclides. However, these baseline fallout values often vary over an order of magnitude. More data is needed on baseline levels of radionuclides in a geographical area around a dumpsite if meaningful estimates are to be made of gradual buildup and movement of any radioactive materials released from the dumpsite.

ACKNOWLEDGEMENTS

The author gratefully acknowledges the support of the National Oceanic and Atmospheric Administration, Woods Hole Oceanographic Institution, the operations teams for the submersibles ALVIN and CURV III, and the many individuals in the Water and Radiation Program Offices of EPA who had the fortitude and vision to support this program through many stormy seas.

REFERENCES

(1) LUEDECKE, A.R., "Disposal of radioactive wastes into the Atlantic Ocean and the Gulf of Mexico", Industrial Radioactive Waste Disposal Hearings Before the Joint Committee on Atomic Energy, 5, Washington, D.C. (1959).

(2) JOSEPH, A.B., A Summary to December 1956 of United States Sea Disposal Operations, WASH-734, U.S. Atomic Energy Commission, Washington, D.C. (1957).

(3) JOSEPH, A.B., GUSTAFSON, P.F., RUSSELL, I.R., SCHUERT, E.A., VOLCHOK, H.L., TAMPLIN, A., "Sources of radioactivity and their characteristics", Ch. 2, Radioactivity in the Marine Environment, National Academy of Sciences, Washington, D.C. (1971).

(4) Ocean Dumping -- A National Policy. A report to the President prepared by the Council on Environmental Quality, U.S. Govt. Printing Office, Washington, D.C. (1970).

(5) FAUGHN, J.L., FOLSOM, T.R., JENNINGS, F.D., MARTIN, DeC., MILLER, I.E., WISNER, R.L., Radiological Survey of the California Disposal Areas, University of California, Scripps Institution of Oceanography, (1957).

(6) Pneumodynamics Corporation, Survey of Radioactive Waste Disposal Sites, USAEC Doc. No. TID-13665 (1961).

(7) JONES, EDMUND L., Special Report on Waste Disposal Program, U.S. Coast and Geodetic Survey Project 10,000-827 under contract to U.S. Atomic Energy Commission, Washington, D.C. (1961).

(8) Handbook of Radiochemical Analytical Methods, (JOHNS, F.B., Ed.), EPA-680/4 Las Vegas, (1975).

(9) TALVITIE, N.A., Radiochemical determination of plutonium in environmental and biological samples by ion exchange, Anal. Chem. 43 13 (1971) 1827.

(10) TALVITIE, N.A., Electrodeposition of actinides for alpha spectrometric determination, Anal. Chem. 44 2 (1972) 280.

(11) Procedures for Radiochemical Analysis at the Eastern Environmental Radiation Facility, (STRONG, A.B., Ed.), U.S. Environmental Protection Agency, Montgomery, (Draft, Unpublished).

(12) A Survey of the Farallon Islands 500-Fathom Radioactive Waste Disposal Si Operations Report, U.S. Environmental Protection Agency, Rep. ORP-75-1, Washington, D.C. (1975).

(13) CHU, N., Plutonium determination in soil by leaching and ion exchange separation, Anal, Chem. 43 3 (1971) 449.

(14) LIVINGSTON, H.D., MANN, D.R., BOWEN, V.T., "Analytical procedures for transuranic elements in sea water and marine sediments", Analytical Methods in Oceanography, Am. Chem. Soc., Advances in Chemistry (1975).

(15) BOWEN, V.T., Personal communication (1976).

(16) NOSHKIN, V.E., Ecological aspects of plutonium dissemination in aquatic environments, Health Physics 22 (1972) 537.

(17) LA BEYRIE, L.D., LIVINGSTON, H.D., BOWEN, V.T., "Comparison of the distributions in marine sediments of the fallout derived nuclides ^{55}Fe an ^{239,240}Pu: A new approach to the chemistry of environmental radionuclide Transuranium Nuclides in the Environment, IAEA, Vienna (1975) in press.

(18) UCHUPI, E., EMERY, K.O., Continental slope between San Francisco, Califor and Cedros Island, Mexico, Deep-Sea Res. 10 (1963) 397.

(19) Fallout from Nuclear Weapons Tests, Hearings Before the Joint Committee o Atomic Energy, 3, Wash., D.C. (1959).

(20) WALDEN, H., WEICHART, G., KAUTSKY, H., Aus der meereskundlichen Forschung, Naturwissenschaften 59 (1972) 19.

(21) ISAACS, J.D., et al., Near-bottom currents measured in 4 kilometers depth off the Baja California coast, J. Geophys. Res. 71 18 (1966) 4297.

(22) JOW, T., Personal communication (1975).

(23) BAKUS, G.J., Personal communication (1975).

(24) May 1974 Baseline Investigation of Deepwater Dumpsite 106, U.S. Departmen of Commerce and U.S. Environmental Protection Agency (1975).

(25) EMERY, K.O., UCHUPI, E., Western North Atlantic Ocean: Topography, Rocks Structure, Water, Life, and Sediments, AAPG Memoir 17 (1972) 417.

(26) VOLKMANN, G., Deep current observations in the Western North Atlantic, Deep-Sea Res. 9 (1962) 493.

(27) ZIMMERMAN, H.R., Bottom currents on the New England continental rise, J. Geophys. Res. 76 24 (1971) 5865.

DISCUSSION

Y. NISHIWAKI: I am greatly impressed by the pictures you showed in your oral presentation of the corrosion of waste packages that have lain for a number of years on the sea bed. What is the main cause of the corrosion? Is it due to chemical action from inside or from outside, and is there any possibility of biocorrosion due to the biological action of marine micro-organisms?

You also showed us a picture of some fish close to the drum lying on the sea bed. Have you noticed any effects on other benthic organisms at the dumpsite?

R.S. DYER: Your first question is a difficult one to answer, since no long-term corrosion studies have been conducted to my knowledge on mild steel in deep water. Assuming that there is no internal chemical corrosion from the wastes themselves, the chief cause of corrosion is most probably localized external pitting in areas of metal stress. But any answer is, of necessity, oversimplified at the present time. The possibility of biocorrosion has only just begun to be investigated. There are indications that both macro- and micro-organisms might be responsible for creating loci for pitting corrosion, in addition to the possibility that there are corrosive secretions from many of the marine organisms themselves. There is also evidence, however, from shallower water situations, where biological fouling is sufficient to create anaerobic barriers over the metal, that oxidative corrosion is retarded.

Regarding your second question, we did not conduct any detailed biological studies, since the primary objective of the surveys was to determine the feasibility of the survey method and to obtain sediment samples for radioanalysis so that we could determine whether the wastes had been released from the packages. Since it is already recognized that marine organisms are relatively radiation-resistant in other than the egg and larval stages, we would not expect to see visible effects from the contamination measured. But it would be interesting to look into possible chromosomal abnormalities arising in benthic organisms resident at these disposal sites.

Y. NISHIWAKI: Would it be possible to obtain detailed information on the design of these old waste packages, the drum material and the content of the waste, so that your very valuable survey data could be utilized more effectively for improving the future design of waste packages for ocean dumping?

R.S. DYER: Yes, it would. We are currently undertaking this rather extensive task, but have encountered some difficulties in obtaining the disposal records. My personal experience has demonstrated a need for one central registry of disposal information.

Y. NISHIWAKI: You stated that one of the reasons for breakage of the drum was the lack of pressure balancing devices. Do you have any other constructive suggestions for improvement of the packages as a result of the survey of the old waste drums? Further, you mentioned that the presence of radioactive contaminants at the dumpsite has not been reflected in any health risk to man or to the marine environment. In that case, what is the main reason for the decision by the United States of America to suspend ocean dumping operations?

R.S. DYER: Although we have considered design improvements for the waste packages utilized in the past, we have not yet conducted the necessary corrosion and matrix leach tests to reach definitive conclusions. However, one of the matrix materials we hope to investigate in the near future is polymer-impregnated concrete.

Regarding suspension of the operation, the United States took this decision in 1962 when land disposal sites became more generally available and, at the same time, more economical to use.

J.M. MATUSZEK: The tables of analytical data in your paper do not include any data on cores at distances comparatively remote from the burial sites. Did you obtain any measurements for such cores that could serve as control samples showing fallout radioactivity transported to the deeper sediments?

R.S. DYER: Yes, we did obtain control cores at some distance from the site; they were collected along the west coast from the State of Washington to southern California. We also compared these fallout values with composite data from other cores collected at the same latitude and depths. The resultant range of expected fallout values for ^{137}Cs, and for $^{239,240}Pu$, are given in the paper.

W. HILD: It would appear from the photographs you have shown us that the drums dumped in the Atlantic are embedded in the sediment to a greater degree than those at the Pacific sites. Would this be due to differences in the physical properties of the sediments, or is it the result of turbidity streams depositing sediment? Can we conclude that after a certain lapse of time the drums would be completely covered by sediment?

R.S. DYER: The drums surveyed at the radioactive waste dumpsite in the Atlantic most probably appeared to be more deeply embedded in the sediment because of physical burial by sediment transport due to a contour current. I have no evidence at this time of the presence of any turbidity currents that might impact the site. While it is conceivable that the drums could become covered over completely by sediment, preliminary evidence indicates that both sediment deposition and scouring action take place around the waste packages. That might leave the waste package exposed until long after the waste materials had been released.

D.W. CLELLAND: What is the maximum depth at which the submersible can work? For example, would 4000 m be possible?

The apparent ease of operating such a device seems to suggest that deep ocean storage with the capability of retrieving and inspecting packages is now a practical possibility.

R.S. DYER: The manned submersible ALVIN is certified to operate down to a depth of approximately 3800 m, with a possibility of increasing the working depth to 4000 m. The United States Navy, however, is developing an unmanned, remote-control submersible, known as the RUWS which will have an operating depth of approximately 5000 m.

With regard to your comment, it is true that the more sophisticated retrieval mechanisms now being developed should make retrieval and inspection of packages a practical possibility in the near future at the depths currently under consideration for low-level waste disposal.

DEVELOPMENTS AND STUDIES FOR MARINE DISPOSAL OF RADIOACTIVE WASTES

C. MACHIDA, A. ITO
Tokai Research Establishment,
Japan Atomic Energy Research Institute,
Ibaraki-ken

A. MATSUMOTO
Oarai Research Establishment,
Japan Atomic Energy Research Institute,
Ibaraki-ken

S. SAKATA
Japan Atomic Energy Research Institute,
Tokyo

T. NAGAKURA, H. ABE
Civil Engineering Laboratory,
Central Research Institute of
Electric Power Industry,
Abiko City, Chiba,
Japan

Abstract

DEVELOPMENTS AND STUDIES FOR MARINE DISPOSAL OF RADIOACTIVE WASTES.

Before realization of experimental sea-dumping operations expected to start in 1977, considerable tasks have been undertaken in various fields related to low-level radioactive wastes and promising areas in the Pacific. The main effort is being put on high performance treatments to reduce the amount to be dumped and demonstrating the integrity of the waste packages. Bituminization and incineration are the preferred processes from the standpoint of reducing the amount. Several developments and improvements to these processes are being explored. A cement solidification method has been utilized to prepare the waste packages from nuclear power stations. To establish provisional guidelines, cement-solidified packages have been examined for their integrity under high hydrostatic pressure tests and free fall tests. Multistage packages have also been investigated. Radioactivity leachability into the aqueous phase is one of the important indices in evaluating package design and specifications. Leaching of solidified radionuclides under normal and higher pressures has been studied on small specimens and full-scale packages.

INTRODUCTION

The need for Japan to take steps to meet the situation that the accumulation of radioactive wastes has reached significant proportions and is rapidly growing has become urgent. In June 1973 the Japan Atomic Energy Commission (JAEC) made a decision concerning radioactive waste management, based on the revised long-term programme for the development and utilization of atomic energy, to promote preparatory investigations followed by experimental sea-dumping operations for marine disposal of low-level radioactive wastes.

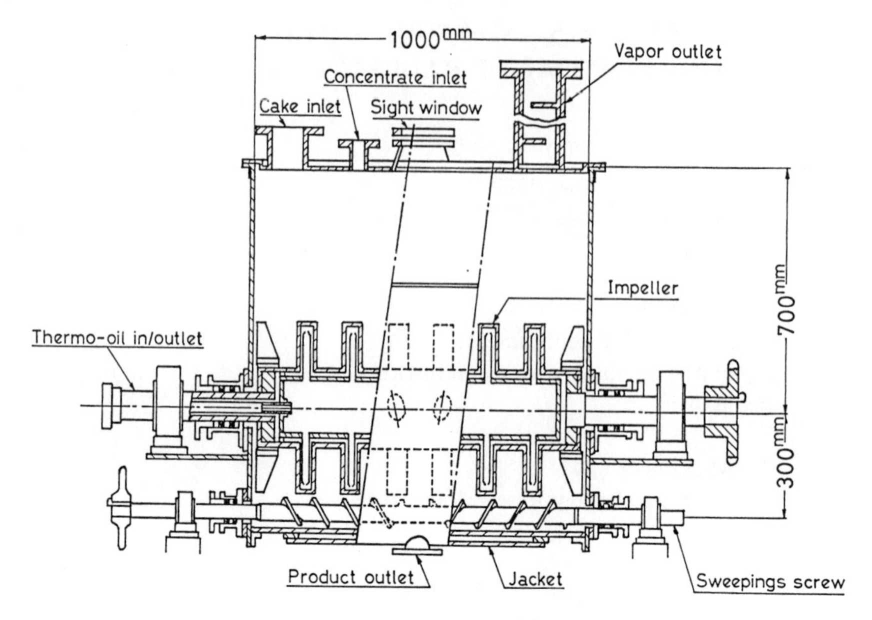

FIG.1. Batch-wise bituminizing unit.

In compliance with the JAEC's decision, several governmental and public organizations, under the leadership of the Science and Technology Agency, undertook a share of the relevant studies including oceanographic investigations on promising areas in the northwestern Pacific and development of deep-sea observation techniques.

It is expected that the experimental sea-dumping operations would be started in 1977 after confirmation of the safety evaluation of the operations through the preparatory investigations.

1. TREATMENT AND CONDITIONING

The main effort in the treatment studies has been concentrated on reducing the difficulties expected in the storage, transport and disposal of the packages. Bituminization and incineration processes are considered effective to minimize the amount concerned.

1.1. Bituminization

The Oarai Research Establishment of the Japan Atomic Energy Research Institute (JAERI) has introduced a batchwise, slow-mixing bituminization unit to treat its chemical sludge and evaporator concentrate. A bathtub-shaped mixing vessel has a horizontal axis impeller with hollow tubes and jacketed walls (Fig. 1). The vessel is heated with thermo-oil (max. temp. 300°C) circulating through the hollow impeller, jacket and a heat source. Its water evaporation rate is about 30 l/h for the chemical sludge and about 50 l/h for the evaporator concentrate. During the last two years the unit (rated capacity 100 l/batch) has successively solidified 50.6 m^3 of the chemical sludge (solid content 2–4%) and 360 l of the evaporator concentrate and proved to be effective in minimizing the resultant solids.

1.2. Incineration

In the Tokai Research Establishment of JAERI the low-level solid wastes were found to be mostly classified as combustible organic (89% in weight). This led to the installation of an incinerator that should be able to withstand the severe operational conditions expected. The design work has been finished in co-operation with NGK Insulators Ltd. for the new incinerator (capacity 100 kg/h) with two-stage high-temperature ceramic filters. The design is based on results from a series of preliminary tests on the ceramic filters and operational performance on the existing incinerator.

The preliminary tests on the two-stage ceramic filters have been carried out to measure the pressure drop, after-burning effect and decontamination characteristics with some radionuclides. A fairly large increase in pressure drop was observed across the secondary filter (from 79 to 131 mm aq.), despite no increase in the primary one, possibly because of tar plugging after 52 days of operation. With regard to the after-burning effect, soot was perfectly burned on the primary filter surface, but most of the tar was observed to be trapped as aerosol in the secondary one. The decontamination factors of the filters depend upon the radionuclide and the chemical form. The observed performance is:

	Primary filter (DF)	Secondary filter (DF)
$^{132}CsCl$	1.7×10^2	2.5×10^2
$^{85}SrCl$	5.5×10	1.0×10^3
$H_3{}^{32}PO_4$	3.7×10	1.1×10^3
$^{58}CoCl$	4.5×10	2.1×10^2

The Oarai Research Establishment adopted a thermal-decomposition type furnace equipped with a dry gas-cleaning system. The rectangular furnace consists of three parts for pyrolysis, main combustion and after-burning. It is characterized by complete combustion and high cleanliness of flue gas. Its dust loading in the flue gas has been observed to be less than 0.1 g/m^3 at the outlet of the furnace during normal operation.

1.3. Multi-stage packaging

The Oarai Research Establishment has applied multistage-type packaging for all processed wastes since 1971. Low-level wastes solidified with cement, concrete or bitumen and compacted solid rags are packed into 200 l steel drums lined with reinforced concrete. Even very low active wastes that raise no significant radiation on the container surface are packed into drums lined with a minimum of 5 cm of concrete in order to improve the resistivity to mechanical and chemical impact in the course of sea dumping. Irradiated metal waste pieces are sorted into a high radio-activity ^{60}Co-rich portion and a moderate radioactive remainder, after which they are compacted and packed into large reinforced concrete vessels. The ^{60}Co-rich portion is in most cases provided with an iron or lead inner shielding. This procedure is effective in decreasing the number of packages. All outer containers are sealed with reinforced concrete poured onto the top.

Most packages are consistent with the guidelines prepared by OECD/NEA in 1974 [1]. Samples were subjected to penetration and free drop tests as defined in the Guidelines and remained intact.

An apparatus that can simulate the conditions of the proposed deep-sea dumping (maximum hydrostatic pressure 600 kgf/cm^2, minimum temperature 2°C) was installed at the Tokai Research Establishment for integrity and radioactivity leaching tests on full-scale packages of 200-l drums in early 1975. A flowsheet of the apparatus is shown in Fig. 2. Multistage packages containing various types of dummy wastes (Fig. 3) were submitted to high pressure tests up to 500 kgf/cm^2. Most packages, which contain compacted solid rags with a rather large void inside and insufficient path for the water ingress required for pressure equalization, were deformed at the top or bottom.

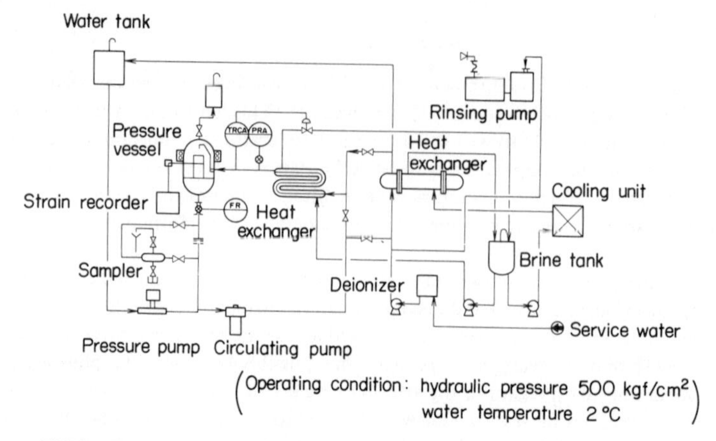

FIG.2. Flowsheet of high-pressure leaching test apparatus for full-scale waste solids.

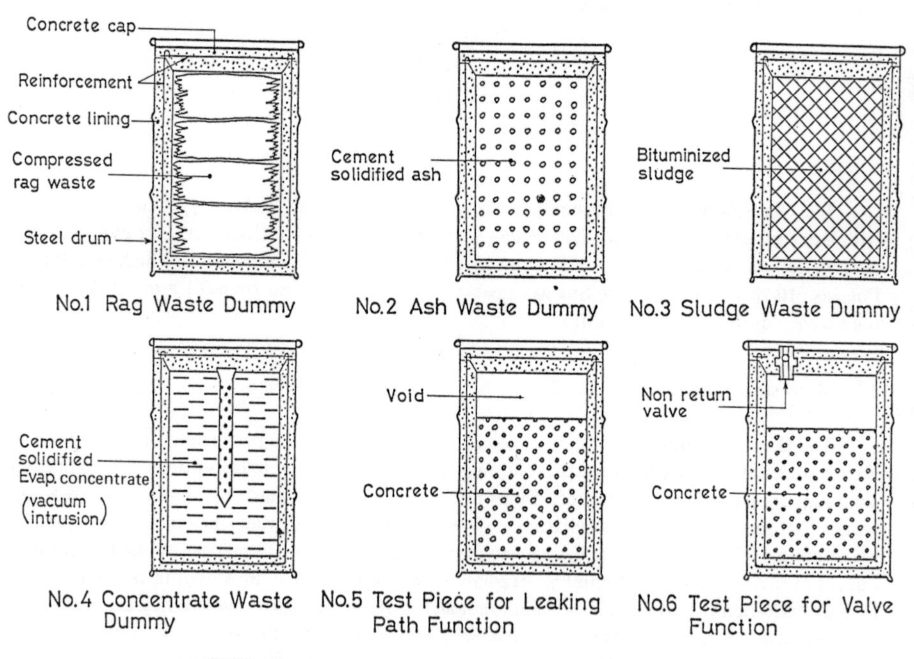

FIG.3. Dummy waste packages submitted to high-pressure tests.

2. CONDITIONING PRACTICE IN NUCLEAR POWER STATIONS

Low-level wastes generated at nuclear power stations have been solidified uniformly in 200 l drums and are expected to be dumped into the deep sea in the near future.

Based on the results of broad experiments [2] carried out in the Civil Engineering Laboratory of the Central Research Institute of Electric Power Industry (CRIEPI), the JAEC prepared provisional guidelines on homogeneously cement solidified packages of low-level wastes for experimental marine disposal. At present BWR-type power stations are adopting the cement solidification method consistent with the guidelines.

The guidelines consist of the requirements for cement solidified waste packaging and specifications for materials, mix proportions, mixing, packaging, quality control and so on. The basic requirements for each package are that:

(1) Solidified wastes shall have a specific gravity greater than 1.2

(2) At least until reaching the surface of the sea bed the package should not suffer any damage that might cause radioactive materials to be scattering

(3) The surface radiation dose rate of the package shall be less than 200 mR/h, in conformity with the Regulations for Transportation of Radioactive Materials by Vehicle.

Particularly, it is emphasized in requirement (2) that the package must be of such durability that leaching of radionuclides from it is prevented.

The Japanese provisional guidelines correspond to the IAEA's recommendation for marine disposal of radioactive wastes and other radioactive materials and the OECD/NEA's guidelines for marine-disposal packages.

2.1. Cement solidification

To prepare homogeneously cement-solidified wastes suitable for marine disposal, the appropriate types of cement and admixture are selected and the proportions of those materials and the mixing method are made clear for each kind of waste, for example, evaporator concentrate, spent resin and spent filter aid. Certain characteristic properties of the solidified products such as specific gravity, mechanical strength and modulus of elasticity were experimentally observed.

(1) A concentrated sodium sulphate solution from BWR-type power stations was solidified with slag cement. The stability of the solidified mixture was much improved by adding artificial light weight sand as admixture. Boric acid concentrates from PWR-type power stations were solidified with Portland cement after neutralization with sodium hydroxide. One cubic metre of the solidified mixture contains from 500 to 550 l of the waste solution, maintaining the required properties.

(2) Spent resin was solidified with slag cement after neutralization. Under proper mixing conditions 300 kg of particulate resin (water content 50%) or 200 kg of powdered resin (water content 50%) was converted to 1 m^3 of solidified waste.

(3) Spent filter aid may be converted into stable solid with either Portland or slag cement, but it was very hard to mix them because of their high viscosity. The content of spent filter aid was about 150 kg (water content 50%) per 1 m^3 of the resultant solid of the required properties.

2.2. Mechanical behaviour under higher hydrostatic pressure

To estimate the soundness of the cement-solidified package in marine dumping, stress analysis and high hydrostatic pressure tests are carried out in the CRIEPI for two types of packages:

Monolithic package, i.e. the homogeneously cement-solidified into 200 l drum

Multistage package, i.e. the cement-solidified into concrete container.

A finite element method was applied for their stress analysis. A unit that can simulate free-falling conditions of the package (pressure increase rate about 0.3 $kgf/cm^2 \cdot s^{-1}$) in seawater, maximum pressure up to 500 kgf/cm^2 and corresponding temperature change, was used for the high hydrostatic pressure test.

(1) The sealed monolithic package that precludes any water ingress remained intact and its strength decreased a little after the maximum load test under 500 kgf/cm^2 of hydrostatic pressure (tri-axial compressive load), when the uniaxial compressive strength (σ) of the cement-solidified waste is greater than 1/4 of the maximum water pressure (σ_p), namely, $\sigma_p/\sigma \leqslant 4$. In view of the

probable quality variation, it is recommended that cement-solidified waste should have a uniaxial compressive strength greater than 150 kgf/cm^2.

(2) The multistage package with a pressure equalization valve on the cap did not suffer any deformation or rupture because the inner and outer pressure of the package was soon equalized but the high hydrostatic pressure inside the package induced micro-structure damage. This phenomenon may cause the materials contained to leach out and may decrease the durability of the package.

2.3. Mechanical behaviour of package under free fall tests

Free fall tests were carried out on the cement-solidified packages to estimate their integrity on striking the sea surface or sea bed during marine dumping and under probable impact during handlin on land. The packages were dropped from various heights up to 10 m on to sand mats, concrete mat or the sea surface.

(1) Cement-solidified waste proved to retain its soundness when colliding against the sea surface or sea bed

(2) In the tests with concrete mats a fall of 1-2 m produced cracks on the package surface, while cracks penetrating through the package have been found in tests carried out from more than 2 m. In tests with sand mats the impact effect was reduced to about half that on concrete mats, and no cracks were found even after dropping from four metres. These results suggest that the packages should be very carefully handled on concrete mats.

2.4. Quality inspection by non-destructive testing

It has been concluded that ultrasonic pulse velocity measurement is one of the suitable method for quality inspection of cement-solidified packages prior to marine dumping. The method is to determine the velocity of an ultrasonic pulse propagated through the homogeneously cement-solidified waste and calculate the dynamic modulus of elasticity from the velocity. The uniaxial compressive strength of the solidified waste can be estimated by considering its proportionality to the dynamic modulus of elasticity.

The measurement unit is under development and expected to become ready for practical use at nuclear power stations.

3. PRELIMINARY EXPERIMENTS WITH RADIOACTIVITY

Radioactivity leachability tests were carried out at JAERI on the solidified waste and packages under normal and higher pressures.

3.1. Leaching test under normal pressure

In accordance with the IAEA's recommendations on 'Determination of leaching rate and application of results' [3], the radioactivity leaching has been examined on specimens (4.5 cm Ø × 4.4 cm) prepared with an evaporator concentrate-cement mixture. Leachability of ^{137}Cs from the solidified waste depends on such factors as waste-cement ratio, salt concentration in waste, type of cement and leachant, temperature and setting period.

The diffusion coefficient D (cm^2/d) of ^{137}Cs has been observed to be:

For slag cement composite	$10^{-4} - 10^{-6}$
For Portland cement composite	$10^{-3} - 10^{-4}$

To lower the high leachability of ^{137}Cs, the effect of the addition of various aggregates and other additives has been examined. It was found that the addition of natural Zeolite (a mixture of

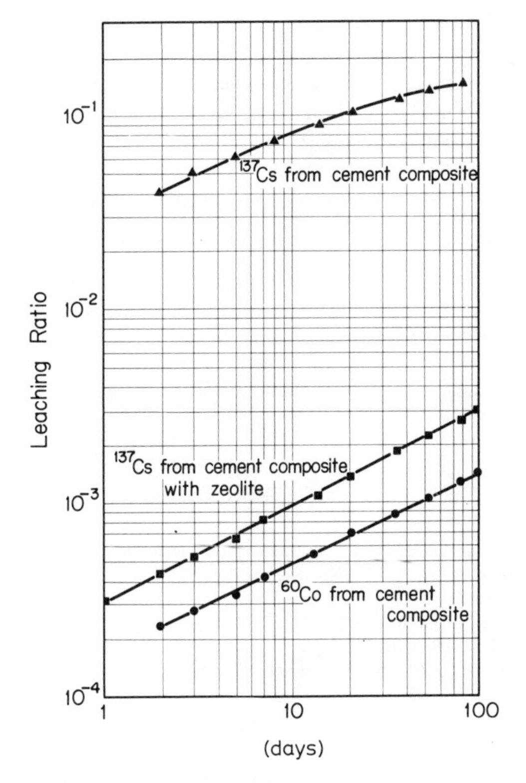

FIG.4. Leaching of ^{137}Cs and ^{60}Co from evaporator concentrate-cement composite.

mordenite and clinoptillolite) is most effective and reduces the diffusion coefficient by a factor of 10^2 to 10^3. A similar examination for ^{60}Co shows that the leachability depends on the type of cement and leachant, temperature and setting period. The diffusion coefficient of ^{60}Co is about 10^{-9} to 10^{-11} (Fig. 4).

Overall leaching tests have been carried out on full-scale multistage packages for more than one year with tap water as leachant. The cumulative radioactivity found in the leachant is about 10^{-4} to 10^{-5} of the total embedded activity per year in the case of 200 l drum-type packages containing compacted rags and 10^{-7} to 10^{-8} for concrete vessel-type package filled with irradiated metal pieces.

3.2. High pressure leaching test

Evaporator concentrate-cement packages were subjected to high hydrostatic pressure tests using the unit described previously in section 1.3. A pressure of 500 kgf/cm² was attained within about 30 min, corresponding to the falling velocity of the package in seawater. The principal results are:

(1) For solidified waste in sealed drums no damage to either the solidified waste or the drum was observed, except that the drum was deformed when a void space existed between the solidified waste and the drum.

(2) For solidified waste with some exposed bare surface, the exposed parts of the solidified waste were cracked. The degree of damage decreased when the pressure increase rate was lowered.

Leaching tests of solidified waste containing ^{137}Cs have been performed under a hydrostatic pressure of 500 kgf/cm^2. The results are:

(1) For solidified waste in sealed drums the radioactivity in the leachant was lower than the detection limit even after seven days. If the radioactivity in the leachant is assumed to be equal to the detection limit, the leaching ratio is 0.03%.

(2) For solidified waste with some exposed bare surface a higher leaching ratio was obtained than under normal pressure, apparently because of an increase in the surface area due to the cracks.

CONCLUSION

Extensive work is still under way in Japan on the marine disposal of low-level radioactive wastes. Recently the JAEC set up a special committee to promote further research and development activities in radioactive waste management and to assess the overall safety of the experimental sea-dumping operations.

REFERENCES

[1] OECD/NEA, Guidelines for Sea Disposal Packages of Radioactive Wastes, OECD/NEA, Paris (1974).

[2] NAGAKURA, T., ABE, H., MAKI, Y., Report of the Central Research Institute of Electric Power Industry, Technical Report C74001 (1974).

[3] IAEA, Treatment of Low- and Intermediate-Level Radioactive Waste Concentrates, Technical Reports Series No. 82, IAEA, Vienna (1968) 101.
HOSPE, E.D., "Leach testings of immobilized radioactive waste solids – A proposal for a standard method", At. Energy Rev. 9 1 (1971) 195.

DISCUSSION

C.F. MARHSALL: You refer in your very interesting paper to the testing of a multi-stage design for a sea disposal package in which a valve is fitted to reduce the differential pressure forces operative during descent to the sea bed. Have you tested or considered any other methods of pressure equalization, such as those described in the NEA guidelines for sea disposal packages?

S. SAKATA: We have tested two types of pressure equalization methods – the non-return valve and a sealing interface between the prefabricated concrete of the vessel and the capping concrete. Some of the packages employing the sealing interface were deformed. We are now attempting to establish a reliable preparation procedure ensuring an adequate water ingress path, using the sealing interface.

J.B. LEWIS: I would like to make two observations on your paper. First, although your leaching experiments are very interesting, it must be remembered that the provisional definition by the IAEA for sea disposal of low-level wastes under the London Convention does not assume containment of the radioactive species once the waste has reached the bottom of the sea.

Second, experiments carried out by us at Harwell, in which we used a pressure chamber to test various pressure relief devices, showed that it was the rate of rise in pressure that was important, and not just the total pressure reached. The device must be capable of rapid pressure equalization as the package sinks. We tested every type of container, using non-radioactive material, by deep sea tests at a depth of 4000 m. The drums were allowed to fall freely, with the camera attached to them. Other drums were lowered onto the sea bottom, and then recovered.

S. SAKATA: In assessing the safety of the experimental sea dumping operation it was assumed that all the radioactivity contained in the packages would be released from them immediately after they reached the sea bed.

Regarding your second point, we have recently started real dumping tests with the dummy packages to establish their integrity when subjected to an actual pressure increase rate.

B.G. PETTERSSON: You mentioned that environmental impact studies were under way. I am wondering whether the environmental impact of the forthcoming sea disposal operations can be quantified in terms of collective dose commitment.

S. SAKATA: I don't know whether such studies have been undertaken.

Y. NISHIWAKI: On the basis of information received from my colleagues in Japan, I should like to add some comments regarding the studies associated with the Japanese deepsea disposal project.

A project to study the feasibility of solidified low-level radioactive waste disposal onto the sea bed was started by the Japanese Atomic Energy Commission in 1969. During the initial period, extensive studies were made with a variety of monitoring systems, including cine camera, still camera, television, illumination for deepsea use, sonar, pinger, recorder, and other such systems. Monitoring systems for canister safety were developed by the Japan Marine Science and Technology Centre.

In order to see how effectively such monitoring systems would work under natural conditions, *in situ* experiments were considered necessary. In the experiments, the emphasis was placed on observation of the behaviour of solidified packaged waste (with accompanying monitoring devices) during descent, and on and after arrival at the sea bottom.

The weight of the experimental unit consisting of the canister containing solidified dummy waste and monitoring devices was adjusted so that its velocity of descent would be equal to the end velocity of freely-falling solidified packaged waste.

For a more precise study of cracks and pinholes in the canister, direct observation and photography by divers and submersibles may be essential. The first series of experiments was therefore conducted in a relatively shallow area where the water was about 100 m deep. The experimental studies will be continued this year, in a deeper area where the depth of the water is about 1000 m, to obtain basic scientific and technical data. Experimental disposal of full-scale dummy canisters at a depth of 6000 m is planned for 1977.

If the tests demonstrate that deepsea disposal can be conducted safely, experimental dumping of solidified low-level waste will be started during the period 1977 to 1978. After two years of trial operations, the results of environmental surveys and monitoring will be evaluated to assess the environmental impact of the waste disposal and to establish whether or not the deepsea disposal operation should be continued at the proposed experimental sites.

In 1971, the following four sites were selected for possible radioactive waste disposal, the criteria for their selection being (a) minimum marine productivity, (b) minimum possibility of upwelling, (c) presence of weak deepsea currents, (d) flatness of the ocean floor, (e) surface coverage by soft sediment, (f) remoteness from seismic zones, (g) convenience of transportation from Japan, (h) absence of submarine cables or equipment, and (i) avoidance of surface vessel sea routes.

Proposed sites			*Estimated depth*
A	26°N	150°E	6000 m
B	30°N	147°E	6000 m
C	30°N	160°E	5000–6000 m
D	36°N	158°E	3000–4000 m

During the years 1972 to 1975, extensive surveys over a considerable area surrounding each of the above four sites were conducted by Japanese Government organizations – the Maritime Safety, Meteorological and Fishery Agencies in co-operation with the Science and Technology Agency – in order to obtain basic data necessary for the assessment of the possible environmental impacts and the collective dose commitment of the proposed deepsea disposal of radioactive wastes at these locations.

BURIAL OF RADIOACTIVE WASTE
(Session XI)

Chairman:

Y.P. MARTYNOV, Union of Soviet Socialist Republics

EXPERIENCE DE SEPT ANNEES DE STOCKAGE DE DECHETS RADIOACTIFS SOLIDES DE FAIBLE ET MOYENNE ACTIVITE EN SURFACE OU EN TRANCHEES BETONNEES

G. BARDET
Département INFRATOME,
PEC-Engineering,
Paris,
France

Abstract–Résumé

SEVEN YEARS' EXPERIENCE OF STORAGE OF SOLID LOW- AND MEDIUM-LEVEL RADIOACTIVE WASTE ON THE SURFACE AND IN CONCRETE-LINED TRENCHES.

The processes for the storage of solid low- and medium-level radioactive waste that have been selected and approved by the public authorities in France are, depending on the activity levels and nature of the contained products: surface storage in bays prepared and drained, and then covered with clay soil and turf; and storage in concrete-lined trenches, with encasement of the drums in a cement grout. Compressible products are compacted and encased in concrete blocks, after which they are stored in bays. Finally, specific facilities are used to contain sealed sources of relatively high activity; likewise, a special facility basically made up of wells enclosed in leak-tight concrete with a metal lining is intended for canisters containing approximately 60 litres with an activity level of up to several hundred rads at contact. After more than seven years' experience one can confirm the safety of the above storage systems, constant monitoring not having made it possible to observe appreciable leakage of the stored activity. The method is one which can be easily applied by developing countries in territory with geological conditions that can be considered satisfactory. Provision has already been made for the future application of these methods in France for the storage of products of the type considered, which covers almost all the waste produced by nuclear power stations.

EXPERIENCE DE SEPT ANNEES DE STOCKAGE DE DECHETS RADIOACTIFS SOLIDES DE FAIBLE ET MOYENNE ACTIVITE EN SURFACE OU EN TRANCHEES BETONNEES.

Les procédés de stockage de déchets radioactifs solides de faible et moyenne activité retenus et acceptés par les pouvoirs publics en France sont, en fonction des niveaux d'activité et de la nature des produits contenus, des stockages en surface, sur aires préparées et drainées avec recouvrement de terre argileuse puis engazonnées, des stockages en tranchées bétonnées avec blocage des fûts dans un coulis de ciment. Les produits compressibles sont compactés et bloqués dans des blocs de béton, puis stockés sur aire. Enfin des installations spécifiques sont utilisées pour le confinement des sources scellées d'activité relativement élevée; de même une installation spéciale constituée essentiellement par des puits enfermés dans une masse de béton étanche par un cuvelage métallique est destinée à recevoir des poubelles d'une soixantaine de litres dont l'activité peut atteindre plusieurs centaines de rads au contact. Après plus de sept ans d'expérience on peut constater la sûreté de tels stockages, les mesures constantes faites n'ayant pas permis de constater de fuites notables de l'activité entreposée. Il s'agit d'une méthode qui peut aisément être mise en œuvre par des pays en voie de développement disposant de terrains dont la nature géologique peut être considérée comme satisfaisante. Il est d'ores et déjà prévu que seront retenues en France à l'avenir les méthodes décrites pour le stockage des produits du type considéré, qui recouvrent en particulier la quasi-totalité des déchets produits par les centrales nucléaires.

HISTORIQUE

Dans le premier trimestre de l'année 1965, le Commissariat à l'énergie atomique a décidé de confier à une S.à.r.l. dénommée INFRATOME, filiale de la Société Potasse et Engrais Chimiques (PEC), «... une mission d'investigation des procédés de traitement, de conditionnement et de stockage des déchets radioactifs, d'inventaire des stocks existants et des productions annuelles du CE en cette matière et de tous autres organismes publics ou privés (EDF, laboratoires, hôpitaux, etc.)...»

En mai 1966 le Commissariat à l'énergie atomique décidait de donner une suite favorable aux conclusions de l'étude faite, qu'il estimait satisfaisante, et demandait à PEC-Engineering d'étudier et de réaliser la création d'une société industrielle chargée de l'application concrète des conclusions d cette étude.

C'est au début de 1967 que fut donc créée la S.A. INFRATOME, qui était invitée par ailleurs par le CEA à installer son premier parc de stockage (Centre de la Manche) sur des terrains appartenan au CEA et situés à proximité immédiate du Centre de La Hague, cette solution paraissant au CEA devoir simplifier les problèmes psychologiques et sanitaires découlant d'une telle création; elle tenait compte également de la vive opposition des élus locaux à la création d'une deuxième installation nucléaire dans le Département de la Manche, totalement indépendante de celle du Centre de La Hagu

Cette implantation sur un terrain appartenant au CEA aussi bien que l'insistance des élus locaux à voir assurer la responsabilité du respect de l'environnement par le CEA lui-même a amené celui-ci à déposer à son nom la demande d'autorisation de l'installation nucléaire de base envisagée; ce décret pris en 1969 définissait les conditions générales d'exploitation de cette installation.

Entrepris en 1968, les travaux de création du Centre permettaient à celui-ci d'entrer effectivem en fonctionnement le 1er janvier 1969.

Précisions enfin qu'en 1974 INFRATOME a fusionné avec sa Société mère: PEC-Engineering.

ELEMENTS DU CHOIX DU SITE

Les éléments déterminant le choix du site de stockage de La Hague sont les suivants:

- population environnante peu nombreuse et peu agglomérée
- utilisation agricole des sols très restreinte et peu développée
- alimentation des populations en eau potable assurée par des forages lointains et des réseaux isolés
- soutien nucléaire et technique assuré par la proximité du Centre de La Hague, permettant en particulier une surveillance très précise de l'environnement
- proximité de la mer
- contrôle aisé des effluents.

DESCRIPTION GENERALE DE L'INSTALLATION ET DES OPERATIONS

L'installation comprend:

Stockage:

- des aires de stockage en surface
- des tranchées bétonnées
- en cours de réalisation, le projet des puits de stockage de décroissance, avec l'installation de déchargement correspondante qui permettrait de recevoir des conteneurs de 50 à 60 litres de déchets très irradiants

Pré-stockage:

- un hall de stockage provisoire pouvant contenir environ 20 000 fûts de 200 litres
- une cuve de stockage des huiles et solvants

Traitement ou conditionnement:

- un bâtiment abritant une presse à compacter de 400 tonnes et un atelier de mécanique
- un bâtiment de décontamination avec récupération des eaux en réservoir
- une centrale à béton

Installations annexes:

- un bâtiment abritant en zone active essentiellement des laboratoires, la salle de comptages, des vestiaires et installations sanitaires; en zone inactive, les magasins, le chauffage électrique, la chaufferie et le réfectoire
- un bâtiment administratif.

Outre ces installations, PEC-INFRATOME utilise un matériel important pour le transfert des déchets sur le site: chariots élévateurs, grues équipées pour une manutention à distance.

Les déchets reçus sont essentiellement des déchets solides de faible et moyenne activité qui sont emballés dans des fûts métalliques de 200 litres; quelques-uns, issus de stocks antérieurs, se présentent en fûts métalliques de 100 litres, parfois de 225 litres. Certains déchets ont été conditionnés en blocs de béton d'un diamètre soit de 95 cm, soit de 130–140 cm, et d'un poids moyen de 2 tonnes ou de 4 à 5 tonnes.

Les activités spécifiques limites acceptables sont:

100 Ci/m^3 pour le ^{90}Sr
1000 Ci/m^3 pour les autres émetteurs $\beta\gamma$
1 Ci/m^3 pour le ^{239}Pu non enrobé
10 Ci/m^3 pour le ^{238}Pu et autres émetteurs α enrobés (excepté le radium).

Il faut cependant noter que ces limites, notamment en ce qui concerne le plutonium, ne sont pas fixées *ne varietur* mais qu'elles pourront être éventuellement modifiées en fonction des résultats d'expérience d'une part, et des possibilités de stockage dont on pourra disposer dans l'avenir d'autre part.

Le conditionnement et le mode de stockage sont déterminés par les caractéristiques des déchets concernés:

Les déchets conditionnés (enrobés dans le ciment, le bitume, etc.) de façon à avoir un taux de lixiviation très faible (au maximum de 10^{-3}) sont déposés sur des aires aménagées.

Les déchets de faible activité ($<1000\ CMA_t$) contenus en fûts ordinaires sont placés dans des «espaces en surface» ménagés entre les déchets conditionnés dont il vient d'être question. L'ensemble est recouvert d'une couche d'argile assurant une couverture imperméable, de terre et de gazon.

Les déchets en fûts autres que ceux de faible activité et les déchets en vrac sont stockés en cases de béton. Ces cases sont remplies par adjonction d'un coulis de ciment fluide et recouvertes par un enduit de bitume. Les premières cases bétonnées avaient comporté un blocage par du sable entre les fûts, le coulis de ciment ne se faisant qu'au-dessus du sable. Cette méthode, qui aurait été plus adaptée à une reprise éventuelle, comportait l'inconvénient de maintenir les fûts de déchets au contact de l'humidité du sable et a été abandonnée.

Les déchets particuliers (grand volume, très haute activité, déchets en décroissance) sont placés sur des aires ou dans des fosses spécialement conçues pour ces cas.

Les déchets très irradiants, c'est-à-dire dont le débit de dose au contact de l'emballage est supérieur à 10 rad/h, sont placés dans des puits spéciaux en attendant que les débits de dose soient suffisamment bas pour permettre un stockage sur aire ou en tranchée, sans risque d'irradiation pour le personnel.

Il existe donc entre les déchets et l'environnement à l'extérieur du site un certain nombre de barrières plus ou moins élaborées: enrobage, revêtement des fosses, terrain. L'efficacité de ces barrières est vérifiée par le contrôle permanent du site.

MESURES DE SURVEILLANCE

Les mesures de surveillance du site sont permanentes, elles sont concrétisées par l'existence d'une série de piézomètres situés en différents points du terrain qui permettent des prélèvements périodiques dans la nappe phréatique en même temps d'ailleurs que la mesure du niveau de celle-ci, prélèvements qui permettent de vérifier la non-contamination des eaux de la nappe.

L'ensemble des eaux de ruissellement en surface sur le site sont recueillies dans un réseau de drains et de fossés et aboutissent à un bassin de décantation où des mesures régulières permettent également de constater une contamination éventuelle de ces eaux. Un système de conduites et de pompage permettrait, si un début de contamination était constaté, une reprise par la station de traitement des effluents liquides du Centre de La Hague afin d'éviter le rejet de ces eaux dans les ruisseaux avoisinants.

Des appareils de prélèvements d'air à l'intérieur du site aussi bien qu'à l'extérieur permettent de déceler des rejets éventuels dans l'atmosphère provenant aussi bien, d'ailleurs, du Centre de La Hague que du site de stockage.

Cette surveillance est directement assurée par du personnel spécialisé d'INFRATOME, elle est régulièrement supervisée par le Service de protection contre les radiations du Centre de La Hague, qui est en mesure d'apporter selon les besoins son concours technique et ses moyens à cette surveillance.

La surveillance médicale du personnel d'INFRATOME est strictement assurée dans les mêmes conditions que pour le personnel du CEA: visites médicales périodiques, port de films et de stylos dosimètres, vérification de non-contamination après les postes de travail, etc.

Il n'est évidemment pas possible à INFRATOME de vérifier le contenu exact de chacun des fûts ou emballages qui lui sont adressés par un producteur de déchets lui ayant demandé son intervention conformément d'ailleurs à la réglementation sur les transports de produits radioactifs il appartient à chaque producteur, lors d'une demande d'enlèvement de déchets, de définir dans un bordereau les caractéristiques de ce produit[1]; ces caractéristiques, par fût ou par lot, sont à l'arrivée soigneusement inventoriées, fichées, et les lieux de stockage de ces fûts ou de ces lots de fûts reportés sur des plans qui permettent et permettront dans le temps de connaître d'une manière précise l'origine et la nature des produits disposés en sous-sol ou en surface du terrain de stockage.

Ainsi les stockages réalisés, s'ils peuvent être considérés comme définitifs, n'ont néanmoins pas un caractère irréversible car, grâce à l'information permanente, valable dans le temps, que donnent le fichier et les plans, il demeure possible de reprendre tel ou tel produit stocké, si cela apparaît commode, utile ou nécessaire.

Pour faciliter le repérage sur le terrain et la reprise éventuelle un système de marquage par bornes ou plaques métalliques est en cours d'étude et sera mis en place au cours de la présente année

VOLUMES TRAITES

Cette société spécialisée a été créée pour répondre non seulement aux besoins des organismes spécialisés dans l'énergie nucléaire, CEA et EDF, mais aussi à ceux des différents producteurs publics ou privés.

C'est ainsi que si le CEA a assuré en 1975 la fourniture de déchets à notre Société pour environ 89,5% de son activité, EDF est également un client important et régulier et tendra évidemment à le devenir de plus en plus, compte tenu du programme de construction de centrales nucléaires; mais aussi de nombreux producteurs: hôpitaux, laboratoires de recherche, industriels divers (plus de 550 clients divers) sont débarrassés de leurs déchets radioactifs par nos soins.

---

[1] A la suite de certaines anomalies constatées dans certains envois il est apparu néanmoins prudent de procéder à des vérifications par sondage, portant notamment sur les produits en provenance de centres de recherche où la diversité des radionucléides contenus est la plus grande.

Pour donner un ordre de grandeur des produits stockés, mentionnons qu'au 31 décembre 1975 les organismes suivants auront demandé l'enlèvement de déchets à destination du centre de stockage d'INFRATOME:
- CEA: 293 900 fûts de 200 litres ou équivalents
- EDF: 25 230 fûts de 200 litres ou équivalents
- clients divers, publics ou privés: 11 220 fûts.

Il convient de noter en ce qui concerne cette dernière activité qu'elle est en progression constante puisque elle a porté:

en 1969	sur	325 colis (sources ou fûts)
en 1970	sur	756
en 1971	sur	823
en 1972	sur	899
en 1973	sur	2212
en 1974	sur	3012
en 1975	sur	3194.

Ces 330 350 fûts (au 31 décembre 1975) représentent un volume d'environ 65 000 m^3 pour un poids approximatif de 87 000 tonnes.

L'activité totale ainsi stockée, estimée d'après les déclarations des expéditeurs, est de l'ordre de 2000 Ci en α et de 189 000 Ci en $\beta\gamma$.

Si l'on examine ces quantités au regard des modes de stockage qui leur ont été appliqués on constate qu'environ 272 000 fûts ont été déposés directement sur aire, 30 000 ont été préalablement compactés à la presse et déposés en blocs de béton, 25 000 environ ont été mis directement en tranchées bétonnées.

Si l'on compare ces quantités aux possibilités de stockage estimées du Centre de La Manche on aboutit aux chiffres suivants:

Le terrain dont dispose INFRATOME a une surface d'un peu plus de 12 ha, c'est-à-dire que compte tenu de l'implantation des bâtiments, des routes, de la presse, environ 10 ha sont disponibles pour le stockage proprement dit.

Compte tenu des pratiques de stockage actuellement en vigueur on peut estimer que la capacité de ce terrain est de l'ordre de 1 200 000 à 1 300 000 fûts.

Il est sans doute intéressant, pour les producteurs de déchets radioactifs, de connaître le coût de revient des stockages réalisés dans les conditions décrites ci-dessus: il est d'environ 115 F hors taxes par fût de 200 litres en moyenne, montant dans lequel les frais de transport représentent 35% en moyenne:

frais fixes	F 61,00
transport	40,00
stockage	14,00
environ	F 115,00 en moyenne pour 1975

TRANSPORT

Il convient de préciser que notre société assure également l'enlèvement et le transport des produits actifs qui lui sont confiés au départ des centres de production, le chargement étant assuré par le producteur sur son propre site.

Ces transports sont en général assurés par route, la plupart des centres de production n'étant pas reliés à la voie ferrée, non plus d'ailleurs que notre centre de stockage. Ceci représente environ 1000 voyages à l'année pour une distance parcourue d'environ 400 000 km.

La réglementation des transports proposée par l'AIEA et adoptée en France est naturellement strictement appliquée, c'est-à-dire que certains produits ne sont transportés qu'en conteneurs ou coques métalliques du type B ayant reçu l'agrément du Ministère des transports.

Mais la sensibilité de l'opinion publique à tout ce qui se rapporte au nucléaire en général, et aux déchets en particulier, nous a conduit à prendre des dispositions plus sûres pour tous les transports de fûts contenant des déchets de très faible activité que la réglementation autorise à déplacer en emballage industriel classique. C'est ainsi que tous les fûts non justiciables d'un emballage du type B sont transportés dans des conteneurs métalliques du type maritime, donc très robustes, conteneurs qui sont fixés sur des semi-remorques.

En cas d'accident grave de la circulation il nous semble que cet emballage, sans donner les garanties totales d'un conteneur ayant les caractéristiques du type B, éviterait la dispersion des fûts e donc des produits qu'ils contiennent, et en conséquence la contamination des sols.

Par ailleurs ces conteneurs sont munis d'une prise d'air qui autorise, avant ouverture des portes, des prélèvements permettant de déterminer qu'aucune contamination ne s'est produite au cours du transport; dans le cas contraire le personnel serait muni d'un équipement adéquat.

Enfin certains transports de produits contenant des émetteurs α ou dont le conditionnement ne donne pas toutes garanties d'homogénéité font l'objet d'un convoyage par un véhicule dont le conducteur possède une qualification élevée en matière de radioprotection et dispose d'appareils de mesure et de contrôle, et du matériel nécessaire pour le balisage éventuel d'une zone contaminée.

Il va sans dire que le respect de la réglementation générale applicable à la circulation des véhicules lourds, tant en ce qui concerne l'état mécanique des véhicules utilisés que le respect de la vitesse maximale autorisée (60 km/h) et la durée de conduite des chauffeurs, est strictement contrôlé

Cet ensemble de précautions ont permis le transport en sept ans d'environ 90 000 tonnes de produits actifs sans incident ou accident.

CONCLUSIONS

Cette expérience industrielle de sept années de transport et de stockage de déchets radioactifs solides de faible et moyenne activité suscite quelques réflexions d'ordre général que l'on ne saurait négliger alors que le développement considérable des installations nucléaires dans le monde, et pas seulement dans des pays ayant atteint un niveau de développement industriel élevé, entraînera inéluctablement, tout au long de la chaîne des combustibles, une augmentation considérable de la production de ces déchets.

Cette expérience démontre d'abord que les produits de cette espèce peuvent être centralisés et confinés suivant des procédés relativement simples, sûrs et d'un coût supportable.

– *Centralisés* car la multiplication des implantations nucléaires, des utilisateurs publics ou privés de produits actifs, de sources et de radioéléments se traduit par une dispersion des déchets, avec la difficulté de contrôle qu'elle entraîne qui constituerait un danger certain pour l'ensemble de la population.

A cet égard les multiples pollutions par des déchets chimiques classiques sont une illustration quotidienne du risque que suppose la dispersion de produits dangereux pour l'homme et son environnement.

Il convient donc de souligner l'intérêt, aussi bien sur le plan international que national, de la création de parcs de stockage à vocation générale et obligatoirement utilisés par tous les producteurs de déchets actifs.

– *Confinés* afin que le passage dans l'environnement de l'activité stockée ne puisse être brutale et susceptible de dépasser les concentrations maximales admissibles fixées par la CIPR. Pour ce faire nous sommes tout naturellement amenés à considérer l'importance des barrières artificielles (conditionnement des déchets) et naturelles (géologie des sols sur lesquels sont déposés les déchets) qui s'opposent ou ralentissent considérablement le passage de l'activité des produits stockés dans les vecteurs, air, eau de surface, nappe phréatique, qui aboutissent à la chaîne alimentaire.

Notre propos n'est pas, dans cette communication, de reprendre en détail l'examen de ces différentes barrières, de meilleurs spécialistes ne manqueront pas au cours de ce Colloque de faire part de leurs études et de leurs conclusions sur ce sujet; il est de toute façon évident que le choix du terrain de stockage présente un intérêt primordial.

Mais il nous paraît indispensable d'insister sur la nécessaire liaison qui doit exister entre le producteur de déchets radioactifs et celui qui assume la charge du stockage, pour ce qui concerne le choix du procédé de conditionnement au lieu de production de ces déchets. En effet il nous apparaît, et c'est une constante résultant de notre expérience, que le conditionnement réalisé au départ détermine dans une large mesure les modalités de manutention et de stockage, puisque c'est en fonction de ce conditionnement (ou en l'absence de tout conditionnement préalable) que devront être prises les dispositions relatives à la radioprotection du personnel de manutention du centre de stockage et au confinement dans le temps des déchets.

Il est certain que l'absence de solution de stockage définitif des déchets radioactifs a longtemps conduit les producteurs à imaginer et à réaliser des conditionnements, tous de bonne valeur, mais qui ne présentent pas tous les mêmes avantages pour la disposition finale du produit.

Dans le domaine des transports on doit constater que les volumes et les poids sont, pour les déchets de faible et moyenne activité, très importants; un maximum de précautions permet de réaliser ces transports par la route en assumant des risques raisonnables; il nous paraît cependant hautement souhaitable qu'un centre de stockage destiné à être utilisé pendant plusieurs décennies soit relié à la voie ferrée, la probabilité d'un accident de transport par cette voie étant nettement plus faible et ses conséquences pratiques et psychologiques beaucoup plus limitées.

Si nous résumons les résultats de notre expérience nous pensons pouvoir affirmer que nous sommes en mesure d'assurer le stockage dans des conditions sûres et pour des coûts acceptables de tous les déchets solides de faible et moyenne activité, à l'exception de ceux comportant plus d'une certaine quantité d'émetteurs α. Pour ce qui est de ces derniers produits, une solution peut être trouvée et bien sûr, compte tenu de la durée de vie de ces produits, davantage à l'échelle des barrières géologiques que du conditionnement, dont la tenue dans le temps peut difficilement être garantie au-delà de quelques centaines d'années.

RADIONUCLIDE DYNAMICS AND HEALTH IMPLICATIONS FOR THE NEW YORK NUCLEAR SERVICE CENTER'S RADIOACTIVE WASTE BURIAL SITE*

J.M. MATUSZEK
Division of Laboratories and Research,
Radiological Sciences Laboratory,
New York State Department of Health,
Albany, New York

F.V. STRNISA
Division of Industrial Sciences and Technologies,
New York State Department of Commerce,
Albany, New York

C.F. BAXTER
New York State Energy Research and
Development Authority,
New York, New York,
United States of America

Abstract

RADIONUCLIDE DYNAMICS AND HEALTH IMPLICATIONS FOR THE NEW YORK NUCLEAR SERVICE CENTER'S RADIOACTIVE WASTE BURIAL SITE.

A commercial radioactive waste burial site has operated since 1963 at the Western New York Nuclear Service Center. Solid low-level radioactive wastes are buried in trenches excavated from a very fine-grained heterogeneous mixture of silt and clay (silty till) and are then covered with the excavated material. Despite many operational precautions, water levels in three burial trenches rose to within a few centimetres of the covering material by late 1973. Activity levels of HTO, ^{90}Sr, and ^{137}Cs in trench water and core samples were measured to obtain preliminary information on the degree of subsurface radionuclide migration from the burial trenches into the surrounding soil. Tritium concentrations measured in void-space water from vertical cores appeared to peak in the cover material 1.5 to 2 m below the ground surface. Concentrations of ^{90}Sr and ^{137}Cs in the silty till were greatest near the surface of the cover material. Concentrations of HTO and ^{90}Sr, measured in a series of slant-hole core samples collected until the trench was intercepted, showed tritium migration to have progressed less than 0.3 m, while ^{90}Sr migration appeared to be somewhat less. The preliminary data suggest that: (a) radionuclide migration from the burial trenches into the undisturbed silty till is slight; (b) radioactivity in the surface soil is not necessarily caused by migration of trench water; (c) groundwater movement is not massive; (d) rainwater infiltration, with settlement and compaction of buried wastes, is the most likely cause of rising trench water levels; and (e) surface contamination may occur from spills during burial operations, from trench digging, and from deposition of stack effluents from a nearby nuclear fuel reprocessing plant. By January 1975 the steadily rising water levels in three trenches were approximately 1 m above the undisturbed soil from which the trenches were excavated, resulting in increased radioactivity levels in local streams draining the site. To lower the trench water levels to below that of the undisturbed soil, water was removed from the trenches, treated to reduce activity levels, and released to local streams. The estimated maximum individual total body dose for the release was 3×10^{-4} mrem for an individual served by the nearest public water supply. The total-body population dose,

* Partially funded by the U.S. Environmental Protection Agency.

integrated for an estimated population of 2×10^6, was 3×10^{-1} person-rem. Statistically, according to the BEIR Committee risk estimates, less than 10^{-4} deaths would be expected to occur from this discharge. Dose estimates indicate that continuous discharge of untreated trench water from the low-level radioactive waste burial site would not produce a statistically significant health effect.

INTRODUCTION

A near-surface burial site at the Western New York Nuclear Service Center (WNYNSC) provides permanent storage of solid, low-level radioactive wastes [1]. Burial trenches are excavated from a fine-grained mixture of silt and clay (silty till) by a bulldozer to form an open 30-m trench segment. Eventually six joined segments form a single trench approximately 180-m long and 6-m deep. The silty till excavated from each trench is piled on the nearest previously filled trench to compact the waste and cover.

The excavated trench bottom is sloped approximately 0.6 m over the 180-m length to permit collection and pumping of rainwater that infiltrates while the trench is open. A stone-filled sump at the end of the last 30-m segment collects seepage in the closed trench. An open standpipe is inserted into the sump for long-term monitoring of the water level in the closed trench.

As each 30-m segment is filled with waste material, the wastes are covered with approximately 1 m of previously excavated silty till. Seepage from precipitation and from the waste containers is pumped from the next open segment into a holding lagoon for treatment and eventual discharge into surface streams draining the area. The excavated silty till is redistributed over the burial area to leave a compacted, seeded cover over each trench approximately 2.5-m thick, graded to facilitate drainage of precipitation from the trench cover into the surrounding streams.

Despite these precautions, water levels began to rise in three of five covered trenches on the north end of the burial site. By late 1973 water levels in these trenches had risen to within a few centimeters of the cover material. A number of small deposits of granular sand and pebbles had been observed during excavation of all thirteen trenches on the site.

Concentrations of tritium and other radionuclides in environmental water samples collected during 1973 from small streams which intercept surface drainage from the burial site were slightly above ambient levels in surface waters statewide [2]. Lack of prior data for these small streams precluded judgment concerning the source of stream contamination. Because the affected stream beds were at or below the deepest portions of the trenches, subsurface radionuclide migration was considered among the possible contributing factors.

These observations led to concern that the burial operations might not be providing the anticipated "perpetual" storage of radioactive wastes. A study to characterize radionuclide dynamics at the burial site was initiated by the New York State Energy Research and Development Authority (NYSERDA), owner of the site, and the U.S. Environmental Protection Agency (EPA). The study was performed in November 1973 and April 1974 by Nuclear Fuel Services, Inc. (NFS), operator of the site, and the New York State Health Department's Radiological Sciences Laboratory (RSL). The radionuclide data reported here were obtained at the RSL, except for a few NFS tritium measurements in one drill hole [1]. Lithologic data were supplied by NFS [1]. Results and conclusions derived from the study are presented as Part I of this paper.

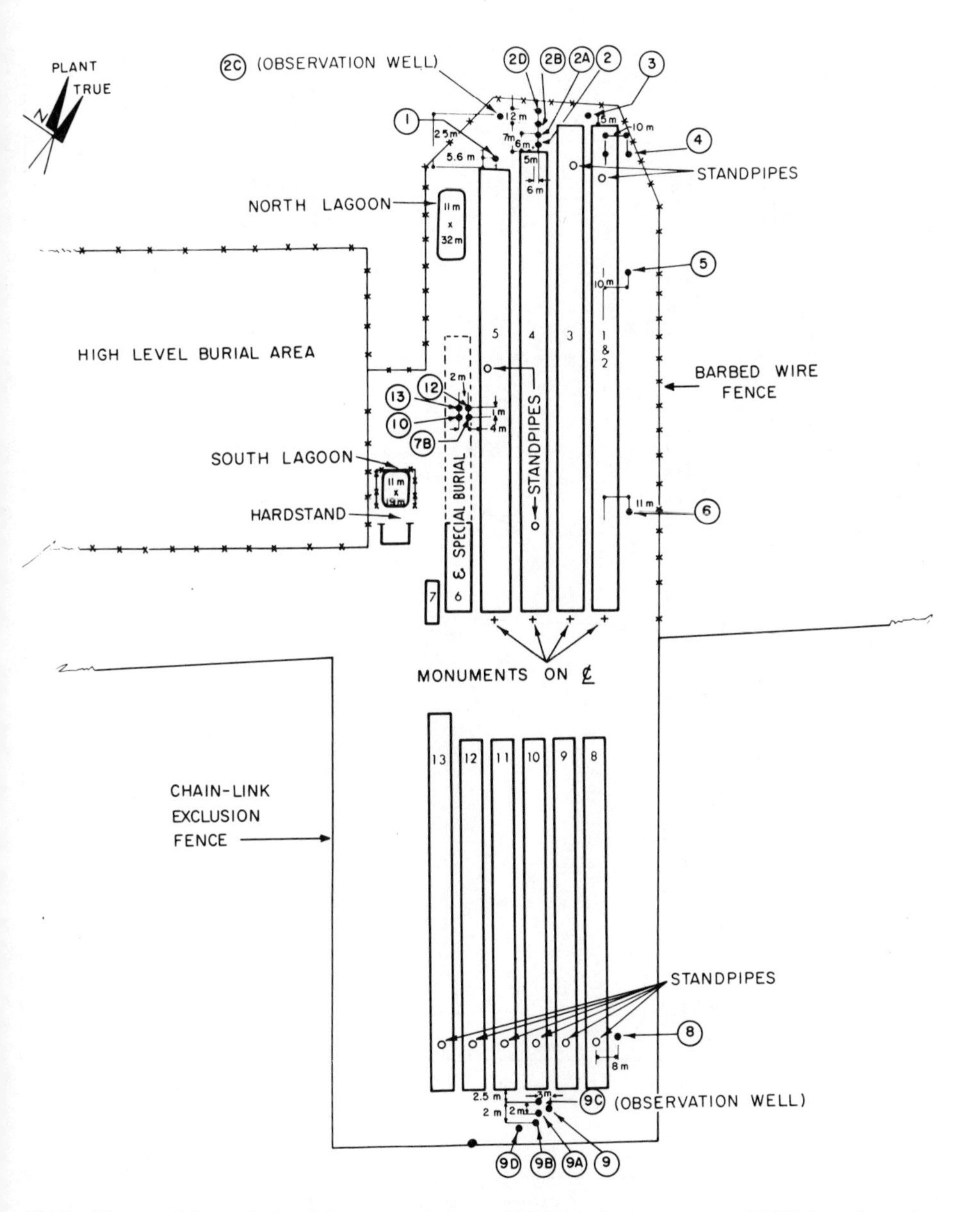

FIG.1. Diagram of the low-level, solid-waste burial area at WNYNSC, indicating locations of drill holes and trenches.

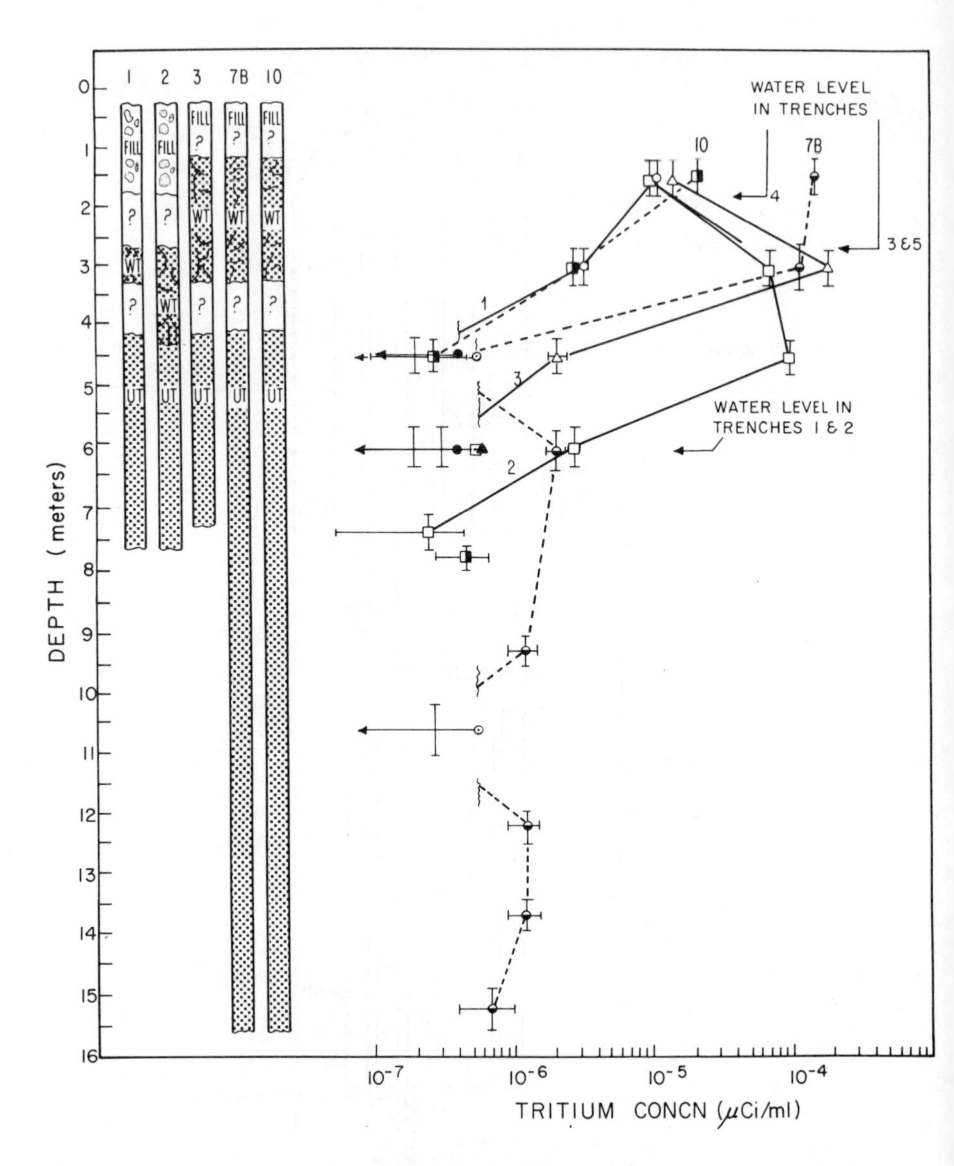

FIG.2. Lithology and concentration of HTO for core samples from Holes 1, 2, 3, 7B, and 10. Completely filled symbols indicate activity levels less than the detectable limit (at 2σ). Water levels in the nearest trenches are also indicated for November 1973. Lithologic abbreviations are described in the text.

By January 1975 the water levels in the three trenches had risen to approximately 1 m above the undisturbed silty till, and increased radioactivity was found in surface water draining from the north end of the burial site. Water was therefore removed from the trenches into a holding lagoon, treated, diluted with reprocessing plant effluents, and then discharged into local streams over a 17-day period. The health effects from that discharge are described in Part II of this paper to provide insight for future burial operations at the site and for burial operations in other similarly wet environments.

PART I. RADIONUCLIDE DYNAMICS IN BURIAL SITE SOIL

The preliminary study conducted to evaluate radionuclide dynamics in the burial site soil provided data concerning the several mechanisms by which contamination can occur at surface burial sites. A review of burial site operations with NFS staff and consideration of environmental data previously obtained by the RSL suggested that the likely sources of surface contamination are: (a) seepage of liquids during burial of waste drums containing dewatered resins, sludges, liquids on vermiculite and animal carcasses; (b) leaks from hoses and pumps during transfer of "rainwater" from operating trenches to holding lagoons; (c) transport of contaminated water and clay to the surface while bulldozing out sections of each new trench; and (d) deposition of effluents discharged from the stack of the fuel reprocessing plant.

SAMPLE COLLECTION

Figure 1 indicates the general burial site layout and the drilling plan used for this study. Trench water samples were collected by lowering glass bottles into the observation standpipes in Trenches 1, 3, 4, 5, 8, 10, 11, and 12.

Soil cores were obtained by drilling with a hollow-core auger, then lowering a hollow-tube sampler through the auger and driving the tube into the soil immediately ahead of the auger. Eighteen vertical holes were drilled around the site, 2 to 20 m from the sides of the trenches and from 1- to 15-m deep. Sampling tubes were approximately 0.5-m long and were either longitudinally separable (split spoon) to permit geological characterization or closed (Shelby tube) to minimize contamination. Samples of both types were submitted for radiochemical analysis.

Two holes were drilled at 25° from the vertical. Hole 12 intercepted the Trench 5 wall near the bottom of the trench; Hole 13 passed within 2 m under Trench 5 and continued to a depth of approximately 15 m. Shelby tube samples from each slant-hole were submitted for analysis.

RADIOCHEMICAL ANALYSES

Void-space water (approximately 15% by weight) was distilled under vacuum from a 100-g portion of each core sample, and 10 ml of the distillate was collected for HTO measurement as a gel suspension in a liquid-scintillation spectrometer. A second portion of the core sample was analyzed as received, using a large-volume Ge(Li) crystal for γ-spectrometry. A third portion was decomposed by $NaOH-Na_2CO_3$ fusion and analyzed for various α- and β-emitting radionuclides, especially ^{90}Sr.

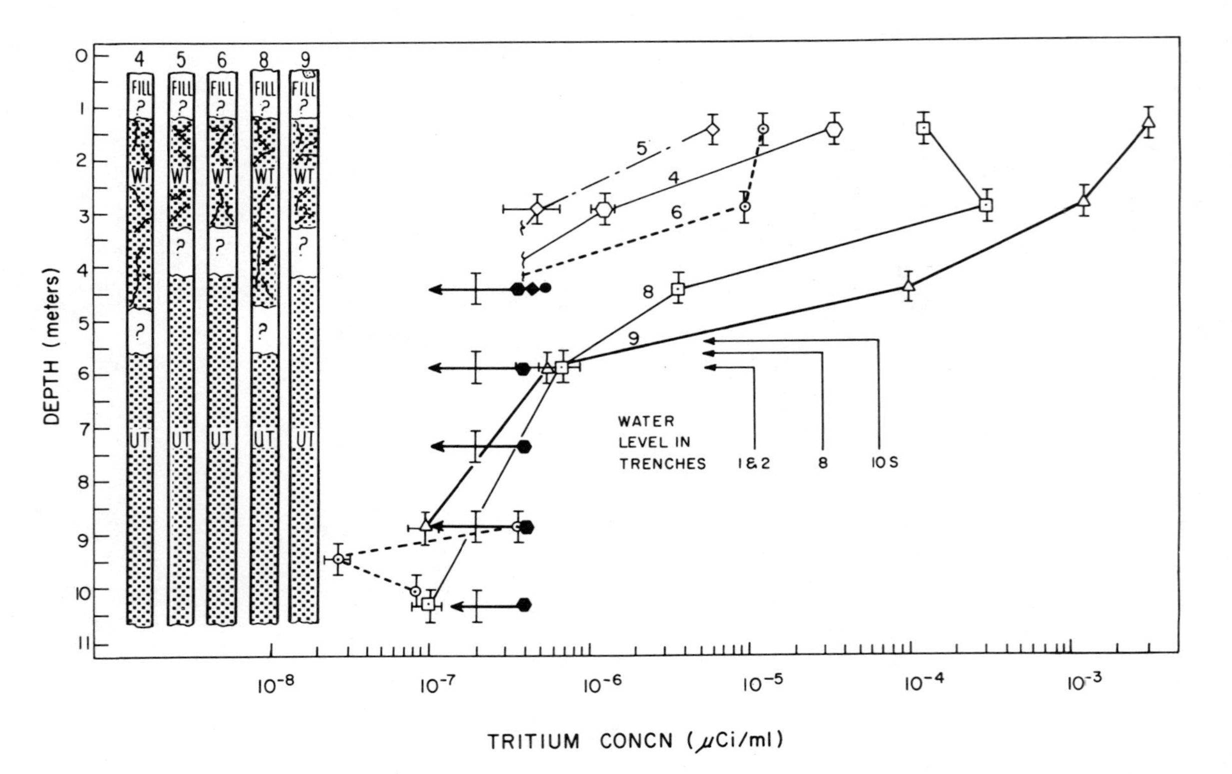

FIG.3. Lithology and concentration of HTO for core samples from Holes 4, 5, 6, 8 and 9 (abbreviations and filled symbols as in Fig.2). Water levels in the nearest trenches are also indicated for November 1973.

Deep core samples, which generally had low HTO concentrations, were analyzed by distilling approximately 300 ml of water from 2 kg of silty till and electrolytically enriching the distillate to a 70-fold greater HTO concentration. The HTO activity in the enriched distillate was measured as a gel suspension on a liquid-scintillation spectrometer or by conversion to hydrogen gas for gas-proportional β-spectrometry.

RESULTS AND INTERPRETATION

Trench water samples

Radionuclide measurements of trench water samples proved inconclusive. Dilution of trench water in the observation standpipes by rainwater appears likely, precluding any direct comparison between radionuclide concentrations in trench water and those in the core samples. The trench water samples did serve to identify HTO, ^{90}Sr, and ^{137}Cs as the predominant long-lived radionuclides for use as tracers.

Core samples

The concentration of HTO in void-space water distilled from the soil cores was the primary indicator of trench water migration. Figures 2-5 summarize the lithology and HTO concentrations in 16 of the vertical holes.

The first layer of soil encountered was a disturbed, partially compacted layer of silty till (FILL in Figs. 2-5) that had been used to cover the burial area to a depth of 1 to 3 m. The second layer was an undisturbed, weathered silty till (WT) generally extending to depths approximately 3 to 5 m from the surface. The third layer was an undisturbed, unweathered silty till (UT) extending to the bottom of all cores taken in this study.

Tritium concentrations in the vertical holes reached a maximum in the cover material near the FILL/WT interface (Figs. 2-5). These results were at first thought to indicate that migration was proceeding through the weathered till. This supposition is particularly plausible for Holes 1, 2, and 3, (Fig. 2) where trench water levels in November 1973 were at, or above, the depth of the maximum HTO concentration. However, the water level in the trench nearest to holes 7B and 10 was nearly 2 m below the depth of the maximum HTO concentration, making direct migration to these holes unlikely. Similarly, the maximum HTO concentrations obtained for Holes 4, 5, 6, 8, and 9 (Fig. 3) were 3 to 5 m above the water levels in neighboring trenches.

A similar inconsistency appears between HTO levels for Holes 2B, 2C, and 2D (Fig. 4) and Holes 9B, 9C, and 9D (Fig. 5), all drilled in April 1974. The HTO concentrations in cores near the surface of these six holes do not appear to vary markedly with distance from the trench.

Tritium concentrations after HTO enrichment of deep samples from Holes 7B, 8, 9, and 10 were less than 1% of the surface activity levels. Unfortunately, with the hollow-core auger and sampling tubes used in this study, each sample could contain as much as 10% by weight of silt or water from the surface layers, and the HTO levels found at depths beyond 5 m are thought to be from contaminating surface material. The levels were low enough, however, to indicate that the void-space water is at least 15 years old and that migration of trench water to these depths is negligibly small.

Slant-holes 12 and 13 were drilled to the west of Trench 5 (Figs. 1 and 6). Tritium results obtained by the RSL and NFS for samples from Hole 12 are

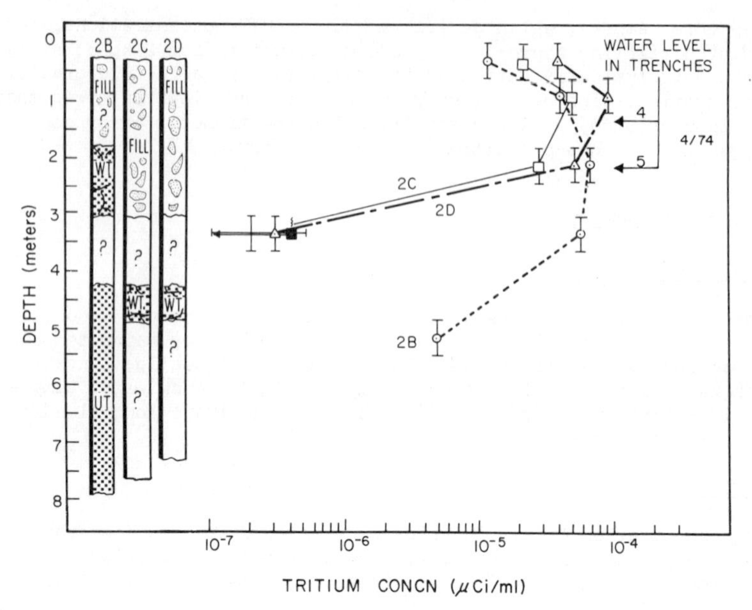

FIG.4. Lithology and concentration of HTO for core samples from Holes 2B, 2C, and 2D (abbreviations and filled symbols as in Fig.2). Water levels in the nearest trenches are indicated for April 1974.

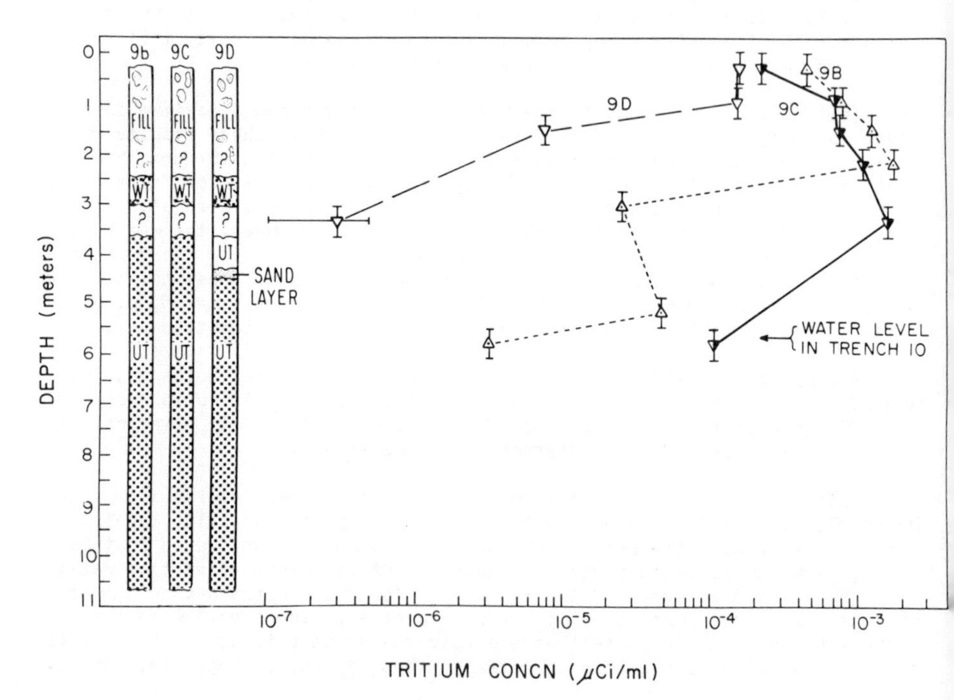

FIG.5. Lithology and concentration of HTO for core samples from Holes 9B, 9C, and 9D (abbreviations as in Fig.2). The water level in Trench 10 is indicated for April 1974.

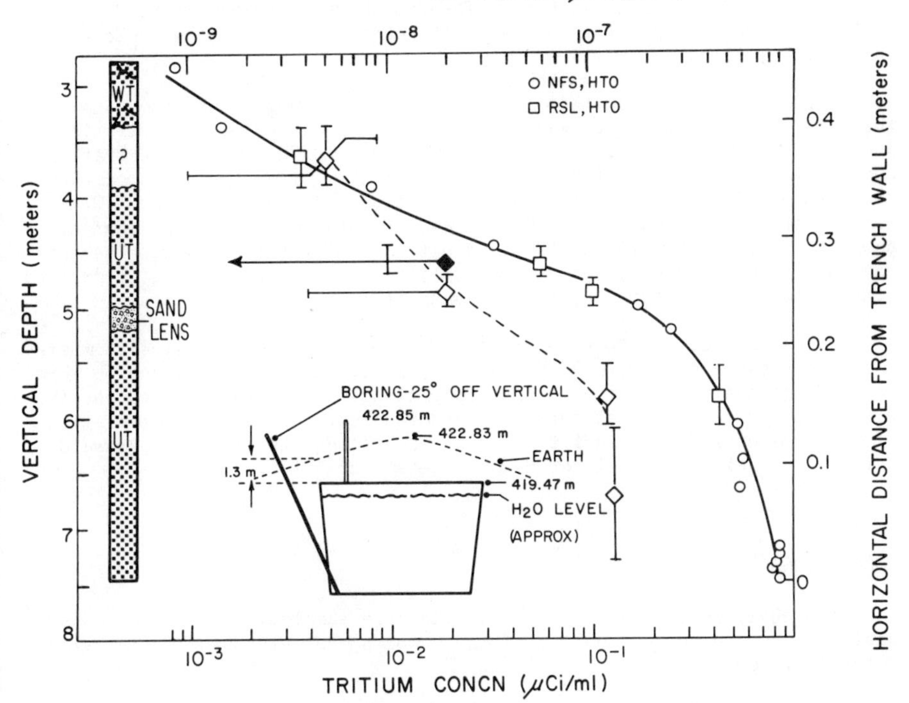

FIG.6. Lithology and concentrations of HTO and ^{90}Sr *for core samples from Hole 12 (abbreviations and filled symbol as in Fig.2). HTO concentrations in units of* μCi/g *soil are approximately 15% of the values in* μCi/ml. *The inset provides a diagram of the slant-hole drilling arrangement.*

combined in Fig. 6 to better define the distance which HTO had migrated from Trench 5 -- approximately 30 cm. Samples from Hole 13 provided little information.

Core samples containing elevated HTO activity levels were analyzed for ^{90}Sr and ^{137}Cs in order to determine the extent of radionuclide migration. The results indicate that in Hole 12 (Fig. 6) ^{90}Sr migration had proceeded to approximately 20 cm. The concentration ratios HTO:^{90}Sr for samples from Hole 12 are 10-fold greater than in the water sample from the standpipe in Trench 5, further indicating that ^{90}Sr migration is being retarded by the silty till.

The ^{90}Sr concentrations in samples from Hole 7B were also very low, and the HTO:^{90}Sr ratios were only 1% of those for water from Trench 5. Since the data from Hole 12 (1 m north of Hole 7B) indicate that ^{90}Sr is not migrating more rapidly than HTO, radionuclide levels in Hole 7B must be presumed to result from a source other than water in Trench 5.

The HTO:^{90}Sr ratios for samples from Hole 9 were similar to those for water from Trench 10, but the water level in Trench 10 is more than 1 m below the deepest sample from Hole 9 and more than 4 m below the sample taken nearest the surface. The mechanism for trench-water transport to Hole 9 must therefore be other than normal water migration through the void-space in the silty till.

A further indication that soil contamination must occur from other than trench water migration was provided by ^{90}Sr and ^{137}Cs concentrations in samples from Holes 2B, 2C, 2D, 9B, 9C, and 9D. The ^{90}Sr concentrations generally were greatest at the surface or within 1 m of the surface. Concentrations of ^{137}Cs were greatest at the surface. The HTO:^{90}Sr and HTO:^{137}Cs ratios for samples from Holes 2B, 2C, and 2D were lowest for the surface samples and increased 10- to 100-fold within 1 m of the surface. The same ratios for Holes 9B, 9C, and 9D varied little with depth, generally proving similar to ratios obtained for water samples from nearby trenches.

Holes 2B, 2C, and 2D were drilled in an area of the burial site where operations had been completed by February 1969. Holes 9B, 9C, and 9D were drilled in an area of active burial operations. The data observed for these six holes indicate initial surface contamination, followed by percolation of the radionuclides by infiltrating precipitation, the rate of downward percolation being greatest for HTO and least for ^{137}Cs.

CONCLUSIONS

The analytical data obtained for the core samples, tempered by considerations concerning the quality of the samples and methods of burial site operation, suggest:

- Migration of radionuclides in trench water into undisturbed silty till is slight,

- Radioactivity in the surface soil is not caused by movement of trench water through void space in the silty till,

- Massive subsurface groundwater movement is not likely,

- Rainwater infiltration through the trench cover, settlement of the covering material, and compaction of buried wastes are the most likely causes of rising trench water levels,

- Surface contamination may occur from spills from waste drums, trench-digging operations or deposition from NFS stack discharges.

PART II. DISCHARGE OF TRENCH WATER

Though the study of radionuclide dynamics indicated that the undisturbed silty till was maintaining reasonable integrity against radionuclide migration, trench water levels continued to rise. By January 1975 the levels in three trenches were approximately 1 m above the undisturbed silty till, and at that time radioactivity levels began to rise markedly in samples from surface streams draining the north end of the burial site. Snow and ice covering the site precluded visual observation of seepage from the trenches until March 1975, when steps were taken by NFS to divert the seepage into a holding lagoon.

To reduce the trench water level below that of the undisturbed silty till, 8.3 x 10^5 liters were removed from the trenches into a holding lagoon for pH adjustment and ferric hydroxide flocculation. The water was then further treated at the NFS low-level waste treatment facility (LLWTF) prior to discharge into local streams. An overall decontamination factor of approximately 2 x 10^2 was obtained for ^{90}Sr. Effluents from decontamination of the fuel-reprocessing facility (which is currently not reprocessing fuel, pending a license application to expand capacity) were also processed through the LLWTF and provided approximately a ten-fold increase in the total volume of composited effluent. Over a 17-day period (June 14-July 1, 1975) 8.7 x 10^6 liters of mixed, treated effluents were discharged into local streams. Radionuclide concentrations in the plant effluent apparently increased the concentrations of all radionuclides in the composited effluent except ^{90}Sr and tritium. Burial-site effluents contributed more than 90% of the total HTO released.

HEALTH EFFECTS FROM 1975 DISCHARGE

First-year individual and population total-body dose values and 50-year individual and population whole-body dose commitments were calculated for the population served by public water supplies on Lakes Erie and Ontario, using dose factors for the average 4-year-old child and for an adult [3]. No public water supplies draw from any of the local streams or from Cattaraugus Creek until it flows into Lake Erie some 61 km downstream from the point of discharge. Dose levels experienced by children were within 10% of those for adults; therefore no distinction is drawn in the following discussion. Similarly, the 50-year dose commitments were within 10% of the first-year dose commitments.

Other dose pathways to large population groups were found to be small in comparison to the values obtained for public water supplies and will not be discussed further here.

Ingestion dose values were also calculated for the hypothetical "maximum individual" (defined as a sportsman who consumes fish from the streams carrying the liquid effluents).

Doses from public water supplies

The individual total-body dose contributed by the discharged trench effluent varied from a maximum 3 x 10^{-4} mrem for an individual served by the nearest public water supply on Lake Erie to 1 x 10^{-4} mrem for an individual served by a public water supply on Lake Ontario.

The population total-body dose, integrated for an estimated population of 2 x 10^6 served by the major public water supplies on Lakes Erie and Ontario, was 3 x 10^{-1} person-rem. According to the BEIR Committee risk estimate [4] for the mortality rate from cancer (50 to 165 deaths/10^6 person-rem), less than 10^{-4} deaths might be expected to occur from this cause in the affected population.

Ingestion dose to the "maximum individual"

The ingestion dose experienced by the hypothetical "maximum individual" was estimated to be less than 4 x 10^{-3} mrem. It was assumed to occur, in this case, entirely from the consumption of fish [5].

Based on a limited creel survey by EPA and the New York State Department of Environmental Conservation, approximately 250 persons fish Cattaraugus Creek regularly [6]. The population total-body dose due to fish ingestion by such a limited population is less than 1×10^{-3} person-rem. The resulting mortality from cancer would be less than 2×10^{-7} deaths. The 17-day discharge of treated trench water thus did not represent a statistically significant health effect.

HEALTH EFFECTS OF DISCHARGING UNTREATED TRENCH WATER

It is reasonable to assume that an amount of water similar to that pumped from the trenches in March 1975 might have to be removed from the trenches each year to maintain the water level below the trench cover. Eventually, however, compaction and settlement of wastes and cover material would be expected to cease, possibly within a few decades after completion of a trench.

A possible solution for regular site maintenance would be to discharge trench water without treatment directly into the local streams. The health implications of that approach were estimated for the population on public water supplies using the maximum isotopic concentrations in samples taken during the trench pumping as a source term. As before, dose levels experienced by children and adults were found to be similar. Discharge of 10^6 liter of untreated water annually from the burial site would result in a first-year individual total-body dose of less than 2×10^{-2} mrem, a 50-year individual total-body dose commitment of less than 4×10^{-2} mrem, a first-year population total-body dose of less than 17 person-rem, and a 50-year population total-body dose commitment of less than 37 person-rem. The major contributor to the increased 50-year dose commitments is the higher level of ^{90}Sr in the untreated discharge. The same population would receive approximately 2×10^5 person-rem annually from natural background radiation. From the BEIR Committee risk estimates, the health impact for continuous discharge of untreated trench water at this level would be less than 6×10^{-3} deaths.

The health impact on the hypothetical maximum individual would also remain small: an annual individual total-body dose less than 4×10^{-2} mrem and an annual population total-body dose less than 10^{-2} person-rem. The resulting mortality among local fishermen would be less than 2×10^{-6}.

On the basis of this analysis, the discharge of untreated trench water from the low-level radioactive waste burial site at the WNYNSC would not present a statistically significant radiological health effect.

ACKNOWLEDGMENTS

The authors are deeply indebted to the NFS staff, particularly Messrs. W. A. Oldham and M. J. Jump, for their cooperation in collecting samples, providing HTO data, and especially providing detailed information concerning burial operations. Mr. K. Anderson served as site supervisor for NYSERDA at the time of the study and provided much valuable information concerning drilling operations.

REFERENCES

[1] DUCKWORTH, J.P., JUMP, M.J., KNIGHT, B.E., Low level radioactive waste management research project - final report, Nuclear Fuel Services, Inc., West Valley, New York (September 15, 1974).

[2] Annual report of environmental radiation in New York State, 1973, New York State Department of Environmental Conservation, Albany, New York (1974).

[3] The potential radiological implications of nuclear facilities in the upper Mississippi River basin in the year 2000, U.S. Atomic Energy Commission, Washington, D.C. (January 1973).

[4] Report of the Advisory Committee on the Biological Effects of Ionizing Radiations, "The effects on populations of exposure to low levels of ionizing radiation," National Academy of Sciences and National Research Council, Washington, D.C. (November 1972).

[5] MARTIN, J.A., Radiation Data and Reports 14, 59 (February 1973).

[6] MAGNO, P.J., KRAMKOWSKI, R., REAVEY, T., WOZNIAK, R., Studies of ingestion dose pathways from the nuclear fuel services fuel reprocessing plant, Report No. EPA-520/3-74-001, U.S. Environmental Protection Agency, Washington D.C. (December 1974).

DISCUSSION

F. GERA: I would have thought that a burial gound where there was extensive contact between groundwater and waste should be considered a failure, since the objective of any type of geological disposal, whether in deep-lying formations or close to the surface, should be to prevent the mobilization of radionuclides through leaching by groundwater.

The question I wish to ask is, do you have an estimate of the time required before the burial area can be made available for unrestricted use? And, what are your plans for the long-term management of the site?

J.M. MATUSZEK: I agree with your comment in part, but the problem is one we can do little about, since most of the north-eastern region of the United States of America is wet. We do not think that, as a low-level waste disposal site, the area presents any radiological hazard.

With regard to the lifting of restrictions on use, the State permit stipulates that the land is to be maintained in perpetuity, which, under State law, is equivalent to a thousand years.

G.L. MEYER: I would like to make a personal comment, namely, that I am not so sure that the available evidence clearly supports your hypothesis regarding the occurrence and movement of radionuclides at West Valley. Rather, I feel there are not yet sufficient data available to confirm your conclusions. The hydrogeological system of the site and the movement of radionuclides through it is not yet known. It is for that reason that the United States Geological Survey is undertaking a five-year hydrogeological study of the site, and the Office of Radiation Programs of the United States Environmental Protection Agency, in co-operation with the State of New York, is sponsoring a radiological, hydrogeological and environmental pathways study at West Valley in order to define the pathways of contamination from the trenches to the environment and man, to develop a model for the movement of radioactivity at the site, and to assess the significance of the ion exchange in the glacial till.

J.B. LEWIS: Mr. Matuszek, what is the authorized limit for disposal at this facility?

J.M. MATUSZEK: 0.2 Ci/ft^3 or, if you wish, 7 Ci/m^3.

M.A.F. AYAD: I notice that you have not mentioned the assumed daily intake of water for individuals. Also, the values of 3×10^{-4} mrem and 1×10^{-4} mrem do not seem to relate to any specified time period, i.e. day, month or year. Would you please clarify.

J.M. MATUSZEK: Tap water consumption rates of 0.7 litre/day for the average child and 1.0 litre/day for the average adult were assumed in this work. These values were taken from Ref. [3] of the paper and are based on water fluoridation studies performed in the United States.

The fifty-year dose commitments for individuals and the population as a whole were found to lie within 10% of the first year dose commitments, and the dose levels calculated for children were found to lie within 10% of those calculated for adults. Consequently, no distinction is drawn between these various values in the paper. The dose levels are essentially the same, whether calculated for children or adults, or for fifty years rather than one year following ingestion.

ENGINEERED STORAGE FOR THE DISPOSAL OF NON-FUEL BEARING COMPONENTS FROM NUCLEAR FACILITIES AND GENERAL RADIOACTIVE WASTE

L.J. ANDREWS
Chem-Nuclear Systems Inc.,
Bellevue, Washington,
United States of America

Abstract

ENGINEERED STORAGE FOR THE DISPOSAL OF NON-FUEL BEARING COMPONENTS FROM NUCLEAR FACILITIES AND GENERAL RADIOACTIVE WASTE.

Non-Fuel Bearing Component (NFBC) waste from nuclear facilities is not a voluminous waste; however, for a commercial burial site, it is a much more hazardous waste by nature of the significantly higher radiation levels. Burial is presently done in excavated slit trenches and a modified form is in use at the CNSI Barnwell facility. This disposal technique creates considerable problems in the handling efficiency, site utilization and personnel exposure. An engineered transport and burial system to reduce the problems and the exposure are discussed. Currently, non-fuel bearing components and irradiated hardware coming from the reactor core are sheared, crushed, containerized and transported to the waste burial site in a top-loading and top-unloading transport cask. At the site, this cask must be removed from the trailer and placed in a vertical position to remove the waste container. The process of removing a highly radioactive container from the cask with site equipment, in the open air, and transferring this container to a slit trench, creates handling problems and excessive exposure. A top-loading bottom-unloading transport cask and trailer unit has been designed so that in combination with an engineered waste disposal will simplify handling, minimize exposure, maximize site utilization and reduce burial site selection problems. The integrated technique discussed sets forth the interfaces, the logic of the transport device, the canister transfer from the cask to the disposal sleeve in a concrete vault with minimal exposure to personnel. Site utilization, cask handling efficiency and exposure control is the motivation for this presentation.

INTRODUCTION

Chem-Nuclear Systems, Inc., (CNSI) is a company involved in waste stream management in the nuclear power industry with a major radioactive waste disposal facility located in Barnwell County, South Carolina. This disposal site is a vital nuclear waste containment processing facility and is in the process of developing an engineered storage for the disposal of non-fuel bearing components (NFBC's) from nuclear fuel facilities and general radioactive waste sources. The remarks in this paper are directed more specifically to the NFBC waste coming from the nuclear reactor core and assemblies. To date, the volume of this waste handled by CNSI at their Barnwell facility has been very small when compared to the total waste generated by a nuclear power plant. The experience which has caused the re-evaluation of current and past disposal techniques has been gained in the handling of sheared components and poison curtains shipped to the facility for disposal from the Commonwealth Edison Dresden plant. This paper represents original work by CNSI engineers and will be addressed briefly to these past experiences, but more specifically to the planned cask handling and transport logic interfaced with an engineered storage facility to be established at the Barnwell disposal facility.

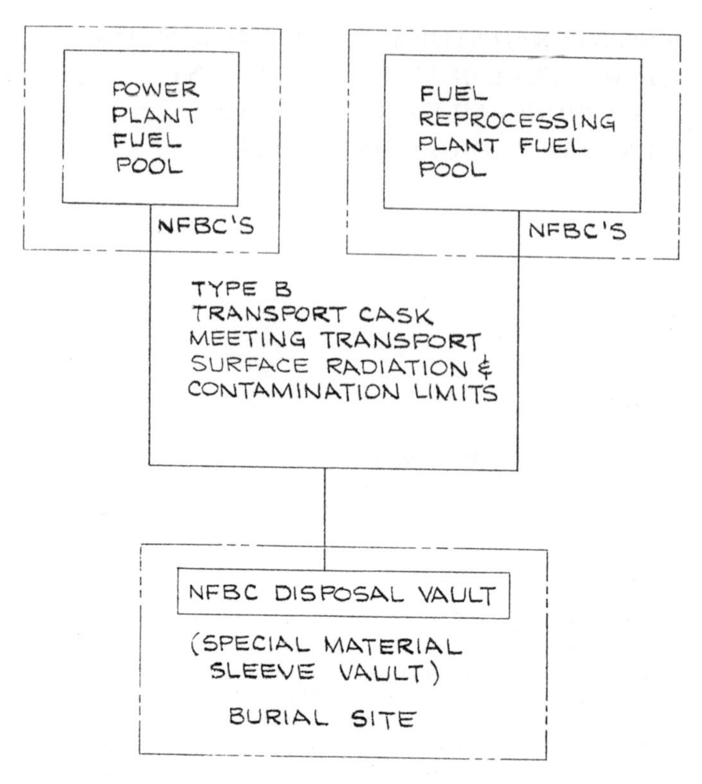

FIG.1. NFBC waste cask handling cycle.

NFBC HANDLING

As defined, the NFBC waste comes from the reactor and as expected is highly radioactive with the major radioactive isotope being cobalt-60. Figure 1 is a brief chart illustrating the source and handling activities for this waste. The in-plant activities of the power plant or fuel reprocessing plant will be the storing of NFBC's in a pool, underwater. Underwater, the waste is loaded into a disposable shipping canister. When the canister is loaded, shipping involves scheduling an appropriately sized and properly licensed cask, removing the cask from the truck trailer or rail car, removing the cask loading lid, opening the cask bottom drain, opening the cask loading lid vent, setting the cask in the storage pool, picking up a filled canister underwater and placing it in the cask, placing the cask loading lid on the cask underwater, lifting the loaded cask out of the pool and holding it above the pool to drain completely, transferring the loaded cask to a decontamination chamber, washing down the cask exterior surfaces and performing the necessary radiation surveys to assure meeting transport criteria (for example, < 200 mR at surface and < 10 mR at 6 feet), replacing the cask loading lid, returning the cask to the truck trailer or rail car and preparing the necessary shipping and inventory control documents.

SHIPMENT

After cask loading, the shipment of NFBC waste will travel from the point of origin or loading to the point of disposal or storage in a Type B cask fabricated to the standards required to support the approved design and analysis requirements. The standard for Type B licensing demands consideration of normal operating and handling conditions (cask removal and placement on the trailer or car, road shock and vibrational loads, ambient temperature conditions and thermal loading associated with NFBC waste decay heat) and withstanding the effects of specific accident events (cask drop, impact on a sharp object and an external fire) while maintaining design shielding and containment integrity.

DISPOSAL

While the licensing standards imposed set a high standard of reliability and safety during operation, these standards do not assure a highly functional unit. Today's typical cask for hauling NFBC waste meets the standard requirements for canister loading underwater and the licensing requirements of normal and accident conditions during transport. However, the typical cask fails to consider and account for the different handling conditions associated with NFBC waste canister removal without benefit of a pool at the typical storage or burial site. Consideration of the final handling requirements and the safest possible NFBC waste canister storage directed Chem-Nuclear to a unique NFBC waste transport cask and cask handling design interfaced with an optimum safe NFBC waste storage vault plan.

PRIOR TECHNIQUE

Prior to elaborating on the optimum plans, a review of previous and intermediate planned NFBC waste disposal activities seems appropriate. The first such activity occurred in 1973 at the Dresden site where NFBC waste was sheared underwater and loaded into 55-gallon drums. The drums were loaded into adequately sized, top-loading/top-unloading transport casks and shipped to the CNSI burial site at Barnwell, S. C. At Barnwell, unloading and trench disposal must be performed in the open air rather than underwater and performed without the relatively sophisticated handling capability normal to a power plant or a reprocessing plant. Needless to say, performing the task of cask lid removal, manual drum handling device grappling and attachment to a mobile crane hook, drum removal and locating in an open slit trench, hook disengagement and drum covering by earth fill, all combined to cause significant site personnel exposure.

IMPROVEMENTS TO PRIOR TECHNIQUES

This year, Chem-Nuclear is again involved in the receipt of NFBC waste. Two major utilities have contracted to remove NFBC waste from their nuclear power plants, and these wastes will be shipped to CNSI's Barnwell, S. C., burial facility. Figure 2 illustrates the intermediate plan for the burial site cask and NFBC waste canister handling improvements over the 1973 Dresden waste handling technique and is designed to yield significantly

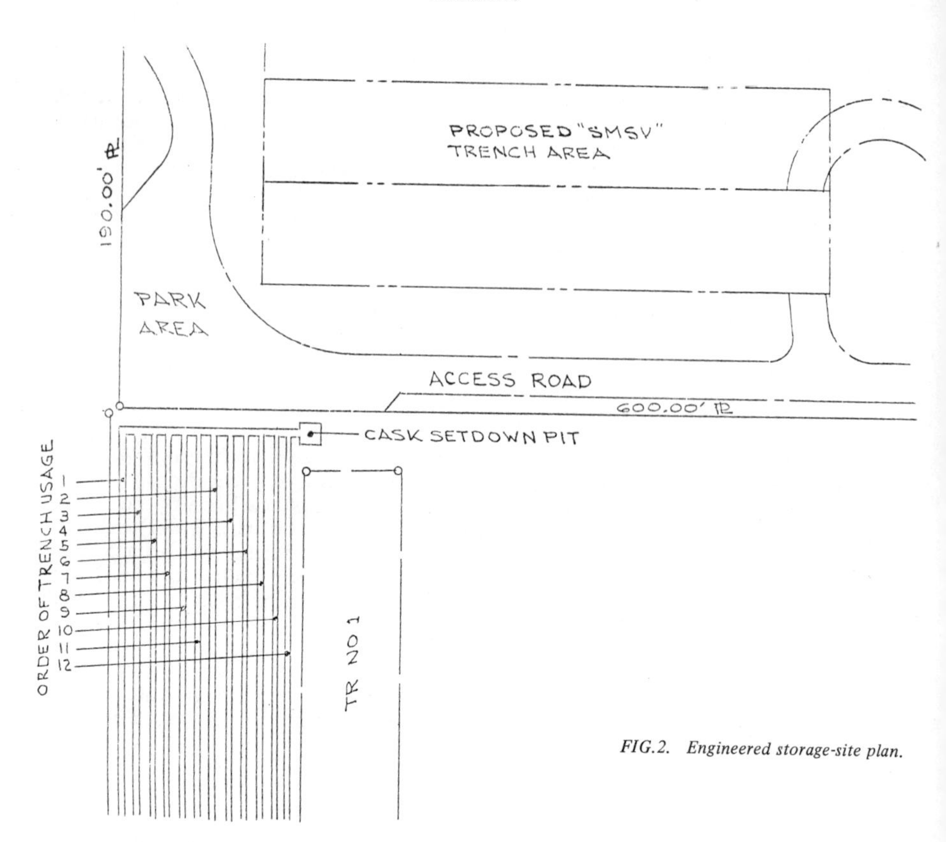

FIG.2. Engineered storage-site plan.

improved burial space utilization. Figure 2 plan incorporates the ability to handle relatively short, 55-gallon type containers as well as long (14 feet) relatively small diameter (14 inches) canisters. Both types of casks are placed in the same cask set down chamber, below ground at the one end of NFBC waste container transfer trench. The unloading tasks are essentially the same as the Dresden activities except that exposure is expected to be controlled significantly better by handling the waste containers below grade in slit type trenches 20 feet deep. While this plan is expected to reduce personnel exposure and improve site space utilization, the relatively long time involved in the unloading cycle, trench transfer technique, canister placement in a burial trench and tasks of covering the containers received at the end of each day all contribute to the potential of considerable personnel exposure conditions that CNSI plans to reduce to a minimum by the implementation of the following optimum plan.

TOP LOADING BOTTOM UNLOADING TRANSPORT CASK

While both systems are interrelated, the improved cask design will be discussed first and then the related improvements in the site burial techniques and site burial space utilization will be reviewed.

Figure 3 shows the combined cask and truck trailer package with the cask in a vertical attitude for either cask removal from the trailer or bottom NFBC waste container discharge. Integral with the cask transport trailer is the mechanisms to raise the cask, rotate it to a vertical attitude, lower the cask to mate with a bottom discharge seal, raise the cask, rotate it to a horizontal attitude and lower the cask to the horizontal transport attitude on support yokes. Figure 4 shows reasonable cask details associated with top and bottom lid seals, top and bottom vent and drain capability, NFBC waste canister bottom support pins and cask vertical support logic. The operational feature of this cask/trailer transportation unit is the ability to relate to the loading point handling conditions previously described. At the power plant or the fuel reprocessing plant, the trailer outrigger supports take the load and level the complete assembly; the cask is rotated to the vertical position; the cask is removed from the circular rotational yoke on the trailer, in the manner described previously; the cask is loaded with a NFBC waste canister and the cask is reinserted in the trailer circular rotational yoke. When the cask has been rotated to the travel position, locked in position on the end support yokes and the trailer outrigger supports retracted, the NFBC waste container is transferred to the storage or burial site. At the burial site and in conjunction with CNSI's long range plan for engineered burial trenches, the cask/trailer unit is located over a vertical disposal sleeve in the Special Material Sleeve Vault (SMSV).

SPECIAL MATERIAL SLEEVE VAULT

The SMSV is a trench designed to primarily receive waste canisters from a NFBC cask and trailer system. Figures 5, 6 and 7 illustrate the basic construction features of a SMSV. Vault construction starts with an excavated trench prepared with standard trench bottom preparation. Clean concrete side walls are then poured. A trench contents monitoring system will be installed as specified by the state license. After the walls and monitoring system are in place, vertical concrete or steel pipe sleeves are placed in the trench and arranged in a square grid pattern with approximately 40 inches between vertical sleeve centerlines. After placing the vertical sleeves, the space between is filled with nuclear and/or chemical waste cement. Sleeve placement and interstitial waste cement pouring would proceed down a walled trench in a progressive schedule using a removable and reusable retaining form. In this manner, blocks of NFBC waste container vertical sleeves will be prepared as NFBC waste or nuclear and/or chemical liquid waste dictates. Over the completed SMSV section will be placed removable and reusable reinforced concrete caps to provide a supporting cover over the empty vertical sleeve section of the SMSV suitable for driving over with the NFBC cask/trailer unit. After a section of the SMSV is filled with NFBC waste containers, these sleeves are either plugged with adequate shielding material or filled with nuclear and/or chemical waste cement level with the top of the sleeve and surrounding cement. After a reasonable section of the vault has been completely filled, the reusable caps are removed and a top layer of clean concrete of the same thickness as the reusable caps is poured. The benefits derived from this storage or burial technique is the highest possible confidence of minimum exposure and environmental contamination, maximum

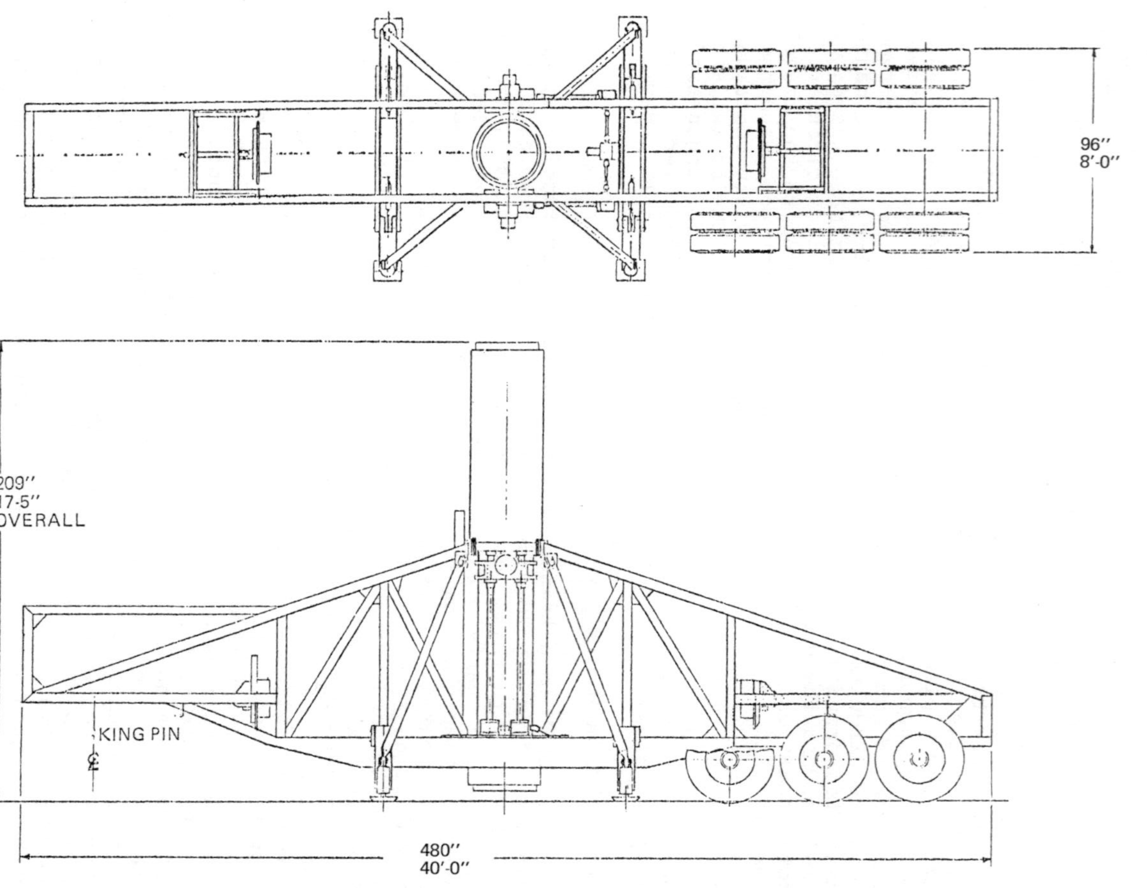

FIG.3. Transport trailer shown in loading/unloading mode.

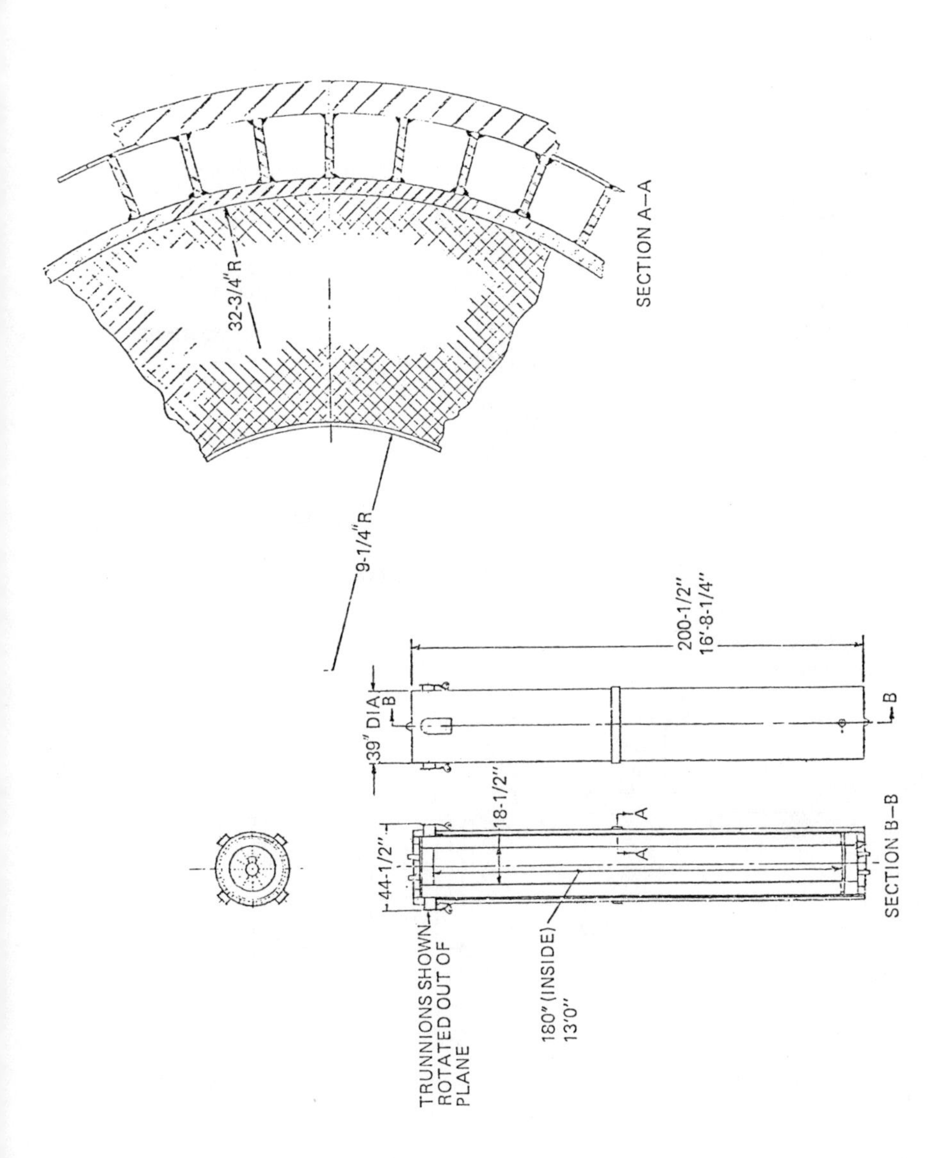

FIG.4. Cask details.

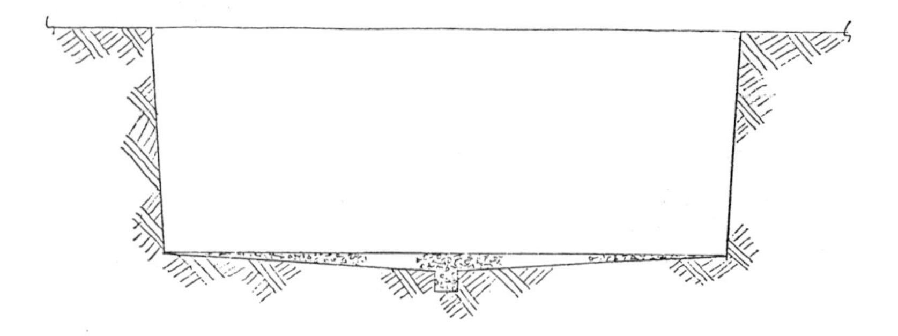

FIG.5. Trench excavation cross-section.

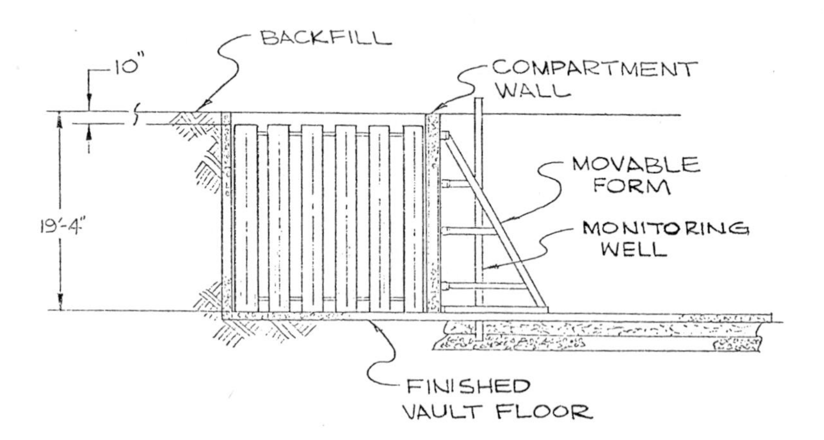

FIG.6. Structural details of trench.

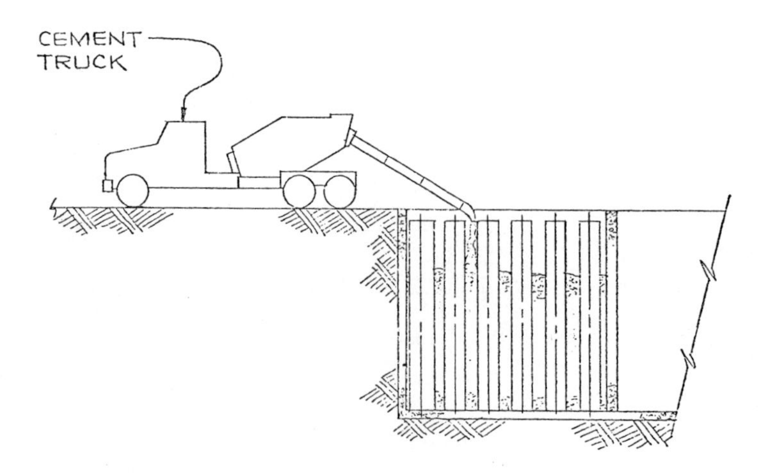

FIG.7. Nuclear and/or chemical cement being added to SMSV.

FIG.8. Cask trailer positioned over SMSV.

trench space utilization for waste disposal and the ability to move from one trench to a new one right adjacent to the one being used, yields maximum site utilization. This site plan imposes the minimum requirements on site selection conditions for maximum containment protection.

NFBC DISCHARGE TO VAULT

Figure 8 illustrates NFBC waste disposal at the site. Disposal activities require positioning of the cask/trailer unit over the selected SMSV disposal sleeve with the reusable cap removed, transferring the cask load to the trailer outrigger, locating the cask to disposal sleeve seal, removing the cask lower lid after confirming that the NFBC waste canister support pins are fully inserted, rotating the cask to a vertical position and lowered onto the sleeve seal spacer to seal the cask cavity and disposal sleeve, and unscrewing the NFBC waste canister support pins until the last pin releases the canister and it drops into the diagonal sleeve. This release mechanism prevents the release of any airborne radioactive particles and the sleeve seal spacer is equipped with a sample connection which will permit cavity air sampling. Provisions are also provided to recirculate and clean up the cavity air should that be required. When the cavity sampling results permit, the cask is raised and rotated to the transport position on the trailer support yokes; the sleeve seal spacer is removed, the reusable sleeve cap is placed over the open sleeve, the cask bottom lid is replaced, the NFBC waste canister support pins are completely inserted, the cask is resecured to the end support yokes on the trailer, and the support outriggers are repositioned for trailer transport conditions. Prior to leaving the site, the cask trailer unit is surveyed and confirmed free of radioactive contamination.

This concludes the cycle associated with handling and disposing of NFBC waste in what CNSI has determined to be the safest, most secure technique economically possible.

CONCLUSIONS

Non-fuel bearing component disposal using past techniques has caused considerable problems and personnel exposure.

CNSI is developing an optimum plan to interface from nuclear plant and reprocessor pool to burial site disposal in an open air environment without the past handling problems and excessive personnel exposure.

DISCUSSION

J.B. LEWIS: What are the authorized activity limits for the facility you describe?

L.J. ANDREWS: At present 10 000 Ci per receipt, but since the NFBC canisters may contain perhaps up to 50 000 Ci, the limits will need to be changed.

P. PATEK: How safe is your storage site in the event of an aircraft crash?

L.J. ANDREWS: Our site is in an area in which there are no flights, but as an extra precaution the engineered storage will have a 12 in concrete cap.

RECENT EXPERIENCE WITH THE LAND BURIAL OF SOLID LOW-LEVEL RADIOACTIVE WASTES

G.L. MEYER
Office of Radiation Programs,
USEPA,
Washington, D.C.,
United States of America

Abstract

RECENT EXPERIENCE WITH THE LAND BURIAL OF SOLID LOW-LEVEL RADIOACTIVE WASTES.

Low-level, nuclear fuel cycle wastes are being disposed of at six commercially operated sites in the United States of America. Similar wastes resulting from Federal activities are being disposed of at five Federally operated sites. The hydrology, geology, climate and operational practices at these sites vary greatly. At three sites in the wetter eastern United States which have low-permeability burial media, it is difficult to keep water from getting into the trenches. Two commercial burial sites in New York and Kentucky have not performed as planned. Authorization to operate these facilities was based on site analyses which, it was believed, demonstrated that the buried radioactive wastes would not migrate from the site during their hazardous lifetime (i.e. for hundreds of years). In ten years or less, however, radioactivity has been detected offsite from these two sites. Radioactivity has migrated offsite from the Federal burial site at Oak Ridge National Laboratory, also. State and Federal authorities have stated that the radioactivity in the environment around the site was not a health hazard at this time. Information is presented on recent disposal practices and experience at these three low-level burial facilities. Based on this experience, the paper (1) briefly describes operations and problems at the sites; (2) suggests factors which led to the problems; (3) identifies problems which appear to be generic to disposal in humid climates; (4) identifies specific problems which could either reduce the ability to predict the impact of disposal operations or reduce the retention capability of the site; and (5) recommends improvements which can be made in site selection, development, and operation to reduce the environmental impact of the site.

1. INTRODUCTION

1.1. Background

In 1972, the Office of Radiation Programs began a program to determine the impact of radioactive wastes from the nuclear fuel cycle on the environment and the public health. EPA's goal for the management of radioactive wastes is to assure that no unwarranted risks are imposed upon present or future generations through the establishment of environmental criteria for all aspects of waste management. [1]

Work began first on "low-level" solid radioactive wastes [2] because they were actively being disposed of throughout the

[1] EPA has divided radioactive wastes into the following general classes or categories because of their markedly differing character, volumes, and treatment/disposition requirements: (1) high-level; (2) low-level; (3) transuranium-contaminated; (4) mill tailings; (5) decommissioning; and (6) naturally occurring.

[2] The AEC divided radioactive waste products into two categories, 'high-level wastes' and 'other than high-level wastes'. 'High-level wastes' are defined as: 'aqueous waste resulting from the operation of the first cycle solvent extraction system, or equivalent, and the concentrated waste of subsequent extraction cycles, or equivalent, in a facility for reprocessing irradiated reactor fuels'. The AEC grouped all other wastes in the category 'other than high-level wastes' [2].

United States, they were being generated in large volumes and their growth was predicted to be exponential for the next 25 years. Six commercial shallow land burial facilities for "low-level" radioactive wastes had been established at Barnwell, South Carolina; Beatty, Nevada; Maxey Flats, Kentucky; Richland, Washington; Sheffield, Illinois; and West Valley, New York. Similar burial facilities were established at five Federally-operated sites located at Idaho National Engineering Laboratory, Idaho; Hanford Facility, Washington; Los Alamos Scientific Laboratory, New Mexico; Oak Ridge National Laboratory, Tennessee; and Savannah River Plant, South Carolina.

A preliminary review of the information available on the sites shows that the hydrology, geology and climate at the different sites vary widely. The operational practices also vary greatly. The wastes are stacked neatly in the trenches at some sites, while at others, the wastes are dumped randomly; and after being covered by earth, they are compacted by earthmoving equipment. In the wetter eastern United States, precipitation presents operational problems. At three sites which have burial media with relatively low permeability, it has been difficult to keep water from accumulating in the trenches.

1.2. Performance Criteria

The performance criteria under which the commercial burial sites such as Maxey Flats were initially licensed were stated by the former USAEC: "The basic objective of shallow land burial is the confinement of radioactive materials on the site over the periods the wastes remain a hazard. Site evaluation and field studies should provide reasonable assurances that the objective will be met." [1, 2] State environmental and health authorities who regulate five of the six commercial sites set similar criteria; for example, "Authorization to operate a commercial land burial facility is based on an analysis of the nature and location of potentially affected facilities; of the site topographical, geographical, meteorological, and hydrological characteristics; and of groundwater and surface water use in the general area which must demonstrate that buried radioactive wastes will not migrate from the site" [3].

These criteria are consistent with EPA's underlying philosophy that waste management means containment of radioactive materials until they have decayed to innocuous levels. The objective is to minimize exposure of present and future populations and to avoid dilution into the biosphere. Containment may involve burial, storage, or some other form of assuring that dispersion into the biosphere does not take place [4].

In 1973, Meyer and O'Connell [5] and EPA [6, 7, 8] in 1974 and 1975 again called attention to the fact that only limited siting-type investigations were available for most of the commercial burial sites and that there was insufficient information available on the sites and the wastes buried therein to evaluate their retention capability. Attention was also called to the fact that there was no method to evaluate quantitatively the retention capability of a site and that there was a general lack of basic data on land burial processes and the transport of radionuclides through the ground under unsaturated and water saturated conditions. Further, EPA expressed concern about present practices used in the disposal of "low-level" radioactive wastes in shallow land burial facilities in humid areas in the United States [4, 36].

1.3. Scope and Purpose

This paper presents information on recent disposal practices and experience at three low-level solid radioactive waste burial facilities in the humid eastern United States. Based on this experience, the paper (1) briefly describes operations and problems at the three sites; (2) suggests factors which led to the problems; (3) identifies problems which appear to be generic to disposal in humid climates; (4) identifies specific problems, any one of which could either reduce the ability to predict the effects of disposal operations or reduce the retention capability of the site; and (5) recommends improvements which can be made in site selection, development, and operation to reduce the environmental impact of the site.

Information is summarized from Maxey Flats and West Valley. Land burial experience at Oak Ridge as described by Duguid [9] has also been included. In addition, experience at general landfills has been introduced because it seemed relevant.

No inferences are made herein about waste burial under arid and semi-arid climatic conditions or about waste buried in a medium with intermediate permeability. Burial under both of these conditions clearly needs more study. Neither does this paper attempt to present detailed data from the sites nor remedial actions already taken because it is concerned with processes and principles. Finally, it should be emphasized that State and Federal scientists have stated that the radioactivity in the environment around the burial sites does not create a public health hazard at this time [15, 16].

2. THE THREE SITES

2.1. Source Term

Most of the wastes buried at Maxey Flats and West Valley are believed to be large volume, low-hazard-potential solid wastes such as paper trash, cleanup materials and sorbed liquids, packing materials, broken glassware, plastics, protective clothing, radioactive carcasses of experimental animals and contaminated equipment. In describing the wastes actually being buried at commercial burial sites, Morton [31] estimated that 70 percent by volume of these wastes would be paper materials and that the density of the total waste would be about 10 lb/ft^3 (0.16 g/cm^3). High-activity wastes such as sealed sources, reactor resins, filters, and irradiated reactor parts have also been buried. A large number of records list the isotopes shipped only as "mixed fission products," "low specific activity," "not specifically identified", and other non-specific identifications [14].

The volume and activity or quantity of wastes which were buried at Maxey Flats and West Valley through 1974 are presented in Table I. Studies by Kentucky, New York, and EPA [14, 26, 16] found that

TABLE I. SUMMARY OF THE VOLUME AND ACTIVITY OR QUANTITY OF RADIOACTIVE WASTES BURIED AT THE MAXEY FLATS AND WEST VALLEY BURIAL FACILITIES THROUGH 1974

	Maxey Flats (Burial began 1973)	West Valley[a] (Burial began 1964)
Volume (m^3)	104 000	64 000
By-Product Material (Ci)[b]	1 638 000	386 000
Special Nuclear Material (g)[b]	349 000	52 000[c]
Source Material (lb)[b]	158 000	986 000
Solidified Liquid Wastes (l)	2 250 000	– 0 –

[a] Burial operations were halted in 1975.
[b] By-product material (radioisotopes produced in reactors), special nuclear material (plutonium and enriched uranium) and source material (uranium and thorium) are defined in USAEC Rules and Regulations, 10 CFR 20.
[c] Further burial of plutonium-239 was prohibited in 1974.

for practical purposes, little information was available on the chemical and physical character of the wastes which were buried at those two sites.

Duguid [9] states that the wastes buried at Oak Ridge consist of anything that was contaminated in normal laboratory operations, including glassware, scrap metal, soil, lumber, contaminated chemicals and, in one case, a small building. Evidence from his monitoring and groundwater investigations suggests that EDTA (ethylenediamine-tetraacetic acid), a widely used agent for cleanup of equipment and decontamination may also have been included in the wastes. Little information is available about the types, concentrations, locations and quantities of radionuclides at one burial ground because all records of wastes buried at that site prior to 1959 were destroyed by fire.

Most wastes are shipped to the burial sites in 55-gallon (208 l) steel drums or in wooden and cardboard boxes and are buried in their shipping containers. These "burial" containers were designed primarily to contain the wastes during shipment and to protect the workers. Their containment capabilities under water- and leachate-saturated conditions are negligible.

2.2. Hydrogeology of the Sites

The geology, the depth to water, and the movement of water through the geologic framework of Maxey Flats, West Valley, and Oak Ridge differ in many details. Yet there are striking similarities in the overall properties and effects of the total hydrogeologic systems of these sites in a land burial situation. From a review of references [9], [11] and [13] through [29] which discuss the characteristics of the three sites, the following similarities were observed:

- All sites receive an average of more than 100 cm precipitation per year and this precipitation is the principal source of recharge to the hydrogeologic system of the site.

- The burial media at all sites generally have very low permeability which causes water infiltrating through the trench caps to fill the trenches and overflow in the "bathtub effect". [3]

- Similar pathways for the migration of contaminants from the trenches and the site exist at all sites; they include (1) surface runoff; (2) interflow or lateral migration through the unsaturated zone; and (3) subsurface migration through joints, fractures and sand lenses.

- The potentially high ion exchange (sorptive) capacity of the burial medium and associated geologic formations may be largely by-passed because the major movement of contamination has occurred directly out of the trenches by overflow and is believed to occur by interflow and by subsurface migration along joints, fractures, and sand lenses rather than by intergranular flow through the fine-grained sediments having high ion-exchange capacity.

- The distance from the trenches to the nearest surface discharge point via the interflow pathway is 15 m or less and the distance via the subsurface migration pathway is 30 to 150 m.

- Burial operations have greatly altered the original hydrogeologic characteristics of the sites. Digging burial trenches at a site, filling them with high porosity wastes, and then covering them with a relatively permeable earthen cover can result in considerably more infiltration from precipitation falling on it than would normally occur under undisturbed conditions.

2.3. Burial Operations at the Sites

General disposal operations are similar at Maxey Flats, West Valley and Oak Ridge even though they vary somewhat in specific detail. The basic technology used until recently to dispose of the "low-level" wastes is simple, consisting primarily of digging a large hole (trench) in the ground, dumping the wastes in the hole and covering the waste with an earthen cover (cap). From a review of references [9], [11] and [13] through [29] which discuss disposal

---

[3] The 'bathtub effect' refers to a situation where precipitation or water infiltrates into a burial trench in which the permeability of the burial medium is so low that water collects in and fills the trench and then overflows at a low side of the trench. This term has been used by several authors but Duguid [9] has an excellent discussion of it.

operations at Maxey Flats, West Valley and Oak Ridge, the following similarities were observed:

- The wastes being buried are largely high-bulk, low-density materials (section 2.1) which when soaked will collapse and undermine the earthen cap.

- The wastes are buried in containers which do not protect them from leaching.

- The earthen covers have until recently, at least, been permeable and have allowed water to infiltrate into the trenches, soak the wastes, and then overflow to spread contamination.

- At times, there have been problems with erosion of the earthen caps and nearby surface materials.

- Until recently, there was insufficient hydrogeological data available to establish an effective monitoring system for the two commercial sites.

In the past three or four years, it was realized that radioactive contaminants might be migrating from the burial sites. Therefore, a number of experiments and changes in operating procedures and techniques have been made in an effort to prevent or mitigate the potential for the migration of radioactive material from the sites. To date, insufficient data has been presented to establish the effectiveness of these improvements.

2.4. Movement of Contamination from the Sites

Two of the commercial burial sites, Maxey Flats and West Valley, have not performed as expected. Authorization to operate the burial facilities was based on analyses of the site hydrology, geology, meteorology, etc., which, it was believed, demonstrated that the buried radioactive wastes would not migrate from the site during their hazardous lifetime (i.e., they would be retained on the site for hundreds or thousands of years). In ten years or less, however, radioactivity has been detected offsite at these two sites. Migration of radioactivity has occurred also from the Federally-operated burial site at Oak Ridge.

Studies supported by EPA's Office of Radiation Programs at the two commercial sites suggest that the following events have occurred: (1) the wastes were buried in large trenches and covered with earthen caps; (2) precipitation infiltrated through the caps, filled the trenches, and soaked the wastes; (3) the water in the trenches interacted with the wastes and formed a leachate, removing radioactive material from the wastes; and (4) the leachate and radioactive material contained therein migrated from the trenches through one or more pathways to the uncontrolled environment. It is believed that studies at the Oak Ridge burial site [9] support the above sequence of events also.

Migration at West Valley apparently occurred because the trenches filled with leachate to ground-level and simply overflowed. Migration may also have occurred along joints in a surficial weathered zone approximately 10 feet thick and along subsurface sand lenses. The latter two modes of migration are possible but have not been confirmed. Migration at Maxey Flats is believed to have occurred by: (1) discharge from the plume of an evaporator used to reduce the volume of the leachate pumped from the trenches; (2) surface transport of contaminants spilled during the burial or trench dewatering operations or leachate which overflowed from the trenches; (3) interflow or lateral migration through the shallow unsaturated zones; and (4) subsurface migration from the trenches along fractures in the burial medium and deeper geologic formations. Based on information from Duguid [9], migration at Oak Ridge is believed to have occurred by: (1) erosion and surface runoff of contaminants overflowing from the trenches; (2) interflow or lateral migration through the shallow soil zones; and (3) subsurface migration from the trenches along fractures in the burial medium.

3. THE GENERIC RADIOACTIVE WASTE DISPOSAL SITE IN A HUMID CLIMATE

3.1. Introduction

In Chapter 2, some practical examples were cited of the problems encountered with current practices in disposing of "low-level" radioactive wastes at Maxey Flats, West Valley and Oak Ridge. This chapter is more hypothetical in nature in that an attempt is made to show a parallel between radioactive waste disposal sites as presently operated in a humid climate and common landfill garbage dumps and to identify specific problems which appear to be common to disposing of radioactive wastes in landfills in a humid climate.

The author would like to state specifically that he does not believe that the problems suggested here regarding the shallow-land of burial of radioactive wastes apply to the deep geologic disposal situation such as has been suggested for the ultimate disposal of high-level and transuranium-contaminated wastes. Rather, the special problems addressed here are associated with the shallow burial radioactive wastes in landfills in humid climates using present methods.

3.2. Common Landfills and Radioactive Waste Disposal Sites

The radioactive waste disposal facilities at Maxey Flats, West Valley, and Oak Ridge have had a number of operational problems such as trenches filling with water, erosion and subsidence of caps, and the movement and possibly migration of leachates from the trenches. These and other problems are normal to most conventional landfills in humid climates--and the above three burial facilities are basically landfills which accept only a special type of waste. Two important causes of these problems are: (1) the landfill process which greatly alters the natural hydrogeologic system of a site and (2) the character of the large amounts of contaminated solid waste which are placed in an environment favorable for its dissolution and dispersal.

To understand the problems at these three sites and how it is possible for radioactivity to migrate, an understanding of what happens to waste after it is buried in a landfill is helpful. Therefore, a brief description of the "life cycle" of a landfill follows: (1) the trench is excavated, and commonly, as at Maxey Flats, in material of relatively low permeability; (2) the trench is filled with high porosity, permeable, compressible wastes which contain organic materials and a wide range of chemical compounds (3) the wastes are covered with an earthen cap which is often more permeable than original pre-trench soil and rock, in effect, allowing increased infiltration; (4) some of the precipitation which falls on the cap infiltrates into the trench and soaks the wastes; (5) the wastes begin to leach aided by the presence of organic matter, bacterial action, the formation of organic and inorganic acids, and chelating agents; (6) the leachate which is formed begins (i) to migrate downward and laterally because of the hydraulic head imposed by the leachate in the trench and/or (ii) to overflow at land surface in springs and seeps at some low point between the cap and the undisturbed earth; and (7) as the wastes continue to soak and leach, they compact, undermine the trench cap, increase the infiltration of water into the trench, and thereby increase leachate generation.

Landfill leachates commonly have 50,000-80,000 mg/1 chemical oxygen demand, 11,000 mg/1 of volatile acids, 6,000 mg/1 of organic acids, and pH ranges of 3.7 to 8.5 [32, 33]. Constituents which possibly might release or mobilize contaminants from the waste include acetic, propionic, isobutyric, butyric, and valeric acids. In brief, landfill leachates may contain agents capable of putting into solution cobalt, strontium, cesium, and other radionuclides including those generally considered relatively insoluble such as plutonium.

Detailed analyses are not available for the non-radioactive constituents of the leachates from Maxey Flats and West Valley. If, however, it is recalled that the wastes entering a commercial radioactive waste burial facility may typically contain 70% by volume paper and plastic material, numerous laboratory animals and organic compounds, and are often packed in wooden crates and cardboard boxes (which are organic material), it can be seen that "low-level" wastes make a good feedstock for generating leachate. In the absence of additional information, it seems reasonable to assign at least some of the characteristics of a "typical" landfill leachate to the Maxey Flats and West Valley leachates similar to that of landfills. Insufficient information is available on the wastes buried at Oak Ridge to make this assumption for those leachates. Preliminary analyses of leachates collected from trenches at West Valley support the hypothesis that they may be similar in character to common landfill leachates [34]. Nineteen different organic peaks were identified including acetic and butyric acids. Radionuclides were detected in both the dissolved and suspended fractions of the West Valley leachates.

The author believes that the evidence described here supports the hypothesis that there is a close similarity between common landfill garbage dumps and the radioactive waste disposal sites at Maxey Flats, West Valley and Oak Ridge. Therefore, it appears that a better understanding of the processes going on within common landfills will be helpful or even necessary for understanding the problems of the radioactive waste burial facilities in humid climates using present disposal methods.

3.3. Common Problems in Disposing of Wastes in Humid Climates

During the past three years while working on the Office of Radiation Programs' "low-level" radioactive waste management program, the author had the opportunity to gain a broad overview of the land disposal of radioactive and other solid wastes. This included a detailed review of studies and reports on burial operations, hydrogeology, and occurrence of radionuclides at Maxey Flats and West Valley; field visits to those sites; extensive discussions with the State officials who regulate the sites and health physicists, hydrologists, and radiochemists conducting studies there. In addition, a review of the literature on the land disposal of solid and other hazardous wastes was made and extensive discussions were held with scientists from EPA's Office of Solid Waste Management Programs, EPA's Solid and Hazardous Waste Research Laboratory, and the U.S. Geological Survey. Finally, an excellent report on burial operations at Oak Ridge by Duguid [9] became available in 1975.

The concept of a typical humid climate burial site was developed from this broad overview of the land disposal of all solid wastes, whether radioactive or non-radioactive, hazardous or non-hazadous. Under present day practices, the fate of a waste appears to be similar regardless of its type once it is buried in a landfill in a humid climate. The wastes become soaked, contaminants are leached from them, and these contaminants migrate or move out of the trenches via one or more water pathways. Further, certain common problems seem to be identifiable at all of the burial sites. Other common weaknesses were apparent in selecting, evaluating and operating the sites.

4. SUMMARY AND CONCLUSIONS

The primary method of disposing of solid "low-level" radioactive wastes today is by burial in landfills. It is simple, relatively inexpensive and potentially an effective method of disposing of the large volumes of wastes resulting from nuclear activities and, in the author's opinion, will continue to be a major method of disposal in the future.

In the humid areas of the United States, however, disposal sites are not performing up to their original criteria -- retaining the wastes for their hazardous lifetimes. The belief in the past that the ground would contain the wastes has to some extent led us into today's present situation of a number of "leaky" sites. The effects of the burial process and of the wastes on the hydrogeologic system of a site in a humid high-rainfall situation were not sufficiently taken into account. Conclusions which were drawn from the information in this paper include:

- Fundamental changes in present disposal methods will be required if landfill disposal of radioactive wastes is to meet the original design criteria.

- New criteria may be required for segregating, treating, and packaging wastes.
- New criteria may be required for the selection, evaluation and operation of new sites.
- The costs of disposing of radioactive wastes may increase if new criteria are implemented.
- The majority of the problems with present disposal methods identified in this paper appear to be correctable, although at additional cost.

5. ACKNOWLEDGEMENTS

I am grateful to the Kentucky Department for Human Resources, Nuclear Engineering Company, New York Department of Environmental Conservation, New York Department of Health, New York Geological Survey, New York Energy Research and Development Authority, Nuclear Fuels Services, Inc., U.S. Geological Survey, EPA's Radiochemistry and Nuclear Engineering Branch - Cincinnati, and EPA's Solid and Hazardous Waste Research Laboratory for their cooperation and assistance in granting access to the sites and in furnishing specific site and general background data.

Investigators and Administrators of the sites and workers in the field of ground disposal including T. J. Cashman, D. T. Clark, K. W. Davis, J. O. Duguid, R. Fakundiny, U. Gat, W. J. Kelleher, H. Kolde, R. LaFleur, A. Lindsey, J. M. Matuszek, D. M. Montgomery, R. Prairie, W. Prell, A. Randall, I. R. Walker, I. J. Winograd, and H. H. Zehner generously gave access to preliminary data from their studies and contributed through stimulating discussions of phenomena they had observed.

REFERENCES

[1] U. S. Atomic Energy Agency, Draft Generic Environmental Statement Mixed Oxide Fuel (GESMO), WASH-1337 (1974).

[2] U. S. Atomic Energy Agency, Proposed Final Environmental Statement for the Liquid Metal Fast Breeder Reactor Program, WASH-1535 (1975).

[3] Conference of State Radiation Control Program Directors, Report to the Conference of the Task Force on Radioactive Waste Management, Annual Report (1974).

[4] U. S. Environmental Protection Agency, Statement of W. D. Rowe, Deputy Assistant Administrator for Radiation Programs, before the Conservation, Energy and Natural Resources Subcommittee of the Committee on Government Operations, United States Congress, on February 23 (1976).

[5] Meyer, G. L. and O'Connell, M. F., Potential Impact of Current Commercial Solid Low-level Radioactive Waste Disposal Practices on the Hydrogeologic Environment, AAPG/USGS/IHAS International Symposium on Underground Waste Management and Artificial Recharge, New Orleans, La., Sept. (1973).

[6] U. S. Environmental Protection Agency, Comments on USAEC WASH-1535, Draft Environmental Statement for the Liquid Metal Fast Breeder Reactor Program, letter, 1974.
[7] U. S. Environmental Protection Agency, Comments on USAEC WASH-1337, Draft Generic Environmental Statement Mixed Oxide Fuel (GESMO), letter, Nov. (1974).
[8] U. S. Environmental Protection Agency, Comments on USEAC WASH-1535 Proposed Final Environmental Statement for the Liquid Metal Fast Breeder Program, letter (1975).
[9] Duguid, J. O., Status Report on Radioactivity Movement from Burial Grounds in Melton and Bethel Valleys, ORNL-5017 (1975).
[10] Number not used.
[11] General Accounting Office, Improvements Needed in the Land Disposal of Radioactive Wastes - A Problem of Centuries, RED-76-54 (1976).
[12] Number not used.
[13] Papadopulos, S. S. and Winograd, I. J., Storage of Low-level Radioactive Wastes in the Ground: Hydrogeologic and Hydrochemical Factors with an Appendix on the Maxey Flats, Kentucky, Radioactive Waste Storage Site: Current Knowledge and Data Needs for a Quantitative Hydrogeologic Evaluation, USEPA Rpt. 520/3-74-009 (1974).
[14] Clark, D. T., A History and Preliminary Inventory Report on the Kentucky Radioactive Waste Disposal Site, Radiation Data and Reports 14 7 (1973) 573.
[15] Kentucky Department for Human Resources, Six Month Study of Radiation Concentrations and Transport Mechanisms at the Maxey Flats Area of Fleming County, Kentucky, open file report, (1974).
[16] Meyer, G. L., Preliminary Data on the Occurrence of Transuranium Nuclides in the Environment at the Radioactive Waste Burial Site Maxey Flats, Kentucky, USEPA, EPA-520/3-75-021, (1976).
[17] Zehner, H. H., Preliminary Hydrogeological Evaluation of the Maxey Flats Radioactive Waste Disposal Site, (study in progress), U. S. Geological Survey.
[18] Montgomery, D., Preliminary Environmental Pathway Study of the Maxey Flats Radioactive Waste Disposal Site, (study in progress, preliminary reports available), USEPA, Radiochemistry and Nuclear Engineering Facility Cincinnati, Ohio, Quarterly Reports, October-December 1974, January-March 1975, April-June 1975.
[19] Kolde, H., Evaporator Study at the Maxey Flats Radioactive Waste Disposal Site, (study in progress, preliminary report available), USEPA, Radiochemistry and Nuclear Engineering Facility, Cincinnati, Ohio, Quarterly Reports, October-December 1974, April-June 1975.
[20] Kentucky Department for Human Resources, Radioactive Waste Burial Ground Field Test, Maxey Flats, Kentucky, report in progress.
[21] U. S. Geological Survey, Study of Principles and Processes of Migration of Radioactive Contaminants at Maxey Flats Burial Facility, in planning.
[22] Walker, I. R., Geologic and Hydrologic Evaluation of Proposed Site for Burial of Solid Radioactive Wastes Northwest of Morehead, Fleming County, Kentucky, unpublished report (1962).
[23] Prairie, R., Preparation of Radioactive Waste Inventory System for Maxey Flats for Operational Status, Southwest Ohio Regional Computer Center, report in progress.
[24] Gat, U., Radioactive Waste Inventory System, USEPA, Unpublished Report (1974).
[25] New York Department of Environmental Conservation - New York Energy Research Development Agency - USEPA, Preliminary Evaluation of Radio-Contaminant Migration at the Radioactive

Waste Burial Facility, West Valley, New York, (a joint report, in preparation).

[26] Kelleher, W. J. and Michael, E. J., Low-level Radioactive Waste Burial Site Inventory for the West Valley Site, Catteragus County, N. Y., New York State Dept. of Environmental Conservation, Albany, N. Y., (1973).

[27] U. S. Geological Survey, Study of Principles and Processes of Migration of Radioactive Contaminants at the West Valley Burial Facility, in progress.

[28] New York State †, Assessment of the Environmental Impact of Shallow Land Burial of Low-level Radioactive Wastes, (in progress).

[29] New York State †, Determination of the Retention of Radioactive and Stable Nuclides by Fractured Soil and Rocks in progress.

[30] Number not used.

[31] Morton, R. J., Land Burial of Solid Radioactive Wastes: Study of Commercial Operations and Facilities, U. S. Atomic Energy Commission. WASH-1143 (1968).

[32] Haxo, H. E., Monthly Progress Report No. 12, Matrecon, Inc. Oakland, California, unpublished report (1975).

[33] USEPA, Summary Report: Gas and Leachate From Land Disposal of Municipal Solid Waste, open file report (1974).

[34] Matuszek, John, New York Department of Health, personal communication (1975).

[36] U. S. Environmental Protection Agency, Statement of Robert Strelow, Assistant Administrator for Air and Wastes Management, before the Joint Committee on Atomic Energy, United States Congress, November 19 (1975).

DISCUSSION

A. SCHNEIDER: Both your data and those presented earlier by Mr. Matuszek (paper IAEA-SM-207/59, these Proceedings, Vol. 2) showed that the migration of radionuclides had occurred at several burial sites in the United States of America to an extent exceeding earlier predictions. Could you assign a quantitative value to the degree of contamination by reference, for example, to MPCs?

G.L. MEYER: The potential health hazard is not an issue in this paper. It is acknowledged that health authorities from the States of Kentucky and New York, the New York Regulatory Commission (NRC) and the Environmental Protection Agency (EPA) have all recommended that the levels of plutonium and other radionuclides at Maxey Flats and West Valley should not be regarded a health hazard at this time. Further, the NRC and the EPA have both stated that the maximum permissible concentration is not a meaningful quantitative measure of contamination when applied to land burial facilities. The important points brought out in my paper are as follows: first, the sites are not performing as predicted and there is no way at present of determining how badly they are performing; second, plutonium and other radionuclides have left the burial trenches; third, the data suggest that plutonium and other radionuclides may be migrating through the ground; and, fourth, these occurrences can and must be prevented in the future.

† This is a multidisciplinary, multiorganizational investigation. The lead agency and coordinator is the New York State Geological Survey and active participants or cooperating agencies include: New York State Department of Environmental Conservation, New York State Department of Health, New York State Energy Research and Development Authority, New York State Department of Commerce, New York State Public Services Commission, U.S. Geological Survey, Nuclear Fuels Services, Inc., Health Research Institute, Inc., and U.S. Environmental Protection Agency. Funding is by USEPA.

K.A. GABLIN: A radioactive waste solidification system developed by Protection Packaging Inc. in the United States includes an inhibitor (formaldehyde) to prevent bacterial growth during storage and disposal. What evidence do you have indicating that the products of anaerobic bacterial breakdown promote the migration of radioactive cations?

G.L. MEYER: We do not yet have direct evidence that the products of anaerobic breakdown promote the migration of radioactive cations. Indirect evidence does suggest this, however, on the basis of the following argument: first, Maxey Flats and West Valley are land fills; second, leachates formed in the trenches; third, the land fill leachates contain acetic, propionic, butyric, isobutyric and valeric acids; fourth, 19 different organic, spectral peaks have been identified in West Valley bacteria, including acetic and butyric acids; and plutonium and other radionuclides were found in the West Valley leachates.

F. GERA: I would like to make a general comment with regard to surface burial of radioactive wastes. This technique can be suitable for the disposal of short-lived wastes, provided the burial grounds possess geohydrological conditions preventing the release of activity into the biosphere.

Some of the older burial grounds cannot be considered successful, since they contain significant amounts of long-lived radionuclides, including plutonium, and the waste is being leached by groundwater. The right approach should be to define the length of time for which burial grounds can be withdrawn from general use and to bury only those wastes which, within that lapse of time, decay to non-hazardous levels of activity. All other wastes should be disposed of in alternative ways.

G.L. MEYER: Thank you for the comment.

CHAIRMEN OF SESSIONS

Session I	P. DEJONGHE	Belgium
Session II	L.E. CARLBOM	Sweden
Session III	Y. TAKASHIMA	Japan
Session IV	R.P. RANDL	Federal Republic of Germany
Session V	Y. SOUSSELIER A.M. PLATT	France United States of America
Session VI	F. GERA	Italy
Session VII	E. DETILLEUX	OECD/NEA
Session VIII	N.J. KEEN	United Kingdom
Session IX	A.F. PERGE	United States of America
Session X	B. VERKERK	Netherlands
Session XI	Y.P. MARTYNOV	Union of Soviet Socialist Republics

SECRETARIAT OF THE SYMPOSIUM

Scientific Secretaries	W.L. LENNEMANN	Division of Nuclear Safety and Environmental Protection, IAEA
	J.P. OLIVIER	Division of Radiation Protection and Waste Management, OECD Nuclear Energy Agency
Administrative Secretary	Gertrude SEILER	Division of External Relations, IAEA
Editor	Brigitte KAUFMANN	Division of Publications, IAEA
Records Officer	J.H. RICHARDSON	Division of Languages and Policy-making Organs, IAEA
Conference Officer	Claire M. BESNYOE	Division of External Relations, IAEA

LIST OF PARTICIPANTS

AUSTRALIA

McDonald, N.R.	Australian Embassy, Mattiellistraße 2–4, A-1040 Vienna, Austria

AUSTRIA

Gattinger, T.E.	Geologische Bundesanstalt, Rasumofskygasse 23, A-1030 Vienna
Halbmayer, H.	Österreichische Elektrizitätswirtschafts AG, (Verbundgesellschaft), Am Hof 6A, A-1010 Vienna
Held, C.	Kernkraftwerk Planungs-GmbH, Jacquingasse 16–18, A-1030 Vienna
Hintermayer, H.	Kernkraftwerk Planungs-GmbH, Jacquingasse 16–18, A-1030 Vienna
Hirsch, H.	Bundesministerium für Handel, Gewerbe und Industrie, Schwarzenbergplatz 1, A-1010 Vienna
Jakusch, H.	Vereinigte Edelstahlwerke AG, A-2630 Ternitz
Knotik, K.	Österreichische Studiengesellschaft für Atomenergie, Forschungszentrum Seibersdorf, A-2444 Seibersdorf
Komurka, M.	Österreichische Studiengesellschaft für Atomenergie, Lenaugasse 10, A-1082 Vienna
Krejsa, P.P.	Österreichische Studiengesellschaft für Atomenergie, Lenaugasse 10, A-1082 Vienna
Kumer, L.	Kernkraftwerk Planungs-GmbH, Jacquingasse 16–18, A-1030 Vienna
Neumann, W.E.	Österreichische Studiengesellschaft für Atomenergie, Lenaugasse 10, A-1082 Vienna
Obermair, G.E.	Bundesministerium für Handel, Gewerbe und Industrie, Schwarzenbergplatz 1, A-1010 Vienna

Oszuszky, F.J.P.	Österreichische Elektrizitätswirtschafts AG, (Verbundgesellschaft), Am Hof 6A, A-1010 Vienna
Patek, P.	Österreichische Studiengesellschaft für Atomenergie, Lenaugasse 10, A-1082 Vienna
Pechacek, F.	Gemeinschaftskernkraftwerk Tullnerfeld GmbH, Marc Aurel Straße 4, A-1010 Vienna
Powondra, F.	Österreichische Elekrizitätswirtschafts AG, (Verbundgesellschaft), Am Hof 6A, A-1010 Vienna
Proksch, E.	Österreichische Studiengesellschaft für Atomenergie, Lenaugasse 10, A-1082 Vienna
Stüger, R.A.	Kernkraftwerk Planungs-GmbH, Jacquingasse 16–18, A-1030 Vienna
Zeger, J.	Österreichische Studiengesellschaft für Atomenergie, Lenaugasse 10, A-1082 Vienna

BELGIUM

Broothearts, J.	SCK/CEN, Boeretang 200, B-2400 Mol-Donk
Cantillon, G.E.	Institut d'hygiène, 14 rue J. Wytsman, B-1050 Brussels
Claes, J.	SCK/CEN, Boeretang 200, B-2400 Mol-Donk
Danguy, J.V.	Société de traction et d'électricité, S.A., 31 rue de la Science, B-1040 Brussels
De Beukelaer, R.C.	Belgonucléaire S.A., 25 rue du Champ de Mars, B-1050 Brussels
Dejonghe, P.	SCK/CEN, Boeretang 200, B-2400 Mol-Donk
Gulinck, M.L.	Service géologique de Belgique, 13 rue Jenner, Brussels
Heremans, R.	SCK/CEN, Boeretang 200, B-2400 Mol-Donk
La Grange, R.L.	Van Leer, 6 Bollaerstraat, B-2500 Lier

Mancini, M.	Westinghouse Nuclear Europe, 73 rue de Stalle, B-1180 Brussels
Mara, G.	Westinghouse Nuclear Europe, 73 rue de Stalle, B-1180 Brussels
Mergan, L.M.	Belgonucléaire S.A., 25 rue du Champ de Mars, B-1050 Brussels
Roofthooft, R.L.A.	Laborelec, B-1640 Rhode St. Genese
Segers, V.H.	Electrobel, 1 place du Trône, B-1000 Brussels
Stallaert, P.F.	Ministère du travail, 53 rue Belliard, B-1040 Brussels
Van de Voorde, N.	SCK/CEN, Boeretang 200, B-2400 Mol-Donk

BULGARIA

Simeonov, S.	Committee for the Peaceful Uses of Atomic Energy, 8 Slavianska Str., Sofia
Stoentchev, I.	Committee for the Peaceful Uses of Atomic Energy, 8 Slavianska Str., Sofia

CANADA

Barnes, R.W.	Ontario Hydro, 700 University Avenue, Toronto, Ontario M5G 1X6
Coady, J.R.	Atomic Energy Control Board, P.O. Box 1046, Ottawa, Ontario K1P 5S9
Didyk, J.P.	Atomic Energy Control Board, P.O. Box 1046, Ottawa, Ontario K1P 5S9
Harms, A.A.	Department of Engineering Physics, McMaster University, Hamilton, Ontario
Mayman, S.A.	Atomic Energy of Canada Ltd., Whiteshell Nuclear Research Establishment, Pinawa, Manitoba ROE 1LO
Merritt, W.F.	Atomic Energy of Canada Ltd., Biology and Health Physics Division, Chalk River Nuclear Laboratories, Chalk River, Ontario K0J 1J0
Muller, E.F.	Department of the Environment, Federal Activities Environmental Branch, Ottawa, Ontario K1A OH3

CZECHOSLOVAKIA

Chyský, J.	Institute of Industrial Hygiene in Uranium Industry, 262 31 Příbram
Kortus, J.	Chemoprojekt, Štěpánská 15, Prague 2
Kyrš, M.	Nuclear Research Institute, 250 68 Řež
Malášek, E.	Czechoslovak Atomic Energy Commission, Slezská 9, Prague 2

DENMARK

Brodersen, K.E.	Danish Atomic Energy Commission, Research Establishment Risø, DK-4000 Roskilde
Hannibal, L.G.	National Health Service of Denmark, State Institute of Radiation Hygiene, Frederikssundsvej 378
Jensen, B.S.	Danish Atomic Energy Commission, Research Establishment Risø, DK-4000 Roskilde
Nielsen, S.O.	Symplexor Engineering Consultants, Inc., Vejlesövej 92, DK-2840 Holte
Singer, K.A.J.	Danish Atomic Energy Commission, Research Establishment Risø, DK-4000 Roskilde

EGYPT

Abdel-Rassoul, A.A.	Atomic Energy Establishment, P.O. Box 1, Cairo
Ayad, M.A.F.	Atomic Energy Establishment, P.O. Box 1, Cairo

FINLAND

Heinonen, J.U.	Technical Research Centre of Finland, SF-02150 Espoo
Kallonen, I.	Imatran Voima Osakeyhtiö, Nuclear Power Project Group, P.O. Box 138, SF-00101 Helsinki
Ruuskanen, A.T.	Institute of Radiation Protection, P.O. Box 268, SF-00101 Helsinki
Söderman, J.K.	Oy W. Rosenlew AB, Engineering Works, SF-28100 Pori 10

FRANCE

Argillier, B. — Electricité de France,
Tour EDF-GDF, Cedex 8,
F-92080 Paris-La Défense

Baer, A. — GAAA,
20 avenue Edouard Herriot,
F-92350 Le Plessis-Robinson

Barbreau, A.F. — CEA, Centre d'études nucléaires de Saclay,
B.P. 2,
F-91190 Gif-sur-Yvette

Bardet, G. — Département Infratome,
PEC-Engineering,
62 rue Jeanne d'Arc,
F-75013 Paris

Beau, P. — Electricité de France,
Département de radioprotection,
6 rue Ampère,
B.P. 120,
F-93203 St. Denis

Bernard, C. — St. Gobain Techniques Nouvelles,
23 boulevard G. Clemenceau,
F-92400 Courbevoie

Bonniaud, R. — CEA, Centre de Marcoule,
B.P. 106,
F-30200 Bagnols-sur-Cèze

Brulé, J.L. — Département Infratome,
PEC-Engineering,
62 rue Jeanne d'Arc,
F-75013 Paris

Candès, P. — Service central de sûreté des installations nucléaires,
13 rue de Bourgogne,
F-75007 Paris

Cousin, Odile — CEA, Centre d'études nucléaires de Fontenay-aux-Roses,
B.P. 6,
F-92260 Fontenay-aux-Roses

Cretey, J. — CEA, Centre d'études de Valduc,
B.P. 14,
F-21120 Is-sur-Tille

Fernandez, N. — CEA, Centre de Marcoule,
B.P. 106,
F-30200 Bagnols-sur-Cèze

Fonné, C.R. — Electricité de France,
2 rue Louis Murat,
F-75008 Paris

Jouan, A. — CEA, Centre de Marcoule,
B.P. 106,
F-30200 Bagnols-sur-Cèze

Laude, F. — CEA, Centre de Marcoule,
B.P. 106,
F-30200 Bagnols-sur-Cèze

Le Bouhellec, J. — CEA, Centre de Marcoule,
B.P. 106,
F-30200 Bagnols-sur-Cèze

Lefillâtre, G.	CEA, Centre d'études nucléaires de Cadarache, B.P. 1, F-13115 St. Paul-lez-Durance
Limongi, A.	CEA, Centre d'études nucléaires de Grenoble, B.P. 85, Centre de tri, F-38041 Grenoble
Macqueron, M.	CEA, Centre d'études nucléaires de Fontenay-aux-Roses, B.P. 6, F-92260 Fontenay-aux-Roses
Migaud, A.J.M.	GAAA, 100 avenue Edouard Herriot, F-92350 Le Plessis-Robinson
Miquel, P.	CEA, Centre d'études nucléaires de Fontenay-aux-Roses, B.P. 6, F-92260 Fontenay-aux-Roses
Mouney, H.	Electricité de France, 1 avenue du Général de Gaulle, F-92141 Clamart
Moutet, P.,	Electricité de France, Tour EDF-GDF, Cedex 8, F-92080 Paris - La Défense
Pelletrat de Borde, B.C.	Ministère de l'intérieur, Direction de la sécurité civile, 18 rue Ernest Cognacq, F-92300 Levallois
Planet, J.	CEA,Centre d'études nucléaires de Fontenay-aux-Roses, B.P. 6, F-92260 Fontenay-aux-Roses
Pomarola, J.	Département Infratome, PEC-Engineering, 62 rue Jeanne d'Arc, F-75013 Paris
Pradel, J.	CEA, Centre d'études nucléaires de Fontenay-aux-Roses, B.P. 6, F-92260 Fontenay-aux-Roses
Scheidhauer, J.G.	CEA, Centre de la Hague, B.P. 209, F-50170 Cherbourg
Schwob, Y.E.	St. Gobain Techniques Nouvelles, 23 boulevard G. Clemenceau, F-92400 Courbevoie
Slizewicz, P.	Ministère de l'industrie et de la recherche, Service central de sûreté des installations nucléaires, 13 rue de Bourgogne, F-75007 Paris
Sousselier, Y.	CEA, Centre d'études nucléaires de Fontenay-aux-Roses, B.P. 6, F-92260 Fontenay-aux-Roses
Tanguy, P.Y.	CEA, Centre d'études nucléaires de Saclay, B.P. 2, F-91190 Gif-sur-Yvette
Thiéry, D.R.	CEA, Centre d'études nucléaires de Grenoble, B.P. 85, Centre de tri, F-38041 Grenoble

Traxler von Schrollheim, Anne-Marie	Groupement atomique alsacienne atlantique (GAAA), 20 avenue Edouard Herriot, F-92350 Le Plessis-Robinson

GERMAN DEMOCRATIC REPUBLIC

Körner, W.	Staatliches Amt für Atomsicherheit and Strahlenschutz der DDR, Waldowallee 117, DDR-1155 Berlin
Richter, D.	Staatliches Amt für Atomsicherheit und Strahlenschutz der DDR, Waldowallee 117, DDR-1155 Berlin

GERMANY, FEDERAL REPUBLIC OF

Andrzejczak, H.J.	Fa. Fichtner Beratende Ingenieure, Haldenäckerstraße 4, D-7000 Stuttgart 30
Anger, W.	Kernforschungsanlage Jülich GmbH, Postfach 1913, D-1517 Jülich
Bähr, W.	Gesellschaft für Kernforschung mbH, Postfach 3640, D-7500 Karlsruhe
Baumgärtner, F.	Gesellschaft für Kernforschung mbH, Postfach 3640, D-7500 Karlsruhe
Bechthold, W.M.	Gesellschaft für Kernforschung mbH, Postfach 3640, D-7500 Karlsruhe
Beisswenger, H.	TÜV, Richard-Wagner-Straße 2, D-6800 Mannheim
Berners, O.F.	Kraftanlagen AG, Im Breitspiel 7, D-6900 Heidelberg
Blume, H.	Programmleitung ASA, Linder Höhe, D-5000 Köln 90
Boden, H.	Maschinenfabrik Werner & Pfleiderer, Theodorstraße 10, D-7000 Stuttgart
Bohnenstingl, J.	Institut für chemische Technologie, Kernforschungsanlage Jülich GmbH, Postfach 3640, D-1517 Jülich
Bokelund, H.	Gesellschaft zur Wiederaufarbeitung von Kernbrennstoffen mbH, D-7514 Leopoldshafen
Born, H.P.	Vereinigte Elektrizitätswerke Westfalen AG, Ostwall 51, D-4600 Dortmund
Braun, R.R.	Hoechst AG, Postfach 800320, D-6230 Frankfurt

Christ, R.	Transnuklear GmbH, Postfach 110030, D-6450 Hanau
De, A.K.	Nuclear Chemistry and Reactor Division, Hahn-Meitner Institut für Kernforschung Berlin GmbH, Glienickerstraße 100, D-1000 Berlin 39 (West)
Diefenbacher, W.	Abteilung Behandlung radioaktiver Abfälle, Gesellschaft für Kernforschung mbH, Postfach 3640, D-7500 Karlsruhe
Dippel, T.	Gesellschaft für Kernforschung mbH, Postfach 3640, D-7500 Karlsruhe
Dobschütz, P.C. von	Institut für Reaktorsicherheit der TÜV, Glockengasse 2, D-5000 Köln 1
Dürr, K.	Institut für Tieflagerung, Gesellschaft für Strahlen- und Umweltforschung, Berlinerstraße 2, D-3392 Clausthal-Zellerfeld
Dyroff, H.	NUKEM GmbH, Postfach 110080, D-6450 Hanau
Engelhardt, G.	Gesellschaft für Kernforschung mbH, Postfach 3640, D-7500 Karlsruhe
Ewest, E.	Hahn-Meitner Institut für Kernforschung Berlin GmbH, Glienickerstraße 100, D-1000 Berlin 39 (West)
Fickel, O.	Babcock Brown Boveri Reaktor GmbH, Postfach 323, D-6800 Mannheim
Friedrich, H.D.	Ciba-Geigy GmbH, Postfach, D-7867 Wehr/Baden
Gattys, F.J.	Gattys-Verfahrenstechnik GmbH, Frankfurter Straße 168–176, D-6078 Neu Isenburg
Goldacker, H.H.	Gesellschaft für Kernforschung mbH, Postfach 3640, D-7500 Karlsruhe
Goralczyk, R.W.C.	Van Leer, B.V., Mijdrecht Engineering Department Centre, Mijdrecht, Energieweg, Netherlands
Grassl, P.	TÜV Bayern, Rüdesheimerstraße 11, D-8000 München
Grziwa, P.	Gelsenberg AG, Rosastraße 2, D-4300 Essen 1
Guber, W.	Gesellschaft für Kernforschung mbH, Postfach 3640, D-7500 Karlsruhe

Halaszovich, S.	Institut für chemische Technologie, Kernforschungsanlage Jülich GmbH, Postfach 1913, D-1517 Jülich
Haug, H.O.	Institut für Heiße Chemie, Gesellschaft für Kernforschung mbH, Postfach 3640, D-7500 Karlsruhe
Heimerl, W.	Gelsenberg AG, Rosastraße 2, D-4300 Essen 1
Hepp, H.	Projektgesellschaft Wiederaufarbeitung von Kernbrennstoffen, Kruppstraße 5, D-4300 Essen 1
Herbrechter, D.	Kraftanlagen AG, Im Breitspiel 7, D-6900 Heidelberg
Hild, W.	Gesellschaft für Kernforschung mbH, Postfach 3640, D-7500 Karlsruhe
Hoffmann, D.	Institut für Reaktorsicherheit der TÜV, Glockengasse 2, D-5000 Köln 1
Höhlein, G.	Gesellschaft für Kernforschung mbH, Postfach 3640, D-7500 Karlsruhe
Hoschützky, A.R.	Bundesministerium des Innern, Husarenstraße 30, D-5300 Bonn
Jelinek-Fink, P.	NUKEM GmbH, Postfach 110080 D-6450 Hanau
Kahl, L.	Gesellschaft für Kernforschung mbH, Postfach 3640, D-7500 Karlsruhe
Kaufmann, F.	Gesellschaft für Kernforschung mbH, Postfach 3640, D-7500 Karlsruhe
Knoglinger, E.	Babcock Brown Boveri Reaktor GmbH, Postfach 323, D-6800 Mannheim
Koch, G.	Gesellschaft für Kernforschung mbH, Postfach 3640, D-7500 Karlsruhe
Kohlbecher, W.	Friedrich Uhde GmbH, D-4600 Dortmund
Kraemer, R.H.	Gesellschaft für Kernforschung mbH, Postfach 3640, D-7500 Karlsruhe
Krause, H.	Abteilung Behandlung radioaktiver Abfälle, Gesellschaft für Kernforschung mbH, Postfach 3640, D-7500 Karlsruhe

Kroebel, R.	Gesellschaft für Kernforschung mbH, Postfach 3640, D-7500 Karlsruhe
Kühn, K.	Institut für Tieflagerung, Gesellschaft für Strahlen- und Umweltforschung mbH, Berlinerstraße 2, D-3392 Clausthal-Zellerfeld
Lahr, H.W.	Gelsenberg AG, Rosastraße 2, D-4300 Essen 1
Laser, M.	Institut für chemische Technologie, Kernforschungsanlage Jülich GmbH, Postfach 1913, D-1517 Jülich
Levi, H.W.	Nuclear Chemistry and Reactor Division, Hahn-Meitner Institut für Kernforschung Berlin GmbH, Glienickerstraße 100, D-1000 Berlin 39 (West)
Lieser, K.H.	Technische Hochschule Darmstadt, D-6100 Darmstadt
Lutze, W.	Hahn-Meitner Institut für Kernforschung Berlin GmbH, Glienickerstraße 100, D-1000 Berlin 39 (West)
Meltzer, A.E.	Institut für Reaktorsicherheit der TÜV, Glockengasse 2, D-5000 Köln 1
Olinger, R.	Friedrich Uhde GmbH, D-4600 Dortmund
Papp, R.	Gesellschaft für Kernforschung mbH, Postfach 3640, D-7500 Karlsruhe
Pascher, K.W.	Maschinenfabrik Werner & Pfleiderer, Theodorstraße 10, D-7000 Stuttgart
Perzl, F.	Gesellschaft für Strahlen- und Umweltforschung mbH, Ingolstädter Landstraße 1, D-8042 Neuherberg
Proske, R.	Gesellschaft für Strahlen- und Umweltforschung mbH, Ingolstädter Landstraße 1, D-8042 Neuherberg
Randl, R.P.	Bundesministerium für Forschung und Technologie, Postfach 120370, D-5300 Bonn
Rönne, P. von	TÜV Hannover, D-3000 Hannover
Saidl, J.	Gesellschaft für Kernforschung mbH, Postfach 3640, D-7500 Karlsruhe
Scheffler, K.	Gesellschaft für Kernforschung mbH, Postfach 3640, D-7500 Karlsruhe
Scheibke, H.G.	Rheinisch-Westfälischer TÜV, Steubenstraße 53, D-4300 Essen

Schikarski, W.O.	Gesellschaft für Kernforschung mbH, Postfach 3640, D-7500 Karlsruhe
Schneider, R.E.P.	Babcock Brown Boveri Reaktor GmbH, Postfach 323, D-6800 Mannheim
Schüller, W.F.J.	Gesellschaft zur Wiederaufarbeitung von Kernbrennstoffen mbH, D-7514 Leopoldshafen
Schwibach, J.	Bundesgesundheitsamt, Ingolstädter Landstraße 1, D-8042 Neuherberg
Seiffert, H.J.	Gesellschaft für Kernforschung mbH, Postfach 3640, D-7500 Karlsruhe
Sitte, R.	Kerntechnischer Ausschuss, Glockengasse 2, D-5000 Köln 1
Stammler, M.	Battelle Institut e.V., Postfach 900 160, D-6000 Frankfurt
Steiner, R.G.	Hoechst AG, Postfach 800320, D-6230 Frankfurt
Stipanits, P.M.	Kernforschungsanlage Jülich GmbH, Postfach 1913, D-1517 Jülich
Storch, S.	Kernforschungsanlage Jülich GmbH, Postfach 1913, D-1517 Jülich
Tägder, K.	TÜV Hannover, D-3000 Hannover
Theenhaus, R.	Kernforschungsanlage Jülich GmbH, Postfach 1913, D-1517 Jülich
Thiele, D.	Kernforschungsanlage Jülich GmbH, Postfach 1913, D-1517 Jülich
Thomas, W.	Laboratorium für Reaktorregelung und Anlagensicherheit, D-8046 Garching
Tolksdorf, P.	TÜV Rheinland, Am Grauen Stein, D-5000 Köln 91
Uerpmann, E.P.	Gesellschaft für Strahlen- und Umweltforschung mbH, D-3392 Clausthal-Zellerfeld
Wenk, E.	Dornier System, Postfach 1360, D-7990 Friedrichshafen
Wilke, M.E.	Kraftanlagen AG, Im Breitspiel 7, D-6900 Heidelberg
Willax, H.O.	Gesellschaft zur Wiederaufarbeitung von Kernbrennstoffen mbH, D-7514 Leopoldshafen

Witte, H.	NUKEM GmbH, Postfach 110080, D-6450 Hanau
Wittenzellner, R.	Gesellschaft für Strahlen- und Umweltforschung mbH, Ingolstädter Landstraße 1, D-8042 Neuherberg

HUNGARY

Lipták, L.	"Eröterv" Power Station and Network Engineering Co., Széchenyi vkp.3, H-1054 Budapest

INDIA

Sunder Rajan, N.S.	Bhabha Atomic Research Centre, Trombay, Bombay 400 085

IRAN

Yamani, S.A.	Atomic Energy Organization of Iran, P.O. Box 12-1198, Teheran

ISRAEL

Levin, I.	NRCN, P.O. Box 9001, Beer-Sheva
Mouyal, M.	NRCN, P.O. Box 9001, Beer-Sheva

ITALY

Antonicci, L.	ENEL, Viale Regina Margherita 137, I-00189 Rome
Bocola, W.	CNEN-CSN Casaccia, C.P. 2400, I-00100 Rome
Brambilla, G.	Agip Nucleare, Centro di Medicina, I-40059 Medicina
Brofferio, Carla	CNEN, Viale Regina Margherita 125, I-00198 Rome
Cassano, G.	CNEN-CRN Trisia, C.P., Policoro
Donato, A.	CNEN-CSN Casaccia, C.P. 2400, I-00100 Rome
Frullani, S.	Istituto Superiore di Sanità, Viale Regina Elena 299, I-00161 Rome

Gera, F.	CNEN-CSN Casaccia, C.P. 2400, I-00100 Rome
Ghilardotti, G.	Agip Nucleare, Corso di Porta Romana 68, I-20122 Milan
La Marca, G.	Consiglio Tecnico Scientifico, Via C. Pascal 6, I-00100 Rome
Lenzi, G.	CNEN-CSN Casaccia, C.P. 2400, I-00100 Rome
Marzullo, T.	ENEL-DCO, Via G.B. Martini 3, I-00198 Rome
Risoluti, P.	Agip Nucleare, Corso di Porta Romana 68, I-20122 Milan
Tabet, E.	Istituto Superiore di Sanità, Viale Regina Elena 299, I-00161 Rome
Tondinelli, L.	CNEN-CSN Casaccia, C.P. 2400, I-00100 Rome
Trenta, G.	CNEN, Viale Regina Margherita 125, I-00198 Rome
Zifferero, M.	CNEN, Viale Regina Margherita 125, I-00198 Rome

JAPAN

Emura, S.	Power Reactor and Nuclear Fuel Development Corporation, 9-13, 1-chome, Akasaka, Minato-ku, Tokyo
Goto, T.	Japan Atomic Industrial Forum, Inc., 1-13, Shimbashi, 1-chome, Minato-ku, Tokyo
Hirai, Y.	D-533, 3-8 Sakuragaoka, Osaka
Ichikawa, R.	National Institute of Radiological Sciences, Anagawa 4-9-1, Chiba
Ihara, S.	Mitsubishi Metal Corporation, Central Research Laboratory, Kitabukuro-machi, Omiya-City, Saitama-Pref.
Ishihama, A.	52-1158 Tomiokamachi, Kanazawa-ku, Yokohama

Kikuchi, A.
Japan Radioisotope Association,
2-28-45, Honkomagome,
Bunkyoku,
Tokyo

Matsuura, H.
Nippon Atomic Industries Group Co. Ltd.,
4-1, Ukishima-cho,
Kanagawa-ken 210,
Kawasaki-City

Meguro, T.
The Enrichment and Reprocessing Group,
6-1, 1-chome,
Ohtemachi, Chiyoda-ku,
Tokyo 100

Nagaike, T.
The Tokyo Electric Power Co,
1-3, 1-chome,
Uchsaiwai-cho, Chiyoda-ku,
Tokyo

Nagakura, T.
Civil Engineering Laboratory,
Central Research Institute of Electric Power Industry,
1646 Abiko, Chiba-ken,
Abiko-shi

Nakamura, S.
The Federation of Electric Power Co.,
9-4, 1-chome,
Ohte-mach, Chiyoda-ku,
Tokyo

Nakayama, Y.
Nippon Atomic Industry Group,
Nuclear Research Laboratory,
4-1 Ukishima-cho,
Kawasaki

Nomi, M.
Ebara Manufacturing Co., Ltd.,
11-1 Asahimachi, Otaku,
Tokyo

Nomura, K.
Sumitomo Shoji Kaisha Ltd.,
Dr. Karl Lueger Ring 10,
A-1010 Vienna,
Austria

Ohtsuka, K.
Plutonium Fuel Division,
Power Reactor and Nuclear Fuel Development Corporation,
Tokai Works, Muramatsu 3371,
Tokai-mura, Ibaraki-ken

Sakata, S.
Japan Atomic Energy Research Institute,
1-1, Shinbashi, Minato-ku,
Tokyo

Tajima, E.
1-12-12 Sasnke,
Kamakura,
Kanagawa

Takashima, Y.
Tokyo Institute of Technology,
Research Laboratory for Nuclear Reactors,
Ookayama, Meguro-ku,
Tokyo

MOROCCO

Belmahi, O.
Faculté des sciences,
Université Mohamed V,
Rabat

NETHERLANDS

Baas, J.L.	Ministry of Health and Environmental Protection, Dokter Reijersstraat 10, Leidschendam
Hamstra, J.	Reactor Centrum Nederland, Westerduinweg 3, Petten N.H.
Joseph, C.J.	Urenco Nederland Operations BV, Planthofsweg 81, Almelo
Shank, E.M.	Comprimo Belgie N.V. Ahlers House, Noorderlaan 139, B-2030 Antwerp, Belgium
Smeets, L.	Reactor Centrum Nederland, Westerduinweg 3, Petten N.H.
Termaat, K.P.	Nukem, Utrechtseweg 310, Arnhem
Van der Plas, T.	Kema, Utrechtseweg 310, Arnhem
Van Erkelens, P.C.	Health Council, J.C. van Markenlaan 5, Rijswijk
Verkerk, B.	Reactor Centrum Nederland, Westerduinweg 3, Petten N.H.

NORWAY

Amundsen, K.	Norwegian Water Resources and Electricity Board, Middelthungsgt. 29, Oslo 3
Berteig, L.	State Institute of Radiation Hygiene, Österdalen 25, N-1345 Österas
Bonnevie-Svendsen, Moj	Institutt for Atomenergi, P.O. Box 40, N-2700 Kjeller
Løken, P.C.	Norwegian Water Resources and Electricity Board, Middelthungsgt. 29, Oslo 3
Michelsen, H.M.	Norwegian Nuclear Energy Safety Authority, Pottemakerveien 4, Oslo 5
Mölsaeter, M.	Norwegian Nuclear Energy Safety Authority, Pottemakerveien 4, Oslo 5
Tallberg, K.	Institutt for Atomenergi, P.O. Box 40, N-2700 Kjeller

PAKISTAN

Siddiqi, A.A.	Karachi Nuclear Power Plant, P.O. Box 3183, Karachi

POLAND

Szumski, W.	Atomic Energy Authority, Palac Kultury i Nauki, XVIII pietro, Warsaw
Tomczak, W.K.	Waste Disposal Department, Institute of Nuclear Research, Swierk K/Otwocka

SOUTH AFRICA

Langford, E.L.	Atomic Energy Board, Private Bag X256, Pretoria
Van der Westhuizen, H.J.	Atomic Energy Board, Private Bag X256, Pretoria

SPAIN

López-Menchero Ordóñez, E.M.	Junta de Energía Nuclear, Avenida Complutense 22, Madrid-3
López-Pérez, B.	Junta de Energía Nuclear, Avenida Complutense 22, Madrid-3
Martínez, A.	Junta de Energía Nuclear, Avenida Complutense 22, Madrid-3
Molina, V.	Nuclenor S.A., Hernan Cortes 26, Santander
Suárez-Estrada, J.A.	Electra de Viesgo S.A., Medio-12, Santander

SWEDEN

Ahlström, P.E.Å.	Statens Vattenfallsverk, Fack, S-162 87 Vällingby
Arnek, R.T.G.	Department of Inorganic Chemistry, Royal Institute of Technology, S-100 44 Stockholm
Boge, R.	National Institute of Radiation Protection, Fack, S-104 01 Stockholm 60
Carlbom, L.E.	AB Atomenergi, Fack, S-611 01 Nyköping

Edwall, B. | South Swedish Power Co., Fack, S-20070 Malmö 5

Hultgren, Å.V. | AB Atomenergi, Fack, S-611 01 Nyköping

Liljenzin, J.O. | Chalmers University of Technology, Fack, S-40220 Göteborg

Mandahl, B. | Oskarshamnsverkets Kraftgrupp AB, Box 1746, S-111 87 Stockholm

Papp, T. | Statens Vattenfallsverk, Fack, S-162 87 Vällingby

Sandklef, S. | Statens Vattenfallsverk, Fack, S-162 87 Vällingby

Thegerström, C.P.A. | AB Atomenergi, Fack, S-611 01 Nyköping

Westermark, T. | Department of Nuclear Chemistry, Royal Institute of Technology, S-100 44 Stockholm

SWITZERLAND

Aeppli, J. | ASK, CH-5303 Würenlingen

Bützer, P. | ASK, CH-5303 Würenlingen

Clausen, A. | Nordostschweizerische Kraftwerke AG, CH-5400 Baden

Gryksa, H.P. | Nuclear Assurance Corporation, Weinbergstraße 9, CH-8001 Zürich

Knutti, P.W. | Kernkraftwerk Kaiseraugst AG, c/o Motor Columbus, Parkstraße 27, CH-5401 Baden

Laske, D. | Eidgenössisches Institut für Reaktorforschung, CH-5303 Würenlingen

Maier, H. | Kernkraftwerk Kaiseraugst AG, c/o Motor Columbus, Parkstraße 27, CH-5401 Baden

Poltier, L. | EOS, P.O. Box 1048, Lausanne

Selinger, J. | Motor Columbus AG, Parkstraße 18, CH-5401 Baden

Weitze, H. | Kernkraftwerk Gösgen-Däniken AG, Postfach 55, CH-4658 Däniken

TURKEY

Göksel, S.A. | Cekmece Nuclear Research Centre, P.O. Box 1, Airport, Istanbul

UNION OF SOVIET SOCIALIST REPUBLICS

Dolgov, V.V.	USSR State Committee on the Utilization of Atomic Energy, Staromonetny pereulok 26, Moscow
Krylova, Nina V.	USSR State Committee on the Utilization of Atomic Energy, Staromonetny pereulok 26, Moscow
Martynov, Y.P.	USSR State Committee on the Utilization of Atomic Energy, Staromonetny pereulok 26, Moscow

UNITED KINGDOM

Allardice, R.	UKAEA, Dounreay Experimental Reactor Establishment, P.O. Box 1, Thurso, Caithness
Bailey, G.	UKAEA, Dounreay Experimental Reactor Establishment, P.O. Box 1, Thurso, Caithness
Clelland, D.W.	British Nuclear Fuels Ltd., Risley, Warrington, Cheshire, WA3 6AS
Corbet, A.D.W.	Fuel Plants Design Office, British Nuclear Fuels Ltd., Risley, Warrington, Cheshire, WA3 6AS
Healy, T.V.	UKAEA, Atomic Energy Research Establishment, Harwell, Didcot, Oxfordshire OX11 0RA
Hesketh, G.E.	Department of the Environment, 28 Broadway, London S.W.1
Jackson, E.W.	Ministry of Defense, AWRE, Aldermaston, Reading RG7 4PR
James, R.H.	Ministry of Defense, AWRE, Aldermaston, Reading, RG7 4PR
Johnson, L.F.	British Nuclear Fuels Ltd., Windscale Works, Sellafield Seascale, Cumbria CA20 1PG
Jones, P.J.	Central Electricity Generating Board, Hartlepool Power Station, Tees Road, Hartlepool
Keen, N.J.	UKAEA, Atomic Energy Research Establishment, Harwell, Didcot, Oxfordshire OX11 0RA

Larkin, M.J.	British Nuclear Fuels Ltd., Windscale and Calder Works, Sellafield, Seascale, Cumbria CA 20 1PG
Lewis, J.B.	UKAEA, Atomic Energy Research Establishment, Harwell, Didcot, Oxfordshire OX11 0RA
Marples, J.A.C.	UKAEA, Atomic Energy Research Establishment, Harwell, Didcot, Oxfordshire OX11 0RA
Marshall, C.F.	Department of the Environment, 2 Marsham St., Westminster, London
Mitchell, N.T.	Ministry of Agriculture, Fisheries and Food, Fisheries Radiobiological Laboratory, Hamilton Dock, Lowestoft, Suffolk
Morris, J.B.	UKAEA, Process Technology Division, Atomic Energy Research Establishment, Harwell, Didcot, Oxfordshire OX11 0RA
Nairn, J.	UKAEA, Atomic Energy Research Establishment, Reactor Group Windscale, Seascale, Cumbria CA20 1PF
Philips, R.J.	Gravatom Industries Ltd., Gosport, Hampshire P013 0AJ
Rawlins, L.V.	Rolls-Royce & Associates Ltd., P.O. Box 31, Derby
Stott, G.	Scottish Development Department, St. Andrews House, Edinburgh EH2 3JY
Taylor, R.F.	UKAEA, Atomic Energy Research Establishment, Harwell, Didcot, Oxfordshire OX11 0RA
Tildsley, F.C.J.	N.I.I.-N.S.E., Thames House North, Millbank, London SW1P 4QL
Varney, E.J.	Nuclear Installations Inspectorate, Silkhouse Court, Tithebarn St., Liverpool L2 2LZ
Webb, G.A.	National Radiological Protection Board, Harwell, Didcot, Oxfordshire OX11 0RQ

UNITED STATES OF AMERICA

Andrews, L.J.	Chem-Nuclear Systems Inc., Yarrow Bay Office Park, 10602 N.E. 38th Place, Kirkland, WA 98033

Angelo, J.A., Jr.	Physics Department, Florida Institute of Technology, Melbourne, FLA 32901
Bishop, W.P.	United States Nuclear Regulatory Commission, Washington, D.C. 20555
Bradley, R.A.	Oak Ridge National Laboratory, P.O. Box X, Oak Ridge, TN 37830
Brubaker, G.L.	Executive Office of the President, Council on Environmental Quality, 722 Jackson Place, Washington, D.C. 20006
Daly, G.H.	US Energy Research and Development Administration, Washington, D.C. 20545
Dyer, R.S.	Office of Radiation Programs, United States Environmental Protection Agency, 401 M Street, S.W., Washington, D.C. 20460
Gablin, K.A.	Protection Packaging Inc., Bluegrass Research & Industrial Park, 328 Production Court, Louisville, KY 40299
Graves, G.A.	National Science Foundation, 1800 G.St., N.W., Washington, D.C. 20550
Matuszek, J.M.	Division of Laboratories and Research, Radiological Sciences Laboratory, New York State Department of Health, New Scotland Avenue, Albany, N.Y. 12201
Meyer, G.L.	Office of Radiation Programs, United States Environmental Protection Agency, Washington, D.C. 20460
Perge, A.F.	Division of Nuclear Fuel Cycle and Production, United States Energy Research and Development Administration, Washington, D.C. 20545
Platt, A.M.	Battelle Pacific Northwest Laboratories, Richland, WA 99352
Rieke, F.E.	Industrial Clinics, 1313 N.W. 19th, Portland, OR 97209
Rieke, Mary W.	House of Representatives, Oregon State Legislature, 5519 S.W. Menefee Drive, Portland, OR 97201
Schneider, A.	Georgia Institute of Technology, School of Nuclear Engineering, Atlanta, GA 30332
Slansky, C.M.	Allied Chemical Corporation, 550 Second St., Idaho Falls, ID 83401
Stevens, P.R.	US Geological Survey Office of Radiohydrology, 12201 Sunrise Valley Drive, Reston, VA 22092

Stuart, R.W.	CTI Nuclear, 266 Second Avenue, Waltham, MA 01880
Wacks, M.E.	Department of Nuclear Energy, University of Arizona, Tucson, AZ 85721
Weart, W.D.	Sandia Laboratories, Albuquerque, NM 87115

YUGOSLAVIA

Pirš, M.	Institute Jožef Stefan, Jamova 39, 61000 Ljubljana

ORGANIZATIONS

Commission of the European Communities (CEC)

Angelini, A.	CCR EURATOM, I-21020 Ispra (Varese), Italy
Bertozzi, G.	CCR EURATOM, I-21020 Ispra (Varese), Italy
Birkhoff, G.	CCR EURATOM, I-21020 Ispra (Varese), Italy
Girardi, F.	CCR EURATOM, I-21020 Ispra (Varese). Italy
Hettinger, H.	CCR EURATOM, I-21020 Ispra (Varese), Italy
Kock, L.W.	European Institute for Transuranium Elements, Postfach 2266, D-7500 Karlsruhe, Federal Republic of Germany
Krischer, W.	200, rue de la Loi, B-1040 Brussels, Belgium
Lanza, F.	CCR EURATOM, I-21020 Ispra (Varese), Italy
Luykx, F.	DG V, Bâtiment Jean Monnet, Luxembourg
Mannone, F.	CCR EURATOM, I-21020 Ispra (Varese), Italy
Mousty, F.	CCR EURATOM, I-21020 Ispra (Varese), Italy
Schmidt, E.M.	CCR EURATOM, I-21020 Ispra (Varese), Italy
Schneider, J.	CCR EURATOM, I-21020 Ispra (Varese), Italy
Simon, R.A.	200, rue de la Loi, B-1040 Brussels, Belgium
Sola, A.J.P.	CCR EURATOM, I-21020 Ispra (Varese), Italy

Council for Mutual Economic Assistance (CMEA)

Tolpygo, V.K.	Prospekt Kalinina 56, Moscow

Economic Commission for Europe (ECE)

Trindade, M. — Palais des Nations, CH-1211 Geneva, Switzerland

International Atomic Energy Agency (IAEA)

Bonhote, P.A. — Division of Nuclear Safety and Environmental Protection, P.O. Box 590, A-1011 Vienna, Austria

Catlin, R.J. — Regional Nuclear Fuel Cycle Centres Study Project, P.O. Box 590, A-1011 Vienna, Austria

Farges, L.F. — Division of Nuclear Safety and Environmental Protection, P.O. Box 590, A-1011 Vienna, Austria

Grover, J.R. — Division of Nuclear Safety and Environmental Protection, P.O. Box 590, A-1011 Vienna, Austria

Jourde, P.H. — Division of Nuclear Safety and Environmental Protection, P.O. Box 590, A-1011 Vienna, Austria

Meckoni, V. — Regional Nuclear Fuel Cycle Centres Study Project, P.O. Box 590, A-1011 Vienna, Austria

Millar, C.H. — Division of Nuclear Safety and Environmental Protection, P.O. Box 590, A-1011 Vienna, Austria

Morozov, V. — Division of Nuclear Safety and Environmental Protection, P.O. Box 590, A-1011 Vienna, Austria

Nishiwaki, Y. — Division of Nuclear Safety and Environmental Protection, P.O. Box 590, A-1011 Vienna, Austria

Pettersson, B.G. — Division of Nuclear Safety and Environmental Protection, P.O. Box 590, A-1011 Vienna, Austria

Organization for Economic Co-operation and Development/Nuclear Energy Agency (OECD/NEA)

Detilleux, E. — Eurochemic, B-2400 Mol, Belgium

Takahashi, M. — 38 bd. Suchet, F-75016 Paris

Van Geel, J. — Eurochemic, B-2400 Mol, Belgium

Wallauschek, E. — 38 bd. Suchet, F-75016 Paris

AUTHOR INDEX

Roman numerals are volume numbers.
Italic-type numerals refer to the first page of a paper by the author concerned.
Arabic numerals denote comments and questions in discussions.
Literature references are not indexed.

TRANSLITERATION INDEX

Басков, Л.И.	Baskov, L.I.
Бельтюков, В.А.	Bel'tyukov, V.A.
Брежнева, Н.Е.	Brezhneva, N.E.
Давыдов, В.И.	Davydov, V.I.
Добрыгин, П.Г.	Dobrygin, P.G.
Долгов, В.В.	Dolgov, V.V.
Константинович, А.А.	Konstantinovich, A.A.
Крылова, Н.В.	Krylova, N.V.
Кузнецов, Д.Г.	Kuznetsov, D.G.
Кулаков, С.И.	Kulakov, S.I.:
Куличенко, В.В.	Kulichenko, V.V.
Мартынов, Ю.П.	Martynov, Y.P.
Минаев, А.А.	Minaev, A.A.
Мусатов, Н.Д.	Musatov, N.D.
Никипелов, Б.В.	Nikipelov, B.V.
Никифоров, А.С.	Nikiforov, A.S.
Озиранер, С.Н.	Oziraner, S.N.
Сергеев, Г.А.	Sergeev, G.A.
Степанов, С.Е.	Stepanov, S.E.
Толпыго, В.К.	Tolpygo, V.K.
Фридрих, Б.	Friedrich, B.

The following conversion table is provided for the convenience of readers and to encourage the use of SI units.

FACTORS FOR CONVERTING UNITS TO SI SYSTEM EQUIVALENTS *

SI base units are the metre (m), kilogram (kg), second (s), ampere (A), kelvin (K), candela (cd) and mole (mol).
[For further information, see International Standards ISO 1000 (1973), and ISO 31/0 (1974) and its several parts]

Multiply		*by*	*to obtain*
Mass			
pound mass (avoirdupois)	1 lbm	= 4.536×10^{-1}	kg
ounce mass (avoirdupois)	1 ozm	= 2.835×10^{1}	g
ton (long) (= 2240 lbm)	1 ton	= 1.016×10^{3}	kg
ton (short) (= 2000 lbm)	1 short ton	= 9.072×10^{2}	kg
tonne (= metric ton)	1 t	= 1.00×10^{3}	kg
Length			
statute mile	1 mile	= 1.609×10^{0}	km
yard	1 yd	= 9.144×10^{-1}	m
foot	1 ft	= 3.048×10^{-1}	m
inch	1 in	= 2.54×10^{-2}	m
mil (= 10^{-3} in)	1 mil	= 2.54×10^{-2}	mm
Area			
hectare	1 ha	= 1.00×10^{4}	m^2
(statute mile)^2	1 mile^2	= 2.590×10^{0}	km^2
acre	1 acre	= 4.047×10^{3}	m^2
yard^2	1 yd^2	= 8.361×10^{-1}	m^2
foot^2	1 ft^2	= 9.290×10^{-2}	m^2
inch^2	1 in^2	= 6.452×10^{2}	mm^2
Volume			
yard^3	1 yd^3	= 7.646×10^{-1}	m^3
foot^3	1 ft^3	= 2.832×10^{-2}	m^3
inch^3	1 in^3	= 1.639×10^{4}	mm^3
gallon (Brit. or Imp.)	1 gal (Brit)	= 4.546×10^{-3}	m^3
gallon (US liquid)	1 gal (US)	= 3.785×10^{-3}	m^3
litre	1 l	= 1.00×10^{-3}	m^3
Force			
dyne	1 dyn	= 1.00×10^{-5}	N
kilogram force	1 kgf	= 9.807×10^{0}	N
poundal	1 pdl	= 1.383×10^{-1}	N
pound force (avoirdupois)	1 lbf	= 4.448×10^{0}	N
ounce force (avoirdupois)	1 ozf	= 2.780×10^{-1}	N
Power			
British thermal unit/second	1 Btu/s	= 1.054×10^{3}	W
calorie/second	1 cal/s	= 4.184×10^{0}	W
foot-pound force/second	1 ft·lbf/s	= 1.356×10^{0}	W
horsepower (electric)	1 hp	= 7.46×10^{2}	W
horsepower (metric) (= ps)	1 ps	= 7.355×10^{2}	W
horsepower (550 ft·lbf/s)	1 hp	= 7.457×10^{2}	W

* Factors are given exactly or to a maximum of 4 significant figures

Multiply			*by*	*to obtain*
Density				
pound mass/inch^3	1 lbm/in^3	=	2.768×10^4	kg/m^3
pound mass/foot^3	1 lbm/ft^3	=	1.602×10^1	kg/m^3
Energy				
British thermal unit	1 Btu	=	1.054×10^3	J
calorie	1 cal	=	4.184×10^0	J
electron-volt	1 eV	≃	1.602×10^{-19}	J
erg	1 erg	=	1.00×10^{-7}	J
foot-pound force	1 ft·lbf	=	1.356×10^0	J
kilowatt-hour	1 kW·h	=	3.60×10^6	J
Pressure				
newtons/metre^2	1 N/m^2	=	1.00	Pa
atmosphere[a]	1 atm	=	1.013×10^5	Pa
bar	1 bar	=	1.00×10^5	Pa
centimetres of mercury (0°C)	1 cmHg	=	1.333×10^3	Pa
dyne/centimetre^2	1 dyn/cm^2	=	1.00×10^{-1}	Pa
feet of water (4°C)	1 ftH_2O	=	2.989×10^3	Pa
inches of mercury (0°C)	1 inHg	=	3.386×10^3	Pa
inches of water (4°C)	1 inH_2O	=	2.491×10^2	Pa
kilogram force/centimetre^2	1 kgf/cm^2	=	9.807×10^4	Pa
pound force/foot^2	1 lbf/ft^2	=	4.788×10^1	Pa
pound force/inch^2 (= psi)[b]	1 lbf/in^2	=	6.895×10^3	Pa
torr (0°C) (= mmHg)	1 torr	=	1.333×10^2	Pa
Velocity, acceleration				
inch/second	1 in/s	=	2.54×10^1	mm/s
foot/second (= fps)	1 ft/s	=	3.048×10^{-1}	m/s
foot/minute	1 ft/min	=	5.08×10^{-3}	m/s
mile/hour (= mph)	1 mile/h	=	4.470×10^{-1}	m/s
			1.609×10^0	km/h
knot	1 knot	=	1.852×10^0	km/h
free fall, standard (= g)		=	9.807×10^0	m/s^2
foot/second^2	1 ft/s^2	=	3.048×10^{-1}	m/s^2
Temperature, thermal conductivity, energy/area·time				
Fahrenheit, degrees −32	°F − 32		$\frac{5}{9}$	°C
Rankine	°R			K
1 Btu·in/ft^2·s·°F		=	5.189×10^2	W/m·K
1 Btu/ft·s·°F		=	6.226×10^1	W/m·K
1 cal/cm·s·°C		=	4.184×10^2	W/m·K
1 Btu/ft^2·s		=	1.135×10^4	W/m^2
1 cal/cm^2·min		=	6.973×10^2	W/m^2
Miscellaneous				
foot^3/second	1 ft^3/s	=	2.832×10^{-2}	m^3/s
foot^3/minute	1 ft^3/min	=	4.719×10^{-4}	m^3/s
rad	rad	=	1.00×10^{-2}	J/kg
roentgen	R	=	2.580×10^{-4}	C/kg
curie	Ci	=	3.70×10^{10}	disintegration/s

[a] atm abs: atmospheres absolute;
atm (g): atmospheres gauge.

[b] lbf/in^2 (g) (= psig): gauge pressure;
lbf/in^2 abs (= psia): absolute pressure.

HOW TO ORDER IAEA PUBLICATIONS

An exclusive sales agent for IAEA publications, to whom all orders and inquiries should be addressed, has been appointed in the following country:

UNITED STATES OF AMERICA	UNIPUB, P.O. Box 433, Murray Hill Station, New York, N.Y. 10016

In the following countries IAEA publications may be purchased from the sales agents or booksellers listed or through your major local booksellers. Payment can be made in local currency or with UNESCO coupons.

ARGENTINA	Comisión Nacional de Energía Atómica, Avenida del Libertador 8250, Buenos Aires
AUSTRALIA	Hunter Publications, 58 A Gipps Street, Collingwood, Victoria 3066
BELGIUM	Service du Courrier de l'UNESCO, 112, Rue du Trône, B-1050 Brussels
CANADA	Information Canada, 171 Slater Street, Ottawa, Ont. K1A 0S9
C.S.S.R.	S.N.T.L., Spálená 51, CS-110 00 Prague Alfa, Publishers, Hurbanovo námestie 6, CS-800 00 Bratislava
FRANCE	Office International de Documentation et Librairie, 48, rue Gay-Lussac, F-75005 Paris
HUNGARY	Kultura, Hungarian Trading Company for Books and Newspapers, P.O. Box 149, H-1011 Budapest 62
INDIA	Oxford Book and Stationery Comp., 17, Park Street, Calcutta 16; Oxford Book and Stationery Comp., Scindia House, New Delhi-110001
ISRAEL	Heiliger and Co., 3, Nathan Strauss Str., Jerusalem
ITALY	Libreria Scientifica, Dott. de Biasio Lucio "aeiou", Via Meravigli 16, I-20123 Milan
JAPAN	Maruzen Company, Ltd., P.O.Box 5050, 100-31 Tokyo International
NETHERLANDS	Marinus Nijhoff N.V., Lange Voorhout 9-11, P.O. Box 269, The Hague
PAKISTAN	Mirza Book Agency, 65, The Mall, P.O.Box 729, Lahore-3
POLAND	Ars Polona, Centrala Handlu Zagranicznego, Krakowskie Przedmiescie 7, Warsaw
ROMANIA	Cartimex, 3-5 13 Decembrie Street, P.O.Box 134-135, Bucarest
SOUTH AFRICA	Van Schaik's Bookstore, P.O.Box 724, Pretoria Universitas Books (Pty) Ltd., P.O.Box 1557, Pretoria
SPAIN	Diaz de Santos, Lagasca 95, Madrid-6 Calle Francisco Navacerrada, 8, Madrid-28
SWEDEN	C.E. Fritzes Kungl. Hovbokhandel, Fredsgatan 2, S-103 07 Stockholm
UNITED KINGDOM	Her Majesty's Stationery Office, P.O. Box 569, London SE1 9NH
U.S.S.R.	Mezhdunarodnaya Kniga, Smolenskaya-Sennaya 32-34, Moscow G-200
YUGOSLAVIA	Jugoslovenska Knjiga, Terazije 27, YU-11000 Belgrade

Orders from countries where sales agents have not yet been appointed and requests for information should be addressed directly to:

Division of Publications
International Atomic Energy Agency
Kärntner Ring 11, P.O.Box 590, A-1011 Vienna, Austria